Food Engineering Series

Springer's *Food Engineering Series* is essential to the Food Engineering profession, providing exceptional texts in areas that are necessary for the understanding and development of this constantly evolving discipline. The titles are primarily reference-oriented, targeted to a wide audience including food, mechanical, chemical, and electrical engineers, as well as food scientists and technologists working in the food industry, academia, regulatory industry, or in the design of food manufacturing plants or specialized equipment.

More information about this series at http://www.springer.com/series/5996

Umesh Hebbar • Shivendu Ranjan
Nandita Dasgupta • Raghvendra Kumar Mishra
Editors

Nano-food Engineering

Volume One

Springer

Editors
Umesh Hebbar
Food Engineering Department
CSIR-CFTRI
Mysuru, Karnataka, India

Nandita Dasgupta
Department of Biotechnology
Institute of Engineering and Technology
Dr. A.P.J. Abdul Kalam Technical University
Lucknow, Uttar Pradesh, India

Shivendu Ranjan
Faculty of Engineering and the Built Environment
University of Johannesburg
Johannesburg, South Africa

Raghvendra Kumar Mishra
EPSRC Fellow
Enhanced Composite Centre
School of Aerospace Transport and Manufacturing
Cranfield University
Bedfordshire, United Kingdom

ISSN 1571-0297
Food Engineering Series
ISBN 978-3-030-44551-5 ISBN 978-3-030-44552-2 (eBook)
https://doi.org/10.1007/978-3-030-44552-2

This Springer imprint is published by the registered company Springer Nature Switzerland AG
The registered company address is: Gewerbestrasse 11, 6330 Cham, Switzerland

Preface

Nanotechnology is currently a buzz word in almost every field, be it be Communication, Material Science, Bioprocessing or Food Processing. Food processing sector is also undergoing radical changes globally and nanotechnology has moved in to different spheres such as emulsion preparation, encapsulation, targeted delivery, packaging, food safety, extension of shelf-life etc. Nanosensors are finding wide range of applications such freshness indicators, detection of spoilage through identification of contaminants in food samples etc. Various research groups are working across the globe in this area and some of the technologies have already been commercialized and well established. This book is an attempt to provide a comprehensive perspective of nanotechnology in food processing; the current state of research, industrial applications and future prospects.

This book on Nanotechnology in Food Engineering Vol 1, contains 12 chapters covering basic information about nanotechnology, various techniques used for the preparation of nanomaterials, nanoencapsulated foods, packaging materials and nanosensors. These chapters have been presented by research groups having rich expertise in this area and hailing for different parts of the world. The book discusses about techniques such as Supercritical Fluid for the preparation of nanoencapsulated food-grade materials, Cavitation Technology for Nano-Foods, Spray drying technique for the preparation of nanomaterial, Inclusion Complexation and Coacervation to fabricate Nanoencapsulated foods and Microfluidization technique in Nano-Food Engineering. Also, the book covers topics on food packaging such as Nano-films for food packaging, Biopolymer Nanocomposite based food packaging and Herbal Bioactive Agents for Edible Food Packaging. It also includes chapters on Nanosensors for foods, nano-sorbents for determination of food contaminants and nanotechnology to improve solubility of bioactive constituents of foods for health-promoting purposes. We hope readers will be benefitted immensely from the vast experience of authors in their respective fields.

Mysuru, Karnataka, India — Umesh Hebbar
Johannesburg, South Africa — Shivendu Ranjan
Lucknow, Uttar Pradesh, India — Nandita Dasgupta

Acknowledgement

We dedicate this book to the research community working towards the betterment of society.

– Editors

Contents

Introduction to Nanofood

Nikita Meghani, Sruja Dave, and Ashutosh Kumar

1 Introduction to Nanotechnology

Richard Feynman, a renowned physicist has first discussed the concept of nanotechnology and the scope for manipulating atoms for synthesis purpose by saying the fact that "There's Plenty of Room at the Bottom", (Feynman 1959).

The term "Nano" is derived from a Greek word that means "dwarf" or "small" thus nanotechnology takes into consideration all the science, engineering and technology operated at the nanoscale (Zhou 2013).

National Nanotechnology Initiative (NNI) is known to be world largest funding source for nanotechnology research and according to them, nanotechnology can be defined as "the understanding and control of matter at dimensions between approximately 1 and 100 nanometers, where unique phenomena enable novel applications not feasible when working with bulk materials or even with single atoms or molecules. Encompassing nanoscale science, engineering, and technology, nanotechnology involves imaging, measuring, modeling and manipulating matter at this length scale" (Zhou 2013).

The field of nanotechnology is developing rapidly as it attracts the attention of everyone by offering novel application and benefits at the nanoscale. Nanoparticles show the different property at a different size, in other words, that nano and macroparticles of the same powder will have different properties (Chaudhry et al. 2008a). Several industries are being affected by this innovative and enabling technology. Starting from cleaning agents to edible items, personal care, sports, medicine, and materials science, there is no field left unchanged by the usefulness of nanotechnology to make things better and susceptible for humans. The food industry is one such area among generals where nanotechnology has already started to

N. Meghani · S. Dave · A. Kumar (✉)
Biological and Life Sciences, School of Arts and Sciences, Ahmedabad University, Ahmedabad, Gujarat, India
e-mail: ashutosh.kumar@ahduni.edu.in

U. Hebbar et al. (eds.), *Nano-food Engineering*, Food Engineering Series,
https://doi.org/10.1007/978-3-030-44552-2_1

inspire and change not only the traditional way of industries existence but also the way research and innovation are being carried out in general. Nanotechnology conceptually has already started to provide the basic framework for the development required to understand the roots of food components into micro and nanoscale, which has led to influence the food structure and it's rheological and functional properties (Sanguansri and Augustin 2006). In a simple way, nanotechnology has the potential to transform, every aspect associated with the food industry. Food packaging, food storage, processing and pesticides sense are some of the aspects of food industry where notable research is in progress using and the unique physico-chemical properties of nanomaterials (Chaudhry et al. 2008a).

2 Nanotechnology in Food Industry – Overview

The food and beverage industry are globally a multi-trillion-dollar industry and looking into the ways to improve in every aspect as possible, starting from production efficiency to food safety. Nanotechnology has emerged as a boon to this industry, as this innovative technology has provided one-stop solution to multiple problems. Food safety, waste reduction and authenticity of the product are the part of food industry dependent on several different industries such as electronics, data storage and advancement of integrated devices. Nanotechnology has tremendously affected this industry and is thus affecting food industries in indirect ways. For e.g. the development of nanosensors has affected both electronics and food industry (Cushen et al. 2012).

Food safety, molecular synthesis of new food products, food packaging, composition, storage etc. are some vital areas concerning food industries which have already being impacted by the incorporation of nanotechnology in food industries which has now let to the emergence of food nanotechnology as an independent field of its own (Chen et al. 2006). Since the physiochemical and biological properties of nanostructures and nanomaterials are different from their bulk product, there is a rise of a new approach to looking into food systems especially the biological and physical occurrence of food. Several studies show a significant improvement in food safety, packaging, processing, and nutrition available in food due to the usage of nanotechnology in the food industry (Chaudhry et al. 2008a; Dasgupta et al. 2015; He and Hwang 2016; Pathakoti et al. 2017; Pradhan et al. 2015; Sekhon 2010; Sekhon 2014).

Food preservation, colouring, flavouring, nutritional additives, usage of antimicrobial agents for food packaging are some of the most noteworthy application of nanotechnology in the food industry. However, food packaging is the area in which nanotechnology is in maximum demands as this has already challenged the conventional methods of packaging by adding an enhanced barrier, and mechanical and heat-resistant properties as merits to its list (Honarvar et al. 2016; Patel et al. 2018; Pathakoti et al. 2017).

Although nanotechnology has shown great potential in the formulation of innovative products and processes in the food sector, there are many challenges left to overcome in food science and technology. One such is the production of edible delivery systems of nutraceuticals using economic approaches that are effective for human consumption and meet the safety norms (Dupas and Lahmani 2007; Patel et al. 2018).

To meet the challenges of the food sector, a great knowledge of food materials and processes at the nanoscale is required, which will be helpful in the innovation of new and improved products. Although the number of nanofood currently available in market is limited, nanotechnology has already thrilled the manufacturers, as its potential seems unlimited. Available technology has so far promised the introduction of new food products in the market while bringing innovation like better food texture, improved taste, higher process ability, and increased shelf life (Gilligan 2008; Rao 2009). Nanotechnology thus has the potential to change the entire food Industry from its roots.

3 Nanofood

Usage of nanotechnology in food Industry has brought back the usage of term nanofood. The term "Nanofood" is now used to describe food that has been cultivated, produced, processed or packaged using nanotechnology techniques or tools, or to which manufactured nanomaterials have been added (Gilligan 2008; Joseph and Morrison 2006). However, nanofood was considered as a part of food industries since centuries as food structure naturally exists at the nanoscale. The idea of nanofood is to have food products in the market with improved safety, enhance nutrition's, flavour and cost-effective. Nanofood though is still a relatively new aspect, one of the earlier applications of nanotechnology is as a carrier to deliver antimicrobial peptides required to stop the antimicrobial decay of food quality in the food industry. The following was achieved by coating starch colloids filled with an antimicrobial agent so that when microorganism grow on the packaged food they will break the coating of starch leading to release of antimicrobial agents (Boumans 2003; Gilligan 2008). The recent application of nanotechnology focuses on detection of food pathogen using nanosensors, which have the merit of being quicker, more sensitive and less labor-intensive procedures than the existing one. On one side where the benefits of nanofood are increasing exponentially like health-promoting additives, longer shelf life, new flavour and smart food packaging the questions are being raised on its safety of the nanomaterials and being used as they may interact with the living system and thereby can cause toxicity (Das et al. 2009).

Sooner or later, the warnings about nanofoods product are going to reach a tipping point in terms of public attentions since nanofood is grabbing media attention worldwide. Questions are going to be raised about the materials and particles used in the products which are already available in the market, which may or may not have FDA approval (Gilligan 2008; Yiannaka 2012). A list of food products currently containing nanoproducts include: Canola Active Oil (Shemen, Haifa, Israel), Nanotea (Shenzhen Become Industry Trading Co. Guangdong, China), Fortified Fruit Juice (High Vive. com, USA), Nanoceuticals Slim Shake (assorted flavors, RBC Lifesciences, Irving, USA), NanoSlim beverage(NanoSlim), Oat Nutritional Drink (assorted flavors, Toddler Health, Los Angeles, USA), and 'Daily Vitamin Boost' fortified fruit juice (Jamba Juice Hawaii, USA) and nanocapsules containing tuna fish oil (a source of omega 3 fatty acids) in "Tip-Top" Up bread (Enfield, Australia) (Gilligan 2008; Kirdar 2015; Sekhon 2010). Thus, along with food industries, a deeper knowledge of nanotechnology is important for innovation.

4 Nanomaterials and Nanostructures

Nanomaterials can either be naturally occurring or be externally added. Most widely used nanostructures in the food industry are engineered nanomaterials, nanoemulsion, nanoliposomes, and nanofibers.

4.1 Engineered Nanomaterials

Nanoemulsion of inorganic and organic substances both are used to form engineered nanomaterials (ENMs), which are used in food industry for multiple purposes. The ENMs are differentiated into three basic categories, inorganic, surface functionalized materials, and organic engineered nanomaterials (Chaudhry et al. 2008a).

Inorganic nanomaterials and surface functionalized nanomaterials are generally used as food additives or in food packaging and storage. While as organic nanomaterials are used to in products to enhance their uptake or absorption or as a carrier for certain nutraceuticals. Usage of inorganic nanomaterials has increased significantly in food packaging and on another hand; surface functionalized nanomaterials are used to enhance properties such as antimicrobial or gas-barrier properties etc. One of the examples of inorganic nanomaterials is nanoselenium which is soon going to be used as an additive to green tea product, in order to enhance its antioxidant effect (Sekhon 2010; Vance et al. 2015) (Table 1).

4.2 Nanoemulsion

The fundamental components of food, the food ingredients are rarely used in their purest form and thus there is a constant need for a better and stable delivery system to deliver nutraceuticals such as vitamins, antioxidants, flavouring etc. The delivery system needs to not only deliver the functional food but also not react with the bio

Table 1 The table lists examples of different type of nanomaterial used in Food Industry. Adapted from FSAI 2008 (Ireland 2008)

Category	Nanomaterials	Application	Reference
Inorganic nanoparticle	Silver	consumer products such as health food and water, food contact surfaces and packaging materials	Sekhon (2010); Vance et al. (2015)
	Iron	as a health supplement, in the treatment of contaminated water	Sekhon (2010)
	Calcium and magnesium	health supplements and use as chewing gum	Gallus et al. (2003); Sekhon (2010)
	Silica	food contact surfaces, food packing	Joseph and Morrison (2006); Sekhon (2010)
Surface functionalized nanomaterial	Nanoclays (Montmorillonite)	improved properties in packaging (barrier, thermal, durability)	Avella et al. (2005)
Organic nanoparticles	Liposomes	bioactive agent nanoencapsulation, improved solubility and bioavailability, cell-specific targeting	Kumari et al. (2014)
	Protein	nanoencapsulation of hydrophobic nutraceuticals. improved functionalities (gelation, heat stability)	Semo et al. (2007)
	Polymeric	nanoencapsulation & improved functionalities (delivery, antimicrobial)	Chen and Subirade (2005)

molecules being delivered, also need to protect the ingredient from various environmental factors and its own degradation (Ravichandran 2010). For this purpose, nanoemulsion and nanoliposomes are already being exploited in the food industry.

Nanoemulsion have been recently exploited in the food industry due to their greater stability, higher optical clarity, increased bioavailability and their efficiency to delivery nutraceuticals (Meghani et al. 2018). Due to above mentioned properties, nanoemulsion are considered as the best emulsion-based delivery systems as they make encapsulation, protection and delivery of hydrophobic nutraceuticals and drugs, very conveniently for both functional food and pharmaceutical application (Kumar et al. 2016; Mirhosseini 2016). However, nanoemulsion so far have been synthesized to decontaminate food packaging equipment, remove pesticide residues from fruits and vegetables, to reduce surface contamination of chicken skin etc (Drusch 2007; Gharsallaoui et al. 2007). Nanoemulsion have also shown great promise to be used in beverage industries (Shukla 2012). Nanoemulsion have been used as an antimicrobial agent against a broad range of food pathogens including some gram-negative bacteria e.g. *Salmonella typhimurium* (Mirhosseini 2016). These are currently being used to deliver nutraceuticals. For e.g. Cinnamon oil nanoemulsion synthesized using ultrasonic emulsification was used to deliver vitamin D (Meghani et al. 2018), while as a corn oil and orange oil were used to deliver polymethoxyflavones (PMFs) extracted from citrus peel (Li et al. 2012), both of the nutraceuticals are hydrophobic compounds (Chakraborty et al. 2009). One of the most successful used of nanoemulsion was to deliver megestrolacetate oral suspension (MAOS) which are used as appetite stimulant in patients with AIDS (Deschamps et al. 2009).

4.3 Nanoliposomes

Like nanoemulsion, nanoliposomes are nanometric version of liposomes, colloidal structures formed by input of energy to phospholipids in an aqueous solution. Nanoemulsion and nanoliposomes both are under extensive research by all different sectors including food industries for their use a colloidal delivery system to deliver hydrophobic bioactive and functional agents. Nanoliposomes technology offers all the merits to industries similar as nanoemulsion such as enhanced stability and the protection of encapsulated material from all the environmental factors, chemical and enzymatic changes, stability from range of pH and ionic strength and most importantly masking unwanted odour or taste (Reza Mozafari et al. 2008). However nanoliposomes offers food industry an extra benefit, a possibility of controlled release of food material, a approach still not feasible with nanoemulsions (Astete et al. 2009). Traditionally food industry has been using liposomes and nanoliposomes to deliver flavours and nutrient, however recently they are being use to deliver antimicrobials to provide protection against microbial contamination (Tumbarski et al. 2018). Since nanoliposomes can be synthesized from natural occurring ingredients such as soy, egg or milk, they have a edge over other in obtaining regulatory approval to be used in food industry (Reza Mozafari et al. 2008). The very first use of liposome was in synthesis of cheese (Law and King 1985). Recently nanoliposomes are used in delivery of concentrated enzyme during food processing and delivering antimicrobials to avoid microbial contamination (Fathi et al. 2012). Coenzyme q10 nanoliposomes are an example of stable nanoliposome used to delivery enzyme in with a desired concentration of the enzyme (Xia et al. 2006) (Fig. 1).

4.4 Nanotubes

Carbon nanotubes (CNT) have been used in numerous application in different industries and thus the idea of using nanotubes emerged in the food industry (Weiss et al. 2006). Nanotubes are beneficial in food industry due to their large surface area which is exploited to immobilize antibodies, detect food pathogens and increase the sensitivity of immunosensors (Yang et al. 2008). Nanotubes having allyl isothiocyanate on their surface area are been used to have an antimicrobial effect and

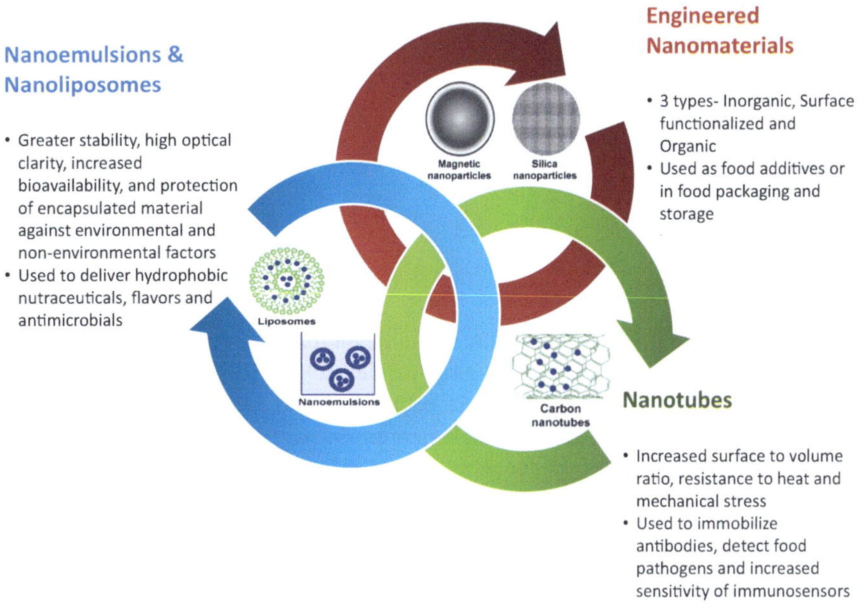

Fig. 1 Schematic for the types of nanomaterials used in food industry

prohibited the growth of *Salmonella* in cellulose-based food packaging (Dias et al. 2013). Nanotubes are more beneficial compared to other particles because of their ability to resist heat and mechanical stress, improved gelation and viscosity and their increased surface to volume ratio which in turn increases the encapsulation or absorption efficiency of the functional food on their surface (Neo et al. 2013; Okutan et al. 2014). The example of nanotubes used in the food industry is the use of CNT assembled with milk protein α-lactalbumin (Graveland-Bikker and De Kruif 2006).

4.5 Others

An association colloid is also a nanosystem used in food industry, is basically a stable dispersion system made up of small particles used to deliver polar, nonpolar and amphiphilic functional ingredients (Flanagan and Singh 2006; Garti et al. 2004; Garti et al. 2005; Golding and Sein 2004). Association colloids have size generally in the range of 5–100 nm but have the disadvantage of compromising flavours and are known to dissociate spontaneously when diluted (McClements et al. 2005). Apart from this, another used nanoscale technique is nanolaminates, which are extremely thin food grade film comprising two or more layers of materials in their nanometer dimensions. Nanolaminates have a layer of 1–100 nm each and are known to have physical or chemical bonded dimensions are used in the preparation of edible films (Cagri et al. 2004; Cha and Chinnan 2004; Morillon et al. 2002; Rhim 2004).

5 Functionality and Applications of Nanoconstructs in the Food Industry

5.1 Functions

Nanotechnology has a potential to play a significant role in the development of food packagings, such as intelligent packaging, active packagings like controlled atmosphere packaging, modified atmosphere packaging, and antimicrobial packaging. Antimicrobial packaging is an area of emerging interest and is evolving with the application of nanotechnology due to its critical role in improving microbial safety and extending the shelf life of the food products (Duncan 2011). It releases the antimicrobials and preservative into the food to elevate the quality of microbes and the safety of food, which is done by incorporating or coating the packaging materials with antimicrobial agents for slow release. This is more efficient as the direct addition of microbes may react with other food components and results in loss of its activity (Mauriello et al. 2005).

There are various nanomaterials used in food industry, which can be classified as organic and inorganic. Inorganic nanoparticles like zinc oxide, silver, copper and titanium oxide are used for improving the physical performance, durability, barrier properties and biodegradation. Nanoparticles are coated on food packaging film or materials which are utilized for preservation and safety, these nanoparticles have the antimicrobial activity against the food spoilage microorganisms and foodborne pathogens (Bradley et al. 2011). Plastics beer bottles are embedded with nanoclays, which increases strength, makes them more shatterproof, increases the shelf life by acting as a barrier to keep oxygen outside the bottle and carbon dioxide inside. ZnO nanoparticles are already integrated into food packaging, ZnO coated food packaging films are developed as ZnO nanoparticles show stability under intensive processing conditions and antibacterial properties (Patel et al. 2014; Taylor et al. 2005). Similarly, silver nanoparticles show antibacterial properties, biosynthesized silver nanoparticles are incorporated into sodium alginate films as it shows antibacterial effect (Mohammed Fayaz et al. 2009). Films are prepared using silver nanoparticles; they are deposited on multi-layered linear low-density polyethylene (LLDPE) by laminating, casting and spraying (Sánchez-Valdes et al. 2008). Nanoparticles of titanium oxide are used widely due to its photocatalytic activity, non-toxicity, and antibacterial activity against a wide spectrum of microorganisms and utilized as self-disinfecting of surfaces (Fujishima et al. 2000). Titanium oxide nanopowder coated on oriented-polypropylene (OPP) film showed a reduction in bacterial cells (Chawengkijwanich and Hayata 2008).

5.2 *Applications*

5.2.1 Nanosensors

Combination of advanced nanoelectronics and suitable functional nanomaterials with smart biological components allows the development of highly specific and selective sensing devices for the detection of viruses, pathogenic microorganisms, detrimental chemicals and physical contaminants in food (Archana H 2012). With the help of nano fluidic technology, miniature sensing devices are developed for field application (Fig. 2).

Nano-biosensors are used to detect gases, pathogens, or toxins in packaged foods. They are also used in food processing plants, alerting consumers, procedures, and distributors on the safety status of food (Baeumner 2004). An electrochemical glucose nano biosensor fabricated by layer by layer self-assembly of polyelectrolyte for detection and quantification of glucose (Rivas et al. 2007). Sensors prepared by liposome nanovesicles are used for detecting peanut allergenic proteins in chocolate and pathogens (Edwards and Baeumner 2006; Wen et al. 2005). An electronic nose is a device, which has a combination of chemical sensors, linked with data processing system; which mimics like the nose, helps to identify different types of odor. These electronic noses are used to detect the fruit odors (J. S Detection of fruit odors using an electronic nose (n.d)).

The electronic nose can be used to measure chemical and physical properties of pears (Zhang et al. 2008), helpful in monitoring strawberry aroma changes during osmotic dehydration (Buratti et al. 2006). The nanosensor can also be used to check the milk quality during the industrial processing. Ultimately, the nanotechnology-based electronic nose can be useful for monitoring and control; more accurate volatiles measurement than measuring temperature and the time is taken, quality assurance and more (J. S Detection of fruit odors using an electronic nose (n.d)). Nanosensors are integrated into food packaging and confirm if the food product is fit enough for the human consumption.

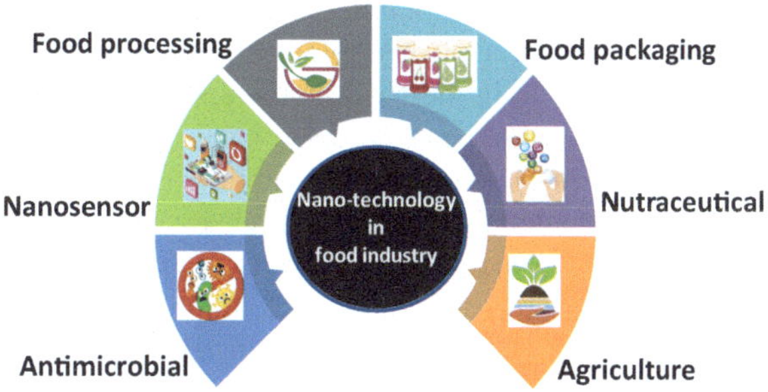

Fig. 2 Applications of nanotechnology in food industry

5.2.2 Food Processing

Food products consist of nanoscaled biomolecules such as proteins, carbohydrates. These nanomaterials are either naturally present in foods or formed during the transformation of preliminary products. In food processing, nanoparticles can be used for improving the nutritional quality of food, flow properties, flavour, colour, and stability or to increase shelf life, to protect functional ingredients such as antimicrobials or vitamins, for instance, chitosan nanoparticles are used as antimicrobial compounds to enhance food safety (MF Hochella 2008).

Food processing includes removal of toxins, prevention from pathogens, preservation, enhancing the consistency of foods. Food processing is equally an important section along with formulation and preparation, as it has a significant impact on the physicochemical properties of food such as texture, flavour release profile, stability and others (A. Wallace and Sahu 2017).Properties of different food-based nanomaterials vary from material to material, like peptide and polysaccharide nanomaterial properties vary from the properties of metal and metal oxide nanoparticles and hence several nanostructured food ingredients are being developed which acts as anticaking agents and gelatin agents and claims to offer an enhanced taste, texture, and consistency (Cientifica Report Nanotechnologies in the Food Industry 2006). It also helps to increase the shelf life of different food materials and also reduces the extent of wastage of food due to microbial infestation (Pradhan et al. 2015). Nanocarriers have been introduced as delivery systems to carry food additives in food products without disturbing their elementary morphology. Nanocapsules as a delivery system play an important role in processing sector as the functional property are maintained by encapsulating simple solutions, colloids, emulsions, biopolymers and others into foods. Self-assembled nanostructured lipids serve as a liquid carrier of healthy components that are insoluble in water and fats called nanodrops, these nanodrops are useful to inhibit transportation of cholesterol from the digestive system into the bloodstream (Abbas et al. 2009; Dingman 2008).

Furthermore, the role of nanotechnology in food processing can be evaluated by considering its part by upgrading the food products in terms of texture, appearance, taste, nutritional value, and food shelf-life. Nanoencapsulation techniques have been implemented to improve the flavour release, retention and to deliver culinary balance (Nakagawa 2014). Multifunctional nanocarriers are fabricated for bioactive molecule protection and delivery, for instance, rutin is one of the basic dietary flavonoids with essential pharmacological activities but due to its poor solubility, it has limited application in the food industry. To eliminate this problem, ferritin nanocages are used which enhances the solubility along with better thermal and UV radiation stability of ferreting trapped rutin (Yang et al. 2015). The main advantage of using nanomaterials is that it can release encapsulated compounds much slower and over longer time periods, carriers of fragrances and flavours (Dekkers et al. 2011), and also provides a favourable improvement in the bioavailability of nutraceuticals due to its subcellular size (Singh et al. 2017).

Various nanomaterials used in food processing are nanocapsules in cooking oils to improve the bioavailability of neutraceuticals, nanotubes, and nanoparticles as gelation and viscosifying agents, nancapsule infusion of plant-based steroids to replace meat cholesterol, nanoemulsions and particles for better availability and dispersion of nutrients, nanoparticles to selectively bind and remove chemicals or pathogens from food.

5.2.3 Food Packaging

Food packaging techniques are designed to maintain the food flavour, quality, and protect from an infestation of microorganisms that leads to food spoilage. Utilization of nanomaterial helps to enhance the shelf life of food by avoiding the spoilage or loss of food nutrients. Nanomaterials as carriers, are used these days as delivery systems in food products which transport food additives without disturbing their basic morphology. An ideal delivery system consists of following properties: (i) targeted delivery of the active compound (ii) ensure availability at a target time and specific rate, and (iii) capability to conserve active compounds at suitable levels for long periods of time. Various nanomaterials in the form of emulsions, encapsulations, biopolymer matrices, colloids offer different and efficient delivery systems with all the above-mentioned properties (Bratovcic et al. 2015).

One of the main reasons behind the deterioration of food is due to oxygen, oxygen inside the food packaging causes oxidation of fats and oils and growth of microorganisms. It also responsible for discoloration, changes in texture, rancidity, and off-odor, and flavour problems. Nanotechnology plays a very significant role here, it has the potential to scavenge oxygen by using moisture absorber sheets and scavenging bags prepared using nanomaterials (Gaikwad et al. 2018). Nanocomposites are used for the controlled release of active substances from the food packaging materials, referred to as ‘active’ food packaging. Active packaging enhances the condition of the packed food, longer shelf life and improves sensory properties while maintaining the freshness and helping the migration of functional additives, such as minerals, probiotics, and vitamins, into food quality of food. Nanocomposites used in ‘active’ packaging consist of polymer composites with antimicrobial nanomaterials, like silver, zinc oxide, copper oxide, magnesium oxide (Chaudhry et al. 2008b). The most common material used in active food packaging is nanosilver in plastics which confers antimicrobial properties to improve food and beverage shelf life (de Azeredo 2013).

Other nanocomposites are used to enhance other physical properties to make the packaging more tensile, durable, or thermally stable, e.g. nanocomposites of titanium dioxide, iron oxides, silica, and alumina are UV absorbers and hence are used to prevent UV degradation of plastic polymers (Beltran et al. 2014). Similarly, nano-titanium nitride is used to advance strength of packaging materials, nano-calcium carbonate-polymer composites, nano-chitosan-polymer composites, biodegradable nanoclay, biodegradable cellulose nano-whiskers, and other gas-barrier coatings (e.g. nano-silica) (Reig et al. 2014). Nanocomposites of nanosilver and nanoclay or other nanomaterials are formulated to enhance the barrier and antimicrobial properties. Nano clay-nylon nanocomposites are used to maintain freshness and block out food odour. Nano-precipitated calcium carbonate is used to improve the mechanical properties, heat resistance and printing quality of polyethylene, nanofilms made up of zinc oxide calcium alginate are used as food preservative, the nanoscale hybrid structure of silica/polymer for better oxygen-diffusion barriers for plastics (Smolander and Chaudhry 2010). Biodegradable nanocomposites are used for packaging, as it has favourable potential towards the environment.

5.2.4 Delivery of Nutraceuticals

Nutraceuticals are a combination of nutrition and pharmaceuticals which have health benefits with their actual function of providing nutrition and hence used to prevent the occurrence of a disease or as its treatment. Principles of nanotechnology have been implemented by various researchers for the efficient delivery of nutraceuticals with the aim to enhance their biological activity. Several formulations have been utilized for the efficient delivery of this nutraceutical, like nanoemulsion, micelles, nanoparticles, nanocapsules, nanocochleats, nanocrystals, etc. (Gupta S 2010). These nanoformulations are benefitted in targeted delivery of encapsulated nutraceuticals with the controlled release and better bioavailability as well as provide protection for bioactive compounds such as vitamins, antioxidants, proteins, carbohydrates and lipids with enhanced functionality and stability (Quintanilla-Carvajal et al. 2010). When a multitude of bioactive and active ingredients are nano-encapsulated, they have an ability to break down and get absorbed by the common food after delivering their active ingredients (Ezhilarasi et al. 2012).

Lipid-based nanoencapsulation systems are used to elevate the performance of antioxidants by producing better solubility and bioavailability, stability, controlled release of food materials, protection of foods and nutraceuticals and prevention against unnecessary interactions with other food components. These lipid-based nanoencapsulation which delivers food and nutraceuticals are mainly nanoliposomes, nanocochleates, and archaeosomes. They act as carrier vehicles of nutrients, nutraceuticals, enzymes, food additives, and food antimicrobials (Mozafari et al. 2008). Preparation of Coenzyme Q nanoliposomes was done with the preferred encapsulation quality and stability (Mozafari et al. 2006). Tiny capsules made of particles one-tenth the size of a human cell known as Colloidsomes, assemble themselves into a hollow shell, where other molecules of any substance such as fat blockers, medicines, and vitamins can be placed inside this shell (Xia et al. 2006). Nanocochleates are nanocoiled particles that wrap around micronutrients and have the ability to stabilize and protect an extended range of micronutrients and can elevate the nutritional value of processed foods (Pathakoti et al. 2017).

Development of nanoencapsulation of probiotics, which blends of bacteria species incorporated in foods in the form of yogurts and yogurt-type fermented milk, cheese, puddings and fruit-based drinks. Encapsulation enables the longer shelf life of the product. These can be useful to deliver the probiotic bacterial formulations to certain parts of the gastrointestinal tract with specific receptors and may act as *de novo* vaccines with the capability of modulating immune responses (Vidhyalakshmi and Subhasree 2009).

5.2.5 Agriculture

Nanomaterials have a significant role in the agriculture sector as it has the potential to boost agricultural production by improving nutrient use efficiency with nanoformulation of fertilizers, boosting the ability of plants to absorb nutrients, agrochemicals for crop enhancement; introducing new tools for detection and treatment of disease, host-parasite interactions at the molecular level using nanosensors, contaminants removal from soil and water, postharvest management of vegetables and flowers, and reclamation of salt-affected soils, precision farming techniques and ultimately leading to highly qualitative and quantitative yield. (Avensblog 2016; Kumar et al. 2018; Mousavi and Rezaei 2011).

Nanomaterials enabled tiny sensors and monitoring systems will have an enormous impact on future of precision farming methodologies. As precision farming makes use of global satellite positioning systems, remote sensing devices through which it can measure highly confined environmental conditions which help determine whether crops are growing at supreme efficiency or precisely identifying the nature and location of problems. Nano-structured smart delivery systems could help in the efficient use of agricultural natural resources like water, nutrients, and chemicals through precision farming. When these nanosensors, are scattered on fields, they are expected to monitor and provide detailed data on crop growth and soil conditions, seeding, fertilizers, usage of chemicals and water. These nanosensors can be useful for diagnosis of the presence of the plant viruses or the level of soil nutrients. At the nanoscale, due to high surface to volume ratio and, ease of access to modify its surface makes it possible to bind selectively with particular biological proteins. Sensors prepared using nanomaterials such as carbon nanotubes or nano-cantilevers can be used to trap and measure individual proteins or even molecules. Nanoparticles can be engineered to trigger an electrical or chemical signal, in presence of crop pathogens like bacteria and viruses (Morrison 2006). Nanostructured catalysts will help elevate the efficiency of pesticides and herbicides, with minimal usage, and also protect the environment indirectly through the use of alternative energy supplies and clean-up existing pollutants (Ditta 2012).

The concept of nano-fertilizers- nano-enabled bulk fertilizers can help enhance crop productivity and reduce nutrient losses. There will thorough and rapid absorption of nano fertilizers by plants. Nano-encapsulated fertilizers with slow release have become a trend to minimize the fertilizer consumption and environmental pollution. Nutrients are released at a slower rate throughout the crop growth and plants are able to take up most of the nutrients without waste, this is done using naturally occurring minerals at nanoscale having a honeycomb-like crystal structure known as zeolites, it has a network of interconnected tunnels and cages which are loaded with nitrogen and potassium, mixed with other gradually dissolving ingredients containing phosphorous, calcium and a complete suite of minor and trace nutrients.

The releases of nutrients from the fertilizer capsule are regulated using a coating and cementing of nano and sub nano-composites (Liu et al. 2006).

Smart seeds are seeds which are imbibed with nano-encapsulations with specific bacterial strain, it helps reduce seed rate, safeguard right field stand and improvement in crop performance. Seeds are coated with nanomembranes, can sense the availability of water and allows seeds to release when ready for germination, aerial broadcasting of seeds embedded with a magnetic particle, detecting the moisture content during storage to take appropriate measure to reduce the damage and use of bio analytical nanosensors to determine the aging of seeds (Manjunatha et al. 2016).

Nanomaterial has a significant role in agriculture machinery as well, like application in machines structure, agriculture tools to increase their resistance to wear and corrosion. Developing strong mechanical tools having nano coating and sensors to control weed growth, to reduce friction in bearings, and in the production of alternative fuels to reduce environmental pollution (DeRosa et al. 2010). Eventually, by implementing the use of smart sensors, productivity in agriculture will be enhanced by providing accurate information and hence help farmers to make better decisions.

Multidisciplinary approaches could potentially improve food production, incorporating new emerging technologies, and disciplines such as chemical biology integrated with nanotechnologies to tackle existing biological bottlenecks that currently limit further developments. The potential benefits of nanotechnology for agriculture, food, fisheries, and aquaculture need to be balanced against concerns for the soil, water, environment, and the occupational health of workers (Sekhon 2014).

6 Toxicity and Safety Aspects

Generally, food and beverage industry are conservative and cautious when the future of nanotechnology and food comes into the picture, despite the evident potential of nanotechnology in the application of food engineering and processing. Nanomaterials, due to their high surface area, might have toxic effects which are not apparent in the bulk materials (Dowling 2004; Kumar et al. 2018). Apart from that, there might be potential and unexpected risks for their use in food industry. It is extremely unlikely that the representatives, scientists from food and beverage corporations and others involved with food and nanotechnology to provide specific details about the level of funding and industry partners. The food industry may be skittish about owning up to R&D on "atomically modified" food products after witnessing widespread rejection of genetically modified foods (Ravichandran 2010) (Fig. 3).

With the application of nanotechnology in food industries, there will be a wide range of food products, like any other new technologies, but there are some serious queries regarding the safety which requires attention by the industry and the policymakers. The generally recognized as safe (GRAS) list of additives universally accepted will have to be reassessed when used at the nanoscale. It was observed that

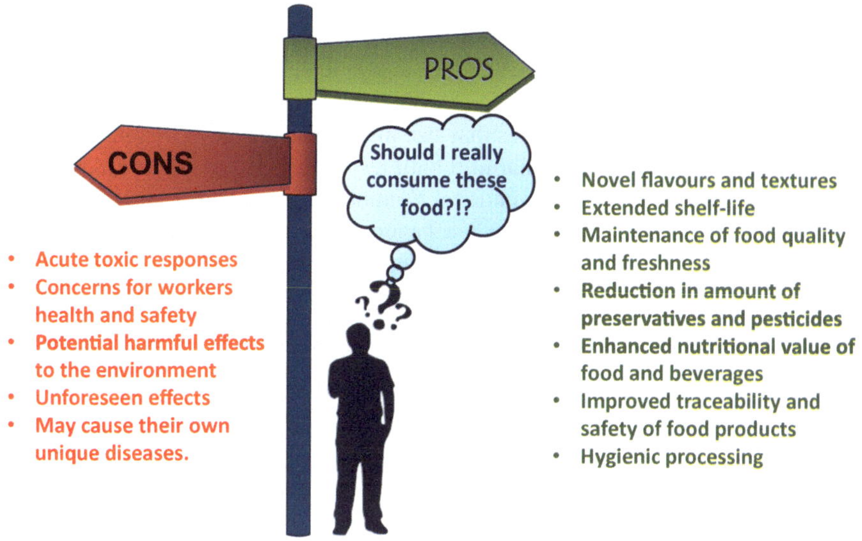

Fig. 3 Nanofood- a dilemma of the decade. The image depicts the positive and negative side of nanotechnology being used in food industry

rats breathing nanoparticles revealed a tendency of collecting nanoparticles in their brain and lungs, leading to increasing in biomarkers for inflammation and stress response. The main concern of any consumer will be about the toxicity and as nanoparticles are more reactive, more mobile and likely to be more toxic which must be addressed. Nanoparticles, when in the body have strong potential to result in increased oxidative stress which can generate free radicals, leading to DNA mutation, cancer, and possible fatality. It is yet to be understood whether enhancing the bioavailability of certain nutrients or food additives might affect human health negatively (Savolainen et al. 2010).

7 Regulations

It is essential for manufacturers to demonstrate that the food ingredients and food products are not injurious to health, according to the Food and Drug Administration (FDA) in the United States. However, this regulation does not explicitly cover nanoparticles, which could become harmful only in nanoscale applications. Therefore, no such distinct regulations exist for the use of nanotechnology in the food industry (Administration.). On the other hand, the European Union has suggested special regulations which are yet to be accepted and imposed. As FDA regulates on a product-by-product basis, it emphasizes that many products that are already under regulation contain particles in the nanoscale range. FDA expects that many products of nanotechnology will come under the jurisdiction of many of its centers; thus, the Office of Combination Products will likely absorb any relevant responsibilities (Chau et al. 2007).

The Institute of Food Science and Technology (IFST), a United Kingdom-based autonomous Professional form for food scientists and technologists has a different opinion of nanotechnology, it states that size matters and suggests that nanomaterials can be treated as potentially harmful until testing proves otherwise (TARVER). IFST recommended that the conventional E-numbering system for labeling should be used along with subscript "n" when nanoparticles are used as food additives (Maynard and Kuempel 2005). The UK government agrees with the Royal Society and the Royal Academy of Engineering that the presence of nanoparticles on ingredient labels is essential for consumers to make informed decisions. Therefore, an updated version of ingredient labeling will be a necessary requirement. Recently, The Swiss center for technology assessment analyzed the circumstances concerning nanotechnologies and food in Switzerland (Ireland 2009).

Certain rules and guiding principles for nanomaterials are established by the regulatory forms around the world that have ramifications for use in food. Due to lack of essential safety data, there is uncertainty over the regulation of nano-based products, which needs to inform regulatory bodies (Sandoval 2009). Efforts are made to facilitate international collaboration and information exchanges are underway to confirm approval and utilization of the advantages of nanotechnology (Magnuson 2009). Hence, organizations across the world are collecting information in an effort to decide how best to proceed (Kahan et al. 2009).

8 Future Aspects

In the past two decades, nanotechnology in the field of food sector has shown 40% increase in publications and 90% patent filings. More than 1000 companies now have an R&D focus on nanotechnology-based products (food?). In the previous few years, it was observed by the food industries that the nanotechnology has integrated into numerous food and food packaging products. Currently, there are around 300 and more nanofood products available in the market worldwide, which has motivated a huge rise of R&D investments in nanotechnology for food industries. Nanotechnology in food has a massive impact, ranging from basic food to food processing, from nutrition delivery to intelligent packaging (Pereira de Abreu et al. 2007) (Fig. 4).

Nanotechnology can benefit exclusively from processed foods in numerous ways. Programmable foods considered the ultimate dream of the consumer, will have designer food features built into it and a consumer can make a product of desired color, flavor, and nutrition using specially programmed microwave ovens. The trick is to formulate the food at the manufacturer's end with millions of nanoparticles of different colors, flavors, and nutrients and under the program in the oven set by the consumer based on his preferences, only selective particles are activated while others stay inert, giving the desired product profile (Kaynak and Tasan 2006).

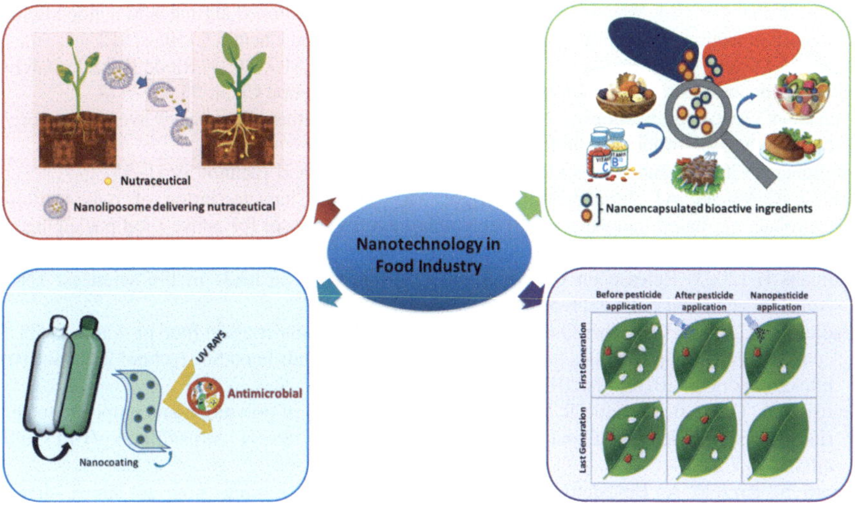

Fig. 4 The image highlights the few areas where nanotechnology is being exploited and how

Nanoscaled resources in food packaging will help amplify nourishment life, upgrade food safety, prepared customers that food is sullied or destroyed, repair tears in packaging, and uniform release added substances to grow the life of the food in the package.

Nanotechnology and nano-bio-info are one of the promising fields to maintain the leadership in food and food-processing industry, in the future. The first step towards it will be improving the safety and quality of food. Nanotechnology permits the change in the existing food systems to ensure products safety, creating a healthy food culture, and enhancing the nutritional quality of food.

Acknowledgement Funding received from the Department of Biotechnology, Government of India under the project "NanoToF: Toxicological evaluation and risk assessment on Nanomaterials in Food" (grant number BT/PR10414/PFN/20/961/2014) and DST SERB Project "Nanosensors for the Detection of Food Adulterants and Contaminants" (grant number EMR/2016/005286) is gratefully acknowledged. Financial assistance by The Gujarat Institute for Chemical Technology (GICT) for the Establishment of a Facility for environmental risk assessment of chemicals and nanomaterials is also acknowledged.

References

Abbas KA, Saleh AM, Mohamed AR, Mohd Azhan N (2009) The recent advances in the nanotechnology and its applications in food processing: a review. J Food Agric Environ 7(3–4):14–17

Administration. UFaD https://www.fda.gov/ScienceResearch/SpecialTopics/Nanotechnology/ucm301114.htm

Archana HCC (2012) Impact of marine nanoparticles for sustained drug delivery. Ind J Innov Develop 1(S8):37–39

Astete CE, Sabliov CM, Watanabe F, Biris A (2009) Ca2+ cross-linked alginic acid nanoparticles for solubilization of lipophilic natural colorants. J Agric Food Chem 57:7505–7512
Avella M, De Vlieger JJ, Errico ME, Fischer S, Vacca P, Volpe MG (2005) Biodegradable starch/clay nanocomposite films for food packaging applications. Food Chem 93:467–474
Avensblog (2016) Application of Nanotechnology in Agriculture, http://www.avensonline.org/blog/application-of-nanotechnology-in-agriculture.html
Baeumner A (2004) Nanosensors identify pathogens in food. Food Technol 58:51–55
Beltran A, Valente AJ, Jimenez A, Garrigos MC (2014) Characterization of poly(epsilon-caprolactone)-based nanocomposites containing hydroxytyrosol for active food packaging. J Agric Food Chem 62:2244–2252. https://doi.org/10.1021/jf405111a
Boumans H (2003) Release on COMMAND: BIO-SWITCH, in leads in life sciences. TNO Nutrition and Food Zeist 22:4–5
Bradley EL, Castle L, Chaudhry Q (2011) Applications of nanomaterials in food packaging with a consideration of opportunities for developing countries. Trends Food Sci Technol 22:601–610. https://doi.org/10.1016/j.tifs.2011.01.002
Bratovcic A, Odobasic A, Catic S, Sestan I (2015) Application of polymer nanocomposite materials in food packaging. Croatian. J Food Sci Technol 7:86–94. https://doi.org/10.17508/cjfst.2015.7.2.06
Buratti S, Rizzolo A, Benedetti S, Torreggiani D (2006) Electronic nose to detect strawberry aroma changes during osmotic dehydration. J Food Sci 71:E184–E189. https://doi.org/10.1111/j.1750-3841.2006.00007.x
Cagri A, Ustunol Z, Ryser ET (2004) Antimicrobial edible films and coatings. J Food Prot 67:833–848
Cha DS, Chinnan MS (2004) Biopolymer-based antimicrobial packaging: a review. Crit Rev Food Sci Nutr 44:223–237
Chakraborty S, Shukla D, Mishra B, Singh S (2009) Lipid–an emerging platform for oral delivery of drugs with poor bioavailability. Eur J Pharm Biopharm 73:1–15
Chau C-F, Wu S-H, Yen G-C (2007) The development of regulations for food nanotechnology. Trends Food Sci Technol 18:269–280. https://doi.org/10.1016/j.tifs.2007.01.007
Chaudhry Q et al (2008a) Applications and implications of nanotechnologies for the food sector. Food Addit Contam 25:241–258
Chaudhry Q et al (2008b) Applications and implications of nanotechnologies for the food sector. Food Addit Contam Part A Chem Anal Control Expo Risk Assess 25:241–258. https://doi.org/10.1080/02652030701744538
Chawengkijwanich C, Hayata Y (2008) Development of TiO2 powder-coated food packaging film and its ability to inactivate Escherichia coli in vitro and in actual tests. Int J Food Microbiol 123:288–292. https://doi.org/10.1016/j.ijfoodmicro.2007.12.017
Chen L, Subirade M (2005) Chitosan/β-lactoglobulin core–shell nanoparticles as nutraceutical carriers. Biomaterials 26:6041–6053
Chen H, Weiss J, Shahidi F (2006) Nanotechnology in nutraceuticals and functional foods. Food Technol
Cientifica Report Nanotechnologies in the Food Industry (2006) Assessment of the potential use of nanomaterials as food additives or food ingredients iin relation to consumer safety and implication for regulatory controls
Cushen M, Kerry J, Morris M, Cruz-Romero M, Cummins E (2012) Nanotechnologies in the food industry–recent developments, risks and regulation. Trends Food Sci Technol 24:30–46
Das M, Saxena N, Dwivedi PD (2009) Emerging trends of nanoparticles application in food technology: safety paradigms. Nanotoxicology 3:10–18
Dasgupta N, Ranjan S, Mundekkad D, Ramalingam C, Shanker R, Kumar A (2015) Nanotechnology in agro-food: from field to plate. Food Res Int 69:381–400
de Azeredo HMC (2013) Antimicrobial nanostructures in food packaging. Trends Food Sci Technol 30:56–69. https://doi.org/10.1016/j.tifs.2012.11.006

Dekkers S et al (2011) Presence and risks of nanosilica in food products. Nanotoxicology 5:393–405. https://doi.org/10.3109/17435390.2010.519836
DeRosa MC, Monreal C, Schnitzer M, Walsh R, Sultan Y (2010) Nanotechnology in fertilizers. Nat Nanotechnol 5:91. https://doi.org/10.1038/nnano.2010.2
Deschamps B, Musaji N, Gillespie JA (2009) Food effect on the bioavailability of two distinct formulations of megestrol acetate oral suspension. Int J Nanomedicine 4:185
Dias MV, Nilda de Fátima FS, Borges SV, de Sousa MM, Nunes CA, de Oliveira IRN, Medeiros EAA (2013) Use of allyl isothiocyanate and carbon nanotubes in an antimicrobial film to package shredded, cooked chicken meat. Food Chem 141:3160–3166
Dingman J (2008) Nanotechnology its impact on food safety. J Environ Health. January/February, 47–50 PMID:18236938
Ditta A (2012) How helpful is nanotechnology in agriculture? Adv Nat Sci Nanosci Nanotechnol 3. https://doi.org/10.1088/2043-6262/3/3/033002
Dowling AP (2004) Development of nanotechnologies. Mater Today 7:30–35. https://doi.org/10.1016/s1369-7021(04)00628-5
Drusch S (2007) Sugar beet pectin: a novel emulsifying wall component for microencapsulation of lipophilic food ingredients by spray-drying. Food Hydrocoll 21:1223–1228
Duncan TV (2011) Applications of nanotechnology in food packaging and food safety: barrier materials, antimicrobials and sensors. J Colloid Interface Sci 363:1–24. https://doi.org/10.1016/j.jcis.2011.07.017
Dupas C, Lahmani M (2007) Nanoscience: nanotechnologies and nanophysics. Springer Science & Business Media, Berlin/Heidelberg
Edwards KA, Baeumner AJ (2006) Liposomes in analyses. Talanta 68:1421–1431. https://doi.org/10.1016/j.talanta.2005.08.044
Ezhilarasi PN, Karthik P, Chhanwal N, Anandharamakrishnan C (2012) Nanoencapsulation techniques for food bioactive components: a review food and bioprocess technology. 6:628–647. https://doi.org/10.1007/s11947-012-0944-0
Fathi M, Mozafari M-R, Mohebbi M (2012) Nanoencapsulation of food ingredients using lipid based delivery systems. Trends Food Sci Technol 23:13–27
Feynman RP (1959) Plenty of room at the bottom: an invitation to enter a new field of physics. In APS annual meeting Caltech
Flanagan J, Singh H (2006) Microemulsions: a potential delivery system for bioactives in food. Crit Rev Food Sci Nutr 46:221–237
food? Witfoni https://wwwfooddivecom/news/what-is-the-future-of-nanotechnology-in-food/446173/
Fujishima A, Rao TN, Tryk DA (2000) Titanium dioxide photocatalysis. J Photochem Photobiol C: Photochem Rev
Gaikwad KK, Singh S, Lee YS (2018) Oxygen scavenging films in food packaging. Environ Chem Lett 16:523–538. https://doi.org/10.1007/s10311-018-0705-z
Gallus S, Carola B, Tilo P, Holger F (2003) Sweet containing calcium
Garti N, Shevachman M, Shani A (2004) Solubilization of lycopene in jojoba oil microemulsion. J Am Oil Chem Soc 81:873–877
Garti N, Spernath A, Aserin A, Lutz R (2005) Nano-sized self-assemblies of nonionic surfactants as solubilization reservoirs and microreactors for food systems. Soft Matter 1:206–218
Gharsallaoui A, Roudaut G, Chambin O, Voilley A, Saurel R (2007) Applications of spray-drying in microencapsulation of food ingredients: an overview. Food Res Int 40:1107–1121
Gilligan B (2008) Nanny, Nano, Boo, Boo Food? http://www.towerofbabel.com/blog/2008/08/28/nanny-nano-boo-boo-food/
Golding M, Sein A (2004) Surface rheology of aqueous casein–monoglyceride dispersions. Food Hydrocoll 18:451–461
Graveland-Bikker J, De Kruif C (2006) Unique milk protein based nanotubes: food and nanotechnology meet. Trends Food Sci Technol 17:196–203

Gupta SCD, Mehla K, Sood P, Nair A (2010) An overview of neutraceuticals: current scenario. J Basic Clin Pharm 1:55–62

He X, Hwang H-M (2016) Nanotechnology in food science: functionality, applicability, and safety assessment. J Food Drug Anal 24:671–681. https://doi.org/10.1016/j.jfda.2016.06.001

Hochella MF Jr (2008) Nanogeoscience: from origins to cutting-edge applications. Elements 4:373–379

Honarvar Z, Hadian Z, Mashayekh M (2016) Nanocomposites in food packaging applications and their risk assessment for health. Electron Physician 8:2531–2538. https://doi.org/10.19082/2531

Ireland F (2008) The Relevance for Food Safety of Applications of Nanotechnology in the Food and Feed Industries The Food Safety and Authority of Ireland (FSAI), Abbey

Ireland FSAo (2009) The Relevance for Food Safety of Applications of Nanotechnology in the Food and Feed Industries. Food Standards Agency UK Web site 2009 http://wwwfoodgovuk/

J. S Detection of fruit odors using an electronic nose (n.d) https://spieorg/newsroom/0137-detection-of-fruit-odors-using-an-electronic nose?highlight&SSO=1

Joseph T, Morrison M (2006) Nanotechnology in agriculture and food. Institute of Nanotechnology, Nanoforum Organization

Kahan DM, Braman D, Slovic P, Gastil J, Cohen G (2009) Cultural cognition of the risks and benefits of nanotechnology. Nat Nanotechnol 4:87–90. https://doi.org/10.1038/nnano.2008.341

Kaynak C, Tasan CC (2006) Effects of production parameters on the structure of resol type phenolic resin/layered silicate nanocomposites. Eur Polym J 42:1908–1921. https://doi.org/10.1016/j.eurpolymj.2006.03.008

Kirdar SS (2015) Current and future applications of nanotechnology in the food industry. In: Conference paper, ISITES2015, Valencia-Spain Google Scholar

Kumar A, Ramalingam C, Dasgupta N, Ranjan S (2016) Nanoemulsions in food science and nutrition. In: Nanotechnology in Nutraceuticals. CRC Press, pp 157–186

Kumar A, Singh S, Shanker R, Dhawan A (2018) Chapter 1 Nanotoxicology: challenges for biologists. In: Nanotoxicology: experimental and computational perspectives. The Royal Society of Chemistry, London, pp 1–16. https://doi.org/10.1039/9781782623922-00001

Kumari A, Singla R, Guliani A, Yadav SK (2014) Nanoencapsulation for drug delivery. EXCLI J 13:265–286

Law BA, King JS (1985) Use of liposomes for proteinase addition to Cheddar cheese. J Dairy Res 52:183–188

Li Y, Zheng J, Xiao H, McClements DJ (2012) Nanoemulsion-based delivery systems for poorly water-soluble bioactive compounds: influence of formulation parameters on polymethoxyflavone crystallization. Food Hydrocoll 27:517–528

Liu X-m, Feng Z-b, Zhang F-d, Zhang S-q, He X-s (2006) Preparation and testing of cementing and coating nano-subnanocomposites of slow/controlled-release fertilizer. Agric Sci China 5:700–706. https://doi.org/10.1016/s1671-2927(06)60113-2

Magnuson BA (2009) Nanoscale materials in foods: existing and potential sources. In: Intentional and unintentional contaminants in food and feed, ACS Symposium Series, pp 47–55. https://doi.org/10.1021/bk-2009-1020.ch004

Manjunatha S, Biradar D, Aladakatti YR (2016) Nanotechnology and its applications in agriculture: a review. J Farm Sci 29:1–13

Mauriello G, De Luca E, La Storia A, Villani F, Ercolini D (2005) Antimicrobial activity of a nisin-activated plastic film for food packaging. Lett Appl Microbiol 41:464–469. https://doi.org/10.1111/j.1472-765X.2005.01796.x

Maynard AD, Kuempel ED (2005) Airborne nanostructured particles and occupational health. J Nanopart Res 7:587–614. https://doi.org/10.1007/s11051-005-6770-9

McClements D, Decker E, Weiss J (2005) Novel procedure for creating nano-laminated edible films and coatings US Patent Application UMA:05–27

Meghani N, Patel P, Kansara K, Ranjan S, Dasgupta N, Ramalingam C, Kumar A (2018) Formulation of vitamin D encapsulated cinnamon oil nanoemulsion: its potential anti-cancerous activity in human alveolar carcinoma cells. Colloids Surf B: Biointerfaces 166:349–357

Mirhosseini M (2016) Evaluation of antibacterial effect of magnesium oxide nanoparticles with nisin and heat in milk. Nanomedicine Journal 3:135–142

Mohammed Fayaz A, Balaji K, Girilal M, Kalaichelvan PT, Venkatesan R (2009) Mycobased synthesis of silver nanoparticles and their incorporation into sodium alginate films for vegetable and fruit preservation. J Agric Food Chem 57:6246–6252. https://doi.org/10.1021/jf900337h

Morillon V, Debeaufort F, Blond G, Capelle M, Voilley A (2002) Factors affecting the moisture permeability of lipid-based edible films: a review. Crit Rev Food Sci Nutr 42:67–89

Morrison TJaM (2006) Nanoforum Report: nanotechnology in agriculture and food A Nanoforum report, available for download from wwwnanoforumorg

Mousavi SR, Rezaei M (2011) Nanotechnology in agriculture and food production. J Appl Environ Biol Sci 1:414–419

Mozafari MR, Flanagan J, Matia-Merino L, Awati A, Omri A, Suntres ZE, Singh H (2006) Recent trends in the lipid-based nanoencapsulation of antioxidants and their role in foods. J Sci Food Agric 86:2038–2045. https://doi.org/10.1002/jsfa.2576

Mozafari MR, Johnson C, Hatziantoniou S, Demetzos C (2008) Nanoliposomes and their applications in food nanotechnology. J Liposome Res 18:309–327. https://doi.org/10.1080/08982100802465941

Nakagawa K (2014) Nano- and micro-encapsulation of flavor in food systems. Nano- and Microencapsulation for foods, Chap 10, ed H-S Kwak Wiley,Oxford, 249–272 doi: https://doi.org/10.1002/9781118292327ch10

Neo YP, Ray S, Jin J, Gizdavic-Nikolaidis M, Nieuwoudt MK, Liu D, Quek SY (2013) Encapsulation of food grade antioxidant in natural biopolymer by electrospinning technique: a physicochemical study based on zein–gallic acid system. Food Chem 136:1013–1021

Okutan N, Terzi P, Altay F (2014) Affecting parameters on electrospinning process and characterization of electrospun gelatin nanofibers. Food Hydrocoll 39:19–26

Patel P et al (2014) Cytotoxicity assessment of ZnO nanoparticles on human epidermal cells. Mol Cytogenet 7:P81

Patel P, Kansara K, Singh R, Shukla RK, Singh S, Dhawan A, Kumar A (2018) Cellular internalization and antioxidant activity of cerium oxide nanoparticles in human monocytic leukemia cells. Int J Nanomedicine 13:39–41. https://doi.org/10.2147/IJN.S124996

Pathakoti K, Manubolu M, Hwang H-M (2017) Nanostructures: current uses and future applications in food science. J Food Drug Anal 25:245–253. https://doi.org/10.1016/j.jfda.2017.02.004

Pereira de Abreu DA, Paseiro Losada P, Angulo I, Cruz JM (2007) Development of new polyolefin films with nanoclays for application in food packaging. Eur Polym J 43:2229–2243. https://doi.org/10.1016/j.eurpolymj.2007.01.021

Pradhan N, Singh S, Ojha N, Shrivastava A, Barla A, Rai V, Bose S (2015) Facets of nanotechnology as seen in food processing, packaging, and preservation industry. BioMed Reserach International 2015:365672. https://doi.org/10.1155/2015/365672

Quintanilla-Carvajal MX, Camacho-Diaz BH, Meraz-Torres LS, Chanona-Perez JJ, Alamilla-Beltran L, Jimenez-Aparicio A, Gutierrez-Lopez GF (2010) Nanoencapsulation: a new trend in food engineering processing. Food Eng Rev 2:39–50. https://doi.org/10.1007/s12393-009-9012-6

Rao M (2009) Nanoscale particles in food and food packaging. J Food Sci 74

Ravichandran R (2010) Nanotechnology applications in food and food processing: innovative green approaches, opportunities and uncertainties for global market. International Journal of Green Nanotechnology: Physics and Chemistry 1:P72–P96. https://doi.org/10.1080/19430871003684440

Reig CS, Lopez AD, Ramos MH, Ballester VAC (2014) Nanomaterials: a map for their selection in food packaging applications. Packag Technol Sci 27:839–866. https://doi.org/10.1002/pts.2076

Reza Mozafari M, Johnson C, Hatziantoniou S, Demetzos C (2008) Nanoliposomes and their applications in food nanotechnology. J Liposome Res 18:309–327

Rhim J-W (2004) Increase in water vapor barrier property of biopolymer-based edible films and coatings by compositing with lipid materials. Food Sci Biotechnol 13:528–535

Rivas GA, Miscoria SA, Desbrieres J, Barrera GD (2007) New biosensing platforms based on the layer-by-layer self-assembling of polyelectrolytes on Nafion/carbon nanotubes-coated glassy carbon electrodes. Talanta 71:270–275. https://doi.org/10.1016/j.talanta.2006.03.056

Sánchez-Valdes S, Ortega-Ortiz H, Ramos-de Valle LF, Medellín-Rodríguez FJ, Guedea-Miranda R (2008) Mechanical and antimicrobial properties of multilayer films with a polyethylene/silver nanocomposite layer. J Appl Polym:NA–NA. https://doi.org/10.1002/app.29051

Sandoval BM (2009) Perspectives on FDA's Regulation of Nanotechnology: Emerging Challenges and Potential Solutions https://onlinelibrarywileycom/action/doSearch?AllField=Perspectives+on+FDA%E2%80%99s+Regulation+of+Nanotechnology%3A+Emerging+Challenges+and+Potential+Solutions

Sanguansri P, Augustin MA (2006) Nanoscale materials development–a food industry perspective. Trends Food Sci Technol 17:547–556

Savolainen K et al (2010) Nanotechnologies, engineered nanomaterials and occupational health and safety – a review. Saf Sci 48:957–963. https://doi.org/10.1016/j.ssci.2010.03.006

Sekhon BS (2010) Food nanotechnology – an overview. Nanotechnol Sci Appl 3:1–15

Sekhon BS (2014) Nanotechnology in Agri-food production: an overview. Nanotechnol Sci Appl 7:31–53. https://doi.org/10.2147/NSA.S39406

Semo E, Kesselman E, Danino D, Livney YD (2007) Casein micelle as a natural nano-capsular vehicle for nutraceuticals. Food Hydrocoll 21:936–942

Shukla K (2012) Nanotechnology and emerging trends in dairy foods: the inside story to food additives and ingredients. Int J Nano Sci & Tech 1:41–58

Singh T, Shukla S, Kumar P, Wahla V, Bajpai VK (2017) Application of nanotechnology in food science: perception and overview. Front Microbiol 8:1501. https://doi.org/10.3389/fmicb.2017.01501

Smolander M, Chaudhry Q (2010) Nanotechnologies in food packaging. Nanotechnologies in Food:86–101. https://doi.org/10.1039/9781847559883-00086

TARVER T Food Nanotechnology Scientific Status Summary synopsis, http://wwwiftorg/~/media/Knowledge%20Center/Science%20Reports/Scientific%20Status%20Summaries/Editorial/editorial_1106_functionalmaterialinfoodpdf

Taylor TM, Davidson PM, Bruce BD, Weiss J (2005) Liposomal nanocapsules in food science and agriculture. Crit Rev Food Sci Nutr 45:587–605. https://doi.org/10.1080/10408390591001135

Tumbarski Y, Lante A, Krastanov A (2018) Immobilization of Bacteriocins from lactic acid Bacteria and possibilities for application in food biopreservation. Open Biotechnol J 12

Vance ME, Kuiken T, Vejerano EP, McGinnis SP, Hochella MF Jr, Rejeski D, Hull MS (2015) Nanotechnology in the real world: redeveloping the nanomaterial consumer products inventory. Beilstein J. Nanotechnol 6:1769–1780. https://doi.org/10.3762/bjnano.6.181

Vidhyalakshmi RBR, Subhasree RS (2009) Encapsulation "the future of probiotics" – a review. Adv Biol Res 3–4:6–103

Wallace Hayes A, Sahu SC (2017) Nanotechnology in the food industry: a short review. Food Safety Magazine

Weiss J, Takhistov P, McClements DJ (2006) Functional materials in food nanotechnology. J Food Sci 71

Wen H-W, Borejsza-Wysocki W, DeCory TR, Baeumner AJ, Durst RA (2005) A novel extraction method for peanut allergenic proteins in chocolate and their detection by a liposome-based lateral flow assay. Eur Food Res Technol 221:564–569. https://doi.org/10.1007/s00217-005-1202-8

Xia S, Xu S, Zhang X (2006) Optimization in the preparation of Coenzyme Q10 nanoliposomes. J Agric Food Chem 54(17):6358–6366

Yang M, Kostov Y, Rasooly A (2008) Carbon nanotubes based optical immunodetection of Staphylococcal Enterotoxin B (SEB) in food. Int J Food Microbiol 127:78–83

Yang R et al (2015) Synthesis of homogeneous protein-stabilized rutin nanodispersions by reversible assembly of soybean (Glycine max) seed ferritin. RSC Adv 5:31533–31540. https://doi.org/10.1039/c5ra03542b
Yiannaka A (2012) Consumer attitudes and labeling regimes as determinants of the market success of food nanotechnology
Zhang H, Wang J, Ye S (2008) Predictions of acidity, soluble solids and firmness of pear using electronic nose technique. J Food Eng 86:370–378. https://doi.org/10.1016/j.jfoodeng.2007.08.026
Zhou G (2013) Nanotechnology in the food system: consumer acceptance and willingness to pay

Supercritical Fluid Techniques to Fabricate Efficient Nanoencapsulated Food-Grade Materials

Umar Garba and Balarabe Bilyaminu Ismail

1 Introduction

In the recent decade, nanotechnology is used to encapsulate bioactive compounds (e.g., micronutrients, antioxidants, enzymes, nutraceuticals, and polyphenols) and protect them from adverse conditions while at the same time ensuring control release of compounds at a targeted site (Ezhilarasi et al. 2013). This fast-emerging area provides great potentials for enhancing the stability, control release and bioavailability of bioactive compounds (Alessa 2014; Suganya and Anuradha 2017). The encapsulation process involves enclosing the desired component within an inert material that provides protection from the environment and control the release of the substance. Consequently, undesirable odours, flavour, and colours can be masked resulting in the improvement of the organoleptic attributes of the end products (Gómez-Estaca et al. 2015).

An important area that combines aspects of nanotechnology with encapsulation is referred to as nanoencapsulation. It is a process, which involves coating substances within another material at nanoscale sizes (between 10 to 1000 nm) and expressed as nanocapsules and nanospheres. Nanoencapsulation differs from micro-encapsulation technique, which involves the production of larger particles at a microscale (between 1 to 1000 μm) (Fathi et al. 2012; Suganya and Anuradha 2017). The basic requirements for a food grade nanoencapsulated compounds are that it should be non-toxic, biodegradable and suitable for human consumption, and must

U. Garba (✉)
Department of Food Science and Technology, Faculty of Agriculture, Bayero University Kano, PMB 3011, Kano state, Nigeria
e-mail: ugarba.fst@buk.edu.ng

B. B. Ismail
Department of Food Science and Technology, Faculty of Agriculture, Bayero University Kano, PMB 3011, Kano state, Nigeria

College of Biosystems Engineering and Food Science, Zhejiang University, Hangzhou, China

U. Hebbar et al. (eds.), *Nano-food Engineering*, Food Engineering Series,
https://doi.org/10.1007/978-3-030-44552-2_2

use an ingredient that is generally regarded as safe (GRAS). Nanoencapsulation technique has initially been used to design a drug delivery system, however, researches in recent time have focused on its application in food systems. A number of methods have been employed for nanoencapsulation including supercritical fluids, self-assembly, nanoemulsification, lipid-based nanoencapsulation nanoprecipitation, high-pressure homogenization, nano complexation, coacervation, among others (Gutiérrez et al. 2013).

The use of supercritical fluid (SCF) technique for nanoencapsulation of heat-sensitive compounds such as bioactive components, micronutrients has become a promising technology with wider applications (Ezhilarasi et al. 2013). The SCF possesses a temperature and pressure greater than its critical points and still remained as a single phase (Kalani and Yunus 2011). The technique is very effective and safe for the production of nanoencapsulated particles. Its widely used due to its low critical temperature and the use of a minimum amount of organic solvent (Ezhilarasi et al. 2013). Carbon dioxide, water, propane, nitrogen are the common compounds that can be brought to the supercritical state. Among these compounds, carbon dioxide is the most commonly used supercritical fluid. This is because it is colourless, odourless, safe, highly pure, cost-effective, non-flammable, non-toxic and recyclable (Girotra et al. 2013; Sohail et al. 2017).

However, despite an incredible growth in the field of nanoencapsulation of bioactive components and the potentials of SCF for active food packaging, its usage is still rare. This chapter aims to fill the gap in knowledge regarding the use of SCF technology for nanoencapsulation by providing a broad review of the current application of this technology as it relates to nanoencapsulation of bioactive compounds in food grade materials, focusing on the techniques, properties, and applications. Topics covered include the properties of SCF, techniques, and application of supercritical fluid techniques to fabricate nanoencapsulated food grade materials. Here we conclude with a brief overview of the future trends in the nanoencapsulation of bioactive materials using SCF.

2 Supercritical Fluids and Their Properties

Supercritical fluids (SCFs) are substances in either liquid or gas state that exist at pressures and temperatures above their critical thermodynamic values (Brunner 2005; Ezhilarasi et al. 2013; King 2014; Subramaniam 2017). SCFs are highly compressive on reaching their supercritical state leading to a dramatic change in their properties (Subramaniam 2017). A number of unique properties can be seen at a stage where pressure and temperature of a substance reached or exceeded its critical levels as such, SCF behaves as either liquid or gas while it is actually neither. The supercritical status of SCF corresponds to a region where the physicochemical properties such as density, viscosity, the coefficient of diffusion, and heat conductivity values are at an intermediate state between those of a liquid and that of a gas (Deshpande et al. 2011). Thus, SCFs exhibit properties such as low viscosity, low density, high solvating power and high diffusivities that are typical of liquids and gases (Ezhilarasi et al. 2013).

King (2014) reported that the use of SCF in the processing of food materials has been on the rise since the late 1960s and to date, this represents undoubtedly the most successful application of SCFs and related compressed fluids. The solvent power of SCF is the highest for non-polar or slightly polar compounds and decreases as the molecular weight increases and can be removed easily by mere expansion to an ambient pressure. Carbon dioxide (CO_2) is particularly useful for the processing of food materials. SCFs are used for batch extractions of solids, multi-stage counter-current separation (fractionation) of liquids, and adsorptive and chromatographic separations. Special food applications of SCFs includes decaffeination of coffee, lipid extraction from foods, enrichment of vitamins from natural sources, removal of alcohol from wine and beer, and for the encapsulation of thermally sensitive compounds (Brunner 2005; Ezhilarasi et al. 2013).

The following are a brief overview of the general properties and thermo-physical properties of SCF as provided by Sadjadi (2016) and Carlès (2010).

1. The compressibility of any pure fluid goes to infinity when it approaches the critical point. Because of high compressibility of SCFs near the critical point, their density and density-dependent solvent properties, such as the dielectric constant and viscosity can be sensitively tuned through small changes in pressure, leading to variations in dissolving power.
2. In its supercritical state, the density of a SCF at a constant temperature increases with pressure and eventually decreases when the temperature rises, (at a constant pressure).
3. The coefficients of diffusion of SCFs are usually in an order of magnitude larger than those obtained in the liquid phase. The diffusivity is directly proportional to the temperature and inversely proportional to the pressure.
4. The viscosity of SCF is near that of normal a gas, which enhances mass transfer. Although temperature has little influence on the viscosity of a liquid, it can have a considerable effect on the viscosity of SCF.
5. There is no surface tension for SCFs because they are not subject to the vapour-liquid boundary.
6. The most interesting characteristics of SCFs which provide the basis for the SCFs processes is their "tunability" with pressure and temperature, particularly the tunability of their solvent power.
7. The thermos-physical properties of SCFs enable them to be dense and at the same time compressible; their coefficient of heat diffusion is generally very small in comparison with the standard gases whereas their coefficients of mass diffusion are generally much larger than in most liquids. Additionally, most of the properties of SCFs can be changed easily by small variations of the average temperature and pressure.

The advantages of using (SCFs) over other solvents are largely due to their physical and chemical properties that are intermediate between those of liquids and gases and their adjustability with temperature and pressure changes. SCFs possess

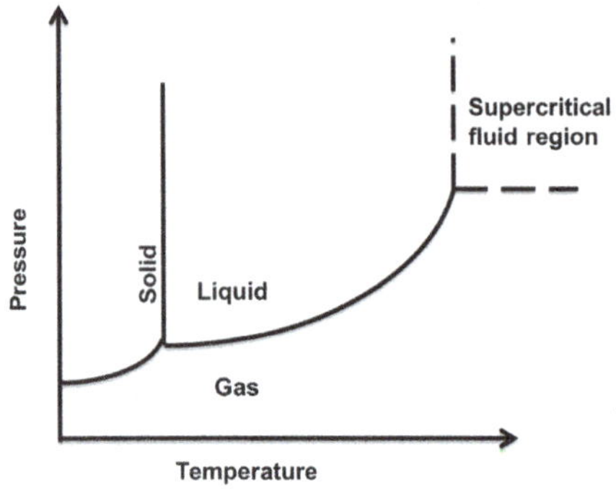

	Density(kg/m^{-3})	Viscosity (mPa/s)	Diffusivity(cm^2/s^{-1})
GAS	0.8-1.3	0.01-0.03	0.1-0.2
LIQUID	800-1200	0.4-1.1	0.00001-0.0001
SCF	300-1000	0.05-0.01	0.0001-0.001

Fig. 1 Phase diagram of a compound and physicochemical properties for liquids, gases, and SCFs

densities comparable to those of liquids, viscosities that are close those of gases and diffusion coefficients that are greater than those of liquids, in addition to a surface tension that is approximately zero (Tabernero et al. 2012). Due to these transport properties as shown in Fig. 1, SCFs are used for a wide range of applications such as extraction of bioactive compounds, chromatographic analyses, and particle formation. The liquid-gas mixing properties are particularly useful in various nanotechnology-related applications (Reverchon and Adami 2006; Tabernero et al. 2012; Silva and Meireles 2014).

3 The Concept of Nanoencapsulation

The concept of nanoencapsulation of food component covers an important application of food nanotechnology. It is a technique used to fabricate nanoparticles by incorporating any intended active ingredient into a desired wall/coating material (Fathi et al. 2012; Shah et al. 2016). The resulting particles after nanoencapsulation process are known as nanoparticles that comprise of nanocapsules and nanosphere (Fig. 2). The nanosphere has matrix type structure in which active materials or drugs are encapsulated within the particle or get absorbed at the surface of the sphere. While nanocapsules are a vesicular system in which an active substance usually dissolve in the inner core or get absorbed on to the surface of the capsules (Pinto et al. 2006). Nanocapsules are also described as hydrophilic or an oil like cavity surrounded by thin wall materials. There are novel delivery systems obtained at a nanoscale to entrap, protect and ensure a controlled release of bioactive compounds (Katouzian and Jafari 2017).

Nanocapsules have a promising application in food delivery systems due to their suitability for the encapsulation of many different bioactive ingredients including antioxidants, antimicrobial agents, phenolic compounds, natural pigments, peptides, essential fatty acids, and minerals (Mittal 2013; Jafari 2017; Katouzian and Jafari 2017). Nanocapsules have various advantages in comparison with

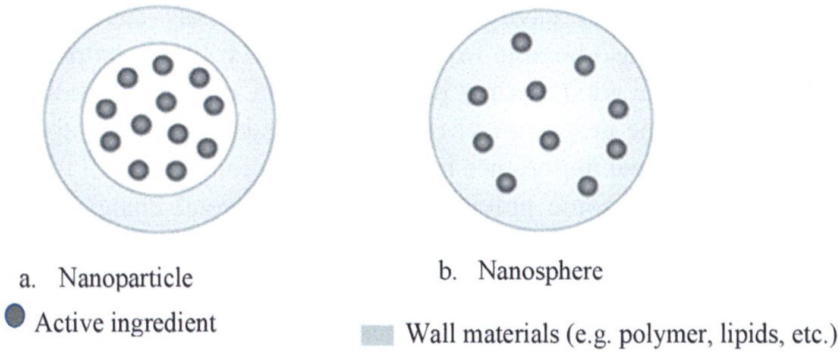

Fig. 2 Structure of nanoparticle (**a**) and nanosphere (**b**)

micron-sized carriers including a larger surface area, enhanced solubility, bioavailability, and controlled release of the encapsulated ingredients. Various wall materials are used to prepare nanocapsules including biopolymers (proteins, carbohydrates), lipids, chemical polymers, surfactants etc. (Jafari et al. 2015; Jafari 2017).

There are several nanoencapsulation technique that can either be (1) top-down approach: such as emulsification and emulsification-solvent evaporation; or (2) bottom-up approach: such as coacervation, nanoprecipitation, inclusion complexation, lipid-based nanoencapsulation, flow-focusing and supercritical fluid (SCF) technique (Garay et al. 2010; Ezhilarasi et al. 2013; Gutiérrez et al. 2013). The "top-down" method is characterised by a decrease in particle size during the encapsulation process due to the utilisation of various mechanical forces. On the other hand, the "bottom-up" process results in the increase of particle size by methods, such as self-assembly. Further, combinations of both approaches are useful in some cases (Jafari 2017).

The nanoencapsulation techniques provide a proper protection of active ingredient from interaction with other food components and degradation, which may lead to the loss of activity. It also enhances the bioavailability of active food ingredient during digestion, uptake in the gastrointestinal tract, transportation to the target sites and retention of volatile compounds (Shah et al. 2016; Trucillo et al. 2018).

4 Requirements for Food-Grade Materials for Nanoencapsulation

Nanoencapsulated compounds that are intended for food application are specifically needed to be non-toxic and biodegradable, and the coating material must have a GRAS status. Further, it is necessary to perform an appropriate toxicological investigation to confirm their safety for human consumption (Gutiérrez et al. 2013). Selection of an appropriate wall/coating material is one vital condition for encapsulation of active ingredients. For any coating material to be used, it must be compatible with the food product, have a good mechanical strength and be able to protect the core material from any environmental factor that may cause degradation. Additionally, it should also be able to limit the loss of volatile compounds, allow control release and have good thermal properties to the designated product (Silver and Meireles 2014). The properties of coating materials particularly their interaction with SCFs are of great importance in the encapsulation process. It is known that when SCF interact with some polymeric materials, several changes in physical properties of the polymer such as swelling, reduction in melting and glass transition temperatures and crystallinity are observed. These properties may enable proper incorporation of active compounds into the carrier materials (Martín et al. 2014).

Polymers are widely used as coating materials due to their ability to maintain the properties and activities of an active ingredient. They are easy to process, nontoxic and biocompatible. Polymers used in the food industry are of food grade and have a GRAS status. Various forms of carbohydrates used for encapsulation purposes include; starch, modified starch (Octenyl succinic anhydride (OSA) starch), maltodextrin, mannitol, etc. Proteins such as whey protein, soy protein isolate, gelatin, or a combination of both proteins and carbohydrates. Many lipids and inorganic materials have commonly been used as wall materials for example in the encapsulation of DHA. Other synthetic polymers such as carboxymethyl cellulose (CMC), hydroxypropyl methylcellulose (HPMC) have also been used (Antonio et al. 2014; Singh et al. 2018). The use of β-cyclodextrin, acacia gum, chitosan, and phospholipids were reported for coating various essential oils (Martín et al. 2014). Cyclodextrin and starches were also used for the encapsulation of various vitamins (Katouzian and Mahdi, 2016). Lipid-based materials have been reported for coating of hydroxycinnamic acids derivatives (HADs) such as ferulic, sinapic, caffeic and coumaric acids (Granata et al. 2018).

The use of carbohydrates and proteins alone has been associated with poor encapsulation efficiency and oxidative instability during storage due to the problem of caking and recrystallization of amorphous carbohydrate matrix, and the denaturation (unfolding) of protein molecules at the droplet-air interface. The use of polymers alone has also been associated with poor loading and water dispersibility (Singh et al. 2018).

5 Supercritical Fluid Technology to Fabricate Nanoencapsulated Food Grade Materials

There are several techniques used for the production of nanocapsules as the end products of nanoencapsulation and some of these include nano spray-drying, spray chilling, jet milling, fluidized bed coating, liposome entrainment, coacervation, thermal and ionic gelation etc. Nonetheless, there are some drawbacks related to the use of these methods including the difficulty of producing capsules with the desired particle size distribution and relatively higher operating temperatures. This leads to the demands for novel techniques to overcome these drawbacks. The use of SCFs serves as a great alternative technique for nanoencapsulation with improvements in the results obtained from other techniques while overcoming their limitations and allowing for the creation of novel formulations and products. Specific properties of SCFs particularly the CO_2 responsible for an increase application in nanoencapsulation include dynamic solvation power, low viscosity, moderately high diffusion coefficient, and near absence of contamination (Osorio-Tobón et al. 2016; Jafari 2017; Katouzian and Jafari 2017). The SCF technique can be categorised according to the function of SCF in the encapsulation process, as solvent, antisolvent, solute or cosolvent, nebulization compound, extractor and antisolvent techniques (Keven et al. 2014). There are various techniques that involve supercritical fluids and these include supercritical antisolvent process (SAS) and its various modifications, rapid expansion of supercritical solutions (RESS), gas antisolvent process (GAS), supercritical fluid extraction of emulsions (SFEE), aerosol solvent extraction system (ASES), precipitation with compressed fluid antisolvent (PCA) etc. However, the SAS has recently received an enormous attention more than other methods because of its feasibility of application (Reverchon and Della 1999; Jin et al. 2009; Nerome et al. 2013; Esfandiari and Ghoreishi 2015). Herewith, these various supercritical processes for nanoparticles formation are discussed in details.

5.1 Supercritical Antisolvent Process (SAS)

SAS is a popular technology in the field of SCF processing which has recently been used as an alternative to liquid antisolvent (LAS) precipitation for the production of nanoparticles with controlled particle size and distribution (Reverchon and Della 1999; Reverchon et al. 2000; De Marco and Reverchon 2008; Campardelli et al. 2015). In a traditional liquid anti-solvent technique, precipitation involves the addition of the antisolvent to a liquid solution of the desired solid compound. The antisolvent is normally miscible with the primary solvent but immiscible with the solute and thus, as the two solvents mix together, the solute precipitated (Reverchon and Della Porta 1999). While in the SAS process, two completely miscible solvents are used. The solute intended for precipitation is miscible with the first solvent but immiscible with the second one. The addition of antisolvent leads to the formation of a solution with two solvents, and a subsequent supersaturation and solute precipitation (Garay et al. 2010). Additionally, the mixing of SAS and the liquid solvent is faster than in the conventional LAS process. This leads to a higher supersaturation level and smaller particle diameters (Nerome et al. 2013). The basic principle of SAS and GAS technique are the same but the SAS process is a semi-continuous process (Silver and Meireles 2014). SAS technique typically uses Carbon dioxide as SCF, which serves as antisolvent for the polymer-solute mixture. (Chong et al. 2009). Depending on the pressure and temperature, the process can be carried out as a one-phase supercritical method or two-phase gas-liquid methods. However both the methods lead to the formation of a particle, in case of SAS, there is no interface between the solution and antisolvent fluid and therefore, the precipitation is achieved primarily by nucleation, and the particles show less agglomeration and smaller particle sizes (Nerome et al. 2013). The challenges associated with this process is that the control of the particle size and distribution of precipitate particle size, as well as the elimination of residual solvent, becomes very difficult (Reverchon and Della Porta 1999).

In a typical SAS process, the SC-CO_2 is fed at a constant flow rate into the precipitation chamber through the nozzles, until the required pressure is reached. The pure solvent is then passed into the chamber through the nozzles with the aim of obtaining a steady state composition condition of fluid phase during precipitation of the solute (De Marco and Reverchon 2008). In a normal operation of SAS process, the stream of SC-CO_2 and solution are co-current, while in modified method (SAS-EM) the streams are brought in contact by a cross pattern. The SAS process is semi-continuous, with CO_2 and organic solution continuously vented from the chamber (Bahrami and Ranjbarian 2007). To avoid blockage of the nozzles during the precipitation of the solute, the solvent is also fed at the initial phase. When a steady state is reached, the flow of the liquid solvent is stopped. The liquid solution is also delivered at a given flow rate through the nozzles. Solute precipitation is achieved when a certain quantity of organic solvent is injected, the liquid pump is stopped. However, the SC-CO_2 continue to flow in order to wash the chamber and eliminate the supercritical solution, which is formed by the liquid, solubilized in the SAS. The final purge with pure CO_2 is done to avoid solvent condensation during depressurization step and modify or solubilize the precipitates. The CO_2 flow is stopped at the end of the washing and the precipitator is depressurised to atmospheric pressure. Various solvents are used for the SAS process, some of the commonly used solvents include: *N*-methyl pyrrolidone (NMP), dimethylsulfoxide (DMSO), methanol (MeOH), ethanol (EtOH), dichloromethane (DCM), chloroform ($CHCL_3$), acetone (AC), isopropanol, formic acid, acetic acid, etc. (Reverchon et al. 2008; Fathi et al. 2012; Tamzeedul et al. 2017). The DMSO and NMP are the most popular solvents for the SAS (Montes et al. 2016). A schematic diagram of a typical SAS process is presented in Fig. 3.

In recent years, several researchers have focused on the synthesis of food and pharmaceutical grade nanocapsules using the SAS technology. Some examples of such studies include the β-carotene nanoparticles with a mean diameter below 400 nm in an aqueous media suspension using SAS precipitation of emulsion (Mattea et al. 2009), and antibiotics (e.g., ampicillin, amoxicillin, griseofulvin, and tetracycline) (Reverchon and Della Porta 1999). Vitamin D_3 nanoparticles were produced through coating with hydrogenated phosphatidylcholine (HPL) and ethanol with n-hexane as solvent. The optimum conditions of this process were a pressure of 8 MPa, temperature of 318 K and mass ratio of HPL to Vitamin D3 of 15%. A solution flow rate of 0.5 mL/min, the CO_2 flow rate of 6.0 kg/h and coaxial nozzle type were used in the process (Xia et al. 2011). Another attempt was made to encapsulate passion fruit seed oil in poly-lactic-co-glycolic acid (PLGA) with dichloromethane as solvent. Nanoparticles of size ranging from 721 to 1498 nm and encapsulation efficiency of 67.8 to 91% were achieved at optimum condition of 35 and 45 °C temperature, 90 and 110 bar pressure, and 92.5 and 95.0% CO_2 mass

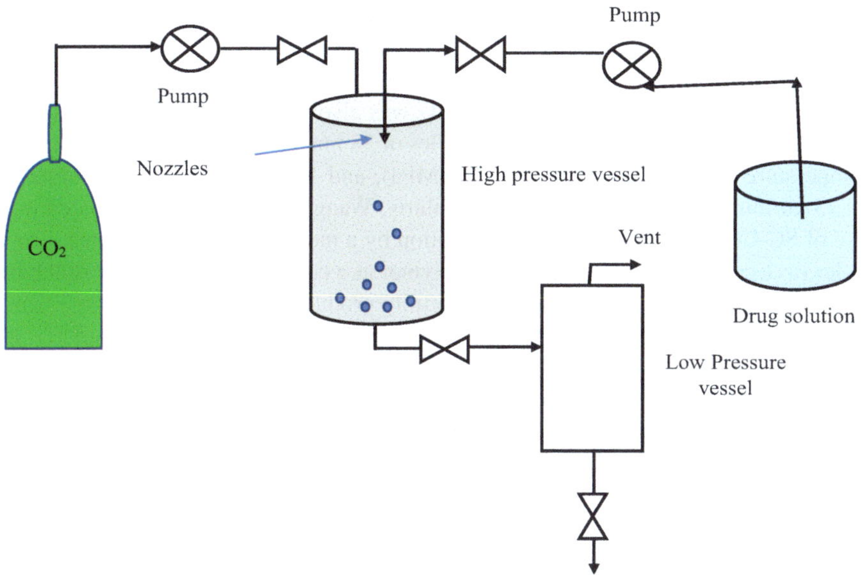

Fig. 3 A diagram for the concept of SAS/PCA process

fraction. Microscopic image analysis revealed that the pressure can affect the nanoparticles morphology and size, consequently, spherical to irregular shape particles were at both low pressure (90 bar) and high pressure (110 bar) (Oliveira et al. 2017). Similarly, Mezzomo et al. (2012) produced nanoparticles of carotenoids extract of shrimp coated using polymer through SAS technique with an encapsulation efficiency of about 74%. Moreso, Guamán-Balcázar et al. (2017) studied the production of mango leaf antioxidant extract by SAS techniques. Nanoparticles of sizes ranging from 40–440 nm were produced, with optimum processing conditions of 150 bar pressure, 50 °C temperature, 20 mg/mL extract concentration, 20 g CO_2/min flow rate and 5 g extract/min flow rate. The study has observed that initial concentration of the solution is a very crucial factor for obtaining nano- or submicron-particles of the leaf antioxidant. Likewise, their result revealed that changing the precipitation condition of temperature from 35 to 50 °C had an influence on the particle size even though the pressure, extract and SC-CO_2 flow rates have remained the same. Another study recommends the use of higher flow rate of CO_2 and liquid solution, higher pressure and lower temperature and lower nozzle diameter to obtain nanoparticles with smaller sizes. At a high flow rate of the liquid solution, the smaller particles are due to an increase in turbulence and effective mixing, which in turns favour the supersaturation. The CO_2 has a great effect on the morphology, and when a higher CO_2 flow rate is used, spherical nanoparticles with a narrower particles size distribution could be obtained (Montes et al. 2014). The bioactivity of lutein was successfully maintained through SAS nanoencapsulation with hydroxypropylmethylcellulose phthalate (HPMCP). The lutein/HPMCP nanocapsules were prevented from thermal and light degradation, and encapsulation efficiency of

88.41% was achieved. The nanocapsules size ranges between 163 to 219 nm at a pressure of 11 Mpa, temperature of 40 °C and concentration of HPMCP: lutein of 5:1 ratio (Jin et al. 2009). A lycopene nanoparticle coated with β-cyclodextrin at a ratio of 0.35:0.74 mg/ml lycopene: β-CyD was also produced through SAS technique. It was observed that smallest particles of 38 nm could be obtained at a higher temperature (50 °C), high pressure (14 MPa), and low flow rate of the solution (0.75 ml/min) (Nerome et al. 2013). Similarly, Wang (2004) has demonstrated the use of SC-CO_2 in SAS for nanoencapsulation by a model system of silica nanoparticles (as a core material) and Eudragit polymer as a coating material. The result has shown that the SAS process could successfully produce nanoparticles of 16–20 nm size. The coating process was found to be independent of surface hydrophilicity. It was also observed that the heterogeneous polymer nucleation where the nanoparticles serve as nuclei and subsequent deposition of polymer on the surface of nanoparticle due to the mass transfer and phase transition. In this study, the thickness of polymer coating was controlled by adjusting the ratio of polymer to core material ratio. Despite the popular application of SAS technique, the major drawback of the process is the inevitability of using an organic solvent (Oliveira et al. 2017).

An advance SAS method with an enhanced mass transfer has been studied to overcome the limitation of SAS process. It is known as a supercritical antisolvent-enhance mass transfer (SAS-EM). This modified technique uses a vibrating ultrasonic processor to atomise the solution into a micro-droplet. High turbulence resulted from this method helps to enhance the mixing operation and the subsequent mass transfer generates particles of smaller size (Miwa et al. 2016; Kankala et al. 2017).

5.2 *Rapid Expansion From Supercritical Solution (RESS)*

The RESS has been used for encapsulation of compounds with a good solubility in SC-CO_2 (Campardelli et al. 2015). The process utilises the high solvating power of SCF (Türk and Lietzow 2004) and it is based on the supersaturation and nucleation of a polymer when it is saturated with SCF at high pressure. Since the polymer is slightly soluble in SCF medium, the resultant solution becomes of very high pressure. When the solution is passed through an orifice or capillary tube, high-pressure drop occurred as a result of which the supersaturation and nucleation take place, and subsequently the formation of micro- or nanoparticles (Chen et al. 2005; Fathi et al. 2012; Girotra et al. 2013). For depressurization to occur, the orifice size of the precipitation chamber is kept at about 150 μm. This process gives smaller and uniform sized particles. Controlling the temperature, pressure, and nozzles orifice geometry can lead to a larger particle size (Girotra et al. 2013). Normally, at the end of RESS process, a solvent-free and dry product is obtained. This is because the solvent is in the gaseous phase and thus no additional drying is required (Chen et al. 2005). Since some materials are not sufficiently soluble in SC-CO_2 which is used by RESS technique, it is often required to modify the process by using organic solvent or cosolvent solid for more efficient particles production. Based on this modification, the RESS process can be classified into rapid expansion of supercritical solution into a liquid solvent (RESOLV), the rapid expansion of supercritical solution into an aqueous solution (RESAS) and rapid expansion of supercritical solution with cosolvent (RESS-SC) (Miwa et al. 2016). A schematic diagram of this process has been presented in Fig. 4.

The RESS techniques offer a solvent-free product because, after expansion, the solvent is in a form of a dilute gas. It also enables the nanoencapsulation of thermally labile materials and the formation of particles with lower than 500 nm diameter (Türk and Lietzow 2004; Calderone and Tallon 2008). It offers a great potential to produce a fine powder with small size distribution, thin films (Matson et al. 1987). This method is simple and more effective when a single nozzle is used. It also minimises the use of organic solvent and allows for reuse of SCF in a continuous process (Parhi and Suresh 2013).

RESS technique was successfully shown to produce a model nanoparticles of SiO_2 coated with poly(vinyl chloride) polymer, using a range of supercritical solvent (Matson et al. 1987). In another attempt, the RESS technique was used to prepare protein-conjugated nanoparticles of silver sulphide using a bovine serum albumin. Monodisperse particles obtained were well coated with the protein (Meziani and Sun 2003). Some disadvantages of the RESS process that could limit its use include the difficulty to scale up and design of nozzle and relatively high-energy cost related with the need to recompress CO_2 to enable its circulation into the system. Consequently, certain modifications are made to overcome some of these drawbacks (Mittal 2013). Another drawback is the fact that many polar and high molecular compounds are insoluble in the SCF especially CO_2 (Chen et al. 2005). Nevertheless, various compounds particularly pharmaceutical products or other materials of high economic value have been encapsulated using the RESS process. The use of the RESS process in the food industry and related products is said to be limited because of the moderate solubility of some food components such as fat-soluble vitamins in CO_2 (Osorio-Tobón et al. 2016).

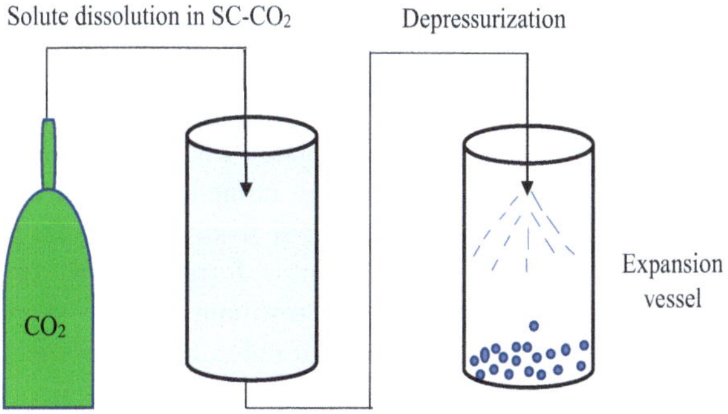

Fig. 4 A schematic diagram of the RESS process

5.3 *Gas Antisolvent (GAS)*

The GAS technique is commonly used for polymers that are insoluble in SCF thus; this process is a suitable alternative to their precipitation or recrystallization to form nanoparticles. In this method, the polymer is initially dissolved in a liquid solvent and a gas is used as antisolvent for the polymer (Garay et al. 2010; Girotra et al. 2013). The antisolvent is used in order to reduce the solubility of compounds that are dissolved in the solution and promote the nucleation and crystals formation (Miwa et al. 2016). The GAS technique involves saturating the polar solvent containing the dissolved substrate with SC-CO_2. By doing so, the solvent power of the polar solvent got reduced, thereby leading to the precipitation of the substrate (Girotra et al. 2013). It is important that the SCF be completely miscible with the liquid solvent, while the solute is insoluble in the SCF. Thus, when the liquid solution comes in contact with the SCF, the supersaturation and precipitation of solute in form of micro- or nanocapsules takes place (Garay et al. 2010). This technique can be batch or semi-continuous process with a good prospect for a scale up (Chen et al. 2005; Bahrami and Ranjbarian 2007). The important parameters in this process are pressure, temperature, flow rate and the initial concentration of SCF. The quality of the product can be controlled by these parameters through their effect on the thermodynamic, hydrodynamic and mass transfer kinetics, and precipitation (Chen et al. 2005). Elvassore et al. (2001) have demonstrated the use of GAS CO_2 precipitation to produce insulin nanoparticles coated with polyethylene glycol (PEG)/poly *l*-lactic acid (PLA). Similarly, a semi-continuous compressed CO_2 GAS precipitation was used to fabricate nanoparticles of insulin/PLA, and more than 70% product recovery was obtained under proper operating conditions of temperature, pressure, CO_2 flow rate and washing time (Caliceti et al. 2004). The major drawback of this technique is the lack of control over the particle formation, which makes it difficult to obtain monodisperse particles (Parhi and Suresh 2013).

5.4 *Supercritical Fluid Extraction of Emulsions (SFEE)*

The SFEE is a particle formulation process that combines the conventional emulsion techniques and the unique properties of SCF in producing nanoparticles (De Aguiar et al. 2016). In this technique, an oil-in-water emulsion is prepared with an active ingredient dissolved in a dispersed organic phase, and the aqueous phase contains the carrier and stabilizing agent. The emulsion is then exposed to SCF which enables the extraction of organic solvent from the organic phase, thereby leading to the precipitation of active ingredient in form of micro- or nanoparticles (Fathi et al. 2012; Gutiérrez et al. 2013). The resultant aqueous suspension can be dried into a powder. As compared to a conventional evaporation, the drying process in this technique avoids the high temperature and possible disruption of the product by gas bubbles formation during the evaporation (Fathi et al. 2012). The concept of this process is presented in Fig. 5.

The SFEE technique has been studied for the production of quercetin nanoparticles to enhance its delivery to the gastrointestinal tract using two different biopolymers (Pluronic L64® poloxamers and soybean lecithin) as coated materials. A quercetin loaded multivesicular liposomes were obtained with an average particle size of 100 nm and encapsulation efficiency of about 70% (Lévai et al. 2015). In a similar attempt, thermo-economic evaluation was carried out to assess the bottleneck of the scale-up of a semi-continuous SFEE technique for the production of quercetin microparticles. According to the study, decreasing CO_2 flow was found to be the major parameter to decrease the energy requirement and cost of the manufacturing process under SFEE technique (Lévai et al. 2017). The SFEE has also been used to fabricate carotenoids (β- carotene and lycopene) nanoparticles using OSA as surfactant and coating materials. The particle size of 344–366 nm with encapsulating efficiency of 34–89% was achieved. This work revealed that emulsion flow rate could influence the production of suspension carotenoids particularly the encapsulation efficiency and degree of isomerisation. The final particle size was influenced by the concentration of the carrier material/surfactant, while the efficiency of the encapsulation efficiency was influenced by the pressure and concentration of the carrier material (Santos et al. 2012). Other attempts were made to produce a composite micro- and nanoparticles of ketoprofen and indomethacin in biodegradable polymers (poly-lactic/glycolic acid (PLGA) and Eudragit RS) by Chattopadhyay et al. (2006) and gene encapsulation in the PLGA by Mayo et al. (2010). The benefit of SFEE over other SCF precipitation methods is that there is a correlation between the distribution of the diameters in emulsion droplets and in the particle suspension. It is thus possible to control the particle size in the final suspension by changing the parameters that directly influence the final emulsion droplets size through their formation (De Aguiar et al. 2016).

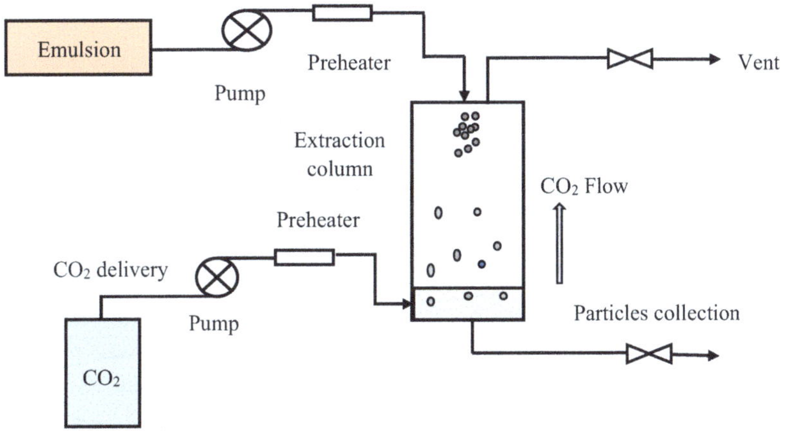

Fig. 5 The concept of SFEE process

5.5 Aerosol Solvent Extraction (ASES)

This technique uses the extraction properties of SCF to produce micro/nanoparticle. Selection of the organic solvent to be used in this method is based on its solubility in SC-CO_2 (Bleich et al. 1993; Girotra et al. 2013). In this process, the SCF is initially pumped to the top pressure vessel until the system reaches a constant pressure and temperature. The polymer and active ingredient are both dissolved in an organic solvent and then sprayed in form of a fine droplet into an SC-CO_2 medium (Figs. 6 and 7) through atomization nozzles (Girotra et al. 2013). The solution of active substance largely expanded in the vessel leading to the dissolution of SCF into droplet followed by supersaturation due to a reduction in the solvent power, which leads to nucleation and ultimately the formation of nanoparticles. Nanoparticles are collected at the bottom of the vessel, and the SCF and organic solvent mixture flow down to the depressurized tank where they are separated based on their suitable temperature and pressure conditions. The major advantage of ASES technique over GAS is that its suitability for continuous operation and has a great potential for a large-scale particles production (Parhi and Suresh 2013).

5.6 Precipitation With Compressed Fluid Antisolvent (PCA)

In PCA technique, the SCF (or compressed gas) is mixed with an organic solvent in a way that is completely in contrast to the GAS method. In this case, the organic solvent is dispersed in SFC, and unlike GAS method, this method does not require

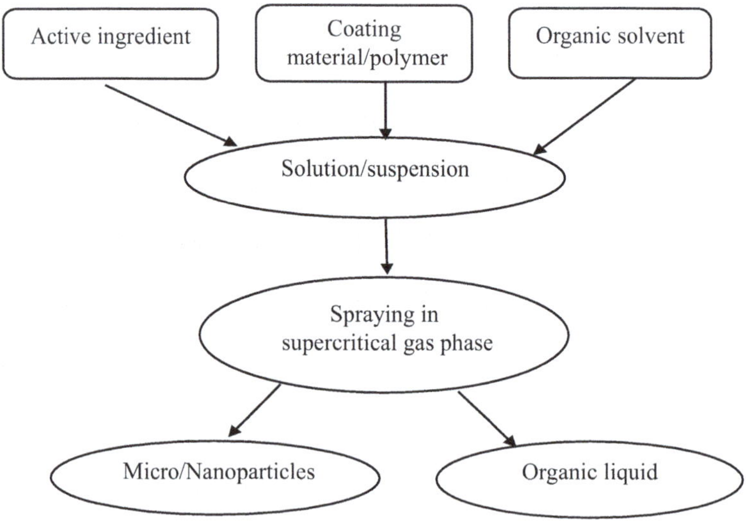

Fig. 6 Flowchart of the ASES process

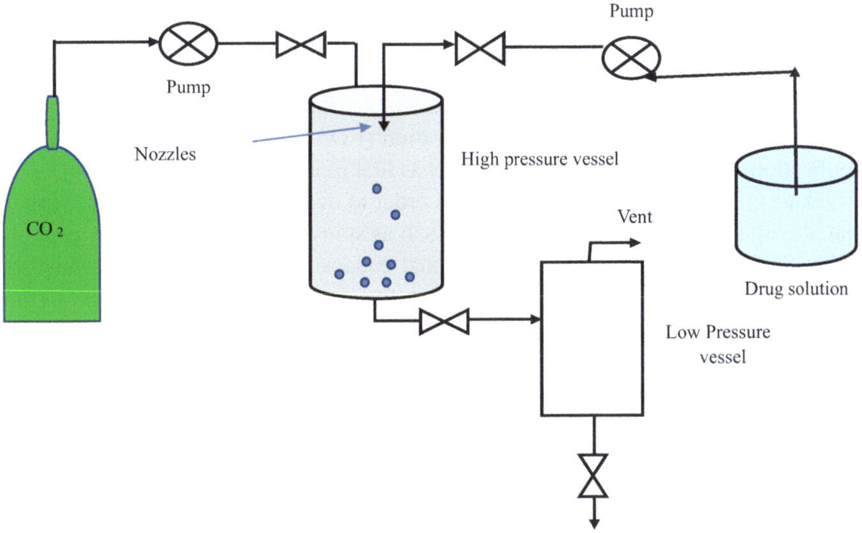

Fig. 7 Schematic diagram of a typical ASES process

a drying step. The SCF phase can be used in continuous or static mode (Bahrami and Ranjbarian 2007). The liquid solution is sprayed through nozzles into a vessel, which was pressurized with a dense gas. A typical schematic diagram of this process is presented in Fig. 3. The PCA technique is more suitable for systems where high supersaturation and quick precipitation are required because it provides a more efficient mass transfer between the solution and antisolvent. The efficiency of the mass transfer can be enhanced by adjusting the operating parameters for full miscibility of solvent and antisolvent, or by substituting the capillary nozzles with more sophisticated devices like coaxial devices, ultrasonic dispersion devices, or two-component jets. The nozzles provide an intimate contact between the antisolvent and solution before the entrance of the precipitator (Fusaro et al. 2005). An important control parameter in this process is the emulsion physical properties rather than the core material characteristics, thus, developing appropriate emulsion formula which will sustain the stable emulsion and ensure particle' suspension is critical (Parhi and Suresh 2013).

6 Application of SCF for Nanoencapsulation of Food Grade Materials

SCF based processes are important for the fabrication of nanoencapsulation products (Caliceti et al. 2004). In recent years, the SCF technique has attracted much attention for the encapsulation of bioactive ingredients (Fathi et al. 2012). Supercritical carbon dioxide (SC-CO_2) is an ideal medium for the particle encapsulation process owing to its relatively mild critical temperature (T_{c} = 304.1 K) and pressure (P_c = 7.38 MPa). The SC-CO_2 is chemically stable, nontoxic, non-flammable, readily available and inexpensive (Wang et al. 2005; Fathi et al. 2012). Generally, CO_2 is the most commonly used supercritical solvent recognised by the United States' Food and Drug Administration (FDA) in food and pharmaceutical applications. The use of SCF technology was first reported by Hannay et al. in 1879 (Kankala et al. 2017). It was developed in order to overcome some challenges associated with the conventional techniques such as spray drying (Chattopadhyay et al. 2006; Singh et al. 2018). In a conventional encapsulation technique, the coating materials are atomised and then dissolve in a fluidized bed reactor together with the active ingredients in form of small solid or liquid particles. Upon atomisation, the droplet of coating material permeates the bed particles and the solvent quickly evaporated. While in SCF technique, production of nanoparticles is achieved through a single precipitation and encapsulation step using SC-CO_2 (Arango-Ruiz et al. 2018). Generally, in SCF technique, the active ingredient (core material) and the polymer (coating material) are both dissolve in SFC, and the solution expanded through nozzles. The SCF is subsequently evaporated by the spraying process and eventually, the particles are precipitated (Reverchon and Della 1999). The SCF technology is based on the principle of co-precipitation of active compound and polymer using antisolvent (non-solvent) properties of the supercritical fluid carbon dioxide (SC-CO_2) (Chattopadhyay et al. 2006). Typically, a polymer-solute is dissolved in a liquid solvent to form a solution, which is sprayed into a chamber that is preconditioned with the SCF, causing a rapid contact between the two media. This form a supersaturated condition leading to a quick nucleation and growth that consequently generates smaller particles (Chong et al. 2009). The major problem associated with the SCF precipitation method is plasticization of polymers in the presence of SC-CO_2. The polymer-CO_2 solution tends to have high viscosity during atomization or solvent extraction, thereby leading to a strong particle agglomeration, inconsistent particle size and incomplete encapsulation (Chattopadhyay et al. 2006).

Oliveira et al. (2017) and Reverchon et al. (2010) stated that during the SCF precipitation process, SC-CO_2 acts like a plasticizer and reduce the glass transition temperature of the polymer, thereby making the particles to aggregate. The SCF nanoencapsulation technique is considered clean because the final precipitated solutes are free from any solvent. The process is also considered to have a suitable technological and biopharmaceutical properties and yield a product of good quality (Pinto et al. 2006). Table 1 provides a summary of some recent application of SCF to fabricate efficient nanoencapsulated food grade materials.

6.1 Advantage and Limitation of SCF Technology

SCF technology is regarded as green technology due to the environmental friendliness of the SCFs such as SC-CO_2. This technology also overcomes the challenges of conventional encapsulation techniques such as thermal and solvent-based

Table 1 Application of supercritical fluid nanoencapsulation using different food-grade coating materials

Core material	Coating material	SCF Technique	Solvent	Efficiency (%)	References
Carotenoid	Polymer	Supercritical anti-solvent		74	Mezzomo et al. (2012)
Astaxanthin	Poly(3-hydroxybutirate-co-hydroxyvalerate) (PHBV)	Supercritical anti-solvent	Dichloromethane	48.25	Machado et al. (2014)
Eugenol acetate	PHBV	Supercritical antisolvent	Dichloromethane	58	Loss et al. (2016)
Trans-resveratrol	PHBV	Supercritical antisolvent	Dichloromethane	99.54	Dal Magro et al. (2017)
Passion fruit seed oil	PLGA(poly(lactic-co-glycolic)Acid	Supercritical antisolvent	Dichloromethane	91	D A Oliveira et al. (2017)
Lycopene	Lecithin and α tocopherol	Supercritical antisolvent	N-dimethylformamide and tetrahydrofuran		Cheng et al. (2017)
Rosemary	Poloxamers	Supercritical antisolvent		90	Visentin et al. (2012)
Curcumin	Lactose	Supercritical antisolvent-fluidized bed process	Free	93	Matos et al. (2018)
Curcumin	Polyvinylpyrrolidone, Eudragit® L100, and Pluronic® F127	Supercritical antisolvent			Arango-Ruiz et al. (2018)
Olive pomace polyphenols	Liposomes	Supercritical assisted liposome formation	Ethanol	58	Trucillo et al. (2018)
Lutein	Hydroxypropylmethylcellulose phthalate (HPMCP)	Supercritical antisolvent	Acetone	88.41	Jin et al. (2009)
Lycopene	β-cyclodextrin	Supercritical antisolvent	N,N-dimethylformamide (DMF)		Nerome et al. (2013)
Insulin	Polyethylene glycol/poly – *l*-lactide	Gas Antisolvent	Dichloromethane and Dimethylsulphoxide		Elvassore et al. (2001)

(continued)

Table 1 (continued)

Core material	Coating material	SCF Technique	Solvent	Efficiency (%)	References
Insulin	Poly *l*-lactide (PLA)	Gas antisolvent	Dimethylsulphoxide (DMSO)		Caliceti et al. (2004)
B-carotene	Modified starch surfactant (n-octenyl succinic anhydrite)	Supercritical antisolvent	Dichloromethane		Mattea et al. (2009)
Curcumin	Polymer (Eudragita L100, Pluronic F127, and Polyvinylpyrrolidone)	Supercritical antisolvent	Absolute ethanol and 5%-pure acetone		(Arango-Ruiz et al. (2018)
Passion fruit seed oil	Poly-lactic co-glycolic acid (PLGA)	Supercritical antisolvent	Dichloromethane	67.8–91	D. A. Oliveira et al. (2017)
Vitamin D3a	Hydrogenated phosphatidylcholine (HPL)	Supercritical antisolvent	Ethanol and h-hexane		Xia et al. (2011)
Capsaicinoids	Hi-cap 100 modified starch	Supercritical fluid extraction of emulsion	Ethyl acetate	99.5	De Aguiar et al. (2016)
Quercetin	Pluronic L64® poloxamers and soybean lecithin	Supercritical fluid extraction of emulsion	Ethyl acetate	70	Lévai et al. (2015)
β-carotene and lycopene	n-octenyl succinic anhydride (OSA)-modified starch	Supercritical fluid extraction of emulsion	Dichloromethane	34–89	Santos et al. (2012)
Mango leaf antioxidant extract		Supercritical Antisolvent			Guamán-Balcázar et al. (2017)

techniques (Girotra et al. 2013). SCF has advantages over other solvents primarily because of their intermediate physical and chemical properties between those of liquids and gasses, and are easily adjustable by changing the temperature and pressure (Silver and Meireles 2014). It has been regarded as a novel technique due to its ability to enhance the solubility of poorly soluble drugs, surface modification, plasticizing of polymers, nanosizing and nanocrystal modification, and chromatographic extraction (Girotra et al. 2013).

One of the limiting factors of using SCF techniques is the solubility of substrate and polymer matrix in SFC (Silver and Meireles 2014). The high-pressure requirement, the high cost of maintenance and need for auxiliary equipment has limited the application of SCF technology for many pharmaceutical industries (Girotra et al. 2013).

7 Conclusion

In the recent years, the use nanoencapsulation process for the encapsulation at a nano scale of bioactive compounds on a wall material is increasingly useful. This is necessary due to their volatile nature and high sensitivity to heat and other environmental conditions. The advantages of nanoencapsulation over other encapsulation techniques include larger surface area, enhanced solubility, bioavailability, and controlled release of the encapsulated ingredients. It is a promising technology with tremendous potentials in food, pharmaceutical, cosmeceutical applications. Several techniques are used for nanoencapsulation of bioactive compounds including emulsification, coacervation, inclusion complexation, emulsification–solvent evaporation, nanoprecipitation, and supercritical fluid technology (SCF). The use of supercritical fluids particularly CO_2 possess several advantages over the traditional encapsulation techniques including mild processing temperatures, generally below 50 °C, an inert atmosphere suitable for unstable bioactive compounds, minimal use of organic solvents and the production of contamination-free products. Among the various techniques that were reviewed, RESS and SAS techniques were particularly most important SCF techniques to fabricate food grade nanoparticles.

References

De Aguiar AC, Silva LPS, De Rezende CA, Barbero GF, Martinez J (2016) Encapsulation of pepper oleoresin by supercritical fluid extraction of emulsions. J Supercrit Fluids 112:37–43

Alessa F (2014) Stability of nano encapsulated Rice bran derived bioactive pentapeptide in apple juice. J Food Proc Technol 05(08). https://doi.org/10.4172/2157-7110.1000356

Arango-Ruiz Á, Martin Á, Cosero MJ, Jiménez C, Londono J (2018) Encapsulation of curcumin using supercritical antisolvent (SAS) technology to improve its stability and solubility in water. Food Chem 258:156–163

Bahrami M, Ranjbarian S (2007) Production of micro- and nano-composite particles by supercritical carbon dioxide. J Supercrit Fluids 40:263–283

Bleich J, Müller BW, Waßmus W (1993) Aerosol solvent extraction system - a new microparticle production technique. Int J Pharm 97:111–117

Brunner G (2005) Supercritical fluids: technology and application to food processing. J Food Eng 67:21–33

Calderone M, Tallon S (2008) Particle formation by rapid expansion from solution using near-critical dimethyl-ether. J Supercrit Fluids 45:245–252

Caliceti P, Salmaso S, Elvassore N, Bertucco A (2004) Effective protein release from PEG/PLA nano-particles produced by compressed gas anti-solvent precipitation techniques. J Control Release 94:195–205

Campardelli R, Baldino L, Reverchon E (2015) Supercritical fluids supercritical fluids applications in nanomedicine. J of Supercrit Fluids 101:193–214

Carlès P (2010) A brief review of the thermophysical properties of supercritical fluids. J Supercrit Fluids 53:2–11

Chattopadhyay P, Huff R, Shekunov BY (2006) Drug encapsulation using supercritical fluid extraction of emulsions. J Pharm Sci 95:667–679

Chen K, Zhang X, Pan J, Zhang W, Yin W (2005) Gas antisolvent precipitation of Ginkgo ginkgolides with supercritical CO_2. Powder Technol 152:127–132

Cheng YS, Lu P-M, Huang C-Y, Wu J-J (2017) Encapsulation of lycopene with lecithin and α-tocopherol by supercritical antisolvent process for stability enhancement. J Supercrit Fluids 130:246–254

Chong GH, Yunus R, Abdullah N, Choog TSY, Spotar S (2009) Coating and encapsulation of nanoparticles using supercritical antisolvent. American J Appl Sci 6:1352–1358

Dal Magro C, Aguir GPS, Veneral J, Santos AE, Chaves LMPC, Oliveira V, Lanza M (2017) Co-precipitation of trans-resveratrol in PHBV using solution enhanced dispersion by supercritical fluids technique. J Supercrit Fluids 127:182–190

Deshpande PB, Kumar AR, Shavi GV, Karthik A, Reddy MS, Udupa N (2011) Supercritical fluid technology: concepts and pharmaceutical applications. PDA J Pharm Sci Technol 65:333–344

Elvassore N, Bertucco A, Caliceti P (2001) Production of insulin-loaded poly(ethylene glycol)/poly(l-lactide) (PEG/PLA) nanoparticles by gas antisolvent techniques. J Pharm Sci 90:1628–1636

Esfandiari N, Ghoreishi SM (2015) Ampicillin nanoparticles production via supercritical CO_2 gas antisolvent process. AAPS Pharm Sci Tech 16:1263–1269

Ezhilarasi PN, Karthik P, Chhanwal N (2013) Nanoencapsulation techniques for food bioactive components: a review. Food Bioproc Technol 6:628–647

Fathi M, Mozafari MR, Mohebbi M (2012) Nanoencapsulation of food ingredients using lipid based delivery systems. Trends Food Sci Technol 23:13–27

Fusaro F, Hänchen M, Mazzotti M (2005) Dense gas antisolvent precipitation: a comparative investigation of the GAS and PCA techniques. Ind Eng Chem Res 44:1502–1509

Garay I, Pocheville A, Madariaga L (2010) Polymeric microparticles prepared by supercritical antisolvent precipitation. Powder Technol 197:211–217

Girotra P, Singh SK, Nagpal K (2013) Supercritical fluid technology: a promising approach in pharmaceutical research. Pharm Dev Technol 18:22–38

Gómez-Estaca J, Gavara R, Hernández-Muñoz P (2015) Encapsulation of curcumin in electrosprayed gelatin microspheres enhances its bioaccessibility and widens its uses in food applications. Innov Food Sci Emerg Technol 29:302–307

Granata G, Consoli GML, Lo NR, Geraci C (2018) Hydroxycinnamic acids loaded in lipid-core nanocapsules. Food Chem 245:551–556

Guamán-Balcázar MC, Montes A, Pereyra C, Martınez de la Ossa E (2017) Precipitation of mango leaves antioxidants by supercritical antisolvent process. J Supercrit Fluids 128:218–226

Gutiérrez FJ, Albillos SM, Casas-Sanz E, Cruz Z, García-Estrad C, García-Guerra A, García-Reverter J et al (2013) Methods for the nanoencapsulation of β-carotene in the food sector. Trends Food Sci Technol 32:73–83

Jafari SM (2017) An introduction to nanoencapsulation techniques for the food bioactive ingredients. Nanoencapsulation of Food Bioactive Ingredients Elsevier Inc. https://doi.org/10.1016/B978-0-12-809740-3.00001-5

Jafari SM, Fathi M, Mandala I (2015) Chapter 13 – emerging product formation. Food Waste Recovery Elsevier Inc. https://doi.org/10.1016/B978-0-12-800351-0.00013-4

Jin H, Fei X, Cuilan J, Yaping Z, Lin H (2009) Nanoencapsulation of lutein with hydroxypropylmethyl cellulose phthalate by supercritical antisolvent. Chin J Chem Eng 17:672–677

Kalani M, Yunus R (2012) Effect of supercritical fluid density on nanoencapsulated drug particle size using the supercritical antisolvent method. Int J Nanomedicine 7:2165–2172

Kankala RK, Zhang YS, Wang S-B, Lee C-H, Chen A-Z (2017) Supercritical fluid technology: an emphasis on drug delivery and related biomedical applications. Adv Healthc Mater 1700433:1–31

Katouzian I, Jafari SM (2017) Nanoencapsulation of vitamins. Nanoencapsulation of food biactive ingredients Elsevier Inc. https://doi.org/10.1016/B978-0-12-809740-3.00004-0

Katouzian I, Mahdi S (2016) Nano-encapsulation as a promising approach for targeted delivery and controlled release of vitamins. Trends Food Sci Technol 53:34–48

Keven ES, Angela M, Meireles A, (2014) Encapsulation of Food Compounds Using Supercritical Technologies: Applications of Supercritical Carbon Dioxide as an Antisolvent. Food and Public Health 4(5):247–258

King JW (2014) Modern supercritical fluid technology for food applications. Annu Rev Food Sci Technol 5:215–238

Lévai G, Martín A, Paz E, Rodríguez-Rojo S, Cocero MJ (2015) Production of stabilized quercetin aqueous suspensions by supercritical fluid extraction of emulsions. J Supercrit Fluids 100:34–45

Lévai G, Albarelli JQ, Santos DT, Meireles MAA, Martin A, Rodríguez-Rojo S, Cocero MJ (2017) Quercetin loaded particles production by means of supercritical fluid extraction of emulsions: process scale-upstudy and thermo-economic evaluation. Food Bioprod Proc 103:27–38

Loss RA, Pereira GN, Boschetto DL, Aguiar GS, Machado JR, Chaves LM, Silva MJ, Oliveira D, Oliveira JV (2016) Encapsulation of eugenyl acetate in PHBV using SEDS technique and in vitro release evaluation. J Food Sci Technol 53:3859–3864

Machado FRS, Reis DF, Boschetto DL, Burkert JFM, Ferreira SRS, Oliveira JV, CAV B (2014) Encapsulation of astaxanthin from Haematococcus pluvialis in PHBV by means of SEDS technique using supercritical CO2. Ind Crop Prod 54:17–21

De Marco I, Reverchon E (2008) Supercritical antisolvent micronization of cyclodextrins. Powder Technol 183:239–246

Martín A, Varona S, Navarrete A, Cocero MJ (2014) Encapsulation and co-precipitation processes with supercritical fluids: applications with essential oils. The Open Chem Eng J 4:31–41

Matson DW, Fulton JL, Petersen RC, Smith RD (1987) Rapid expansion of supercritical fluid solutions: solute formation of powders, thin films, and fibers. Ind Eng Chem Res 26:2298–2306

Matos RL, Lu T, McConville C, Leeke G, Ingram A, (2018) Analysis of curcumin precipitation and coating on lactose by the integrated supercritical antisolvent-fluidized bed process. The Journal of Supercritical Fluids 141:143–156

Mattea F, Martin A, Matias-Gago A, Cocero MJ (2009) Supercritical antisolvent precipitation from an emulsion: β-Carotene nanoparticle formation. J Supercrit Fluids 51:238–247

Mayo AS, Ambati BK, Kompella UB (2010) Gene delivery nanoparticles fabricated by supercritical fluid extraction of emulsions. Int J Pharm 387:278–285

Meziani MJ, Sun YP (2003) Protein-conjugated nanoparticles from rapid expansion of supercritical fluid solution into aqueous solution. J Am Chem Soc 125:8015–8018

Mezzomo N, Paz E, Maraschin M, Martin A, Cocero MJ, Ferreira SRS (2012) Supercritical antisolvent precipitation of carotenoid fraction from pink shrimp residue: effect of operational conditions on encapsulation efficiency. J Supercrit Fluids 66:342–349
Mittal V (2013) Encapsulation Nanotechnologies, Encapsulation nanotechnologies. Wiley
Miwa VMHY, Balcao VMCF, Villa MMDC, Junior JMO, Gramiao MPD, Chaud MV (2016) Supercritical fluid and pharmaceutical applications. Part I: process classification. Afr J Pharm Pharmacol 10:132–144
Montes A, Gordillo MD, Clara P, Ossa EJM (2014) New insights into acrylic polymer precipitation by supercritical fluids. Chem Eng Technol 37:141–148
Montes A, Wehner L, Pereyra C, Ossa EJM (2016) Mangiferin nanoparticles precipitation by supercritical antisolvent process. J Supercrit Fluids 112:44–50
Nerome H, Machmudah S, Wahyudiano FR, Higashiura T, Youn Y-S, Lee Y-W, Gato M (2013) Nanoparticle formation of lycopene/β-cyclodextrin inclusion complex using supercritical antisolvent precipitation. J Supercrit Fluids 83:97–103
Oliveira DA, Mezzomo N, Gomes C, Ferreira SRS (2017) Encapsulation of passion fruit seed oil by means of supercritical antisolvent process. J Supercrit Fluids 129:96–105
Osorio-Tobón JF, Silva EK, Meireles MAA (2016) Nanoencapsulation of flavors and aromas by emerging technologies. Nanotech Agri-Food Ind 2:89–126
Parhi R, Suresh P (2013) Supercritical fluid technology: a review. Adv Pharm Sci Technol 1:13–36
Pinto Reis C, Neufeld RJ, Ribeiro AJ, Veiga F (2006) Nanoencapsulation I. methods for preparation of drug-loaded polymeric nanoparticles. Nanomed: Nanotechnol, Bio, Med 2:8–21
Reverchon E, Porta GD, Rosa ID, Subra P, Letourneur D (2000) Supercritical antisolvent micronization of some biopolymers. J Supercrit Fluids 18:239–245
Reverchon E, Adami R, Caputo G, Marco I (2008) Spherical microparticles production by supercritical antisolvent precipitation: interpretation of results. J Supercrit Fluids 47:70–84
Reverchon E, Torino E, Dowy S, Braeuer A, Leipertz A (2010) Interactions of phase equilibria, jet fluid dynamics and mass transfer during supercritical antisolvent micronization. Chem Eng J 156:446–458
Reverchon E, Adami R (2006) Nanomaterials and supercritical fluids. J Supercrit Fluids 37:1–22
Reverchon E, Della Porta G (1999) Production of antibiotic micro- and nano-particles by supercritical antisolvent precipitation. Powder Technol 106:23–29
Sadjadi S (2016) Supercritical fluids in nanoreactor technology. In: Organic nanoreactors: from molecular to supramolecular organic compounds. Elsevier Inc., pp 373–419
Santos DT, Martin A, Meireles MAA, Cocero MJ (2012) Production of stabilized sub-micrometric particles of carotenoids using supercritical fluid extraction of emulsions. J Supercrit Fluids 61:167–174
Shah MA, Mir SA, Bashir M (2016) Nanoencapsulation of food ingredients. Adv Med Technol Clin Prac (AMTCP). United State: IGI Global Medical Information Science Reference, pp. 132–152
Silva EK, Meireles MAA (2014) Encapsulation of food compounds using supercritical technologies: applications of supercritical carbon dioxide as an antisolvent. Food Pub Health 4:247–258
Singh H, Kumar C, Singh N, Paul S, Jain SK (2018) Nanoencapsulation of docosahexaenoic acid (DHA) using a combination of food grade polymeric wall materials and its application for improvement in bioavailability and oxidative stability. Food Funct 9:2213–2227
Sohail M, Rakha A, Butt MS, Iqbal MJ, Rishid S (2017) Rice bran nutraceutics: a comprehensive review. Crit Rev Food Sci Nutri 57:3771–3780
Subramaniam B (2017) Sustainable processes with supercritical fluids. Elsevier, Encyclopedia of Sustainable Technologies
Suganya V, Anuradha V (2017) Microencapsulation and nanoencapsulation: a review. Int J Pharm Clin Res 9:233–239
Tabernero A, Martín del Valle EM, Galán MA (2012) Supercritical fluids for pharmaceutical particle engineering: methods, basic fundamentals and modelling. Chem Eng Proc: Proc Intens 60:9–25

Tamzeedul F, Ghafoor K, Ferdosh S, Al-Juhaimi F, Ali E, Yunus K, Hamed H, Islam A, Asif M, Sakar MZ (2017) Microencapsulation of fish oil using supercritical antisolvent process. J Food Drug Anal 25:654–666

Trucillo P, Campardelli R, Reverchon E (2018) Production of liposomes loaded with antioxidants using a supercritical CO_2 assisted process. Powder Technol 323:155–162

Türk M, Lietzow R (2004) Stabilized nanoparticles of phytosterol by rapid expansion from supercritical solution into aqueous solution. AAPS Pharm Sci Tech 5:56

Visentin A, Rodríguez-Rojo S, Navarrete A, Maestri D, Cocero MJ, (2012) Precipitation and encapsulation of rosemary antioxidants by supercritical antisolvent process. Journal of Food Engineering 109 (1):9–15

Wang Y, Pfeffer R, Dave R, Enick R (2005) Polymer encapsulation of fine particles by a supercritical antisolvent process. Am Inst Chem Eng J 51:440–455

Wang Y, Dave RN, Pfeffer R (2004) Polymer coating/encapsulation of nanoparticles using a supercritical anti-solvent process. J Supercrit Fluids 28:85–99

Xia F, Jin H, Zhao Y, Guo X (2011) Supercritical antisolvent-based technology for preparation of vitamin D3 proliposome and its characteristics. Chin J Chem Eng 19:1039–1046

Preparation Methods and Advantages of Nano-Sorbents for Food Contaminants Determination

Adrián Gutiérrez-Serpa, Javier González-Sálamo, and Javier Hernández-Borges

1 Introduction

The presence of contaminants in foods is nowadays a topic of special relevance as a result of the security/hazard issues in which they are involved. Pesticides, pharmaceutical compounds, polycyclic aromatic hydrocarbons (PAHs), mycotoxins, hormones or heavy metals are some examples of food contaminants of concern that have already been found in a wide variety of foods at different concentrations (Farré and Barceló 2013; González-Sálamo et al. 2016b; Socas-Rodríguez et al. 2017), thanks to the analytical methods that have been developed so far and whose application and development still represents a challenge (Farré and Barceló 2013; González-Sálamo et al. 2016b; Socas-Rodríguez et al. 2017). Such analytical information has also played and, in fact, is currently playing, a very important role in the implementation of food safety actions in all stages of food production -harvest, processing, storage and preparation-, since trace levels of those contaminants can inadvertently enter the food chain at different stages and, as a result, represent serious health risks. In particular, analytical information concerning food contaminants has been essential for the establishment of maximum residue limits (MRLs) by different regulatory

A. Gutiérrez-Serpa
Departamento de Química, Unidad Departamental de Química Analítica, Facultad de Ciencias, Universidad de La Laguna (ULL), San Cristóbal de La Laguna, España

Instituto Universitario de Enfermedades Tropicales y Salud Pública de Canarias, Universidad de La Laguna (ULL), San Cristóbal de La Laguna, España

J. González-Sálamo · J. Hernández-Borges (✉)
Departamento de Química, Unidad Departamental de Química Analítica, Facultad de Ciencias, Universidad de La Laguna (ULL), San Cristóbal de La Laguna, España

Instituto Universitario de Enfermedades Tropicales y Salud Pública de Canarias, Universidad de La Laguna (ULL), San Cristóbal de La Laguna, España
e-mail: jhborges@ull.edu.es

U. Hebbar et al. (eds.), *Nano-food Engineering*, Food Engineering Series,
https://doi.org/10.1007/978-3-030-44552-2_3

authorities worldwide or even to ban the commercialization or use of certain products in order to guarantee a safe consumption.

In a context with an increased globalization in the food supply and with a constant flow of chemicals being released to the environment, the development of accurate, sensitive and reliable methodologies for food safety and food quality assurance (particularly including identification, monitoring and quantification of newly-recognized food contaminants) is still a hot topic of research nowadays. For that purpose, especially when organic analytes are studied, most conventional analytical methods are chromatographic-based techniques, mainly liquid chromatography (LC) and gas chromatography coupled with mass spectrometry (GC-MS), combined with different sample pre-treatment methods. In this sense, it should be highlighted that sample preparation is still a key step in any analytical method. As it is widely known, its main objective is the selective extraction of the target analytes and, in many cases, their preconcentration in order to increase the sensitivity of the method. Contrary to what it may seem, the development of new analytical methods is a field under a constant evolution in which new alternative sample preparation methods are being proposed, based also on the developments of other areas.

Nanoscience and nanotechnology, which produce materials for their applications well below the sub-microscopic level (< 100 nm), have reached an extraordinary level of development. At the nm level, surface physics and chemistry start to dominate the material's properties and, therefore, such fact, if properly considered and managed, could be of high utility. Such physico-chemical properties depend on their three-dimensional morphology, the surrounding media and their arrangement in space (Adams and Barbante 2013). Among the different fields in which nanomaterials have found a good level of application, Analytical Chemistry could be highlighted (González-Sálamo et al. 2016b, Socas-Rodríguez et al. 2017). In particular, their use as nano-sorbents for sample preparation purposes currently provides one of the most promising avenues for developments in Analytical Chemistry. Concerning the food safety field, they have been applied with success for the extraction and preconcentration of a wide variety of contaminants and matrices.

Among the nanomaterials most commonly used as sorbents it could be highlighted the use of carbon-based materials such as graphene, fullerenes, carbon nanotubes (CNTs), carbon nanohorns or nanodiamonds (all of them functionalized or not) (Liang et al. 2014; Ravelo-Pérez et al. 2010), as well as metal-organic frameworks (MOFs) (Wang et al. 2019; Hashemi et al. 2017) and a wide variety of nanoparticles (NPs) (González-Sálamo et al. 2016a; Płotka-Wasylka et al. 2016) coated or combined with different materials such as those previously commented (MOFs and carbon-based materials) and polymers (including polymeric ionic liquids (PILs) (Yu et al. 2013a) or molecular imprinted polymers (MIPs)) (Speltini et al. 2017), among others. All of them have been mostly used with success as solid-phase extraction (SPE) and solid-phase microextraction (SPME) sorbents, under their different modalities.

This book chapter is aimed at providing a general description of the different nano-sorbents that have been used in food safety for the extraction and preconcentration of food contaminants as well as at describing the most general preparation methods of each nanomaterial. Particular emphasis has been paid to describe the most recent applications in this field.

2 Nano-Sorbents Synthesis/Preparation Methods

As it has already been commented, the features of the nanomaterials used for extraction purposes are of special concern in order to achieve a suitable and selective extraction of the target analytes. Thus, especial attention should be paid to the development of preparation strategies which provide the most suitable structures and the highest purity possible. This section is aimed at providing a general overview of the most common synthesis procedures used in this field.

2.1 Carbon-Based Nano-Sorbents

Carbon-based nanomaterials are the ones mostly applied in sample preparation methodologies as a result of their extraordinary properties which have transformed them in exceptional sorbents for the extraction of a wide number of analytes from matrices of different nature. Among them, CNTs and graphene have been the most widely used and, as a consequence, numerous synthetic routes have been proposed for their obtention (Bhuyan et al. 2016; Herrera-Herrera et al. 2012; Ravelo-Pérez et al. 2010).

Since the first report of the preparation of CNTs using an arc discharge evaporation method in 1991 by Sumio Iijima (Iijima 1991), several synthetic methodologies have allowed the preparation of both single- (SWCNTs) and multi-walled CNTs (MWCNTs). In this sense, chemical vapor deposition (CVD), arc discharge and laser-ablation are the three most established (Eatemadi et al. 2014). Among them, the former technique has been the preferred one, probably because it requires lower temperatures and it offers a better control of the final structural features of the nanomaterial. This method makes use of carbon-based gases (e.g. methane or ethylene) to provide a carbon source in the presence of metallic NPs (e.g. Au, Pt, Fe, Co, etc.) at high temperature (350–1000 °C). In this case, the carbon atoms deposited on the surface of the NPs, which can be floating or supported on a substrate, precipitate as CNTs. This methodology has given place to multiple variants of the thermochemical CVD traditionally used, including aerosol assisted CVD (AA-CVD), alcohol catalyst CVD (AACC-CVD) or plasma enhanced CVD, among others (Crevillen et al. 2019; Rümmeli et al. 2010). Besides CVD, arc discharge constitutes a simple methodology for obtaining CNTs. In this high temperature procedure, the ignition of an arc placed between two high-purity graphite electrodes in a chamber filled with helium (argon or hydrogen can also be used) produces the evaporation of carbon. When carbon cools, it condenses forming CNTs. However, this technique shows as an important disadvantage the fact that not all carbon condenses as nanotubes, being necessary the inclusion of purification steps at the end of the process

(Eatemadi et al. 2014; Rümmeli et al. 2010). Finally, laser-ablation allows obtaining SWCNTs with controlled diameters by suitably changing the conditions influencing the reaction during the synthesis process. Briefly, CNTs are obtained by applying an intense pulse of a laser on a carbon surface in a stream of helium gas and in the presence of a metal catalyst. The evaporated material condenses on the surface of the catalyst particles forming CNTs (Rümmeli et al. 2010).

It is important to highlight that the dimensions (length and diameter), morphology (closed or open ends), the number of walls and the carbon atoms alignment (zig-zag, armchair or chiral -see Fig. 1-) of the CNTs obtained, directly depend on the synthetic strategy followed to obtain them. Moreover, it should also be considered the possibility of ulterior treatments of the CNTs for their covalent or non-covalent functionalization in order to provide a higher selectivity during the extraction process (Herrera-Herrera et al. 2012).

As it has been mentioned, CNTs can be functionalized or combined with other materials, normally in order to improve their selectivity or their extraction capacity, although in certain cases it has been carried out to improve their solubility, which is very low in most solvents as a consequence of the strong inter-tube van der Waals interactions (Socas-Rodríguez et al. 2014). In this sense, the process usually includes an initial oxidative or acidic treatment of pristine CNTs followed by different procedures for their covalent (Meng et al. 2009) or non-covalent functionalization (Hu et al. 2009). In the first case, it is possible to carry out a direct or an indirect functionalization by covalent linkage between the functional groups and the CNT skeleton or the carboxylic groups present on its surface, respectively. The second one is based on physical adsorption processes of molecules onto the CNT sidewall by van der Waals, hydrophobic or electrostatic forces, among others (Ravelo-Pérez et al. 2010). Thus, this versatility has made possible their application for the extraction of a wide variety of organic and inorganic analytes from matrices of different nature (Socas-Rodríguez et al. 2014).

Regarding graphene synthesis, a broad spectrum of methods can be found in the literature, some of them very similar to those used for CNTs obtention. All these methodologies can be classified in two well-differenced groups: *top-down methods*,

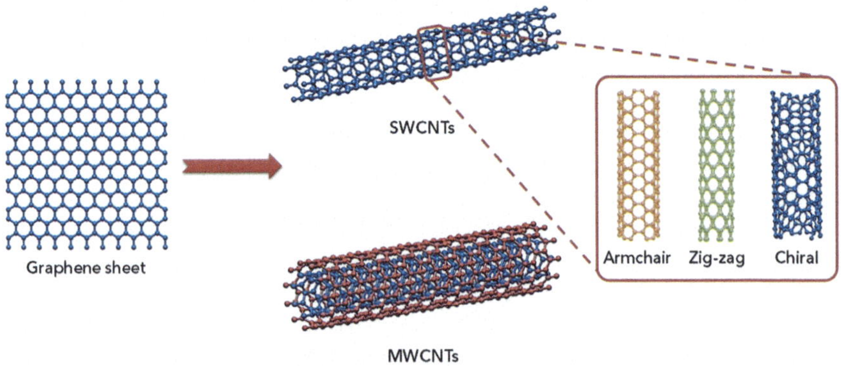

Fig. 1 SWCNTs and MWCNTs structures and torsions. Reprinted from (González-Sálamo 2018) with permission

which are those based on the separation of the stacked layers of graphite to obtain single two-dimensional sheets; and *bottom-up methods*, which are those based on the synthesis graphene sheets from different carbon sources (Crevillen et al. 2019).

In the first group, different processes can be found such as mechanical, electrochemical or solvent-based exfoliation or graphene oxide (GO) reduction, among others. In relation to mechanical exfoliation, the first application of this kind of methods, developed by Novoselov and co-workers in 2004 (Novoselov et al. 2004), used scotch tape to cleave the layers apart. Although this approach provided high quality layers, the process is slow and laborious, which makes it unsuitable for large-scale production. Since then, the following methods have emerged as the two main alternatives to obtain graphene sheets from monocrystalline graphite based on the application of ultrasounds and electric fields (Bhuyan et al. 2016; Crevillen et al. 2019). Electrochemical exfoliation is based on the application of a certain voltage between two electrodes, both or at least one made of graphite, in an electrolyte solution, producing the exfoliation of graphite and delivering the graphene sheets from the sacrificial electrode. In this case, surfactants or sulfuric acid have been added to the electrolyte solution in order to avoid the re-agglomeration of graphene sheets (Shams et al. 2015). The use of ultrasounds has also been explored for obtaining graphene. In this case, it is really important to make a good selection of the solvent used, since it determines the number of layers of the graphene obtained. N-methyl pyrrolidone, N,N-dimethyl formamide, cyclopentanone or even aqueous surfactant solutions have been used in this sense (Crevillen et al. 2019; Shams et al. 2015).

Besides these methods, the most commonly used to obtain graphene sheets are based on the reduction and exfoliation of graphite oxide. In such methods, the exfoliation process produces GO layers, which are subsequently reduced using different reducing agents such as hydrazine, hydroquinone or $NaBH_4$, among others (Bhuyan et al. 2016; Crevillen et al. 2019).

Concerning bottom-up methods, CVD is the most widely used. This approach results very similar to the above-mentioned for CNTs obtention, with the difference that a catalytic metal foil (Ni, Pd, Cu, Ru, etc.) is used instead of catalyst metal NPs (Bhuyan et al. 2016; Crevillen et al. 2019). Besides CVD, other methodologies have also been proposed, including epitaxial growth or pyrolysis. Pyrolysis of graphene is a solvothermal process in which graphene sheets can be detached using a mixture of sodium:ethanol 1:1 (v/v) under sonication in a closed recipient. Regarding epitaxial thermal growth, it uses single-crystalline SiC substrates. In this process, the substrate is heated above 1000 °C under ultra-high vacuum, resulting in the sublimation of Si atoms, which originate a realignment of the carbon atoms producing graphene sheets. Controlling the process conditions, it is possible to control the number of graphene layers (Bhuyan et al. 2016; Crevillen et al. 2019).

2.2 *Metal-Organic Frameworks (MOFs)*

Nowadays, there is an increasing interest on the use of MOFs in different fields, including Analytical Chemistry. In the particular case of sample preparation procedures (Hashemi et al. 2017; Pacheco-Fernández et al. 2019), MOFs gather more than 2800 scientific publications and 370 patents in the last five years (Scopus Database n.d.) as sorbent materials in different techniques such as SPME (Gutiérrez-Serpa et al. 2019; González-Hernández et al. 2019; Rocío-Bautista et al. 2016b, 2018), SPE (He et al. 2017; Rocío-Bautista et al. 2016a; Sajid 2017), and dispersive solid-phase extraction (dSPE), including its miniaturized form, micro-dSPE (μ-dSPE) (Rocío-Bautista et al. 2016a, 2017), or even as stationary phases in chromatography (Qian et al. 2018; Wang and Ye 2017).

MOFs are three-dimensional nanomaterials consisting on an inorganic secondary building unit (SBU) of at least one metal ion (or metal cluster), and one or more organic linkers, connected by coordinated bonds in a specific spatial disposition forming a highly well-organized crystalline structure full of cavities and porous feasibly tuneable. Some of them are even flexible, changing their pore size without affecting their crystallinity (Kalmutzkl et al. 2018). These materials are known to possess the highest surface area known so far with values as high as 10,000 $m^2 \cdot g^{-1}$ (Kalmutzkl et al. 2018; Moghadam et al. 2017; Rowsell and Yaghi 2004). Depending on the selected precursors (inorganic SBU and linkers), the way in which building units are connected, and the synthesis conditions, the network will present a specific structure, pore volume and surface area enrolled with the material properties - see Fig. 2- (Kalmutzkl et al. 2018; Yuan et al. 2018). Therefore, there is an enormous number of possible combinations of all those factors being theoretically possible the design of specific MOFs with concrete physical characteristics in order to use it for a particular application such as CO_2 capture, catalysis, environmental monitoring studies, among others (Furukawa et al. 2013; Yuan et al. 2018). Currently, there are more than 70,000 different MOF structures characterized and registered in the Cambridge Structural Database (CSD) (Cambridge Crystallographic Data Center n.d.; Piri-Moghadam et al. 2017).

Conventional MOFs synthesis is performed by crystallization procedures that can be divided into non-solvothermal and solvothermal methods (Rowsell and Yaghi 2004; Stock and Biswas 2012).

Non-solvothermal methods are those that imply soft synthesis conditions of pressure and temperatures under the solvent boiling point. Examples of that are the direct precipitation and modulated precipitation methods, evaporation approaches, re-crystallization and other strategies like slow diffusion (Stock and Biswas 2012).

Precipitation methods consist on obtaining the MOF by the simple mixing of the precursors. However, the direct mixing can result in a quick precipitation (Yuan et al. 2018) which is not favourable for obtaining a highly ordered crystalline powder. One way of obtaining a more ordered structure once the non-ordered precipitate has been formed is to perform a recrystallization by a better control of the crystal reformation (Stock and Biswas 2012). The other option is to perform modulated precipitations instead of direct precipitations. The main difference between them is the introduction of a third compound into the reactive solution: a modulator. Modulators are substances that prevent the quick precipitation by the regulation of the precipitation equilibrium by competitive coordination with the metal or by blocking the organic linker reducing the nucleation rate and slowing the crystal growth time, allowing a more controlled crystal growth (Yuan et al. 2018).

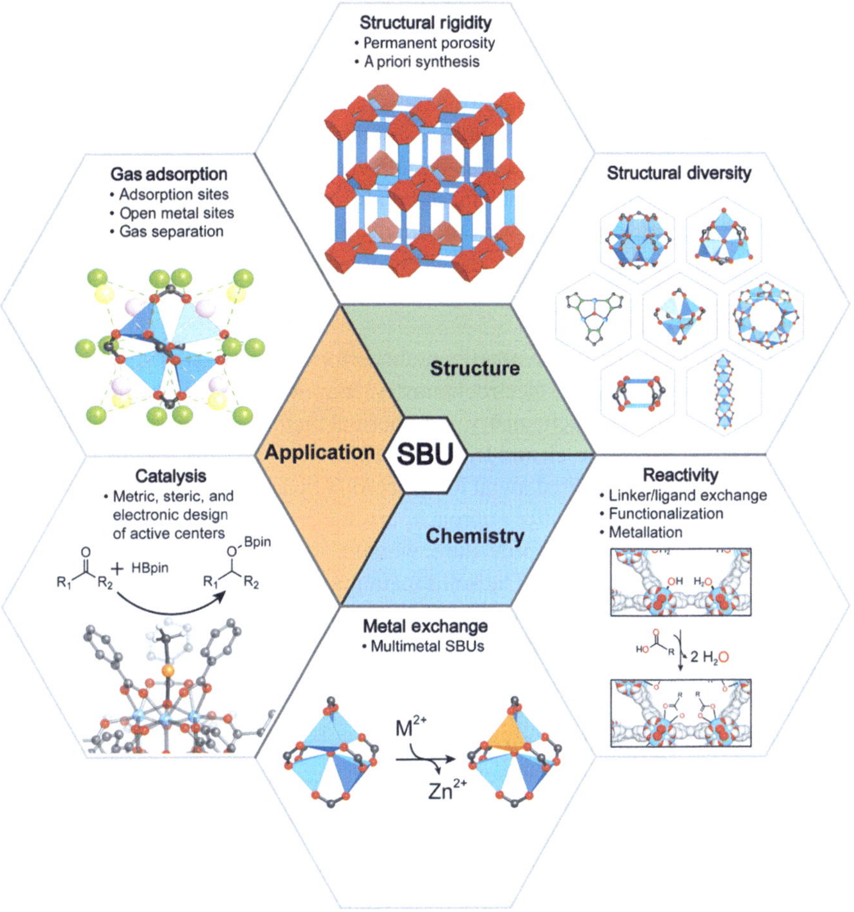

Fig. 2 Influence of the SBU election in the MOF chemical and structural characteristics, and its applicability. Reprinted from (Kalmutzkl et al. 2018) with permission from the American Association for the Advancement of Science

Regarding the evaporation approach, it consists on the removal of the solvent by soft heating (though it can also take place at room temperature), increasing the MOFs precursors concentration in solution and forcing the crystal formation when the solution reaches oversaturation. However, long times are required for the formation of the crystalline structures. This long time wasting inconvenient is very frequent in the diffusion synthesis strategy (Dey et al. 2014). In that case, two non-miscible liquids, one of them containing the inorganic SBU and the other the organic linker, get slowly in contact by diffusion and the MOF is formed. Whenever the reaction advances, the precursors concentrations decrease in the two original solutions until no more crystals are formed. These two methods often provide large crystals highly ordered and stable compared with the other methods (Dey et al. 2014; Stock and Biswas 2012).

Solvothermal methods consist on performing the synthesis under high temperature and pressure conditions. Generally, MOFs precursors are dissolved with a polar solvent and mixed. Then, the resulting solution is placed into a Teflon-lined stainless-steel autoclave, which is sealed and heated at high temperatures over the boiling point of the solvent. Under these conditions, regular crystals are formed. Nevertheless, the extreme conditions required for their obtaining, and the organic solvents used has brought to light the need of developing novel strategies more environmentally friendly using non-organic solvents and lower synthesis temperatures (Rowsell and Yaghi 2004).

There are other alternative synthetic methods for obtaining MOFs such as solid matters (mechanochemistry), electrochemistry, microwave assisted synthesis and, in a less extended way, sonochemistry, ionothermal methods and surfactant thermal process (Li et al. 2018b; Stock and Biswas 2012).

Solid matters are considered metal ion reservoirs that can be used for the synthesis of MOFs. In the solid matter approach, crystals synthesis can take place by a solvent-based or a solvent-free procedure, which is related with mechanochemistry reactions (Zhan & Zeng 2017). The solid matter and the organic linker are mixed, then external energies are applied in order to promote the reaction due to the low diffusion rate that the ions present into the solid state. The external energy can be mechanical or supported by heating using an oven or by microwaves (Zhan & Zeng 2017).

Electrochemical procedures take place by introducing progressively metal ions into a solution containing a conducting salt and the organic linker dissolved in a protic solvent in order to avoid the crystal formation over the cathode surface. Applying a potential to the solution, the MOF is formed as microcrystals. Such electrochemical procedures are widely used in the industry and are specially followed for obtaining thin films coatings (Stock and Biswas 2012; Yue et al. 2015).

Microwave assisted synthesis is based on the interactions between MOFs promoters and electromagnetic waves. Applying the appropriate wave frequency, the number of collisions that take place into the solution increases exponentially, giving as result a homogeneous system with high temperatures and high kinetic energies, allowing a fast synthesis of small crystals (Stock and Biswas 2012).

Sonochemistry synthesis is based on a cavitation process that takes place when an intense ultrasound radiation is applied to a solution. Ultrasound waves produce the collapse of bubbles in the solution creating short-time local hot spots with thermal and pressure conditions favourable for the promotion of the chemical reactions responsible of the nucleation points formation (Dey et al. 2014; Stock and Biswas 2012).

Regarding ionothermal methods and surfactant thermal processes, their use has turn up as an alternative to conventional methods using less aggressive substances with low toxicity and cost as well as high biodegradability and recyclability. That implies the use of ionic liquid (IL) solvents, deep eutectic solvents or even surfactants (Li et al. 2018b).

2.3 Nanoparticles (NPs)

Apart from carbon-based materials and MOFs, NPs of different composition have also been used as nano-sorbents for food contaminants analysis. Among them, those that show a permanent magnetism (ferromagnetism) or those that can be attracted by a magnetic field without retaining residual magnetism after such field is eliminated (superparamagnetism) have been the most applied in this field. Such fact greatly facilitates and simplifies the extraction process.

The m-NPs most commonly used are those composed of metals like iron, nickel and cobalt or their oxides, spinel-type ferromagnets such as $MnFe_2O_4$, $MgFe_2O_4$ and $CoFe_2O_4$ as well as alloys such as $CoPt_3$ and FePt. In particular, the iron oxides magnetite (Fe_3O_4) and maghemite (γ-Fe_2O_3), which can be easily obtained (He et al. 2014), are the ones with the highest applications, also in the food analysis field, since they can be easily produced. Concerning their synthetic methods, they can be obtained by different procedures: chemical co-precipitation, solvothermal/hydrothermal reactions, thermal decomposition, microemulsion/nanoemulsion based synthesis, metal reduction, flow injection methods and aerosol/vapor phase procedures (He et al. 2014; Lu et al. 2007). Most of these methods can also be used to obtain m-NPs different from iron by just changing the precursor salt. It should be mentioned that in the particular case of maghemite it can also be easily obtained from the oxidation of magnetite (Jolivet and Tronc 1988; Lyon et al. 2004). Despite the wide variety of methods currently used, it should be highlighted that the obtained m-NPs do not have a specific shape and size; in fact, a wide variety of shapes and sizes can be obtained depending on the method. Besides, they tend to aggregate and are also easily oxidizable in air and, as a result, they may lose part of their magnetism and dispensability. To overcome this problem, and to increase their selectivity, they are normally coated or grafted with an inorganic (silica, carbon –including graphene and CNTs–, alumina, etc.) or an organic (surfactants or polymers) layer (Xie et al. 2014).

Concerning the synthetic procedures most widely used for iron m-NPs, the co-precipitation method, which is the most common, consists on the mixing of Fe^{2+} and Fe^{3+} salts (mainly their chlorides, although sulphates and nitrates can also be used) and precipitation of the oxides in the presence of an alkaline solution (i.e. aqueous ammonia) (Giakisikli and Anthemidis 2013), since pH should be in the range 8–14. Appropriate control of the NP size and magnetic response can be obtained by varying the experimental conditions (molar ratio of reagents and temperature, pH, etc.), although the co-precipitation method tends to produce particles with a wide size distribution. m-NPs made of cobalt (Román et al. 2011), nickel-zinc ferrite (Mahmoodi 2013) or manganese (Kumar et al. 2014) have also been obtained with this procedure.

Another method that has been used with success is thermal decomposition, which is also applicable to Ni and Co m-NPs. In this case, a metal carbonyl, a low-valent metallic alkene, polyene metal complexes or metal oleates precursors are mixed with high-boiling point organic solvents containing stabilizing surfactants at high temperatures and pressures in reactors or autoclaves (He et al. 2014). The fact that the cost and toxicity of the reagents is high has precluded its wider application, although m-NPs of a controlled size distribution and good crystallinity are obtained.

Regarding the solvothermal approach, also widely applied, it is mainly carried out by mixing $FeCl_3$, ethylene glycol (which acts as solvent and reductant) and sodium acetate (as assistant reductant and electrostatic stabilizer) (Faraji et al. 2010; Kaur et al. 2014) and heating above 200 °C. In some cases, the synthesis has also been assisted with microwaves to accelerate it (Li et al. 2013).

Metal reduction consists on the electrochemical reduction of metallic salts (nitrates, chlorides, and acetates, among others) or on their treatment with appropriate reducing agents in the presence of a surfactant (Lu et al. 2007), while flow injection requires the continuous (or segmented) mixing of reagents under a laminar flow (He et al. 2014). The last methods applied are aerosol/vapor phase techniques in which the precursors are continuously reduced in an aerosol or vapor state in a reactor (He et al. 2014).

As previously indicated, and concerning the application of m-NPs as sorbents, there currently exists a wide variety of inorganic and organic coating materials being the most popular in food sample analysis those consisting of silica (Abbaszadeh and Tadjarodi 2016; Chen and Zhu 2015, 2016; Chen et al. 2016; Du et al. 2014; Hou et al. 2013; Keramat and Zare-Dorabei 2017; Li et al. 2018c, Mahpishanian et al. 2015; Miao et al. 2015; Noormohamadi et al. 2018; Wang et al. 2013; Zhan et al. 2013; Zheng et al. 2014b), carbon-based coatings (Du et al. 2013; González-Sálamo et al. 2016a; Ruan et al. 2014), MOFs (Abbaszadeh and Tadjarodi 2016; Ghorbani-Kalhor 2016; Liu et al. 2015b, 2016; Shi et al. 2018; Xia et al. 2017), and polymers (González-Sálamo et al. 2017; McCullum et al. 2014; Xu et al. 2011; Ríos and Zougagh 2016; Safdarian et al. 2013; Zhao et al. 2013).

Concerning silica functionalization, this strategy is widely used to avoid agglomerates formation and also if further functionalization is desired. Silanol groups existing on silica are versatile and allow performing an easy functionalization of the material surface in order to coat them with another nanomaterial. The silanization reaction is carried out by a single step Störber process (Stöber et al. 1968) introducing slight differences. The general procedure is based on the silica precursor -tetraethyl orthosilicate (TEOS)- hydrolysis followed by a crosslinking condensation in an alcoholic solution using ammonia as catalyst (Stöber et al. 1968). Once the SiO_2 layer is obtained, the possible combinations of the m-NPs with other materials are infinite. Regarding food sample analysis, the materials attached to this kind of NPs have been CNTs (Du et al. 2014), GO (Chen et al. 2016; Hou et al. 2013; Keramat and Zare-Dorabei 2017; Mahpishanian et al. 2015), MOFs (Abbaszadeh and Tadjarodi 2016), ILs (Chen and Zhu 2015, 2016), and polymers (Li et al. 2018c; Miao et al. 2015; Noormohamadi et al. 2018; Wang et al. 2013; Zhan et al. 2013; Zheng et al. 2014b).

Regarding polymers, they are by far the nanomaterials most extensively combined with NPs, due to their good versatility and stability as well as the good retention of multiple compounds in comparison with other materials. In this sense, different natural (e.g. chitosan (Xu et al. 2011) or agarose (Safdarian et al. 2013)) and synthetic (e.g. polypyrrole (Zhan et al. 2013) or polydopamine (McCullum et al. 2014)) polymers have been applied as coatings in diverse extraction procedures. Concerning the synthesis processes, the method most commonly used consists on the dispersion of the NPs in the same media in which the polymerization takes place. Thus, the polymer grows around the NPs, resulting in a core-shell type structure. In this sense, it is important to mention that in some cases, NPs are previously coated with silica in order to provide a better surface for polymer binding as it has been previously indicated (Ríos and Zougagh 2016). However, polymeric coatings have also been synthesized directly on the NPs surface in other occasions. As an example of this, González-Sálamo et al. (González-Sálamo et al. 2017) synthesized Fe_3O_4@polydopamine by dispersing the m-NPs (previously synthetized by a co-precipitation method) in a phosphate buffered saline media at pH 8.3. Then, the dopamine was added, and the polymerization process was maintained for 6 hours under magnetic stirring at room temperature. This coated m-NPs showed good stability for at least one week after their preparation, which is more than acceptable to their application in any extraction procedure. Finally, it is important to mention that polymers and NPs have not only been combined by coating procedures. In some occasions these two materials are simply mixed, creating a composite in which NPs are embedded in the polymeric structure (Zhao et al. 2013), although its use is reduced in comparison with coated NPs.

Besides polymers, the combination of NPs, especially m-NPs, with nanomaterials to form composites have also gained importance in the last years. This kind of new nanocomposites gather the magnetic properties of m-NPs with the extraordinary characteristics of materials such as CNTs, graphene or MOFs, among others (González-Sálamo et al. 2016a), which makes them excellent alternative sorbents for their application in m-dSPE. Among them, the use of carbon-based nanomaterials can be highlighted, as a result of the well-known properties of CNTs or graphene. Besides that, the combination of m-NPs with this kind of nanomaterials is also of special concern because CNTs and graphene show a low solubility in water and organic solvents and their isolation from the sample matrix results quite difficult. However, their combination with m-NPs solves this problem, since they can be easily isolated with a simple magnet (González-Sálamo et al. 2016a). Regarding the synthesis of this kind of nanocomposites, the most usual consists on the deposition of m-NPs onto the surface of the carbon-based material, normally by electrostatic forces (Ruan et al. 2014). However, in the case of CNTs, they can also be filled with m-NPs (Du et al. 2013). As an example, Du et al. (Du et al. 2013) carried out the synthesis of cobalt ferrite filled MWCNTs for the extraction of organochlorine pesticides (OCPs) from tea and honey samples. For this purpose, MWCNTs were firstly cut and oxidized with nitric acid at 80 °C for 4 hours. Then, they were dried and mixed with a solution containing iron and cobalt nitrates under ultrasound. The particles were dried overnight at 60 °C and then heated at 100 °C in air. Finally, they were heated at 550 °C with charcoal or activated carbon as reductant.

MOFs-based materials employed in m-dSPE for the analysis of contaminants in food samples have been obtained by combination with other materials forming a composite, being this approach the most commonly used so far (Abbaszadeh and

Tadjarodi 2016; Liu et al. 2015b; Shi et al. 2018; Xia et al. 2017), or by a direct synthesis without the need of another magnetic material to induce the magnetism (Liu et al. 2015b, 2016). Composite nanomaterials are obtained by the combination of m-NPs and MOFs. There are several ways to obtain them, being the in-situ growth of the MOF over the surface of the m-NP the strategy most extensively used (Abbaszadeh and Tadjarodi 2016; Liu et al. 2015b; Shi et al. 2018; Xia et al. 2017). In order to achieve that crystal growth over the particle surface, there must be first a functionalization step of the m-NP in order to establish the seed for the MOFs formation and growth. Some examples of that, are the dithiocarbamate functionalized Fe_3O_4 (Abbaszadeh and Tadjarodi 2016), mercapto acetic acid functionalized Fe_3O_4 (Xia et al. 2017), amine functionalized Fe_3O_4 (Shi et al. 2018), and 4-(thiazolylazo) resorcinol (TAR) functionalized Fe_3O_4 (Ghorbani-Kalhor 2016). This strategy has been successfully applied to obtain, for example, magnetic HKUST-1 (Abbaszadeh and Tadjarodi 2016; Ghorbani-Kalhor 2016), MIL-101 (Shi et al. 2018) and JUC-48 (Xia et al. 2017) nanocomposites, which have been applied in food sample analysis.

Regarding the direct synthesis of magnetic nanomaterials, it is becoming trendy the use of carbonized MOFs in sample preparation. Some of these nanomaterials have magnetic properties and are obtained by carbonization of the MOF once it has been synthetized (Li et al. 2016; Liu et al. 2015b). Although these nanomaterials have been prepared out of a crystalline structure, when carbonized, the three-dimensional structure is broken and is no more a perfect organized network. Thus, a careful study of what the carbonized MOFs are and how they should be classified (as carbonaceous nanomaterial or as framework) must be done.

3 Nano-Sorbent Applications in Food Contaminant Analysis

3.1 Solid-Phase Extraction (SPE)

SPE, introduced in the 1970s (Fritz 1999), is probably the sorbent-based extraction and clean-up procedure most widely used nowadays in Analytical Chemistry (Anderson et al. 2015). It consists on the use of solid chromatographic materials normally packed into a column or cartridge. Such materials are initially activated with an appropriate solvent and then the liquid sample/extract containing the target analytes is passed through while they are retained. Afterwards, the cartridge can be washed with another solvent, dried with air or under vacuum and finally eluted with an appropriate solvent. Although this is the general procedure, gaseous samples can also be extracted, and it can be used for clean-up purposes when the components of the matrix are retained onto the sorbent and not the target analytes. In all cases, the technique is characterized by its simplicity, an easy handling and a much lower consumption of solvents compared, for example, with liquid-liquid extraction.

The sorbent is selected according to the target analytes or sample components and to the type of modality to be used: extraction or clean-up. In this sense, it should be highlighted that there are four main characteristics of a chemical compound that can be used to achieve the extraction/retention in SPE: polarity, electrical charge, molecular recognition and molecular size (Anderson et al. 2015). In all cases, a wide variety of sorbents are nowadays commercially available which are also accompanied by feasible and effective procedures already studied.

One of the main drawbacks of SPE is probably the long times required for the sample-loading step, which is indeed its bottleneck. To overcome this limitation, SPE has evolved in two different ways: towards its miniaturization (which is frequently called μ-SPE, mainly when less than 100 mg of sorbent are needed) and also towards the complete dispersion of the material in the sample/extract (also called dSPE) (Anderson et al. 2015). In some cases, dSPE also implies miniaturization, which is frequently designated as μ-dSPE.

Miniaturization is one of the current trends in the Analytical Chemistry field (Anderson et al. 2015), which contributes to the simplification of the procedures and time-saving as well as to a low consumption of organic solvents and reagents. The use of nano-sorbents has greatly contributed to its development, as a result of their high extraction capacity, their ability to establish different types of interactions and their high surface-to-volume ratio.

Regarding dSPE, such modality has several advantages over traditional SPE. First of all, there is no need to activate the cartridge, which clearly reduces the consumption of organic solvents and simplifies the procedure. Instead, the sorbent is directly dispersed in the liquid sample/extract and the equilibrium is quickly reached. When samples containing microparticles or microorganisms that could block traditional SPE cartridges are analysed, dSPE is more suitable. Furthermore, if magnetic materials are used (which is also a current trend in dSPE (Anderson et al. 2015)), the procedure, which is called magnetic-dSPE (m-dSPE), is even easier since it is developed with the help of an external magnet. The application of magnetic sorbents for m-dSPE was not developed until 1996, when Towler et al. (Towler et al. 1996) reported the recovery of different metals from seawater samples using manganese dioxide coated magnetite as the magnetic sorbent. However, the term m-SPE was only introduced three years later by Safariková et al. (Šafaříková and Šafařík 1999).

Concerning the use of nano-sorbents for food contaminants analysis, both dSPE and m-dSPE have been the SPE modalities most widely applied and their application in both procedures using nano-sorbents will be discussed below. However, and regarding the use of SPE, Table 1 compiles different works in which nano-sorbents have been used in SPE for the extraction of different types of contaminants from food samples. As can be in the table, such works include the use of MWCNTs (Qin et al. 2016), GO (Chen et al. 2013b, c; Wu et al. 2012), combinations of graphene and MWCNTs (Yuan et al. 2019) or combinations of GO with other compounds or polymers (Hou et al. 2018; Qi et al. 2014; Zhai et al. 2015; Zhang et al. 2015), coated silica NPs (Ahmadi et al. 2019; Sun et al. 2016; Li et al. 2014a, b) or even a MOF combined with a MIP (Liu et al. 2017), among others. Regarding the food matrices selected, such nanomaterials have been mainly applied to the extraction of fish (Ahmadi et al. 2019; Chen et al. 2013b, c; He et al. 2019; Wu et al. 2012; Zhang et al. 2015), fruits and vegetables (Barreto et al. 2010; Li et al. 2014a; Duo et al. 2018; Qi et al. 2014; Zhai et al. 2015), milk and its derivatives (Deng et al. 2018; He

Table 1 Recent examples of the application of nano-sorbents in SPE for food contaminants analysis

NM	NM amount (mg)	Analyte (n)	Sample	Extraction T (°C) / time (min)	Desorption T (°C) / time (min)	Eluent	LOD	Intraday RSD (%)	Interday RSD (%)	RR (%)	Ref.
MWCNTs	10	Malachite green, Leucomalachite green, diethylstilbestrol and dienestrol (4)	Carp, striped bass and giant salamander	- / -	- / -	ACN (10 mL)	< 0.15 μg·kg^{-1}	< 14	-	73–106	Qin et al. (2016)
GO	20	Chloramphenicol (1)	Fish and muscles	- / 5	- / -	MeOH (10 mL)	0.036 μg·kg^{-1}	< 4.9	< 7.4	92.3–103	Wu et al. (2012)
GO	10	Malachite green and leucomalachite green (2)	Fish	- / 7	- / 4	NH_3/MeOH (4.0 mL)	< 0.12 μg·kg^{-1}	< 10.6	< 6.29	82.0–103	Wu et al. (2012)
GO	15	SAs (13)	Carp fish	- / 11	- / 3	Ammoniacal acetone (2%) (3.0 mL)	< 2.68 μg·kg^{-1}	< 5.7	< 7.6	76.2–91.5	Chen et al. (2013c)
G/MWCNTs	1	17β-estradiol (−)	Milk and milk powder	- / -	- / -	MeOH/NH_3 (95/5, v/v)	0.7 μg·kg^{-1}	≤ 5.5	–	88.4–108.9	Yuan et al. (2019)
GO@MWCNTs-MIP	150	Tetracyclines (3)	Milk, chicken and fish	- / 2	- / -	ACN/Formic acid (4/1, v/v)	< 128 μg·kg^{-1}	< 9.55	–	85.6–116.9	He et al. (2019)
GO@Acrylamide	670	Heterocyclic amines (6)	Duck, fish and fried chicken	- / -	- / -	-	< 2.5 ng·L^{-1}	< 8.7	-	65.9–118	Zhang et al. (2015)
GO@MIP	2	Phloxine B (1)	Coffee beans	- / -	- / -	MeOH/HAc (0.040 mL)	0.075 ng·mL^{-1}	< 4.7	< 4.1	89.5–91.4	Zhai et al. (2015)
GO@PBMED	-	Pesticides (4)	Apple, grape, orange, tomato, cucumber and rape	- / 3	- / 2	MeOH/H_2O (0.10 mL)	< 3.0 μg·kg^{-1}	<3.4	< 6.2	70.2–95.0	Qi et al. (2014)

NM	NM amount (mg)	Analyte (n)	Sample	Extraction T (°C) / time (min)	Desorption T (°C) / time (min)	Eluent	LOD	Intraday RSD (%)	Interday RSD (%)	RR (%)	Ref.
SiO_2@GO@IL	50	Acidic herbicides (5)	*Lycium barbarum*	- / 60	- / 0.5	MeOH/Formic acid (1.5 mL)	< 2 $\mu g{\cdot}L^{-1}$	< 7.1	-		Hou et al. (2018)
(La0.9Eu0.1)$(DPA)_3(H_2O)_3$	500	Pesticides (5)	Lettuce	- / 3	- / -	ACN (30 mL)	< 0.05 $mg{\cdot}kg^{-1}$	< 8.0	-	78–107	Barreto et al. (2010)
MIL-101@MIP	60	Pyrraline	Milk, and milk powder	- / -	- / -	MeOH (3.0 mL)	40.7 $\mu g{\cdot}L^{-1}$	< 6.87	-	92.2–103	Liu et al. (2017)
PCN-224	250	SAs (6)	Milk and honey	- / -	- / 2.5	MeOH (10 mL)	< 0.47 $ng{\cdot}L^{-1}$	< 6.7	< 5.1	87.3–115	Deng et al. 2018
UiO-67	15	Acidic herbicides (4)	Tomatoes, cucumbers and wax gourds	- / 25	- / 3	MeOH/formic acid (1 mL)	< 0.5 $\mu g{\cdot}L^{-1}$	< 3.92	< 7.65	89.4–113	Duo et al. (2018)
SiO_2@DSMIP	100	Bisphenol A (1)	Canned drinks. apple, pear and pitaya	- / 20	- / -	MeOH (6.0 mL)	3.0 $ng{\cdot}mL^{-1}$	< 7	-	98–105	Li et al. (2014)
SiO_2@fumed@Pr-INAH	150	Heavy metals (1)	Fish	- / 25	- / -	EDTA	0.018 $\mu g{\cdot}L^{-1}$	2.8	-	95.5–100	Ahmadi et al. (2019)
SiO_2@NiO	200	Benzimidazoles (9)	Eggs and milk	- / -	- / 3	Ammonium hydroxide/acetone (3.0 mL)	< 2.2 $ng{\cdot}mL^{-1}$	< 5.8	-	70.8–119	Sun et al. (2016)

ACN: Acetonitrile
DPA: Pyridine-2,6-dicarboxylic acid
DSMIP: Dummy surface molecularly imprinted polymer
GO: Graphene oxide
IL: Ionic liquid
LOD: Limit of detection
MeOH: Methanol

MIP: Molecularly imprinted polymer
MWCNTs: Multiwalled carbon nanotubes
NM: Nanomaterial
PBMED: Poly(butyl methacrylate–ethylene dimethacrylate
RR: Recovery
RSD: Relative standard deviation
SAs: Sulfonamides

et al. 2019; Liu et al. 2017; Sun et al. 2016; Yuan et al. 2019), meat (He et al. 2019) and honey (Deng et al. 2018).

3.1.1 Dispersive Solid-Phase Extraction (dSPE)

As pointed before, dSPE has turn up as a reliable alternative to conventional/static SPE, also for contaminants analysis in foods, due to the reduction in the amount of the extractant material required (Jia et al. 2018), the organic solvents needed (less than 5 mL in many cases (Zeng et al. 2015)) and the extraction and desorption times (less than 30 and 10 minutes, respectively (Xu et al. 2016; Zeng et al. 2015)). However, the number of steps followed in order to apply this extraction technique are high and could be considered as tedious. The general extraction procedure is as follows. A certain amount of the nanomaterial (in many cases less than 100 mg) is added to the sample. Then, the extractant material has to be dispersed into the solution in order to maximize the interaction between the analytes and the sorbent. Such dispersion can be performed by soft shaking (Li et al. 2014b; Mirabi et al. 2017; Wu et al. 2015; Yu et al. 2013b) or stirring (Xu et al. 2016), or by another more energetic procedure such as sonication (Jia et al. 2018; Kashanaki et al. 2018; Li et al. 2015b, c; Lu et al. 2019; Stanisz and Krawczyk 2017) or vortex (Eftekhari et al. 2019; Tokalioglu et al. 2017; Zeng et al. 2015). Once the extraction has ended up, the sorbent has to be separated from the solution in order to remove the sample and to continue with the desorption of the analytes. This separation is often performed by centrifugation (see Fig. 3A) instead of decantation or filtration to reduce the sample treatment time. In some cases, the sorbent is retained into a column or cartridge (see Fig. 3B). Then, the solid sorbent containing the analytes is mixed with a suitable solvent to elute the analytes. This desorption requires also an external agitation to favour the interactions between the adsorbed analytes and the selected solvent that is preferably performed by an energetic procedure such as vortex (Eftekhari et al. 2019; Tokalioglu et al. 2017; Wu et al. 2015; Yu et al. 2013b; Zeng et al. 2015) or sonication (Jia et al. 2018; Kashanaki et al. 2018; Li et al. 2014b, 2015b, c; Lu et al. 2019; Xu et al. 2016) to ensure the maximum interactions in a short period. Finally, the nanomaterial and the elution solvent containing the analytes are separated by centrifugation. Sometimes, further steps such as filtration, solvent evaporation and reconstitution with a solvent compatible with the equipment used for the analytical determination is also required for the analysis.

Regarding the nanomaterials most commonly used in dSPE so far for contaminants determination in food samples, Table 2 summarizes their applications in this field. As can be seen, CNTs (as MWCNTs) (Lu et al. 2019; Mirabi et al. 2017; Wu et al. 2015; Xu et al. 2016; Zeng et al. 2015) are the ones mostly applied, followed by MOFs (Jia et al. 2018; Kashanaki et al. 2018; Li et al. 2014b, 2015b, c; Tokalioglu et al. 2017) and, in a less extended way, NPs (Gao et al. 2010; Stanisz and Krawczyk 2017) and GO (Yu et al. 2013b; Eftekhari et al. 2019).

Commercial MWCNTs, synthesized in the previous indicated ways, can be used directly without further functionalization for most applications. Examples of that

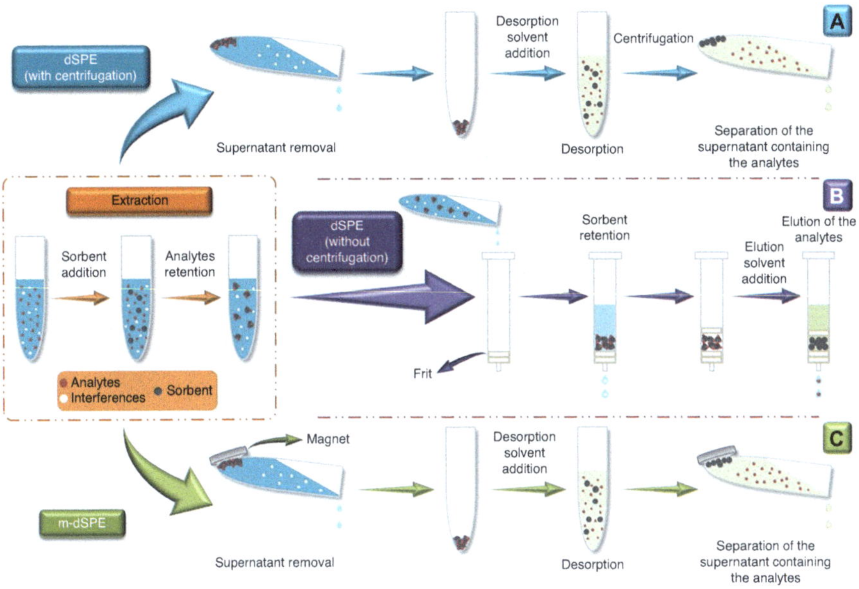

Fig. 3 Schematic representation of (**A**) dSPE with centrifugation, (**B**) dSPE without centrifugation (the sorbent in retained into a column) and (**C**) m-dSPE procedures. Reprinted from (González-Sálamo 2018) with permission

are the determination of tocopherol and tocotrienols (Lu et al. 2019), adulterants (Zeng et al. 2015) and fungicides (Wu et al. 2015) in tea, fruits, cereals, nuts and vegetables samples obtaining excellent analytical quality parameters. Nevertheless, in other cases a functionalization is required. An example is the extraction of metal ions, which requires the presence of a complexant able to form complexes with the metal ions dissolved on the solution. For that purpose, Mirabi et al. modified the surface of the MWCNTs with dithiooxamide. Making this modification excellent results are obtained with limits of detection (LODs) of 0.6 $\mu g \cdot L^{-1}$ and with recovery values around 97% (Mirabi et al. 2017).

Regarding MOFs, MIL-101(Cr) has been widely used for the determination of herbicides in peanuts (Li et al. 2014b), different kinds of oils (Li et al. 2015c), and soybeans (Li et al. 2015b) combined with high-performance liquid chromatography coupled to a diode array detector (HPLC-DAD), obtaining in all the cases similar LODs (under 2.0 $\mu g \cdot kg^{-1}$) and relative standards deviations (RSDs) lower than 7%, proving the reliability of the use of this kind of materials for the determination of contaminants at trace levels in complex samples. In the particular case of the work of Li et al., the extraction using MIL-101(Cr) is performed by dynamic microwave assisted extraction reducing considerably the extraction times (Li et al. 2015b). Other MOFs used in food sample analysis are Al-Fu MOF (Kashanaki et al. 2018) and MOF-545 (Tokalioglu et al. 2017) for the analysis of metal ions in fruits, rice and vegetables. More recently, Jia et al. has described the use of a MOF-125(Ti) core-shell surface modified with Indium (III) sulphur for the analysis of nitro-PAHs

Table 2 Recent examples of the application of nano-sorbents in dSPE for food contaminants analysis

NM	NM amount (mg)	Analyte (n)	Sample	Extraction T (°C) / time (min) / agitation	Desorption T (°C) / time (min) / agitation	Eluent	LOD	Intraday RSD (%)	Interday RSD (%)	RR (%)	Ref.
MWCNTs	20	Adulterants (6)	Tea	- / 30 / vortex	- / 2 / vortex	MeOH (5.0 mL)	< 0.2 μg·mL^{-1}	< 5.2	< 5.9	71.0–105	Sun et al. (2016)
MWCNTs	10	Fungicides (16)	Cabbage, celery, strawberry, and grape	- / 1 / shake	- / 1 / vortex	Acetone (10 mL)	< 3 μg·kg^{-1}	< 10	< 5.4	72.4–98.5	Wu et al. (2015)
MWCNTs	20	Tocopherols and tocotrienols (7)	Cereals and nuts	- / 3 / ultrasounds	- / 1 / ultrasounds	MeOH/2% HCl (3 mL)	< 0.36 μg·L^{-1}	< 2.00	< 4.26	86.1–102.5	Lu et al. (2019)
MWCNTs@DTAm	20	Metal ions (1)	Tomato, potato, wheat flour, red beans and oat	- / 20 / shake	- / - / -	HCl (1.0 mL)	0.6 ng·mL^{-1}	[illegible]	-	97.9	Mirabi et al. (2017)
MWCNTs-OH	7	Fungicide (1)	Fruit juice	25 / 4 / stir	- / 10 / ultrasounds	MeOH / C_6MIMPF_6 (1.2 mL)	< 9.0 ng·mL^{-1}	< 4.9	< 6.5	92.9–104	Xu et al. (2016)
GO	5	Aflatoxins (4)	Peanuts	- / 10 / shake	- / 0.5 / vortex	MeOH (1.0 mL)	< 0.6 ng·g^{-1}	< 4.5	< 9.6	85.1–101	Yu et al. (2013b)
GO@pectic acid	5	Cu^{2+} (–)	Black tea, red lentil, wheat and sesame	- / 10 / vortex	- / 5 / vortex	1 mol·L^{-1} HNO_3 (0.4 mL)	2.22 μg·L^{-1}	[illegible]	–	90–102	Eftekhari et al. (2019)

Al-Fu MOF	30	Metal ion (1)	Fruit juice, rice and green tea	- / 10 / ultrasounds	- / 2 / ultrasounds	THF (1.0 mL)	< 0.2 ng·mL^{-1}	< 6.5	-	90–96	Kashanaki et al. (2018)
MIL-101	7	Herbicides (7)	Peanuts	- / 5 / shake	- / 3 / ultrasounds	ACN (2.0 mL)	< 2.0 μg·kg^{-1}	< 5.1	< 5.7	89.5–103	Li et al. (2014b)
MIL-101	7	Herbicides (7)	Soybean oil, corn oil and peanut oil	- / 15 / ultrasounds	- 1/ ultrasounds	MeOH (1.5 mL)	< 1.0 μg·L^{-1}	< 4.46	< 3.39	87.3–107	Li et al. (2015c)
MIL-101	5	Herbicides (7)	Soybeans	- / 5 / ultrasounds	- / 2 / ultrasounds	MeOH (2.0 mL)	< 2.0 μg·kg^{-1}	< 6.7	< 6.4	91.1–107	Li et al. (2015b)
MIL-125(Ti)@ In_2S_3	1	Nitro-PAHs (16)	Grilled pork, smoked sausages	40 / 34 / ultrasounds	- / 10 / ultrasounds	CH_2Cl_2 (2.0 mL)	< 83 ng·L^{-1}	< 9.2	< 10	71.3–112	Jia et al. (2018)
MOF-545	10	Metal ions (1)	Chick pea, beans, wheat, lentils and cherry juice	- / 15 / vortex	- / 15 / vortex	HCl (2.0 mL)	< 1.8 μg·L^{-1}	2.6	-	125	Tokalioglu et al. (2017)
SiO_2@MIP	100	SAs^{n} (6)	Milk and egg	- /45 / -	- / - / -	MeOH / HAc (−mL)	< 14.6 μg·L^{-1}	< 6.3	-	69.8–89.1	Gao et al. (2010)
ZnO	20	Metal ions (1)	Tea, goji, ginger and aloe vera juice	- / 0.6 / ultrasounds	- / - / -	H_2O (0.2 mL)	0.003 ng·mL^{-1}	< 23	-	89.0–102	Stanisz and Krawczyk (2017)

NM: Nanomaterial
LOD: Limit of detection
RSD: Relative standard deviation
RR: Recovery
MWCNTs: Multiwalled carbon nanotubes
MeOH: Methanol
DTAm: Dithioamide
GO: Graphene oxide
THF: Tetrahydrofuran
ACN: Acetonitrile
PAH: Polycyclic aromatic hydrocarbon
MIP: Molecularly imprinted polymers
SAs: Sulfonamides

in cooked meats (Jia et al. 2018). The combination of both materials (MOF-125 and In_3S_2) increases significantly the extraction properties of the resulting material due the high porosity of the MOF and the Indium nanosheet activity.

Concerning NPs, these materials have been less used in dSPE for food sample analysis (Gao et al. 2010; Stanisz and Krawczyk 2017) due to their properties, which are more interesting in other sample preparation techniques such as m-dSPE. As an example of their applications in dSPE, Gao et al. have proposed the use of silica modified NPs with a layer of a MIP for the determination of sulfonamides (SAs) in milk and egg samples (Gao et al. 2010). In that case, the main responsible of analytes extraction is not the NP itself, but the MIP coating which is characterized by presenting a highly selective extraction capability. Silica NPs act as the support for the extractant material. Their reduced dimensions ensure a high surface area to volume ratio suitable for the attachment of the extractant material and to perform dSPE using this kind of polymers that are more commonly used in static extraction procedures such as SPME and SPE (Turiel and Martín-Esteban 2010, Zhang et al. 2013c). However, NPs can also be used as extractant material and not just as support. As an example, Stanisz et al. have used ZnO NPs for the determination of metal ions in tea, goji berries, fresh ginger root and aloe vera pulp juice by electrothermal atomic absorption spectrometry (ETAAS) (Stanisz and Krawczyk 2017). The NPs are added to the digested samples and an ultrasound assisted extraction is performed. Then, the NPs with the metal ions retained are separated and suspended in a more reduced volume of water (200 μL), achieving, as a result, a slurry that is introduced in the ETAAS system.

3.1.2 Magnetic Dispersive Solid-Phase Extraction (M-dSPE)

As it has already been mentioned, the use of m-NPs, either as sorbents or as part of them, has provided an important simplification of the dSPE procedure. In that case, in which the procedure is called m-dSPE, the initial extraction step is developed as described for dSPE in the previous section. However, in m-dSPE the sorbent is easily retained and isolated from the sample matrix using an external magnetic field (see Fig. 3C), which allows avoiding any centrifugation or sorbent retention and results in a significant reduction of the extraction time (Socas-Rodríguez et al. 2015; Płotka-Wasylka et al. 2015). After analytes retention, a magnet is placed on the outer wall of the recipient to retain the magnetic sorbent, which allows removing the supernatant by decantation. Then, the sorbent is re-dispersed in the desorption solvent and isolated again with the magnet in order to recover the supernatant containing the analytes by decantation. It is important to highlight that the strength of the magnet, together with the paramagnetic behaviour of the m-NPs, determine the time needed to settle the m-NPs before decantation.

The inherent advantages derived from the use of m-dSPE have made this technique one of the most used in last years for the extraction of a wide variety of inorganic and organic compounds from matrices of very different nature (Khezeli & Daneshfar 2017). Regarding its application to the analysis of food contaminants, Table 3 compiles some of the most relevant works in this sense. As can

Table 3 Recent examples of the application of nano-sorbents in m-dSPE for food contaminants analysis

NM	NM amount (mg)	Analyte (n)	Sample	Extraction T (°C) / time (min) / agitation	Desorption T (°C) / time (min) /agitation	Eluent	LOD	Intra-day RSD (%)	Inter-day RSD (%)	RR (%)	Ref.
YS-Fe_3O_4@ GC	6	Sulfonamides (4)	Milk, pork and chicken	- / 15 / stir	- / 5 / ultrasouns	MeOH 5% 1 $mol{\cdot}L^{-1}$ NaOH (600 μL)	< 0.26 $\mu g{\cdot}L^{-1}$ < 2.25 $\mu g{\cdot}kg^{-1}$	< 7.6	< 9.2	77.2–118.0	Liu et al. (2020)
Fe_3O_4@ MWCNTs @ DMMIPs	50	PCBs (6)	Fish	- / 90 / stir	- / 90 / -	n-hexane / CH_2Cl_2 (14 mL)	< 0.0068 $\mu g{\cdot}L^{-1}$	< 5.7	< 5.3	73.41–114.2	Du et al. (2014)
Fe_3O_4@ MWCNTs @ MIP	15	Sulfamethoxazole (1)	Milk and honey	- / 30 / -	- / - / -	MeOH/ HAc(– mL)	6.04 $\mu g{\cdot}L^{-1}$	< 6.8	-	68.3–80.1	Zhao et al. (2015)
PAMAN@ Fe_3O_4-CNTs	100	Alkaloids (15)	Pork and green-vegetables	- / 2 / -	- / - / -	ACN (5.0 mL)	< 0.317 $ng{\cdot}g^{-1}$	< 5.4	< 5.3	83.4–125	Cheng et al. (2018)
Fe_3O_4@GO	10	Aflatoxins (4)	Rice, wheat and sesame	- / 15 / shake	- / 5 / -	Acetone (4.0 mL)	< 0.075 $ng{\cdot}g^{-1}$	< 3.2	-	64.38–122.2	Es'haghi et al. (2014)
Fe_3O_4@GO	20	Aflatoxins (4)	Bread	- / - / -	- / - / -	MeOH (0.7 mL)	< 0.11 $ng{\cdot}g^{-1}$	< 5.7	< 8.6	84–105	Sereshti et al. (2019)
Fe_3O_4@GO	20	Pyrethroid pesticides (6)	Orange and lettuce	- /30 / shake	- / 1 / vortex	Acetone (1.0 mL)	–	< 11	-	90.0–104	Hou et al. (2013)
Fe_3O_4@ GO/2-PTSC	8	Metal ions (1)	Fish, rice, tea and milk	- / 3 / shake	- / 5.4 / ultrasounds	HCl (0.5 mL)	0.0079 $\mu g{\cdot}L^{-1}$	1.6	-	92.75–100	Keramat and Zare-Dorabei (2017)
Fe_3O_4@GO@ DVB-VA	100	Metal ions (5)	Chilli, black pepper and hemp	- / 5 / shake	- / - / ultrasounds	HCl (1.0 mL)	< 2.39 $\mu g{\cdot}L^{-1}$	< 3.1	-	94–107	Khan et al. (2016)
Fe_3O_4@GO@ MIP	20	17β-estradiol (1)	Milk powder	25 / 60 / shake	- / - / -	MeOH (HAc (2.0 mL)	0.035 $\mu mol{\cdot}L^{-1}$	< 4.7	-	72.63–84.20	Ning et al. (2014)

(continued)

Table 3 (continued)

NM	NM amount (mg)	Analyte (n)	Sample	Extraction T (°C) / time (min) / agitation	Desorption T (°C) / time (min) /agitation	Eluent	LOD	Intra-day RSD (%)	Inter-day RSD (%)	RR (%)	Ref.
Fe_3O_4@GO@MIP	50	PCBs (6)	Fish	- / 75 / stir	- / 120 / -	n-hexane / CH_2Cl_2 (13 mL)	< 0.007 μg·L^{-1}	< 3.7	< 5.0	79.90–94.23	Lin et al. (2015a)
Fe_3O_4@GO@PIL	80	Preservatives (20)	Puple cabbage	- / 2 / vortex	- / 30 / stir	Acetone (1.0 mL)	< 6.64 μg·kg^{-1}	< 12	-	81.7–118	Chen et al. (2016)
Fe_3O_4@GO@PTh	25	Zinc (1)	Flour, celery, and egg	- / 7 / ultrasounds	- / - / -	HNO_3 (0.5 mL)	0.08 ng·mL^{-1}	< 5.8	-	87.4–96.2	Babaei et al. (2018)
Fe_3O_4@GO-PEA	30	OPPs (6)	Apple, grape, pear, bell pepper and celery	- / 5 / shake	- / 2 / shake	Acetone (1 mL)	< 0.1 μg·L^{-1}	< 4.8	< 6.8	94.6–104	Mahpishanian et al. (2015)
GO@Fe_3O_4	15	Chlorophenols (4)	Honey	- / 25 / shake	- / 0.3 / vortex	Alkaline methanol (0.3 mL)	< 1.5 ng·g^{-1}	< 4.6	-	93.2–98.9	Liu et al. (2015c)
Fe_3O_4@HKUST	50	Metal ions (3)	Fish, shrimp, broccoli, apple, lettuce and mushroom	- / 10 / shake	- / 15 / vortex	EDTA (5.0 mL)	< 0.8 ng·mL^{-1}	< 8.5	-	-	Ghorbani-Kalhor (2016)
Fe_3O_4@HKUST@DTC	14.5	Metal ions (2)	Rice and canned tuna	- / 13 / stir	- / 15 / shake	Thiourea / HCl (2.5 mL)	1.2 ng·L^{-1}	< 3.4	-	93.1–96.2	Abbaszadeh and Tadjarodi (2016)
Fe_3O_4@JUC-48	25	SAs (5)	Chicken, pork and shrimp	- / 8 / shake	- / 11 / stir	MeOH /HAc (0.8 mL)	< 5.23 ng·g^{-1}	< 5.4	< 4.5	76.1–103	Xia et al. (2017)
Fe_3O_4@MIL-101	3	Sudan dyes (6)	Tomato sauce	- / 2 / shake	- /10 / ultrasounds	Ethyl acetate (1.0 mL)	< 2.5 μg·kg^{-1}	< 5.3	< 9.2	69.6–91.6	Shi et al. (2018)

Fe_3O_4@ MIL-101	5	Herbicides (7)	Rice	- / 25 / ultrasounds	- / 5 / ultrasounds	ACN (2.0 mL)	< 0.08 μg·kg^{-1}	< 5.9	< 7.8	83.9–103.5	Liang et al. (2018)
Fe_3O_4@ MIL-101	11	Herbicides (7)	Rice	- /5 / ultrasounds	- / 4 / ultrasounds	MeOH (4.0 mL)	< 18.1 pg·g^{-1}	< 4	< 9.6	79.3–117	Jiang et al. (2019)
Fe_3O_4@ MOF235	5	Insecticides (5)	Honey and fruit juice	- / 5 / vortex	- / 0.5 / ultrasounds	ACN	< 0.5 μg·L^{-1}	< 4.8	< 4.9	79.9–99.62	Duo et al. (2019)
ZIF 8 (Zn/Co)	15	Chlorophenols (4)	Tea	- / 20 / shake	- / 10 / vortex	MeOH (0.2 mL)	< 0.2 ng·mL^{-1}	< 8.7	-	83.0114	Li et al. (2016)
ZIF-67 based magnetic nanoporous carbon sorbent	10	Phenylurea herbicides (4)	Grapes and bitter gourd	- / 25 / shake	- / - / -	Acetone (0.1 mL)	< 0.46 ng·g^{-1}	< 6.7	-	88.9–105	Liu et al. (2015b)
Co-Fe_3O_4@ PEI	10	Tartrazine (1)	Saffron spray and cotton candy	- / 2 / ultrasounds	- / 1 / vortex	EtOH/NaOH (0.15 mL)	5.09 ng·mL^{-1}	< 2.5	< 5.5	99.5–106	Noormohamadi et al. (2018)
Fe_3O_4	500	Sudan dyes (4)	Chilli oil, chilli powder and tomato paste	25 / 1 + 20 / ultrasounds + no agitation	- / 1 / shake	MeOH (5.0 mL)	< 10 μg·kg^{-1}	< 5	-	92 .9–111	Zhang et al. (2013b)
Fe_3O_4	200	SAs (2)	Milk	- / 5 / ultrasounds	- / 1 / ultrasounds	ACN (2.0 mL)	< 0.01	< 8.1	< 7.3	92.9–102	Karami-Osboo et al. (2015)
Fe_3O_4@AMT	100	Aflatoxin (1)	Milk	- / 5 / ultrasounds	- / 1 / shake	$(CH_3)_2CO$/ ACN/CH_2Cl_2 (2.0 mL)	10 ng·mL^{-1}	< 4.5	< 5.3	91.6–96.1	Hashemi et al. (2014)
Fe_3O_4@ Aptamer	-	Ochratoxin (1)	Wheat, cereals and coffee	- / 2 + 30 / vortex + no agitation	- / 3 / vortex	ACN /HAc (2.5 mL)	< 0.5 μg·kg^{-1}	< 7.8		67–90	Wu et al. (2011)
Fe_3O_4@C_{14}	200	Bisphenol A (1)	Soft drinks	- / 20 / stir	- / 5 / vortex	MeOH (2.0 mL)	0.01 ng·mL^{-1}	< 8	-	91–105	Yazdinezhad et al. (2013)

(continued)

Table 3 (continued)

NM	NM amount (mg)	Analyte (n)	Sample	Extraction T (°C) / time (min) / agitation	Desorption T (°C) / time (min) /agitation	Eluent	LOD	Intra-day RSD (%)	Inter-day RSD (%)	RR (%)	Ref.
Fe_3O_4@C_{18}	17	OCPs (20) and PCBs (6)	Milk	- / 1 + 6 / vortex + ultrasounds	- / 10 / vortex	n-hexane (2.3 mL)	< 0.3 μg·L^{-1}	< 12	< 14	79–116	Synaridou et al. (2014)
Fe_3O_4@ CNDs	25	Fungicide (1)	Rice and cracked wheat	- / 2 / vortex	- / 1 / ultrasounds	HNO_3 / Acetone (1.0 mL)	5.3 μg·L^{-1}	-	-	93–100	Yilmaz and Soylak (2016)
Fe_3O_4@IL	100	Herbicide (1)	Apple and lettuce	RT / 15 / shake	25 / 10 / ultrasounds	EtOH (4.0 mL)	5.0 ng·mL^{-1}	< 8	-	95.0–98.5	Chen and Zhu (2015b)
Fe_3O_4@IL		Rhodamine B (1)	Chilli pepper	30 / 10 / shake	- / 5 / -	EtOH (4.0 mL)	0.08 μg·L^{-1}	< 5.1	-	99.0–101	Chen and Zhu (2016)
Fe_3O_4@IL	30	PAHs (5)	Rice	- / 30 / shake	- / 15 / shake	ACN (0.2 mL)	< 0.18 μg·Kg^{-1}	< 5.3	< 7.1	80.4–112	Boon et al. (2019)
Fe_3O_4@MIP	20	Chloramphenicol (3)	Pork and honey	25 / 50 / -	- / - / -	- (– mL)	10 μg·L^{-1}	< 1.9	< 2.8	95.31–106.9	Li et al. (2018c)
Fe_3O_4@MIP	20	Drug (1)	Health food	- / 15 / ultrasounds	- / - / -	MeOH /H_2O (0.5 mL)	5.49 ng·mL^{-1}	< 9	-	81.46–93.53	Wang et al. (2013)
Fe_3O_4@MIP	20	Fluoroquinolones (3)	Milk	- / 1 / vortex	- / 2 / vortex	MeOH /NH_3 (1.0 mL)	< 3.2 ng·g^{-1}	< 9.5	< 13	94.0–124	Zheng et al. (2014)
Fe_3O_4@MIP	10	Herbicides (5)	Rice	- / 60 / -	- / - / -	- (– mL)	-	< 8.2	-	73.2–91.5	Miao et al. (2015)
Fe_3O_4@MIP	50	Metronidazole	Milk and honey	25 / 90 / no agitation	- / 30 / ultrasounds	MeOH /HAc (2.0 mL)	0.05 μmol·L^{-1}	< 3.5	-	88.5–97	Chen et al. (2013)
Fe_3O_4@MIP	100	Orange II (1)	Red pepper and meat	- / 5 / shake	- / 0.5 / ultrasounds	MeOH (2.0 mL)	< 17.4 ng·L^{-1}	< 3.3	-	70–94	Zhan et al. (2013)
Fe_3O_4@MIP	10	Steroids (1)	Milk	- / 10 / shake	- / 3 / -	MeOH /HAc (10 mL)	0.01 ng·mL^{-1}	< 5.9	-	88.9–92.1	Gao et al. (2016)
Fe_3O_4@MIP	5	Sulfadimethoxine (1)	Fish	- / 5 + 120 / vortex + shake	- / - / -	MeOH /HAc (– mL)	< 52 μg·kg^{-1}	–	-	89.3–107	Feng et al. (2016)

Fe_3O_4@MIP	15	SAs (22)	Chicken breast	- / 10 / shake	- / - / -	MeOH /NH_3 (1.5 mL)	< 0.030 ng·g^{-1}	< 6.0	< 8.9	85.0–112	Zhao et al. (2014)
Fe_3O_4@MIP	50	SAs (4)	Chicken	- / 70 / stir	- / 120 / stir	MeOH (500 mL)	< 0.5 μg·L^{-1}	< 5.4	-	92.0–106	Karimi et al. (2013)
Fe_3O_4@MIP	50	Tetracycline antibiotics (6)	Milk	- / 20 / stir	- / - / -	MeOH /HAc (2.0 mL)	< 19.4 μg·kg^{-1}	< 5.2		74.5–93.8	Lv et al. (2013)
Fe_3O_4@MIP	20	Antibiotics (1)	Pork and milk	RT / 60 / shake	- / - / ultrasounds	MeOH /HAc (3.0 mL)	5.0 μg·kg^{-1}	< 3.6	< 4.3	85.5–94.0	Chen et al. (2019)
Fe_3O_4@oleic acid	60	Aflatoxins (4)	Pistachio	- / 1 / vortex	- / 3 / vortex	ACN (2.0 mL)	21 ng·L^{-1}	< 3.3	< 4.6	91.6–99.6	Taherimaslak et al. (2015)
Fe_3O_4@ stearic acid	98.2	Lead (–)	Red pepper	- /0.25 + 0.25 / shake + ultrasounds	- / - / -	HNO_3 65% (0.2 mL)	10 μg·L^{-1}	11	–	102.6–106.6	Zaman et al. (2020)
γ-Fe_2O_3@ imino-pyridine	148	Rhodamine B (–)	Chilli powder	- / 4 / shake	- / - / -	MeOH (0.5 mL)	1.6 μg·L^{-1}	2.3	–	90.8–94.2	Khani et al. (2019)
Fe_3O_4@OTC	-	Antibiotics (2)	Pork, milk and honey	37 / 40 / -	37 / 40 / -	PBS (– mL)	< 0.92 μg·L^{-1}	< 9.8	-	76.5–94.7	Liu et al. (2015)
Fe_3O_4@PIL	60	OPPs (4)	Tea	- / 1 / vortex	- / 1 / vortex	MeOH (2.0 mL)	< 0.01 μg·L^{-1}	< 11	-	81.4–113	Zheng et al. (2014b)
Fe_3O_4@PIL	10	Pyrethroids (8)	Apple	- /1- / shake	- / 1 / shake	MeOH /HAc (1 mL)	< 1.98 ng·g^{-1}	< 21	< 13	85.7–103	Zhao et al. (2019)
Fe_3O_4@Si-C_8/C_{18}	10	Chlorophenols (4)	Fish	- / 3 / vortex	- / 5 / ultrasounds	ACN (1.0 mL)	-	< 10		92–95	Zhang et al. (2013)
mPPY	2	Pesticides (11)	Juice and tea	- / 5 /vortex	- /2 / vortex	Acetone (0.2 mL)	< 0.09 μg·L^{-1}	< 14	< 16	63–129	Zhao et al. (2013)

(continued)

Table 3 (continued)

NM	NM amount (mg)	Analyte (n)	Sample	Extraction T (°C) / time (min) / agitation	Desorption T (°C) / time (min) /agitation	Eluent	LOD	Intra-day RSD (%)	Inter-day RSD (%)	RR (%)	Ref.

ACN: Acetonitrile
AMT: 2-amino-5-mercapto-1,3,4-thiadiazole
CNDs: Carbon nanodiamonds
CNTs: Carbon nanotubes
DMMIP: Dummy template magnetic molecularly imprinted polymer
DTC: Dithiocarbamate
DVB-VA: poly(vinylacetate-co-divinylbenzene
EDTA: Ethylenediaminetetraacetic acid
EtOH: Ethanol
GC: Graphitic carbon
GO: Graphene oxide
ILs: Ionic liquid
LOD: Limit of detection
MeOH: Methanol
MIP: Molecularly imprinted polymer
mPPY: Magnetic polypyrrole nanowires
MWCNTs: Multiwalled carbon nanotubes
NM: Nanomaterial
OCPs: Organochlorine pesticides
OPPs: Organophosphorus pesticides
OTC: Oxytetracycline
PAHS: Polycyclic aromatic hydrocarbons
PAMAN: Polyamidoamine
PBS: Phosphate buffered saline
PCBs: Polychlorinated biphenyls
PEA: Phenylethyl amine
PEI: Polyethyleneimine
PIL: Polymeric ionic liquid
PTh: Polythionine
PTSC: 2-pyridinecarboxaldehyde thiosemicarbazone
PVA: Polyvinil acrylate
RR: Recovery
RSD: Relative standard deviation
SAs: Sulfonamides
YS: Yolk-shell

be seen in the table, magnetic sorbents have been successfully used for the extraction of a wide variety of analytes, which includes from metal ions (Abbaszadeh and Tadjarodi 2016; Babaei et al. 2018; Ghorbani-Kalhor 2016; Keramat and Zare-Dorabei 2017; Khan et al. 2016; Zaman et al. 2020) to multiple organic contaminants such as antibiotics (Chen et al. 2013a; Chen et al. 2019; Karami-Osboo et al. 2015; Karimi et al. 2013; Li et al. 2018c; Liu et al. 2015a; Lv et al. 2013; Xia et al. 2017; Zhao et al. 2014, 2015; Zheng et al. 2014a), pesticides (Hou et al. 2013; Mahpishanian et al. 2015; Synaridou et al. 2014; Zhao et al. 2013; Zheng et al. 2014b), PCBs (Du et al. 2014; Lin et al. 2015a; Lin et al. 2015b; Synaridou et al. 2014), chlorophenols (Li et al. 2016; Liu et al. 2015c; Zhang et al. 2013a), PAHs (Li et al. 2018a; Boon et al. 2019), mycotoxins (Es'haghi et al. 2014; Hashemi et al. 2014; Sereshti et al. 2019; Taherimaslak et al. 2015; Wu et al. 2011), steroidal compounds (Gao et al. 2016; Ning et al. 2014) or dyes (Shi et al. 2018; Zhan et al. 2013, b), among others. Such compounds have been extracted from food matrices of different nature, including animal origin food, such as fish (Abbaszadeh and Tadjarodi 2016; Du et al. 2014; Feng et al. 2016; Ghorbani-Kalhor 2016; Keramat and Zare-Dorabei 2017; Li et al. 2018a; Lin et al. 2015a; Lin et al. 2015b; Zhang et al. 2013a), milk (Chen et al. 2013a; Chen et al. 2019; Gao et al. 2016; Hashemi et al. 2014; Karami-Osboo et al. 2015; Keramat and Zare-Dorabei 2017; Liu et al. 2015a; Liu et al. 2020; Lv et al. 2013; Synaridou et al. 2014; Zhao et al. 2015; Zheng et al. 2014a), milk powder (Ning et al. 2014), honey (Chen et al. 2013a; Li et al. 2018c; Duo et al. 2019; Liu et al. 2015c; Zhao et al. 2015), meat (Chen et al. 2019; Cheng et al. 2018; Karimi et al. 2013; Li et al. 2018a, c; Liu et al. 2015a; Liu et al. 2020; Xia et al. 2017; Zhan et al. 2013; Zhao et al. 2014), eggs (Babaei et al. 2018) or shrimps (Ghorbani-Kalhor 2016; Xia et al. 2017), among others, food of plant origin like vegetables (Babaei et al. 2018; Chen and Zhu 2015, 2016; Chen et al. 2016, 2018; Ghorbani-Kalhor 2016; Hou et al. 2013); Liu et al. 2015b; Mahpishanian et al. 2015; Zaman et al. 2020; Zhan et al. 2013), rice (Abbaszadeh and Tadjarodi 2016; Es'haghi et al. 2014; Boon et al. 2019; Jiang et al. 2019; Keramat and Zare-Dorabei 2017; Li et al. 2015c; Liang et al. 2018; Yilmaz and Soylak 2016), cereals (Es'haghi et al. 2014; Wu et al. 2011; Yilmaz and Soylak 2016), fruits (Chen and Zhu 2015; Ghorbani-Kalhor 2016; Hou et al. 2013; Liu et al. 2015b; Mahpishanian et al. 2015; Zhao et al. 2019), spices (Khan et al. 2016; Khani et al. 2019; Zhang et al. 2013b) or flour (Babaei et al. 2018) as well as different types of beverages (Keramat and Zare-Dorabei 2017; Li et al. 2016; Yazdinezhad et al. 2013; Zhao et al. 2013; Zheng et al. 2014b). The high number and the varied nature of the compounds and matrices to which magnetic sorbents have been applied, as well as the good results provided in all cases, highlight the good extraction performance and versatility of this kind of sorbents.

Regarding the composition of the sorbents, multiple combinations of materials have been used. In this sense, and as it has been previously mentioned, Fe_3O_4 NPs have been by far the preferred material to provide magnetic properties to the sorbents. It is noteworthy to mention that, although this kind of m-NPs are usually functionalized/modified, Fe_3O_4 NPs have shown a good extraction capacity in certain applications without further modifications, also for the analysis of contaminants in food samples, although in very few occasions (Karami-Osboo et al. 2015; Zhang et al. 2013b). As an example of their good performance in this field, Zhang et al. (Zhang et al. 2013b) carried out the ultrasound-assisted m-dSPE of Sudan I, II, III and IV dyes from chilli oil, chilli powder and tomato paste as well as tap and river water, and their subsequent determination by HPLC-UV. These NPs allowed obtaining recovery percentages in the range 93–111%, which show that the application of uncoated m-NPs is an option to consider in some cases in this field.

However, and as it is usual, m-NPs have been normally superficially modified in order to obtain a higher selectivity during the extraction of the target analytes. In this sense, and although a wide variety of materials have been used with this goal, polymeric materials have been the most common alternative by far (Chen et al. 2013a; Feng et al. 2016; Gao et al. 2016; Karimi et al. 2013; Li et al. 2018c; Lv et al. 2013; Miao et al. 2015; Noormohamadi et al. 2018; Wang et al. 2013; Zhan et al. 2013; Zhao et al. 2014; Zheng et al. 2014a). Regarding the multiple polymers which can be used, in the particular case of their application in food analysis, m-NPs have been combined with MIPs in almost all works (Chen et al. 2013a; Feng et al. 2016; Gao et al. 2016; Karimi et al. 2013; Li et al. 2018c; Lv et al. 2013; Miao et al. 2015; Wang et al. 2013; Zhan et al. 2013; Zhao et al. 2014; Zheng et al. 2014a), as a result of the great selectivity and versatility shown by this kind of coatings. At this point, it is important to mention that m-NPs combined with polymers have been used in two different ways: they have been coated with the polymeric material resulting in a core-shell type structure; or they have been simply mixed with the polymer resulting in a composite material in which m-NPs have been embedded in the polymeric structure. In both cases, the polymer can be isolated from the sample matrix by applying an external magnetic field. As an example of the former, Karimi et al. (Karimi et al. 2013) coated Fe_3O_4 m-NPs with a MIP using sulfamethazine as template and applied them as sorbent for the selective m-dSPE of four SAs from chicken meat. The analytes were finally separated and determined by HPLC-UV. The m-NPs coating resulted in an important increment of the active surface, which allowed obtaining recovery percentages higher than 95% and LODs lower than 0.6 $\mu g \cdot L^{-1}$, without losing the characteristic simplicity and rapidity of dSPE. Another interesting article that should be commented in this sense was developed by Zhao and coworkers (Zhao et al. 2014). In this work, the authors synthesized what they called a core–shell nanoring amino-functionalized superparamagnetic MIP, abbreviated as CS-NR-Mag-MIP. This particular sorbent, whose synthesis process is shown in Fig. 4, does not show a classical core-shell structure. Instead, several m-NPs arranged in a circular distribution are coated with the MIP providing multiple selective cavities for the target analytes retention. This sorbent, designed for the selective extraction of SAs, was applied for the m-dSPE of 22 SAs from chicken breast muscle, allowing the obtention of recovery values in the range 85–112%.

In relation to composite materials (m-NPs combined with polymers -non-coated NPs-), although their use is reduced in comparison to the above-commented, some examples can be found. An interesting application of this kind of sorbents was proposed by Zhao et al. (Zhao et al. 2013), in which the authors prepared polypyrrole nanowires by in situ chemical oxidative polymerization directed by

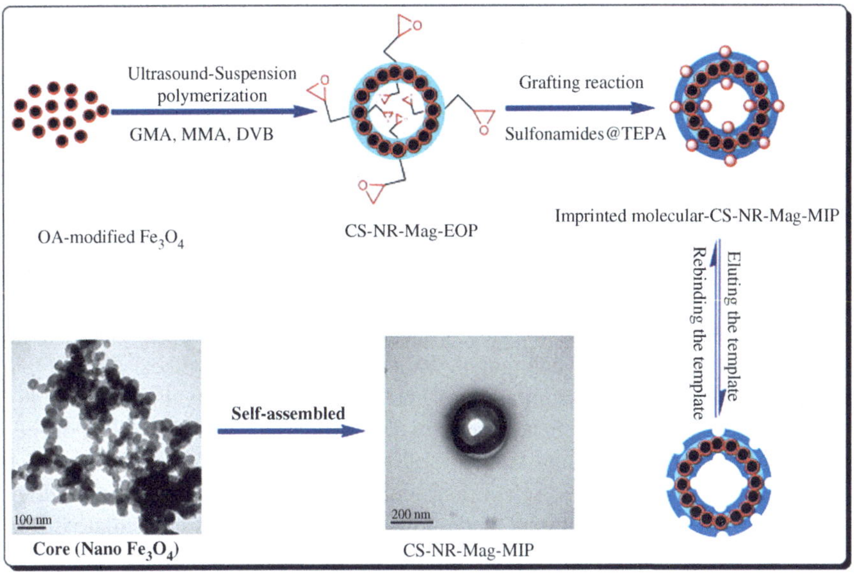

Fig. 4 Preparation procedure of a CS-NR-Mag-MIP. Reprinted from (Zhao et al. 2014) with permission from Elsevier

cetyltrimethylammonium bromide. These polymeric nanowires were magnetized by mixing them with Fe_3O_4 m-NPs with ultrasound agitation. This sorbent was further applied for the m-dSPE of 11 pesticides from juice and tea samples (water samples were also analysed) obtaining recovery percentages in the range 63–129%.

Besides polymeric coatings, and as it has been previously mentioned, other materials have also been used as superficial modifiers of m-NPs in order to achieve a high selectivity. In this sense, aptamers (Liu et al. 2015a; Wu et al. 2011), hydrocarbon chains of different lengths (Synaridou et al. 2014; Yazdinezhad et al. 2013; Zhang et al. 2013a), ILs (Chen and Zhu 2015, 2016) and PILs (Zheng et al. 2014b) have been some of the most common coating materials used in food analysis. However, one of the most relevant magnetic sorbents used in this field have been synthesized by combining m-NPs with nanomaterials, creating nanocomposites which gather the extraordinary extraction capacity of materials such as CNTs, graphene, GO or MOFs with the advantages derived from the magnetic properties of such NPs. In the particular case of CNTs (Cheng et al. 2018; Du et al. 2014; Zhao et al. 2015), MWCNTs have been used in almost all cases. However, regarding food analysis, CNTs have not been used as sorbents but as a support for other materials which interact with the analytes. As an example of it, Zhao et al. (Zhao et al. 2015) combined Fe_3O_4 m-NPs with CNTs and used them as support for a MIP. This sorbent was applied for the selective extraction of sulfamethoxazole from milk and honey, with recovery values between 68 and 80%. Such materials combination provides a high imprinting factor, an improved binding capacity and a fast adsorption kinetics, as well as an easy separation of the sorbent from the sample matrix thanks to its magnetic properties.

As it has already been mentioned, m-NPs have also been combined with graphene (Es'haghi et al. 2014) or GO (Babaei et al. 2018; Chen et al. 2016; Hou et al. 2013; Keramat and Zare-Dorabei 2017; Khan et al. 2016; Lin et al. 2015a; Liu et al. 2015c; Mahpishanian et al. 2015; Ning et al. 2014; Sereshti et al. 2019). Similarly to CNTs, GO has also been used as support for further modifications with other materials. In this sense, polymeric materials have been the most common coatings once more (Khan et al. 2016; Lin et al. 2015b; Ning et al. 2014), although phenylethyl amine (Mahpishanian et al. 2015) has also been used, among others. As an example, Es'haghi et al. (Es'haghi et al. 2014) carried out the synthesis of Fe_3O_4@graphene by chemical coprecipitation of Fe(II) and Fe(III) in an alkaline media in the presence of graphene. This nanocomposite was applied for the m-dSPE of four aflatoxins from rice, wheat and sesame, followed by the HPLC with fluorescence detection (FD) for analytes separation and determination. This methodology resulted a promising pre-treatment method for the fast trace analysis of aflatoxins in complicated matrices, with recovery values in the range 64–122% and LODs lower than 0.075 $\mu g \cdot kg^{-1}$.

Besides the above-mentioned combinations, m-NPs have also been combined with MOFs such as HKUST-1, MIL-101, among others, in a good number of applications (Abbaszadeh and Tadjarodi 2016; Ghorbani-Kalhor 2016; Li et al. 2016; Liang et al. 2018; Shi et al. 2018; Xia et al. 2017). The extremely porous structure of MOFs makes them an excellent alternative for their combination with m-NPs. Thus, they have been applied for the extraction of different families of compounds from matrices of different nature. As it has been previously commented for CNTs and graphene, the extraordinary properties of MOFs have allowed them not only to be used as sorbents but also as supports for other materials with higher selectivity or affinity for the target analytes. However, and despite their good performance in these two applications, they have also been used to create new nanoporous sorbents. That is what Liu et al. proposed in their work (Liu et al. 2015b). In this case, the authors carried out the synthesis of MOF ZIF-67 which was later calcined at 700 °C under nitrogen atmosphere, providing a magnetic black carbon material powder that the authors denoted as magnetic nanoporous carbon. This magnetic material was then applied for the extraction of phenylurea herbicides from grape and bitter gourd samples, providing recovery values in the range 89–105%. Besides MOFs, covalent organic materials (COFs) have emerged in recent years as a novel kind of porous materials assembled reversibly through strong covalent bonds. These new materials have also been used in combination with m-NPs in food analysis. That is the case of the work of Li et al. (Li et al. 2018a), who designed Fe_3O_4@COF hybrid microspheres for the m-dSPE of PAHs from pork, fish, grilled fish, smoked bacon and coffee. These materials present some advantages against MOFs, such as better stability, greater specific area or better π-π interactions, among others. Such good features have made them excellent sorbents, providing recovery percentages between 84 and 107% in this particular case.

As can be seen, the great versatility of Fe_3O_4 m-NPs to be combined with multiple materials or to be used alone as sorbent, and the advantages derived from their magnetic properties have made m-dSPE one of the most extended pre-treatment techniques in the last years in food analysis, as a result of its simplicity, rapidity, and low solvents and sorbents consumption against other classical methods.

3.2 Solid-Phase Microextraction (SPME)

SPME is a non-exhaustive sample preparation technique based on the establishment of partitioning equilibriums between the analytes contained in a sample and an extractant material exposed to it (Arthur et al. 1992). Generally, the extractant material is located at the extreme of a fused silica wire (Ø 110 nm) coating at least one-centimetre of the surface with a variable thickness forming a fibre. In addition, there are other SPME variants where the extractant is not onto the surface of a fibre but onto a magnetic stir bar, for example, as it happens in stir bar sorptive extraction (SBSE) devices (Baltussen et al. 1999; Nogueira 2012; Rykowska and Wasiak 2013), in hollow fibres (HF) (Li et al. 2015a), or onto the inner walls of tubes in the so called in-tube SPME (Fernández-Amado et al. 2016; Moliner-Martínez et al. 2015).

Analytes extraction can take place in two different ways: by the direct exposition of the fibre to the sample matrix (direct immersion -DI- modality) or by the exposition of the fibre to the gas phase situated over the sample (headspace -HS- modality). In both cases, an equilibrium between the donor phase and the extractant phase has to be reached, being just one in the DI modality (sample/fibre) case and two for the HS modality (sample/gas phase/fibre). Thus, the extraction capability of the extractant material, its volume and the possible interactions between the fibre and the sample matrix will play an important role in the extraction and preconcentration procedure. After a certain time (the technique can be applied under equilibrium or non-equilibrium conditions), the adsorbed/absorbed analytes are submitted to a desorption process. In most of the cases, a thermal desorption into a gas chromatograph injection port is performed avoiding the use of organic solvents and further sample treatment steps prior to the analysis. Nevertheless, this option is not always feasible due to the thermal instability of certain analytes or to the detection requirements, and an off-line desorption using organic solvents might be needed.

Since its commercialization (Reyes-Garcés et al. 2018), SPME fibres popularity has experienced an increase, and the number of users has raised up considerably due to its advantages over conventional sample preparation procedures. Some of these advantages are the reduction of extraction time, the high pre-concentration factor obtained using low sample volumes, the reduction of organic solvents volumes required in the sample preparation step, and the reusability of the fibres among others (Reyes-Garcés et al. 2018).

Currently, there are six commercial SPME fibres available (Shirey 2012). Three of them are single polymeric films of divinylbenzene (DVB), polyacrylate (PA), or polydimethylsiloxane (PDMS), and the other three are a mixture of the polymers with another compound: carboxen/PDMS (CAR/PDMS), PDMS/DVB, and DVB/CAR/PDMS. Fibre selection will depend mainly on the nature of the analytes and the desired analysis. For non-polar volatile or semi-volatile compounds, PDMS fibres are the best option, while for polar compounds, PA and CAR fibres are recommended. The employment of mixed fibres (CAR/PDMS, PDMS/DVB, and DVB/CAR/PDMS) is appropriated for the extraction of more complexed analytes that need a higher adsorption efficiency, and be released faster, due to their low trace concentration levels compared with other compounds with a higher molecular weight and size.

Although the commercially available fibres have demonstrated to possess an indubitable excellent extraction capability and analytical performance quality, polymeric coatings present in some cases a low selectivity, making difficult the analysis of the target analytes due to the extraction of non-desired compounds that are also in the sample (Shirey 2012). Such fact should be added to their chemical, mechanical and thermal stability, which in some cases are not enough to ensure analyte extraction without breaking or damaging the fibre under the required extraction conditions. In this sense, current investigations have focused their efforts on the development of novel coatings using fashion coatings such as nanomaterials, which can offer those properties (Piri-Moghadam et al. 2017; Reyes-Garcés et al. 2018).

Regarding the nanomaterial-based coatings developed for the extraction of pollutants in food samples (Souza-Silva et al. 2015), the most studied have been MOFs (Jia et al. 2016; Lan et al. 2014; Liang et al. 2016; Lin et al. 2014, 2016; Lirio et al. 2016; Liu et al. 2019; Mondal et al. 2019; Wu et al. 2016; Zhang et al. 2015; Zhang et al. 2017), followed by graphene (Hou et al. 2016; Shi et al. 2019; Sun et al. 2014; Zhang et al. 2014), and in a less number of occasions CNTs (Saraji et al. 2016), and certain types of NPs (Sehati et al. 2014). Table 4 summarizes some of the nanomaterials-based coatings developed and their application for the analysis of emerging and persistent contaminants in food samples.

SPME-nanomaterial-based coatings can be divided into neat nanomaterials coatings (Liang et al. 2016; Lin et al. 2016; Zhang et al. 2014, 2015) and hybrid/composites obtained from the combination with another material as a protection layer or if a more specific and selective extraction is desired (Hou et al. 2016; Jia et al. 2016; Lan et al. 2014; Lirio et al. 2016; Lin et al. 2014; Saraji et al. 2016; Shi et al. 2019; Sun et al. 2014; Wu et al. 2016; Zhang et al. 2017).

There is a wide variety of procedures for the preparation of SPME fibres coatings. They are divided mainly into three differentiated groups taking into account the existence of a chemical bonding or not, and if the nanomaterial is already prepared or not (Pacheco-Fernández et al. 2018; Rocío-Bautista et al. 2018). These groups are *physical adhesion* methods (Jia et al. 2016; Lan et al. 2014; Zhang et al. 2017), *in-situ growth* methods (Lin et al. 2016; Wu et al. 2016; Zhang et al. 2015),

Table 4 Recent examples of the application of nano-sorbents in SPME for food contaminants analysis

Coating	Analyte (n)	Sample	Extraction T (° C) / time (min)	Desorption t. (° C) / time (min)	LOD	Intraday RSD (%)	Inter-fiber RSD (%)	RR (%)	Ref.
MWCNTs@ SiO_2	Pesticides (4)	Pear, grape and eggplant	45 / 30	260 / 5	< 0.02 μg·L^{-1}	< 7.8	< 12.6	79–99	Saraji et al. (2016)
Go@IL	PAHs (6) and PAEs (3)	Potatoes	70 / 40	260 / 3	< 0.025 μg·L^{-1}	< 10.8	< 14.6	78.3–101.7	Hou et al. (2016)
GO	PAHs (6)	Grilled meat	- / 120	- / 5	< 0.50 pg·mL^{-1}	< 4.97	< 6.45	88.5–113.6	Zhang et al. (2014)
GO-C_{18}	Chlorophenols (4)	Honey	- / 30	- / 2	< 2.0 ng·g^{-1}	< 6.8	–	88–108	Sun et al. (2014)
PS/G@SiO_2	Fungicides (2)	Mung bean	25 / 10	260 / 5	< 0.09 μg·L^{-1}	< 8.9	< 12.2	72–99	Shi et al. (2019)
Aptamer-MOF-5	PCBs (6)	Fish	50 / 50	- / 20	< 0.004 ng·mL^{-1}	< 2.6	< 4.3	89.2–97.1	Lin et al. (2016)
Fe_3O_4@ ZIF-8@MIP	Estrogens (4)	Fish and pork	- / 15	- / 10	< 6.2 ng·g^{-1}	< 5.19	–	73.8–96.7	Lan et al. (2014)
MAF-66	PAHs (7)	Potato and roast pork	25 / 40	250 / 5	< 7.5 ng·L^{-1}	< 4.2	< 8.3	89–118	Liu et al. (2019)
MIL-101@ MON	PAHs (16)	Beef, pork and lamb	45 / 35	290 / 5	< 0.30 ng·L^{-1}	< 8.4	< 8.7	76.3–85.6	Jia et al. (2016)
MIL-101-NH_2	Antibiotics (6)	Fish	- / 10	- / 40	< 4.6 ng·L^{-1}	< 6.8	< 9.5	-	Mondal et al. (2019)
MIL-125	Phthalates (5)	Tea	RT / 40	260 / 6	< 3.0 ng·L^{-1}	< 10.2	–	82–106	Liang et al. (2016)
MOF-5@ IL-GO	Chloramphenicols (2)	Milk and honey	50 / 40	300 / 3	< 19.5 ng·L^{-1}	< 3.7	< 6.0	87.4–103.2	Wu et al. (2016)

(continued)

Table 4 (continued)

Coating	Analyte (n)	Sample	Extraction T (° C) / time (min)	Desorption t. (° C) / time (min)	LOD	Intra-fiber RSD (%)	Inter-fiber RSD (%)	RR (%)	Ref.
ZIF-90 (Zn)	PCBs (6)	Vegetable oil	50 / 40	230 / 15	< 0.053 ng·L^{-1}	< 5.5	< 9.1	85.9–105.8	Zhang et al. (2015)
ZIF-90 (Zn) - NPC	Pesticides (8)	Peach, cucumber and cabbages	40 / 40	280 / 10	< 0.5 ng·g^{-1}	< 7.6	< 10.9	86.0–103.5	Zhang et al. (2017)
Fe_3O_4@ MOF-5	PCBs (6)	Fish	- / 30	- / 3	< 0.096 ng·L^{-1}	< 3.6	–	94.3–97.5	Lin et al. (2014)
MIL-53@MIP	Penicillin (6)	Milk	- / 37	- / -	< 0.26 μg·L^{-1}	< 3.5	< 3.4	81.1–100.7	Lirio et al. (2016)

LOD: Limit of detection
RR: Recovery
MWCNTs: Multiwalled carbon nanotubes
G: Graphene
GO: Graphene oxide
IL: Ionic liquid

PAH: Polycyclic aromatic hydrocarbons
PAEs: Phthalate esters
PCBs: Polychlorinated biphenyls
PS: Polystyrene
MIP: Molecularly imprinted polymer

and *chemical bonding* methods (Hou et al. 2016; Liang et al. 2016; Saraji et al. 2016; Zhang et al. 2014).

Physical adhesion preparation methods are based on the material attachment to the SPME wire by physic interactions to avoid the previous support chemical functionalization, which is time consuming and tedious due to the numerous steps required (Rocío-Bautista et al. 2018). One example is the use of sealants such as epoxy glue or silicone to fix the nanomaterial to the support surface (Jia et al. 2016; Zhang et al. 2017). Zhang et al. have proposed the employment of a carbonized ZIF-90 based coating for the analysis of pyrethroid pesticides in fruits and vegetables prepared by physical adhesion (Zhang et al. 2017). Briefly, an active stainless-steel wire is coated with a silver mirror by reduction using glucose as reductant agent in order to increase the surface area. Then, the wire is immersed into a silicone sealant to impregnate the surface with the adhesive. Finally, before the silicone is dried, the wire is dipped into the MOF carbonized powder. Fibre fabrication takes less than three hours ensuring porous crystalline structures homogeneously disposed all over the fibre coating surface. Nevertheless, sealants have an appreciable content of PDMS, which provides a non-selective extraction. Thus, comparison studies using just the sealant coating and the material attached coating using the sealant must be performed. Another option to avoid the influence of the sealant is to prepare the coating by an electromagnetic bonding. Ferrite NPs are synthetized and coated with the desired material. Then, these magnetic particles are exposed to the stainless-steel wire and are physically adsorbed onto the wire surface without the use of any external chemical substance (Lan et al. 2014).

In-situ growth methods consist on the nanomaterial synthesis in the presence of the support where it is going to be attached. In order to obtain those coatings, the support must be pre-treated to increase the roughness and to ensure the presence of active places or nucleation points where the nanomaterial can grow. The easiest option is to expose the wire to an aggressive procedure in order to remove the passivation film and to activate the silanol groups of the wire surface. As an example, Wu et al. employed that strategy for the development of MOF-5 and MOF-5-IL-GO composite-based coatings for the determination of chloramphenicol and thiamphenicol in milk and honey samples (Wu et al. 2016). A stainless-steel wire was activated using hydrofluoric acid. Then, it was introduced into a suspension containing MOF-5 precursors, and the IL modified GO particles. MOF-5 synthesis took place by the solvothermal method. Finally, the fibre coating was activated under nitrogen atmosphere at 300 ° C for 10 h. A similar strategy was followed by Lin et al. to obtain a MOF-5 based coating but instead of the solvothermal synthesis of the MOF, an electrochemical in-situ growth was performed by cyclic voltammetry obtaining the coating in less time than the solvothermal option (Lin et al. 2016). In addition, an aptamer was immobilized over the fibre surface for the selective extraction of polychlorinated biphenyls. With application of the cyclic voltammetry, the experimental parameters that influence the deposition of the material can be better controlled (Lin et al. 2016). Although these procedures have demonstrated to pose good inter-fibre reproducibility (under 6.0% in both cases) and good extraction efficiency, in some cases there are not enough activated silanol groups to ensure the in-situ growth. Thus, a chemical functionalization procedure is previously required for the surface modification. In this sense, Zhang et al. coated the wire with a silver mirror, after the acid pre-treatment, and then this thin silver layer was functionalized with 3-mercaptopropyl-trimethoxysilane (3-MTMS). This compound poses three silanol groups and a thiol that is easily bonded to silver due to the strong interactions between the metal and the group forming a high stable metal-sulphur bond, allowing to obtain a surface totally functionalized and with silanol groups available for the in-situ growth of the MOF (Zhang et al. 2015).

Chemical procedures are half on the way between physical and in-situ growth coatings preparation methods. The main idea is to perform a wire pre-treatment and functionalization to ensure a chemical bonding and then nanomaterial is attached to the surface. However, in spite of the development of an in situ growth of the nanomaterial it is already synthesized and ready for the deposition. A clear example of that is the layer-by-layer procedure, which consists on the successive deposition of a material by the functionalization of the surface. For that purpose, two different parts are needed: the selected nanomaterial and a linker that must be able to establish a union between the layers. Following this kind of strategies, the coating thickness is controlled because the number of layers that are deposited over the fibre determines the coating thickness. Nevertheless, the layer-by-layer method is time consuming and, in most cases, for obtaining large thickness coatings an elevated number of layers must be deposited (Hou et al. 2016; Liang et al. 2016; Zhang et al. 2014).

All these coatings have been applied for a wide range of food contaminants such as organophosphorus pesticides (OPPs) (Saraji et al. 2016), fungicides (Shi et al. 2019), PAHs (Hou et al. 2016; Jia et al. 2016; Liu et al. 2019; Zhang et al. 2014), PCBs (Lin et al. 2014, 2016; Zhang et al. 2015, Zhang et al. 2018) and phthalates (Hou et al. 2016; Liang et al. 2016) (see Table 4).

Concerning the food samples, they have been applied to different kinds of matrices such as fruits and vegetables (Liu et al. 2019; Saraji et al. 2016; Shi et al. 2019; Zhang et al. 2017), meat (Jia et al. 2016; Lan et al. 2014; Liu et al. 2019; Zhang et al. 2014), fish (Lan et al. 2014; Lin et al. 2014, 2016; Mondal et al. 2019), milk (Lirio et al. 2016; Wu et al. 2016), honey (Sun et al. 2014; Wu et al. 2016), among others (Liang et al. 2016; Zhang et al. 2015).

Regarding MOFs based coatings, MOF-5 is the most commonly used mainly as a neat MOF (Lin et al. 2016) but also as functionalized MOF with other materials/compounds such as aptamers (Lin et al. 2014) and ILs combined with GO (Wu et al. 2016), specially for the study of PCBs in fish samples (Lin et al. 2014, 2016). The study of these compounds using this kind of fibres has shown excellent analytical results (reproducibility, LODs and recovery values), improving the results obtained by the commercially available PDMS, PDMS/DVB, CAR/PDMS and PA fibres. Similar results have been obtained by Zhang et al. for this group of analytes in vegetable oils using a ZIF-90 based coating instead of MOF-5 -see Fig. 5- (Zhang et al. 2015). The improved results obtained compared to commercial fibres acclaim the excellent extraction capability that characterize these porous materials.

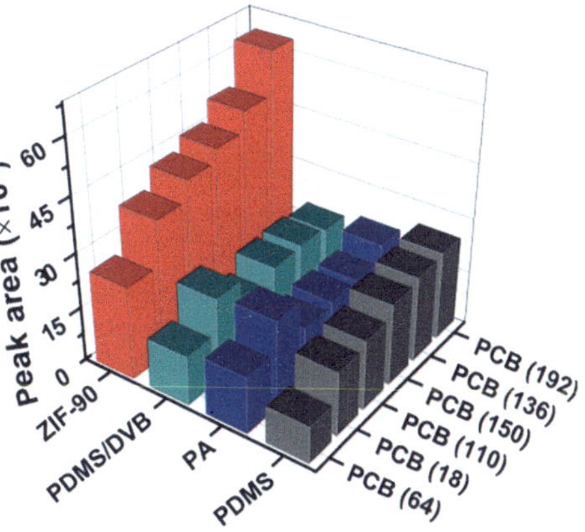

Fig. 5 Comparison of the extraction capability between a ZIF-90 based SPME fibre and PDMS, PA, and PDMS/DVB commercial fibres for the analysis of six PCBs. Reprinted from (Zhang et al. 2015) with permission from Elsevier

Concerning GO based coatings, they have been mainly used for determining PAHs in grilled meat (Zhang et al. 2014) and potatoes (Hou et al. 2016), this may be due to the high interactions between the GO and the π electrons of these compounds favouring the extraction of these analytes (Hou et al. 2016; Zhang et al. 2014).

4 Conclusions and Future Trends

Nanomaterials have been applied with success for the selective extraction and preconcentration of contaminants from a wide variety of food matrices, which has notably increased the number of analytical methods devoted to food contaminants analysis. Their high porosity, high volume-to-surface ratio and their ability to establish different types of interactions with a wide variety of analytes are the main reasons for their success.

Among the different types of nano-sorbents currently applied in this field, the use of carbon-based nanomaterials, in particular, CNTs, graphene and GO, MOFs and a wide variety of NPs in which different materials are combined can be highlighted. Most of them have been used in SPE, μ-dSPE, m-dSPE and SPME procedures. In particular, GO has been the most widely applied in SPE, MOFs and CNTs in dSPE, MOFs in SPME and a wide variety of m-NPs in m-dSPE. The low amounts at which they are frequently used and the low volumes of organic solvents normally necessary for the desorption process, has greatly contributed to the development of the so-called "Green Analytical Chemistry" processes (Armenta et al. 2015).

Notwithstanding the high number of publications in this field, there exists a lot of scope for further investigation. In particular, there are still groups of contaminants and food matrices that have not been studied and, therefore, the full potential of those materials currently available has not yet been established. Besides, nanoscience and nanotechnology fields are in continuous development and novel nanosorbents, including modifications of previous ones, will surely appear in the near future and will also be surely applied in this and other fields.

Acknowledgements A. G.-S. thanks the Canary Agency of Research and Innovation (ACIISI), co-funded by the European Social Fund, for his FPI PhD fellowship.

References

Abbaszadeh A, Tadjarodi A (2016) Speciation analysis of inorganic arsenic in food and water samples by electrothermal atomic absorption spectrometry after magnetic solid phase extraction by a novel MOF-199/modified magnetite nanoparticle composite. RSC Adv 6:113727–113736

Adams FC, Barbante C (2013) Nanoscience, nanotechnology and spectrometry. Spectrochim Acta B 86:3–13

Ahmadi I, Bahar S, Ziarani GM, Badiei A (2019) Formation of functionalized silica based nanoparticles and their application for extraction and determination of Hg (II) ion in fish samples. Food Chem 300:125180

Anderson JL, Bethod A, Pino V, Stalcup A (2015) (Ed) Analytical separation science. Sample preparation, method validation, and analytical applications. Wiley-VCH, Verlag ISBN: 978-3-527-33374-5.

Armenta S, Garrigues S, de la Guardia M (2015) The role of the green extraction techniques in green analytical chemistry. TrAC-Trends Anal Chem 71:2–8

Arthur CL, Killam LM, Buchholz KD, Pawliszyn J (1992) Automation and optimization of solid-phase microextraction. Anal Chem 64:1960–1966

Babaei A, Zeeb M, Es-haghi A (2018) Magnetic dispersive solid-phase extraction based on graphene oxide/Fe3O4@polythionine nanocomposite followed by atomic absorption spectrometry for zinc monitoring in water, flour, celery and egg. J Sci Food Agric 98:3571–3579

Baltussen E, Sandra P, David F, Cramers C (1999) Stir bar sorptive extraction (SBSE), a novel extraction technique for aqueous samples: theory and principles. J Microcolumn Sep 11:737–747

Barreto AS, Luiz da Silva R, Santos Silva SCG, Rodríguez MO, Simone CA, Sá GF, Júnior SA, Navickiene S, Mesquita ME (2010) Potential of a metal–organic framework as a new material for solid-phase extraction of pesticides from lettuce (Lactuca sativa), with analysis by gas chromatography-mass spectrometry. J Sep Sci 33:3811–3816

Bhuyan MSA, Uddin MN, Islam MM, Bipasha FA, Hossain SS (2016) Synthesis of graphene. Int Nano Lett 6:65–83

Boon YH, Zain NNM, Mohamad S, Osman H, Raoov M (2019) Magnetic poly(β-cyclodextrin-ionic liquid) nanocomposites for micro-solid phase extraction of selected polycyclic aromatic hydrocarbons in rice samples prior to GC-FID analysis. Food Chem 278:322–332

Chen J, Zhu X (2015) Ionic liquid coated magnetic core/shell Fe3O4@SiO2 nanoparticles for the separation/analysis of linuron in food samples. Spectrochim Acta A 137:456–462

Chen J, Zhu X (2016) Magnetic solid phase extraction using ionic liquid-coated core-shell magnetic nanoparticles followed by high-performance liquid chromatography for determination of Rhodamine B in food samples. Food Chem 200:10–15

Chen D, Deng J, Liang J, Xie J, Huang K, Hu C (2013a) Core-shell magnetic nanoparticles with surface-imprinted polymer coating as a new adsorbent for solid phase extraction of metronidazole. Anal Methods 5:722–728

Chen L, Lu Y, Li S, Lin X, Xu Z, Dai Z (2013b) Application of graphene-based solid-phase extraction for ultra-fast determination of malachite green and its metabolite in fish tissues. Food Chem 141:1383–1389

Chen L, Zhou T, Zhang Y, Lu Y (2013c) Rapid determination of trace sulfonamides in fish by graphene-based SPE coupled with UPLC/MS/MS. Anal Methods 5:4363–4370

Chen Y, Cao S, Zhang L, Xi C, Li X, Chen Z, Wang G (2016) Preparation of size-controlled magnetite nanoparticles with a graphene and polymeric ionic liquid coating for the quick, easy, cheap, effective, rugged and safe extraction of preservatives from vegetables. J Chromatogr A 1448:9–19

Chen S, Fu J, Li Z, Zeng Y, Li Y, Su X, Jiang X, Yang H, Huang L, Zou L, He L, Liu S, Ao X, Yang Y (2019) Preparation and application of magnetic molecular imprinted polymers for extraction of cephalexin from pork and milk samples. J Chromatogr A 1602:124–134

Cheng H-L, Wang F-L, Zhao Y-G, Zhang Y, Jin M-C, Zhu Y (2018) Simultaneous determination of fifteen toxic alkaloids in meat dishes and vegetables dishes using double layer pipette tip magnetic dispersive solid phase extraction followed by UFLC-MS/MS. Anal Methods 10:1151–1162

Crevillen AG, Escarpa A, García CD (2019) Carbon-based nanomaterials in analytical chemistry. In: García CD, Crevillén AG, Escarpa A (ed) Carbon-based nanomaterials in Analytical Chemistry. Royal Society of Chemistry

Deng ZH, Xu GJ, Wang X, Wang ML, Lin JM, Zhao RS (2018) A Zr(IV)-based porphyrinic metal-organic framework as a solid-phase sorbent for extraction of sulfonamides prior to their quantitation by LC-MS. Microchim Acta 185:450–458

Dey C, Kundu T, Biswal BP, Mallick A, Banerjee R (2014) Crystalline metal-organic frameworks (MOFs): synthesis, structure and function. Acta Crystallogr Sect B-Struct Sci Cryst Eng Mat B70:3–10

Du A, Liu M, Li G (2013) Novel magnetic SPE method based on carbon nanotubes filled with cobalt ferrite for the analysis of organochlorine pesticides in honey and tea. J Sep Sci 36:3387–3394

Du X, Lin S, Gan N, Chen X, Cao Y, Li T, Zhan P (2014) Multi-walled carbon nanotube modified dummy-template magnetic molecularly imprinted microspheres as solid-phase extraction material for the determination of polychlorinated biphenyls in fish. J Sep Sci 37:1591–1600

Duo H, Wang Y, Wang L, Lu X, Liang X (2018) Zirconium-based metal-organic frameworks (UiO-67) as solid-phase adsorbents for extraction of phenoxyacetic acid herbicides from vegetables. J Sep Sci 41:4149–4158

Duo H, Lu X, Wang S, Wang L, Guo Y, Liang X (2019) Synthesis of magnetic metal-organic framework composites, Fe_3O_4-NH_2@MOF-235, for the magnetic solid-phase extraction of benzoylurea insecticides from honey, fruit juice and tap water. New J Chem 43:12563–12569

Eatemadi A, Daraee H, Karimkhanloo H, Kouhi M, Zarghami N, Akbarzadeh A, Abasi M, Hanifehpour Y, Joo SW (2014) Carbon nanotubes: properties, synthesis, purification, and medical applications. Nanoscale Res Lett 9:393–406

Eftekhari A, Shakerian M, Majeed HJ, Eftekhari M (2019) Pectic acid–graphene oxide nanocomposite as an adsorbent in vortex-assisted dispersive solid-phase extraction for preconcentration of copper ion followed by flame atomic absorption spectrometry. Polym Bull

Es'haghi Z, Beheshti HR, Feizy J (2014) Extraction of aflatoxins from food samples using graphene-based magnetic nanosorbents followed by high-performance liquid chromatography: a simple solution to overcome the problems of immunoaffinity columns. J Sep Sci 37:2566–2573

Faraji M, Yamini Y, Rezaee M (2010) Magnetic nanoparticles: synthesis, stabilization, functionalization, characterization, and applications. J Iran Chem Soc 7:1–37

Farré M, Barceló D (2013) Analysis of emerging contaminants in food. TrAC-Trends Anal Chem 43:240–253

Feng M, Li H, Zhang L, Zhang J, Dai J, Wang X, Zhang L, Wei Y (2016) Preparation and application of novel magnetic molecularly imprinted composites for recognition of sulfadimethoxine in feed samples. Analytical Sci 32:517–521

Fernández-Amado M, Prieto-Blanco MC, López-Mahía P, Muniategui-Lorenzo S, Prada-Rodríguez D (2016) Strengths and weaknesses of in-tube solid-phase microextraction: a scoping review. Anal Chim Acta 906:41–57

Fritz JS (1999) Analytical solid-phase extraction. Wiley-VCH, New York isbn:0–471–24667–0

Furukawa H, Cordova KE, O'Keeffe M, Yaghi OM (2013) The chemistry and applications of metal-organic frameworks. Science 341:1230444

Gao R, Zhang J, He X, Chen L, Zhang Y (2010) Selective extraction of sulfonamides from food by use of silica-coated molecularly imprinted polymer nanospheres. Anal Bioanal Chem 398:451–461

Gao R, Cui X, Hao Y, Zhang L, Liu D, Tang Y (2016) A highly-efficient imprinted magnetic nanoparticle for selective separation and detection of 17β-estradiol in milk. Food Chem 194:1040–1047

Ghorbani-Kalhor E (2016) A metal-organic framework nanocomposite made from functionalized magnetite nanoparticles and HKUST-1 (MOF-199) for preconcentration of Cd(II), Pb(II), and Ni(II). Microchim Acta 183:2639–2647

Giakisikli G, Anthemidis AN (2013) Magnetic materials as sorbents for metal/metalloid preconcentration and/or separation. A review. Anal Chim Acta 789:1–16

González-Hernández P, Gutiérrez-Serpa A, Rocío-Bautista P, Pasán J, Ayala JH, Pino V (2019) Micro-solid-phase extraction using metal-organic frameworks. In: Mittal V (ed) Metal organic frameworks. Central West Publishing, Australia

González-Sálamo J, (2018) PhD Thesis, New extraction protocols for the determination of endocrine disrupting compounds in samples of environmental and agri-food interest . University of La Laguna, Spain

González-Sálamo J, Herrera-Herrera AV, Fanali C, Hernández-Borges J (2016a) Magnetic nanoparticles in solid-phase extraction. LCGC Europe 29:180–193

González-Sálamo J, Socas-Rodríguez B, Hernández-Borges J, Rodríguez-Delgado MÁ (2016b) Nanomaterials as sorbents for food sample analysis. TrAC-Trends Anal Chem 85:203–220

González-Sálamo J, Socas-Rodríguez B, Hernández-Borges J, Rodríguez-Delgado MÁ (2017) Core-shell poly(dopamine) magnetic nanoparticles for the extraction of estrogenic mycotoxins from milk and yogurt prior to LC-MS analysis. Food Chem 215:362–368

Gutiérrez-Serpa A, Pacheco-Fernández I, Pasán J, Pino V (2019) Metal-organic frameworks as key materials for solid-phase microextraction devices – a review. Separations 6:47

Hashemi M, Taherimaslak Z, Rashidi S (2014) Enhanced spectrofluorimetric determination of aflatoxin M1 in liquid milk after magnetic solid phase extraction. Spectrochim Acta A 128:583–590

Hashemi B, Zohrabi P, Raza N, Kim K-H (2017) Metal-organic frameworks as advanced sorbents for the extraction and determination of pollutants from environmental, biological, and food media. TrAC-Trends Anal Chem 97:65–82

He J, Huang M, Wang D, Zhang Z, Li G (2014) Magnetic separation techniques in sample preparation for biological analysis: a review. J Pharmaceut Biomed 101:84–101

He M, Huang L, Zhao B, Chen B, Hu B (2017) Advanced functional materials in solid phase extraction for ICP-MS determination of trace elements and their species - a review. Anal Chim Acta 973:1–24

He J, Huan Y, Zhao T (2019) Well-designed high selective carbon molecularly imprinted polymer nanocomposite based on a green synthesis strategy for solid-phase extraction of tetracyclines residues in food samples. Food Anal Methods 12:2601–2613

Herrera-Herrera AV, González-Curbelo MÁ, Hernández-Borges J, Rodríguez-Delgado MÁ (2012) Carbon nanotubes applications in separation science: a review. Anal Chim Acta 734:1–30

Hou M, Zang X, Wang C, Wang Z (2013) The use of silica-coated magnetic graphene microspheres as the adsorbent for the extraction of pyrethroid pesticides from orange and lettuce samples followed by GC-MS analysis. J Sep Sci 36:3242–3248

Hou X, Guo Y, Liang X, Wang X, Wang L, Wang L, Liu X (2016) Bis(trifluoromethanesulfonyl) imide-based ionic liquids grafted on graphene oxide-coated solid-phase microextraction fibre for extraction and enrichment of polycyclic aromatic hydrocarbons in potatoes and phthalate esters in food-wrap. Talanta 153:392–400

Hou X, Tang S, Guo X, Wang L, Liu X, Lu X, Guo Y (2018) Preparation and application of guanidyl-functionalized graphene oxide-grafted silica for efficient extraction of acidic herbicides by box-Behnken design. J Chromatogr A 1571:65–75

Hu C-Y, Xu Y-J, Duo S-W, Zhang R-F, Li M-S (2009) Non-covalent functionalization of carbon nanotubes with surfactants and polymers. J Chin Chem Soc 56:234–239

Iijima S (1991) Helical microtubules of graphitic carbon. Nature 354:56–58

Jia Y, Su H, Wang Z, Wong YLE, Chen X, Wang M, Chan TWD (2016) Metal-organic framework@microporous organic network as adsorbent for solid-phase microextraction. Anal Chem 88:9364–9367

Jia Y, Zhao Y, Zhao M, Wang Z, Chen X, Wang M (2018) Core–shell indium (III) sulfide@metal-organic framework nanocomposite as an adsorbent for the dispersive solid-phase extraction of nitro-polycyclic aromatic hydrocarbons. J Chromatogr A 1551:21–28

Jiang Y, Piao H, Qin Z, Li X, Ma P, Sun Y, Wang X, Song D (2019) One-step synthesized magnetic MIL-101(Cr) for effective extraction of triazine herbicides from rice prior to determination by liquid chromatography-tandem mass spectrometry. J Sep Sci 42:2900–2908

Jolivet J-P, Tronc E (1988) Interfacial electron transfer in colloidal spinel iron oxide. Conversion of Fe3O4-γFe2O3 in aqueous medium. J Colloid Interf Sci 125:688–701

Kalmutzkl MJ, Hanlkel N, Yaghi OM (2018) Secondary building units as the turning point in the development of the reticular chemistry of MOFs. Sci Adv 4:1–16

Karami-Osboo R, Miri R, Javidnia K, Shojaee MH, Kobarfard F (2015) Extraction and determination of sulfadiazine and sulfathiazole in milk using magnetic solid phase extraction-HPLC-UV. Anal Methods 7:1586–1589

Karimi M, Aboufazeli F, Zhad HRLZ, Sadeghi O, Najafi E (2013) Determination of sulfonamides in chicken meat by magnetic molecularly imprinted polymer coupled to HPLC-UV. Food Anal Methods 7:73–80

Kashanaki R, Ebrahimzadeh H, Moradi M (2018) Metal–organic framework based micro solid phase extraction coupled with supramolecular solvent microextraction to determine copper in water and food samples. New J Chem 42:5806–5813

Kaur R, Hasan A, Iqbal N, Alam S, MKR S, Raza SK (2014) Synthesis and surface engineering for magnetic nanoparticles for environmental cleanup and pesticide residue analysis: a review. J Sep Sci 37:1805–1825

Keramat A, Zare-Dorabei R (2017) Ultrasound-assisted dispersive magnetic solid phase extraction for preconcentration and determination of trace amount of Hg (II) ions from food samples and aqueous solution by magnetic graphene oxide (Fe3O4@GO/2-PTSC): central composite design optimization. Ultrason Sonochem 38:421–429

Khan M, Yilmaz E, Sevinc B, Sahmetlioglu E, Shah J, Jan MR, Soylak M (2016) Preparation and characterization of magnetic allylamine modified graphene oxide-poly(vinylacetate-co-divinylbenzene) nanocomposite for vortex assisted magnetic solid phase extraction of some metal ions. Talanta 146:130–137

Khani R, Sobhani S, Yari T (2019) Magnetic dispersive micro solid-phase extraction of trace Rhodamine B using imino-pyridine immobilized on iron oxide as nanosorbent and optimization by Box–Behnken design. Microchem J 146:471–478

Khezeli T, Daneshfar A (2017) Development of dispersive micro-solid phase extraction base on micro and nano sorbents. TrAC-Trends Anal Chem 89:99–118

Kumar S, Nair RR, Pillai PB, Gupta SN, Iyengar MAR, Sood AK (2014) Graphene oxide-MnFe2O4 magnetic nanohybrids for efficient removal of lead and arsenic from water. ACS Appl Mater Interfaces 6:17426–17436

Lan H, Gan N, Pan D, Hu F, Li T, Long N, Shen H, Feng Y (2014) Development of a novel magnetic molecularly imprinted polymer coating using porous zeolite imidazolate framework-8 coated magnetic iron oxide as carrier for automated solid phase microextraction of estrogens in fish and pork samples. J Chromatogr A 1365:35–44

Li C, Wei Y, Liivat A, Zhu Y, Zhu J (2013) Microwave-solvothermal synthesis of Fe3O4 magnetic nanoparticles. Mat Lett 107:23–26

Li J, Zhou H, Liu YX, Yan XY, Xu YP, Liu SM (2014a) Solid-phase extraction for selective determination of bisphenol a in drinks and fruits by dummy surface molecularly imprinted polymer with direct synthetic method. Food Addit Contam 31:1139–1146

Li N, Wang Z, Zhang L, Nian L, Lei L, Yang X, Zhang H, Yu A (2014b) Liquid-phase extraction coupled with metal–organic frameworks-based dispersive solid phase extraction of herbicides in peanuts. Talanta 128:345–353

Li J, Wang YB, Li KY, Cao YQ, Wu S, Wu L (2015a) Advances in different configurations of solid-phase microextraction and their applications in food and environmental analysis. TrAC-Trends Anal Chem 72:141–152

Li N, Wu L, Nian L, Song Y, Lei L, Yang X, Wang K, Wang Z, Zhang L, Zhang H, Yu A, Zhang Z (2015b) Dynamic microwave assisted extraction coupled with dispersive micro-solid-phase extraction of herbicides in soybeans. Talanta 142:43–50

Li N, Zhang L, Nian L, Cao B, Wang Z, Lei L, Yang X, Sui J, Zhang H, Yu A (2015c) Dispersive micro-solid-phase extraction of herbicides in vegetable oil with metal-organic framework MIL-101. J Agric Food Chem 42:5806–5813

Li M, Wang J, Jiao C, Wang C, Wu Q, Wang Z (2016) Magnetic porous carbon derived from a Zn/co bimetallic metal-organic framework as an adsorbent for the extraction of chlorophenols from water and honey tea samples. J Sep Sci 39:1793–2012

Li N, Wu D, Hu N, Fan G, Li X, Sun J, Chen X, Suo Y, Li G, Wu Y (2018a) Effective enrichment and detection of trace polycyclic aromatic hydrocarbons in food samples based on magnetic covalent organic framework hybrid microspheres. J Agric Food Chem 66:3572–3580

Li P, Cheng FF, Xiong WW, Zhang Q (2018b) New synthetic strategies to prepare metal-organic frameworks. Inorg Chem Front 5:2693–2708

Li Z, Lei C, Wang N, Jiang X, Zeng Y, Fu Z, Zou L, He L, Liu S, Ao X, Zhou K, Chen S (2018c) Preparation of magnetic molecularly imprinted polymers with double functional monomers for the extraction and detection of chloramphenicol in food. J Chromatogr B 1100-1101:113–121

Liang X, Liu S, Wang S, Guo Y, Jiang S (2014) Carbon-base sorbents: carbon nanotubes. J Chromatogr A 1357:53–67

Liang W, Wang J, Zang X, Wang C, Wang Z (2016) A porous carbon derived from amino-functionalized material of Institut Lavoisier as a solid-phase microextraction fibre coating for the extraction of phthalate esters from tea. J Sep Sci 39:1331–1338

Liang L, Wang X, Sun Y, Ma P, Li X, Piao H, Jiang Y, Song D (2018) Magnetic solid-phase extraction of triazine herbicides from rice using metal-organic framework MIL-101(Cr) functionalized magnetic particles. Talanta 179:512–519

Lin S, Gan N, Qiao L, Zhang J, Cao Y, Chen Y (2015a) Magnetic metal-organic frameworks coated stir bar sorptive extraction coupled with GC–MS for determination of polychlorinated biphenyls in fish samples. Talanta 144:1139–1145

Lin S, Gan N, Zhang J, Chen X, Cao Y, Li T (2015b) A novel reductive graphene oxide-based magnetic molecularly imprinted poly(ethylene-co-vinyl alcohol) polymers for the enrichment and determination of polychlorinated biphenyls in fish samples. J Mol Recognit 28:359–368

Lin S, Gan N, Zhang J, Qiao L, Chen Y, Cao Y (2016) Aptamer-functionalized stir bar sorptive extraction coupled with gas chromatography–mass spectrometry for selective enrichment and determination of polychlorinated biphenyls in fish samples. Talanta 149:266–274

Lirio S, Liu WL, Lin CL, Lin CH, Huang HY (2016) Aluminum based metal-organic framework-polymer monolith in solid-phase microextraction of penicillins in river water and milk samples. J Chromatogr A 1428:236–245

Liu C, Lu C, Tang Z, Chen X, Wang G, Sun F (2015a) Aptamer-functionalized magnetic nanoparticles for simultaneous fluorometric determination of oxytetracycline and kanamycin. Microchim Acta 182:2567–2575

Liu X, Wang C, Wu Q, Wang Z (2015b) Metal-organic framework-templated synthesis of magnetic nanoporous carbon as an efficient absorbent for enrichment of phenylurea herbicides. Anal Chim Acta 870:67–74

Liu X, Zhou X, Wang C, Wu Q, Wang Z (2015c) Magnetic three-dimensional graphene solid-phase extraction of chlorophenols from honey samples. Food Addit Contam A 32:40–47

Liu H, Mu L, Chen X, Wang J, Wang S, Sun B (2017) Core-shell metal-organic frameworks/molecularly imprinted nanoparticles as absorbents for the detection of pyrraline in milk and milk powder. J Agric Food Chem 65:986–992

Liu M, Liu J, Guo C, Li Y (2019) Metal azolate framework-66-coated fiber for headspace solid-phase microextraction of polycyclic aromatic hydrocarbons. J Chromatogr A 1584:57–63

Liu X, Tong Y, Zhang L (2020) Tailorable yolk-shell Fe_3O_4@graphitic carbon submicroboxes as efficient extraction materials for highly sensitive determination of trace sulfonamides in food samples. Food Chem 303:125369
Lu A-H, Salabas EL, Schüth F (2007) Magnetic nanoparticles: synthesis, protection, functionalization, and application. Angew Chem Ed 46:1222–1244
Lu D, Yang Y, Sun C, Wang S (2019) Determination of tocopherols and tocotrienols in cereals and nuts by dispersive solid-phase microextraction-gas chromatography-mass spectrometry. Anal Methods 11:5439–5446
Lv Y-K, Zhao C-X, Li P, He Y-D, Yang Z-R, Sun H-W (2013) Preparation of doxycycline-imprinted magnetic microspheres by inverse-emulsion suspension polymerization for magnetic dispersion extraction of tetracyclines from milk samples. J Sep Sci 36:2656–2663
Lyon JL, Fleming DA, Stone MB, Schiffer P, Williams ME (2004) Synthesis of Fe oxide core/Au shell nanoparticles by iterative hydroxylamine seeding. Nano Lett 4:719–723
Mahmoodi NM (2013) Magnetic ferrite nanoparticle-alginate composite: synthesis, characterization and binary system dye removal. J Taiwan Inst Chem Eng 44:322–330
Mahpishanian S, Sereshti H, Baghdadi M (2015) Superparamagnetic core-shells anchored onto graphene oxide grafted with phenylethyl amine as a nano-adsorbent for extraction and enrichment of organophosphorus pesticides from fruits, vegetable and water samples. J Chromatogr A 1406:48–58
McCullum C, Tchounwou P, Ding L-S, Liao X, Liu Y-M (2014) Extraction of aflatoxins from liquid foodstuff samples with polydopamine-coated superparamagnetic nanoparticles for HPLC-MS/MS analysis. J Agric Food Chem 62:4261–4267
Meng L, Fu C, Lu Q (2009) Advanced technology for functionalization of carbon nanotubes. Prog Nat Sci 19:801–810
Miao SS, Wu MS, Zuo HG, Jiang C, Jin SF, Lu YC, Yang H (2015) Core-shell magnetic molecularly imprinted polymers as sorbent for sulfonylurea herbicide residues. J Agric Food Chem 63:3634–3645
Mirabi A, Rad AS, Abdollahi M (2017) Preparation of modified MWCNT with dithiooxamide for preconcentration and determination of trace amounts of cobalt ions in food and natural water samples. ChemistrySelect 2:4439–4444
Moghadam PZ, Li A, Wiggin SB, Tao A, Maloney AGP, Wood PA, Ward SC, Fairen-Jimenez D (2017) Development of a Cambridge structural database subset: a collection of metal–organic frameworks for past, present, and future. Chem Mat 29:2618–2625
Moliner-Martinez Y, Herráez-Hernández R, Verdú-Andrés J, Molins-Legua C, Campíns-Falcó P (2015) Recent advances of in-tube solid-phase microextraction. TrAC-Trends Anal Chem 71:205–213
Mondal S, Xu J, Chen G, Huang S, Huang C, Yin L, Ouyang G (2019) Solid-phase microextraction of antibiotics from fish muscle by using MIL-101(Cr)NH_2-polyacrylonitrile fiber and their identification by liquid chromatography-tandem mass spectrometry. Anal Chim Acta 1047:62–70
Ning F, Peng H, Li J, Chen L, Xiong H (2014) Molecularly imprinted polymer on magnetic graphene oxide for fast and selective extraction of 17β-estradiol. J Agric Food Chem 62:7436–7443
Nogueira JMF (2012) Novel sorption-based methodologies for static microextraction analysis: a review on SBSE and related techniques. Anal Chim Acta 757:1–10
Noormohamadi HR, Fat'hi MR, Ghaedi M (2018) Fabrication of polyethyleneimine modified cobalt ferrite as a new magnetic sorbents for the micro-solid phase extraction of tartrazine from food and water samples. J Colloid Interf Sci 531:343–351
Novoselov KS, Geim AK, Morozov SV, Jiang D, Zhang Y, Dubonos SV, Grigorieva IV, Firsov AA (2004) Electric field effect in atomically thin carbon films. Science 306:666–669
Pacheco-Fernández I, Gutiérrez-Serpa A, Afonso AM, Pino V (2018) Metallic coatings in solid-phase microextraction: environmental applications. In: Zhang J, Jung YG (eds) Advanced ceramic and metallic coating and thin film materials for energy and environmental applications. Springer Nature, Switzerland. isbn:978-3-319-59905-2

Pacheco-Fernández I, González-Hernández P, Pasán J, Ayala JH, Pino V (2019) The rise of metal-organic frameworks in analytical chemistry. In: de la Guardia M, Esteve-Turrillas FA (eds) Handbook of smart materials in analytical chemistry. Wiley, New York. isbn:978-1-119-42262-4

Piri-Moghadam H, Alam MN, Pawliszyn J (2017) Review of geometries and coating materials in solid phase microextraction: opportunities, limitations, and future perspectives. Anal Chim Acta 984:42–65

Płotka-Wasylka J, Szczepańska N, de la Guardia M, Namieśnik J (2015) Miniaturized solid-phase extraction techniques. TrAC-Trends Anal Chem 73:19–38

Płotka-Wasylka J, Szczepańska N, de la Guardia M, Namieśnik J (2016) Modern trends in solid phase extraction: new sorbent media. TrAC-Trends Anal Chem 77:23–43

Qi R, Jiang H, Liu S, Jia Q (2014) Preconcentration and determination of pesticides with graphene-modified polymer monolith combined with high performance liquid chromatography. Anal Methods 6:1427–1454

Qian HL, Yang CX, Wang WL, Yang C, Yan XP (2018) Advances in covalent organic frameworks in separation science. J Chromatogr A 1542:1–18

Qin Y, Zhang J, Li Y, Han Y, Zou N, Jiang Y, Shan J, Pan C (2016) Multiplug filtration cleanup method with multi-walled carbon nanotubes for the analysis of malachite green, diethylstilbestrol residues, and their metabolites in aquatic products by liquid chromatography–tandem mass spectrometry. Anal Bioanal Chem 408:5801–5809

Ravelo-Pérez LM, Herrera-Herrera AV, Hernández-Borges J, Rodríguez-Delgado MÁ (2010) Carbon nanotubes: solid-phase extraction. J Chromatogr A 1217:2618–2641

Reyes-Garcés N, Gionfriddo E, Gómez-Ríos GA, Alam MN, Boyaci E, Bojko B, Singh V, Grandy J, Pawliszyn J (2018) Advances in solid phase microextraction and perspective on future directions. Anal Chem 90:302–360

Ríos Á, Zougagh M (2016) Recent advances in magnetic nanomaterials for improving analytical processes. TrAC-Trends Anal Chem 84:72–83

Rocío-Bautista P, González-Hernández P, Pacheco-Fernández I, Pino V, Ayala JH, Afonso AM (2016a) Metal-organic frameworks as promising sorbents in solid-phase microextraction. In: Reeves P (ed) Metal-organic frameworks (MOFs): chemistry, technologies and applications. Nova Science Publishers, New York. isbn:978-1-63485-051-3

Rocío-Bautista P, Pacheco-Fernández I, Pasán J, Pino V (2016b) Are metal-organic frameworks able to provide a new generation of solid-phase microextraction coatings? Anal Chim Acta 939:26–41

Rocío-Bautista P, González-Hernández P, Pino V, Pasán J, Afonso AM (2017) Metal-organic frameworks as novel sorbents in dispersive-based microextraction approaches. TrAC-Trends Anal Chem 90:114–134

Rocío-Bautista P, Gutiérrez-Serpa A, Pino V (2018) Solid-phase microextraction coatings based on tailored materials: metal-organic frameworks and molecularly imprinted polymers. In: Li L, Yang Q (eds) Advanced coatings materials. Scrivener Publishing, Massachusetts, p 9781119407560

Román IP, Chisvert A, Canals A (2011) Dispersive solid-phase extraction based on oleic acid-coated magnetic nanoparticles followed by gas chromatography–mass spectrometry for UV-filter determination in water samples. J Chromatogr A 1218:2467–2475

Rowsell JLC, Yaghi OM (2004) Metal-organic frameworks: a new class of porous materials. Microporous Mesoporous Mat 73:3–14

Ruan X-L, Qiu J-J, Wu C, Huang T, Meng R-B, Lai Y-Q (2014) Magnetic single-walled carbon nanotubes-dispersive solid-phase extraction method combined with liquid chromatography-tandem mass spectrometry for the determination of paraquat in urine. J Chromatogr B 965:85–90

Rümmeli MH, Ayala P, Pichler T (2010) Carbon nanotubes and related structures: production and formation. In: Guldi DM, Martin N (eds) Carbon nanotubes and related structures. Wiley VCH

Rykowska I, Wasiak W (2013) Advances in stir bar sorptive extraction coating: a review. Acta Chromatogr 25:27–46

Šafaříková M, Šafařík I (1999) Magnetic solid-phase extraction. J Magn Magn Mater 194:108–112
Safdarian M, Hashemi P, Adeli M (2013) One-step synthesis of agarose coated magnetic nanoparticles and their application in the solid phase extraction of Pd(II) using a new magnetic field agitation device. Anal Chim Acta 774:44–50
Sajid M (2017) Porous membrane protected micro-solid-phase extraction: a review of features, advancements and applications. Anal Chim Acta 965:36–53
Saraji M, Jafari MT, Mossaddegh M (2016) Carbon nanotubes@silicon dioxide nanohybrids coating for solid-phase microextraction of organophosphorus pesticides followed by gas chromatography–corona discharge ion mobility spectrometric detection. J Chromatogr A 1429:30–39
Scopus Database.: https://www.scopus.com/search/form.uri?display=basic
Sehati N, Dalali N, Soltanpour S, Dorraji MSS (2014) Extraction and preconcentration of tylosin from milk samples through functionalized TiO2 nanoparticles reinforced with a hollow fibre membrane as a novel solid/liquid-phase microextraction technique. J Sep Sci 37:2025–2031
Sereshti H, Khodayari F, Nouri N (2019) Simultaneous determination of aflatoxins in bread by in-syringe dispersive micro-solid phase extraction using magnetic three-dimensional graphene followed by HPLC-FLD. Food Anal Methods 12:2273–2281
Shams SS, Zhang R, Zhu J (2015) Graphene synthesis: a review. Mater Sci 33:566–578
Shi X-R, Chen X-L, Hao Y-L, Li L, Xu H-J, Wang M-M (2018) Magnetic metal-organic frameworks for fast and efficient solid-phase extraction of six Sudan dyes in tomato sauce. J Chromatogr B 1086:146–152
Shi Z, Chen D, Chen T-T, Wei G, Yin C-Y, Xu H, Yang G-F (2019) In vivo analysis of two new fungicides in mung bean sprouts by solid phase microextraction-gas chromatography-mass spectrometry. Food Chem 275:688–695
Shirey RE (2012) SPME commercial devices and fibre coatings. In: Pawliszyn J (ed) Handbook of solid phase microextraction. Elsevier, Amsterdam. isbn:978-0-12-416017-0
Socas-Rodríguez B, Herrera-Herrera AV, Asensio-Ramos M, Hernández-Borges J (2014) Recent applications of carbon nanotube sorbents in analytical chemistry. J Chromatogr A 1357:110–146
Socas-Rodríguez B, Herrera-Herrera AV, Asensio-Ramos M, Hernández-Borges J (2015) Dispersive solid-phase extraction. In: Anderson JL, Berthod A, Pino V, Stalcup AM (eds) Analytical separation science. Sample preparation method validation, and analytical applications. Wiley-VCH Verlag, Weinheim/Germany
Socas-Rodríguez B, González-Sálamo J, Hernández-Borges J, Rodríguez-Delgado MÁ (2017) Recent applications of nanomaterials in food safety. TrAC-Trends Anal Chem 96:172–200
Souza-Silva EA, Gionfriddo E, Pawliszyn J (2015) A critical review of the state of the art of solid-phase microextraction of complex matrices II. Food analysis. TrAC-Trends Anal Chem 71:236–248
Speltini A, Scalabrini A, Maraschi F, Sturini M, Profumo A (2017) Newest applications of molecularly imprinted polymers for extraction of contaminants from environmental and food matrices: a review. Anal Chim Acta 974:1–26
Stanisz E, Krawczyk-Coda M (2017) ZnO nanoparticles as an adsorbent in ultrasound-assisted dispersive micro solid-phase extraction combined with high-resolution continuum source electrothermal atomic absorption spectrometry for determination of trace germanium in food samples. Microchem J 132:136–142
Stöber W, Fink A, Bohn E (1968) Controlled growth of monodisperse silica spheres in the micron size range. J Colloid Int Sci 26:62–69
Stock N, Biswas S (2012) Synthesis of metal-organic frameworks (MOFs): routes to various MOF topologies, morphologies,and composites. Chem Rev 112:933–969
Sun M, Cui P, Ji S, Tang R, Wu Q, Wang C, Wang Z (2014) Octadecyl-modified graphene as an adsorbent for hollow fibre liquid phase microextraction of chlorophenols from honey. Bull Kor Chem Soc 35:1011–1015
Sun H, Yu QW, He HB, Lu Q, Shi ZG, Feng YQ (2016) Nickel oxide nanoparticle-deposited silica composite solid-phase extraction for benzimidazole residue analysis in milk and eggs by liquid chromatography-mass spectrometry. J Agric Food Chem 64:356–363

Synaridou M-ES, Sakkas VA, Stalikas CD, Albanis TA (2014) Evaluation of magnetic nanoparticles to serve as solid-phase extraction sorbents for the determination of endocrine disruptors in milk samples by gas chromatography mass spectrometry. J Chromatogr A 1348:71–79

Taherimaslak Z, Amoli-Diva M, Allahyari M, Pourghazi K, Manafi MH (2015) Low density solvent based dispersive liquid-liquid microextraction followed by vortex-assisted magnetic nanoparticle based solid-phase extraction and surfactant enhanced spectrofluorimetric detection for the determination of aflatoxins in pistachio nuts. RSC Adv 5:12747–12754

The Cambridge crystallographic data center.: https://www.ccdc.cam.ac.uk/

Tokalioglu S, Yavuz E, Demir S, Patat S (2017) Zirconium-based highly porous metal-organic framework (MOF-545) as an efficient adsorbent for vortex assisted-solid phase extraction of lead from cereal, beverage and water samples. Food Chem 237:707–715

Towler PH, Smith JD, Dixon DR (1996) Magnetic recovery of radium, lead and polonium from seawater samples after preconcentration on a magnetic adsorbent of manganese dioxide coated magnetite. Anal Chim Acta 328:53–59

Turiel E, Martín-Esteban A (2010) Molecularly imprinted polymers for sample preparation: a review. Anal Chim Acta 668:87–99

Wang X, Ye N (2017) Recent advances in metal-organic frameworks and covalent organic frameworks for sample preparation and chromatographic analysis. Electrophoresis 38:3059–3078

Wang R, Wang Y, Xue C, Wen T, Wu J, Hong J, Zhou X (2013) Selective separation and enrichment of glibenclamide in health foods using surface molecularly imprinted polymers prepared via dendritic grafting of magnetic nanoparticles. J Sep Sci 36:1015–1021

Wang P-L, Xie L-H, Joseph EA, Li J-R, Su X-O, Zhou H-C (2019) Metal-organic frameworks for food safety. Chem Rev 119:10638–10690

Wu X, Hu J, Zhu B, Lu L, Huang X, Pang D (2011) Aptamer-targeted magnetic nanospheres as a solid-phase extraction sorbent for determination of ochratoxin a in food samples. J Chromatogr A 1218:7341–7346

Wu J, Chen L, Mao P, Lu Y, Wang H (2012) Determination of chloramphenicol in aquatic products by graphene-based SPE coupled with HPLC-MS/MS. J Sep Sci 35:3586–3592

Wu YL, Chen RX, Zhu Y, Zhao J, Yang T (2015) Simultaneous determination of sixteen amide fungicides in vegetables and fruits by dispersive solid phase extraction and liquid chromatography–tandem mass spectrometry. J Chromatogr B 989:11–20

Wu M, Ai Y, Zeng B, Zhao F (2016) In situ solvothermal growth of metal-organic framework–ionic liquid functionalized graphene nanocomposite for highly efficient enrichment of chloramphenicol and thiamphenicol. J Chromatogr A 1427:1–7

Xia L, Liu L, Lv X, Qu F, Li G, You J (2017) Towards the determination of sulfonamides in meat samples: a magnetic and mesoporous metal-organic framework as an efficient sorbent for magnetic solid phase extraction combined with high-performance liquid chromatography. J Chromatogr A 1500:24–31

Xie L, Jiang R, Zhu F, Liu H, Ouyang G (2014) Application of functionalized magnetic nanoparticles in sample preparation. Anal Bioanal Chem 406:377–399

Xu Z, Zhang J, Cong L, Meng L, Song J, Zhou J, Qiao X (2011) Preparation and characterization of magnetic chitosan microsphere sorbent for separation and determination of environmental estrogens through SPE coupled with HPLC. J Sep Sci 34:46–52

Xu X, Long N, Lv J, Wang L, Zhang M, Qi X, Zhang L (2016) Functionalized multiwalled carbon nanotube as dispersive solid-phase extraction materials combined with high-performance liquid chromatography for thiabendazole analysis in environmental and food samples. Food Anal Meth 9:30–37

Yazdinezhad SR, Ballesteros-Gómez A, Lunar L, Rubio S (2013) Single-step extraction and cleanup of bisphenol a in soft drinks by hemimicellar magnetic solid phase extraction prior to liquid chromatography/tandem mass spectrometry. Anal Chim Acta 778:31–37

Yilmaz E, Soylak M (2016) Preparation and characterization of magnetic carboxylated nanodiamonds for vortex-assisted magnetic solid-phase extraction of ziram in food and water samples. Talanta 158:152–158

Yu H, Ho TD, Anderson JL (2013a) Ionic liquid and polymeric ionic liquid coatings in solid-phase microextraction. TrAC-Trends Anal Chem 45:219–232
Yu L, Li P, Zhang Q, Zhang W, Ding X, Wang X (2013b) Graphene oxide: an adsorbent for the extraction and quantification of aflatoxins in peanuts by high-performance liquid chromatography. J Chromatogr A 1318:27–34
Yuan S, Feng L, Wang K, Pang J, Bosch M, Lollar C, Sun Y, Qin J, Yang X, Zhang P, Wang Q, Zou L, Zhang Y, Zhang L, Fang Y, Li J, Zhou HC (2018) Stable metal–organic frameworks: design, synthesis, and applications. Adv Mater 30:1704303–1704338
Yuan Y, Wang M, Jia N, Zhai C, Han Y, Yan H (2019) Graphene/multi-walled carbon nanotubes as an adsorbent for pipette-tip solid-phase extraction for the determination of 17β-estradiol in milk products. J Chromatogr A 1600:73–79
Yue Y, Mehio N, Binder AJ, Dai S (2015) Synthesis of metal–organic framework particles and thin films via nanoscopic metal oxide precursors. CrystEngComm 17:1728–1735
Zaman BT, Erulaş AF, Chormey DS, Bakirdere S (2020) Combination of stearic acid coated magnetic nanoparticle based sonication assisted dispersive solid phase extraction and slotted quartz tube-flame atomic absorption spectrophotometry for the accurate and sensitive determination of lead in red pepper samples and assessment of green profile. Food Chem 303:125396
Zeng L, Wu X, Li Y, Lu D, Sun C (2015) Multiwalled carbon nanotube-dispersive solid-phase extraction followed by high performance capillary electrophoresis for simultaneous determination of six adulterants in antihypertensive functional foods. Anal Methods 7:543–550
Zhai H, Su Z, Chen Z, Liu Z, Yuan K, Huang L (2015) Molecularly imprinted coated graphene oxide solid-phase extraction monolithic capillary column for selective extraction and sensitive determination of phloxine B in coffee bean. Anal Chim Acta 865:16–21
Zhan G, Zeng HC (2017) Alternative synthetic approaches for metal-organic frameworks: transformation from solid matters. Chem Commun 53:72–81
Zhan J, Fang G, Yan Z, Pan M, Liu C, Wang S (2013) Preparation of a semicovalent, molecularly surface imprinted polymer for the rapid determination of trace acid orange II in food and environmental samples. Anal Bioanal Chem 405:6353–6363
Zhang H, Liu Q-S, Yang C-L, Lv J-Z, Xie L-Q, Tang M-J, Yue Z-F, Wan Z-G (2013a) C18/C8-functionalized magnetic silica nanospheres (Fe3O4@Si-C8/C18) as capture probes for highly efficient and rapid purification of veterinary drug residues. Food Anal Methods 6:933–940
Zhang J, Shao J, Guo P, Huang Y (2013b) A simple and fast Fe3O4 magnetic nanoparticles-based dispersion solid phase extraction of Sudan dyes from food and water samples coupled with high-performance liquid chromatography. Anal Methods 5:2503–2510
Zhang M, Zeng J, Wang Y, Chen X (2013c) Developments and trends of molecularly imprinted solid-phase microextraction. J Chromatogr Sci 51:577–586
Zhang W, Zhang Z, Zhang J, Meng J, Bao T, Chen Z (2014) Covalent immobilization of graphene onto stainless steel wire for jacket-free stir bar sorptive extraction. J Chromatogr A 1351:12–20
Zhang Q, Li G, Xiao X (2015) Acrylamide-modified graphene for online micro-solid-phase extraction coupled to high-performance liquid chromatography for sensitive analysis of heterocyclic amines in food samples. Talanta 131:127–135
Zhang S, Yang Q, Yang X, Wang W, Li Z, Zhang L, Wang C, Wang Z (2017) A zeolitic imidazolate framework based nano-porous carbon as a novel fibre coating for solid-phase microextraction of pyrethroid pesticides. Talanta 166:46–53
Zhang N, Huang C, Feng Z, Chen H, Tong P, Wu X, Zhang L (2018) Metal-organic framework-coated stainless-steel fibre for solid-phase microextraction of polychlorinated biphenyls. J Chromatogr A 1570:10–18
Zhao Q, Lu Q, Feng Y-Q (2013) Dispersive microextraction based on magnetic polypyrrole nanowires for the fast determination of pesticide residues in beverage and environmental water samples. Anal Bioanal Chem 405:4765–4776
Zhao Y-G, Zhou L-X, Pan S-D, Zhan P-P, Chen X-H, Jin M-C (2014) Fast determination of 22 sulfonamides from chicken breast muscle using core-shell nanoring amino-functionalized super-

paramagnetic molecularly imprinted polymer followed by liquid chromatography-tandem mass spectrometry. J Chromatogr A 1345:17–28

Zhao Y, Bi C, He X, Chen L, Zhang Y (2015) Preparation of molecularly imprinted polymers based on magnetic carbon nanotubes for determination of sulfamethoxazole in food samples. RSC Adv 5:70309–70318

Zhao B, Wu D, Chu H, Wang C, Wei Y (2019) Magnetic mesoporous nanoparticles modified with poly(ionic liquids) with multi-functional groups for enrichment and determination of pyrethroid residues in apples. J Sep Sci 42:1896–1904

Zheng H-B, Mo J-Z, Zhang Y, Gao Q, Ding J, Yu Q-W, Feng Y-Q (2014a) Facile synthesis of magnetic molecularly imprinted polymers and its application in magnetic solid phase extraction for fluoroquinolones in milk samples. J Chromatogr A 1329:17–23

Zheng X, He L, Duan Y, Jiang X, Xiang G, Zhao W, Zhang S (2014b) Poly(ionic liquid) immobilized magnetic nanoparticles as new adsorbent for extraction and enrichment of organophosphorus pesticides from tea drinks. J Chromatogr A 1358:39–45

Inclusion Complexation and Coacervation to Fabricate Nanoencapsulated Foods

Shivani Uppal, Aashima, Khushwinder Kaur, and S. K. Mehta

1 Introduction

From the time immemorial, man has tried to mimic nature and create novel materials for the advancement, improvement, and betterment of humanity. With this desire, the nanotechnology platform was discovered and a new nomenclature for small things evolved. Richard Feynman (Feynman 1961) introduced the concept of "Nano" to the world in 1959 that has revolutionized the scientific world. He proposed a technological direction of minimizing the material's size, manipulating their shape and size at nanoscale level (Sahoo et al. 2007). Small-scale, small things and miniaturization of materials are the terms which were replaced by nanotechnology, nanoengineering, and nanoscale in the modern concept of Feynman (Feynman 1961). In 1974 at Tokyo Science University, Norio Taniguchi first elaborated upon the term "nanotechnology" explaining the separation procedure, consolidation and deformation of materials by one atom or one molecule.

Despite the hype created for nanotechnology in recent years, it is not a new technology. The examples can be traced back from ancient times e.g. Lycurgus glass cup in British Museum appears jade green in natural light and red color when bright light shines through it due to the presence of nanoparticles of gold and silver and carbon nanoparticles were utilized in tire manufacturing. The first observation and size measurements of nanoparticles were carried out using an ultra-microscope by Richard Zsigmondy in 1902. It is an innovative technology with the potential to change the world (Verma et al. 2011). It has emerged as one of the most promising

S. Uppal · Aashima · K. Kaur (✉) · S. K. Mehta
Department of Chemistry and Centre of Advanced Studies in Chemistry, Panjab University, Chandigarh, India

U. Hebbar et al. (eds.), *Nano-food Engineering*, Food Engineering Series,
https://doi.org/10.1007/978-3-030-44552-2_4

fields with multifaceted applications starting from production and processing to enhancement in the inherent mechanical, electrical and optical properties of the materials.

1.1 Nanotechnology and Food

Food, water, shelter, and clothing are the four essential needs of mankind. The progress and development of human race has transformed the food needs. The advent of nanotechnology has revolutionized the entire food system from manufacturing to handling and packaging. After the "Green" and "White" Revolution in the food industry, related to agricultural and food respectively, now it is nanotechnology that has the potential to change the face of the food industry. There are new innovations strengthening the taste, color and texture of food which indirectly augments the shelf life and enhances the stability of food materials. The applications of nanotechnology (Fig. 1) in the food industry has metamorphosed food to give it a new outlook as "intelligent food" with controlled diffusion, improved quality, strengthened nutrient content, low contamination, and efficient packaging. The entire cycle from the procurement of raw materials to the processing of food, it's packaging and selling of the final product to the market has been modernized by nanotechnology.

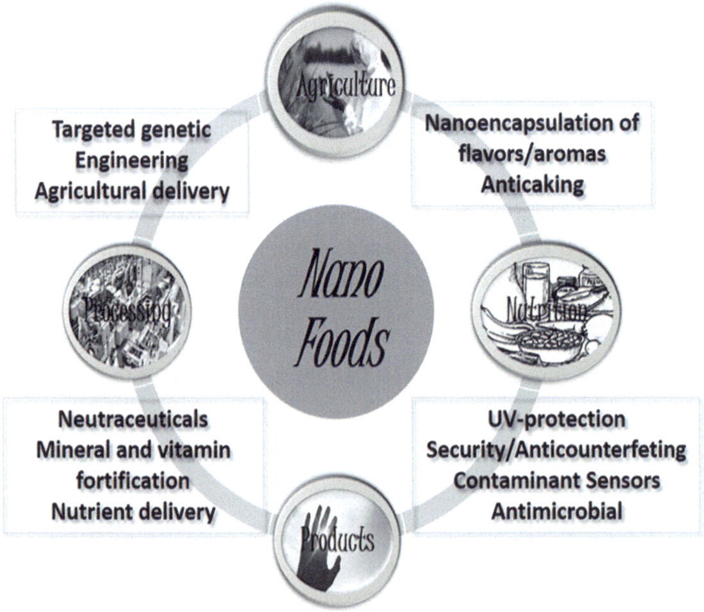

Fig. 1 Various aspects of nanotechnology in the food industry

It also has the potential to customize the food making it handier and customer friendly; meeting with the demands of an ever-changing busy world.

This technology not only improves the water solubility, thermal stability and bioavailability of the bioactive component of the food but has proven its potential in increasing the number of new intelligent products in the market that enhances the consumer benefits. Thus, the introduction of nanotechnology in the food sector has ushered new aspirations and unfolded myriad of arenas for this emerging industry. Yaktine and Pray (Yaktine and Pray 2009) reported numerous advantages of nanotechnology and recognized it it as a food friendly technique. They elaborated upon intelligent nano-packaging and eco-friendly processes for the synthesis of food products with tunable properties.

The use of medicinal herbs (Fig. 2) in traditional medicine practices has been known since prehistoric times. Numerous phytochemicals with potential therapeutic functions have been identified. However, since a single plant can contain diverse phytochemicals, the effects of using a whole plant as medicine are uncertain. The place of plants in medicine was radically altered in the nineteenth century by chemical analysis and technological intrusion. Nanotechnology has helped to bridge the gap between the ancestral medicine where bioactive foods were used to cure the ailments, and the modern-day dietary supplements. It has taken the plant therapeutics to the advanced treatment level. The holy basil, a sacred super herb was used in the Indian system of medicine to harmonize the mind, body, and spirit. The unique ability of basil is empowered by nanotechnology and there are numerous syrups and tablets for a cough, fever, respiratory disorders using Basil. Currently, the market of nanotechnology products in the food industry approaches US$1 billion (most of this

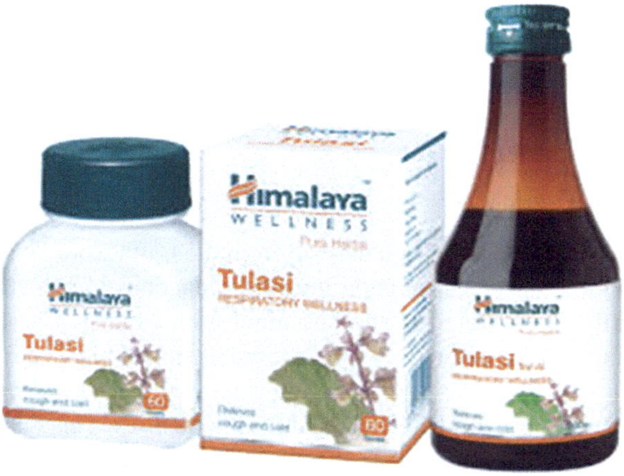

Fig. 2 Commercial products of natural bioactive

on nanoparticle coatings for packaging applications, health-promoting products, and beverages) and it has the potential to grow to more than US$20 billion in the next decade (Chau et al. 2007).

1.2 *Merits and Demerits of Employing Nanotechnology to the Food Industry*

The dynamic entry of nanotechnology within the food sector has generated new hopes and is expected to bring various benefits to the food industry's concerns. Nevertheless, as technology brings risks with benefits for consumers, it may also do the same. It may bring various kinds of opportunities such as developing nano-processes, nano eco-friendly fabricating techniques, intelligent nano-packaging, providing consumers with low-calorie food. It aims at changing the unhealthy food habits of the busy world into healthy ones (Silvestre et al. 2011).

However, there are questions in the mind of researchers that how nano food components are formed, disintegrated, and ingested into our body. Without this knowledge, it will not be possible to overcome the potential risks and uncertainties of this technology. Consumption of excessive nutrient will leave toxic effects on human body and create an imbalance in the functioning of the body in the future. Talking about the risks and disadvantages of applying this technology within the food industry, most of the nanoparticles enter the gut through oral administration and gets absorbed via intestine cells (enterocytes). It is designed in a way that they do not allow large or foreign particles to pass through them, nevertheless, the nano-sized ingredients are able to cross these barriers. Therefore, there is a potential risk in bringing up gastric diseases which should be investigated through *in vivo* and clinical studies. Before analyzing these uncertainties, it is not possible to overcome the potential risks within this nanotechnology food combination.

2 Encapsulation

The word "Encapsulation" has been derived from the Latin word "Caspa" meaning "box". In a literal sense, encapsulation means "to encompass something in a capsule". The word first appeared in the late nineteenth century. Encapsulation is a technological method for entrapping active components within a carrier to improve its delivery (Fig. 3). It is an expeditionary developing technology with many prospective applications in pharmaceutical and food industries. This budding methodology is successfully used to bestow the sensitive bioactive material in safe core thus protecting it from environmental decay. It preserves its taste and aroma, masks the foul odor, enhances stability and increases its bioavailability. Some of the prime reasons for encapsulation of active ingredients are listed in Fig. 2. Shahidi and Han

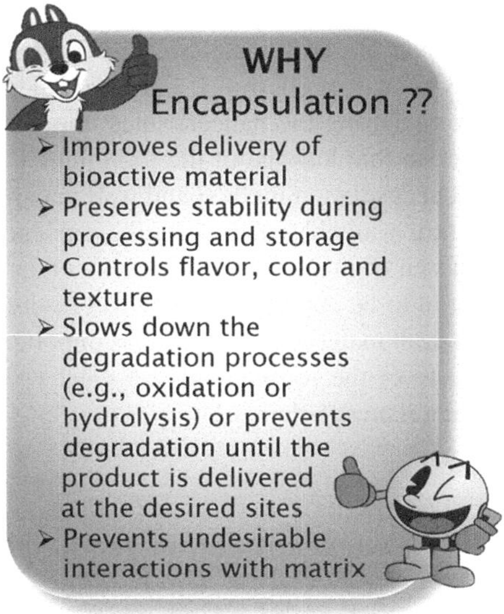

Fig. 3 Reasons for encapsulation

reported that most of the times, the core material is sensitive towards moisture, sunlight so the outer layer in encapsulation prevents the core from damage and increases its shelf-life, leading to a controlled release of encapsulate (Shahidi and Han 1993). Norton and Frith (Norton and Frith 2001) discussed upon the improved facilitation of micronutrient bioavailability due to the nanometric size of the assembly which is one of the major advantages of the nanoencapsulation technique. The improved bioavailability is due to two main mechanisms (i) by increasing bioactive stability during digestion (ii) by enhancing the solubility in intestinal fluids and accelerating the transport within the body, absorption through or between the intestinal epithelial cells, and decreasing first pass metabolism (Yao et al. 2015). Carotenoids represent a large group of tetraterpenoid organic pigments and one of them is β-carotene, a well-known active phytochemical possessing promising health benefits. The bio-accessibility of β- carotene in starch hydrogels containing no fat was less than 1% due to its crystalline nature and lack of mixed micelles to solubilize it (Mun et al. 2015). The bio-accessibility increases from 3.1 to 35.6% through nanoencapsulation (Liang et al. 2013). Thus, with encapsulation, the bioactive component could be kept as fully efficient. Curcumin, another neuro-protective and anticancer potential agent containing phytochemical properties, lack retention in our body. But, when complexed with β- cyclodextrin (β-CD) using co-precipitation, freeze-drying and solvent evaporation methods, they exhibit higher pH, storage and heating stability (Mangolim et al. 2014). In addition to the above, encapsulation can be applied for modification of physical characteristics of the original bioactive to (a) allow easier handling, (b) help separate the components of the mixture that would otherwise react with one another, (c) provide an adequate concentration and uniform dispersion of an active agent (Desai and Park 2005).

Nanoencapsulation is used to produce nano-capsules from food ingredients and wall materials. Wall material is the outer covering which protects the food ingredient by means of controlled release. Generally, some precautions need to be taken while preparing the nanocapsules: the wall around the food ingredient should be proper so that leakage of the food ingredient does not occur. Further, selection of nanoencapsulation method and wall materials is interdependent. Based on the wall material, a suitable method for the nanoencapsulation is chosen. Wall materials can be chosen from a variety of synthetic or natural polymers which is dependent on the material to be coated and properties desired in the final formulation. The selection of the wall material composition is important and responsible for the final functional property of the formulation. An ideal wall material should exhibit the following characteristics (Scheme 1).

2.1 Nanoencapsulation Techniques

The encapsulation process was originally established in biotechnology to build production processes more efficient. This process starts with the matrix formation around the cells which allows for rapid and efficient separation of the producer cells and the metabolites. These technologies developed approximately 60 years ago, are of significant interest to the pharmaceutical sector (especially for drug and vaccine delivery), but also have relevance for the food industry. The reason behind this is the food industry demands for complex functional compounds in products. The encapsulated food has a multitude of possible benefits in the food industry. Encapsulation aims to preserve the stability of the bioactive compounds during processing and storage and to prevent undesirable interactions with the food matrix. During the encapsulation process, bio-actives are preserved for profit since the procedure

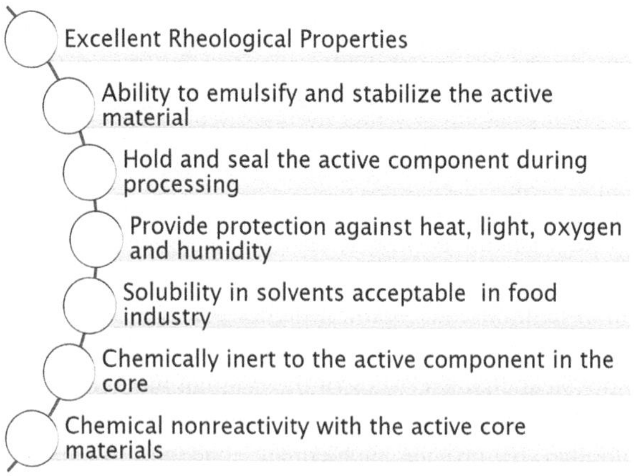

Scheme 1 Characteristics of an ideal wall material

reduces the degradation processes (e.g., hydrolysis or oxidation) or halts the degradation process until the product is delivered at the desired sites (McClements et al. 2009). This is how the bio-active component is kept fully functional. Also, this technology develops a barrier between sensitive bioactive materials and the environment, thus, allowing taste and aroma differentiation, mask bad tasting or smelling, stabilize food ingredients or increase their bioavailability. In addition to the above, encapsulation can be applied for modification of physical characteristics of the original material to (a) allow easier handling, (b) help separate the components of the mixture that would otherwise react with one another, (c) provide an adequate concentration and uniform dispersion of an active agent.

The encapsulation technique and delivery system have a very important impact on the physicochemical properties. In general, physicochemical properties such as surface area, shape, size distribution, particle size, solubility and encapsulation efficiency, and releasing mechanisms have been reported to be varied via the encapsulation technique. Depending on which kind of application someone is focusing, the nanoencapsulation technique is chosen. Therefore, it is important to choose an effective technique based on physicochemical properties. The prevailing techniques for this purpose include emulsification, coacervation, inclusion complexation, emulsification–solvent evaporation, nanoprecipitation, and supercritical fluid technique (Risch 1995). They produce capsules of nanometer range due to which they are considered as nanoencapsulation techniques.

Nanoencapsulation technique uses either top-down or bottom-up approaches for the development of nano-encapsulated formulation (Fig. 4). The size reduction and structure shaping for the desired application can be developed by employing precise tools in a top-down approach. In the bottom-up approach, formulations are developed by self-assembly and self-organization of molecules. The self-assembly is affected by many factors including pH, temperature, concentration, and ionic strength (Augustin and Sanguansri 2009). Techniques such as emulsification and emulsification–solvent evaporation are used under the top-down approach whereas bottom-up approach includes supercritical fluid technique, inclusion complexation, coacervation, and nanoprecipitation (Fig. 4) (Sanguansri and Augustin 2006, Mishra

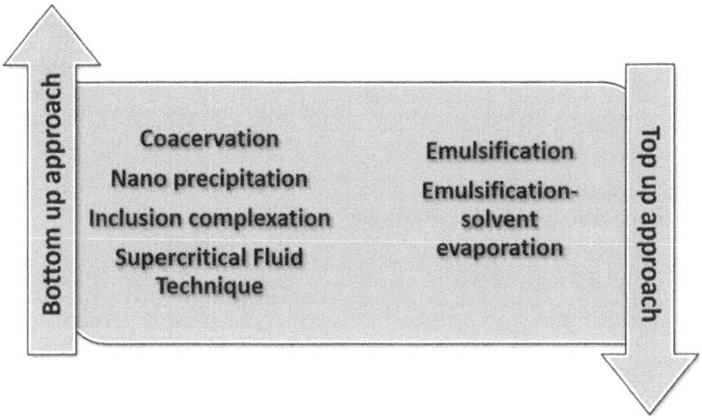

Fig. 4 Top-down and bottom-up approaches in nanoencapsulation techniques

Table 1 Physical methods for the preparation of nano encapsulated food

<table>
<tr><th colspan="3">Physical Methods</th></tr>
<tr><td>Spray drying</td><td>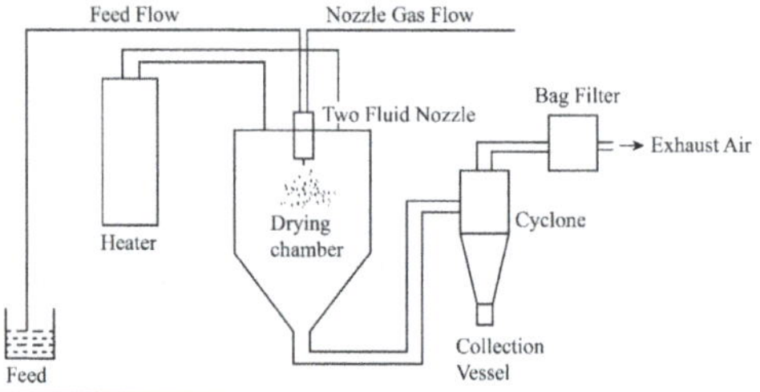
</td><td>Oldest encapsulation method capable of encapsulating liposoluble vitamins and producing encapsulated powders with different particle size.</td></tr>
<tr><td>Fluid bed coating</td><td>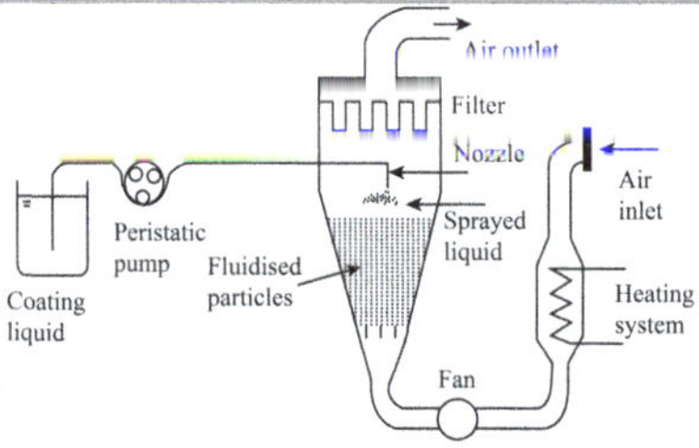
</td><td>In this method, a coating is applied onto powder particles in a batch process or a continuous set-up and 5–50% of the coating is applied depending on the particle of the core material.</td></tr>
<tr><td>Extrusion Spheronization</td><td>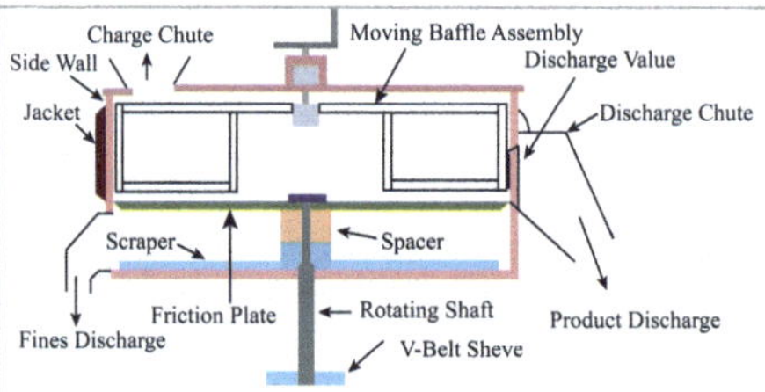
</td><td>It is a multistep process which starts with dry mixing of active ingredients with the addition of a binder to form wet mass and extrusion and finally spherization to form extrudate into spheroids of uniform size</td></tr>
<tr><td>Centrifugal extrusion</td><td>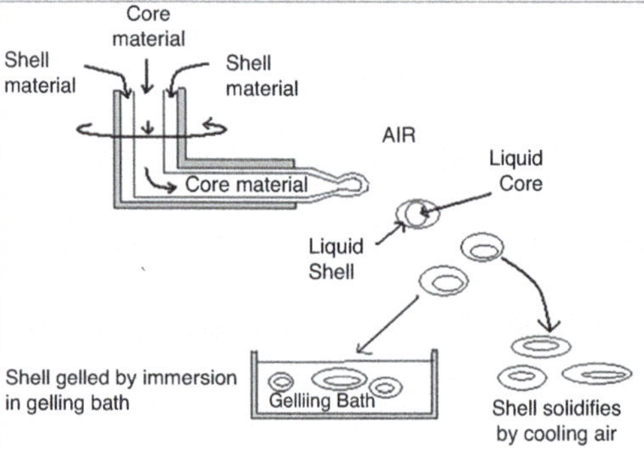
</td><td>It is a process in which liquids are encapsulated using a rotating extrusion head containing concentric nozzles to form particles size of 400–2000 μm in diameter.</td></tr>
</table>

Table 1 (continued)

<table>
<tr><th colspan="3">Physical Methods</th></tr>
<tr><td>Spray drying</td><td>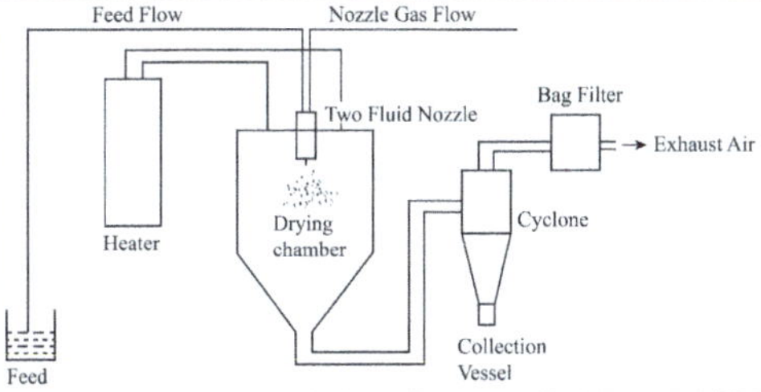
</td><td>Oldest encapsulation method capable of encapsulating liposoluble vitamins and producing encapsulated powders with different particle size.</td></tr>
<tr><td>Supercritical liquid modelling</td><td>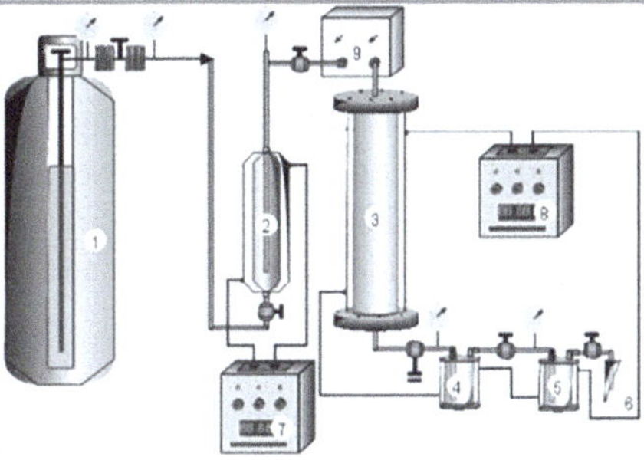</td><td>It involves method like rapid expansion from a supercritical solution, gas antisolvent, aerosol solvent extraction. It is used for encapsulation of thermally sensitive compounds.</td></tr>
<tr><td>Spray freezing</td><td>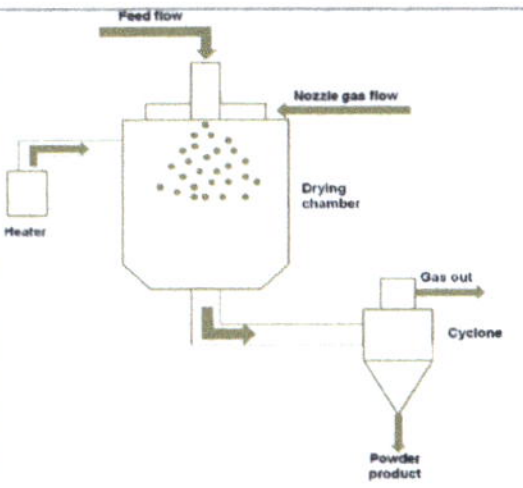
</td><td>They are used for encapsulation heat sensitizing flavors.</td></tr>
<tr><td>Emulsification</td><td>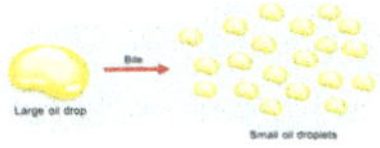
</td><td>It protects the droplets from re-coalescence and improves their stability for use in food and pharmaceutical industry.</td></tr>
<tr><td>Coacervation</td><td>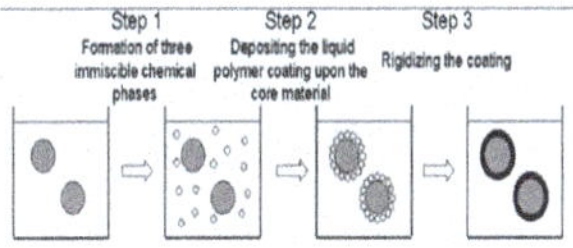
</td><td>It starts with the formation of a chemical phase and depositing the liquid polymer coating upon the core material and rigidizing the coating.</td></tr>
<tr><td>Inclusion Complexation</td><td>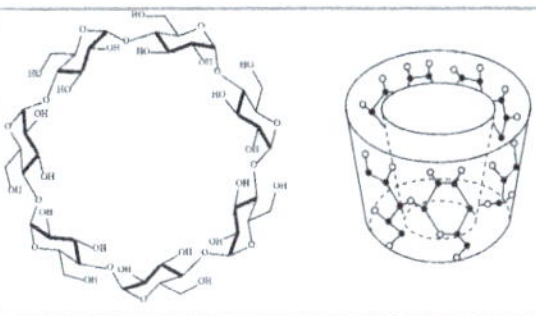</td><td>This technique works on the principal of host-guest chemistry using the lock and key mechanism.</td></tr>
</table>

et al. 2010). Hydrophilic and lipophobic compounds are generally encapsulated via emulsification, coacervation, and supercritical fluid technique (McClements et al. 2009, Cheong et al. 2008, Leong et al. 2009). But, emulsification–solvent evaporation inclusion complexation, and nano-precipitation techniques are employed for lipophilic compounds (Lakkis 2007). Table 1 summarizes the details of the different techniques used for the nanoencapsulation process.

3 What Is Coacervation?

Coacervation technique is based on the principle of phase separation or desolvation of macromolecules. It is a physicochemical process involving a thermodynamic transition in which a uniform colloidal solution separates into two immiscible liquid phases which are thermodynamically compatible - (a) Dense Coacervate and (b) Dilute Liquid Phase. The former thick liquid is concentrated and rich in colloidal particles while the latter is an equilibrium solution. Structurally, it lies between the interface of crystallinity and liquid phase. Thus, it can bear the intermediate range of structural order and can be referred to as mesophase.

Gouin (Gouin 2004) elaborated the concept of the coacervation encapsulation phenomenon which is based on phase separation of one or many hydrocolloids form the initial solution along with deposition of the newly formed coacervate phase around the ingredient suspended or emulsified in the reaction media. Ambinder et al. described the production of coacervates using calcium alginate and calcium alginate chitosan having active ingredient yerba mate extract. Chitosan-coated beads displayed lower loading capacity than the alginate beads. The results implied that wall materials can affect the release of natural antioxidants of yerba mate (Ambinder et al. 2011). A new type of protein/polyphenol microcapsule-based on (−)-epigallocatechin gallate (EGCG) and gelatin containing glycine, proline, and 4-hydroxyproline residues, has been prepared using layer-by-layer assembly (LbL) method (Shutava et al. 2009). The outermost (first layer) was a gelatin layer over the $MnCO_3$ microcores, over which an EGCG layer was formed by adding EGCG solution to the gelatin-coated microparticles. The EGCG content of the protein/polyphenol film material was as high as 30% w/w, while the EGCG in the LbL assemblies retained its antioxidant activity.

3.1 *The Process of Coacervation*

The process starts with the deposition of liquid polymer coating on the core material and then rigidizing the coating usually by different techniques like thermal, cross-linking and de-solvation. This step is the most difficult step and important too. Each step involves a distinct technology in physical chemistry to obtain a coacervate for

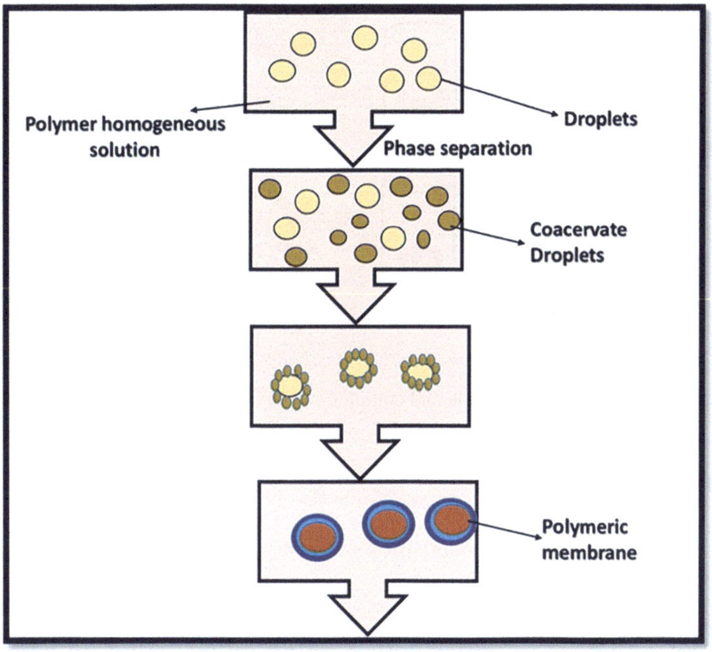

Fig. 5 Various steps involved in the process of coacervation

improved encapsulation. The coacervation and phase separation from organic and aqueous media depend on many properties i.e. materials and processes such as phase inducing agents, core to wall ratio, polymer characteristics, cooling rates and rates of addition, stirring rates, core characteristics (wettability, solubility).

The technology for the fabrication of coacervates was first developed by National Cas Register Co. in the year 1950. It was the first commercial product of microencapsulates and was the rationale behind the carbonless copy paper. The process of coacervation involves three steps (Fig. 5):

3.1.1 Emulsion Formation

In the first step, three immiscible chemical phases are formed by the addition of the carrier/vehicle to the core (the active component) in the presence of the liquid coating under continuous agitation (thermal and mechanical).

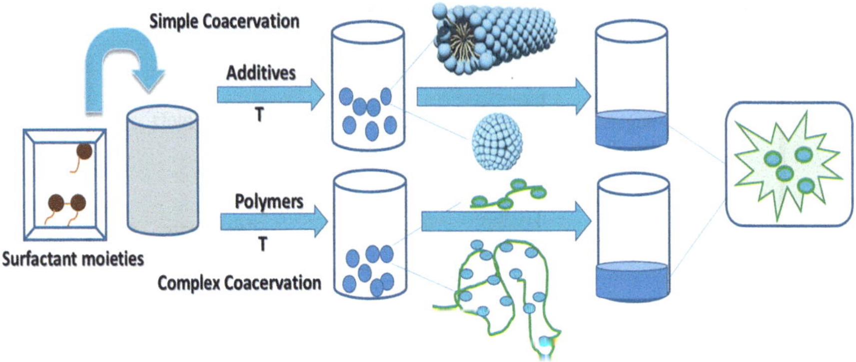

Fig. 6 Comparative analysis of simple and complex coacervation

3.1.2 Induction of Coatin

The second step comprises the deposition of the liquid polymer which acts a protective layer upon the core material.

3.1.3 Stabilization

The prepared nanocapsules can be stabilized and made rigid by (a) Crosslinking; (b) De-solvation; (c) Thermal treatment. The liquid coating of an immiscible polymer is made rigid by temperature changes and add salt or non-solvent (if needed). Inducing polymer-polymer interactions by addition of copolymer also leads to the generation of liquid coating. Phase separation occurs due to reduced solubility caused by an interaction of oppositely charged polyelectrolytes.

3.2 Types of Coacervation

3.2.1 Coacervation Is Generally of Two Types (Fig. 6)

3.2.1.1 Simple Coacervation

It involves only one polymer which is de-solvated or dehydrated resulting in phase separation. There are various factors which effect phase separation phenomenon. It can be affected by mixing two colloidal dispersions, one having a high affinity for water. It changes by addition of strongly hydrophilic substance such as alcohol or sodium sulphate. It can be driven via the addition of salts or alcohols or by increasing the temperature. It demands either more water-soluble polymer or an aqueous non-solvent for the macromolecule or surfactant. Singh et al. (Singh et al. 2016) reported that two phases including polymer rich and polymer poor phases are produced during the processing when liquid-liquid phase separation happens. Since a macromolecule is difficult to solubilize so to bring out the coacervation equilibrium, the mixing of various ratios of water and alcohol is done. This brings the macromolecular solute to coacervation phase separation. Separation of a water-soluble macromolecule into two phases at a special water/alcohol ratio results in the formation of one phase with a high concentration of macromolecule in equilibrium with the second phase containing a higher water concentration. This phenomenon is known as "simple coacervation".

Jincheng et al. (Jincheng et al. 2010) reported the nanoencapsulation of capsaicin with natural polymeric materials shell, gelatin. Gelatins made nanocapsules are compatible with the applied polymer matrix and are helpful for the users. They concluded that gelatin is a good candidate for fabrication of nanospheres, nanocapsules for controlled release applications of drugs and food components. The main reason to use gelatin is an excellent safety, biocompatibility, and biodegradation. The combination of gelatin and encapsulated capsaicin is for enhanced bio-safety. These gelatin-based systems can effectively release the components to a specific target along with protecting them from vulnerable molecules. They also help them form degradation so that they can't lose their bio-efficacy while interacting with the external environment. Taylor and Taylor (Taylor and Taylor 2010) showed that by dissolving kafirin in glacial acetic acid and adding water in a controlled manner, kafirin precipitated out of solution as organized structures, a process of simple coacervation. It does not change the properties of kafirin and sustained its bioactivity which can be explored for further applications.

3.2.1.2 Complex Coacervation

Complex coacervation is defined as a phenomenon of the interaction of two or more oppositely charged macromolecular colloidal systems in a single solvent i.e. it is induced in those systems which has two dispersed hydrophilic colloids of opposite charge. It is primarily driven by the electrostatic interactions between two or more colloidal components. The established interactions between oppositely charged biopolymers in aqueous media lead to the formation of an electrostatic equilibrium in the concentrated phase. The charged macromolecules in a single solvent separate itself into two or more phases with each phase having both the polymers. The formation of a dense phase called coacervate; coexist with the polymer poor phase; called the supernatant. Secondary hydrogen bonding and hydrophobic interactions also facilitate the formation of coacervates, although, less is known about these forces (Liu et al. 2015). This process of neutralization of overall positive charges on one of the colloids by negative charges on the other is used to bring about the separation of polymer-rich complex coacervate phase. The gelatin-gum Arabic (gum acacia) system is the extensively studied complex coacervation system (Lv et al. 2014). Coacervation relies on the mixing ratio of the two polymers.

The mixing of oppositely charged species leads to the formation of a complex named "electrostatic complex". This complex can be soluble or insoluble in the phase. The insoluble part condenses and leads to the formation of coacervate droplets due to Ostwald ripening. This causes mesophase separation resulting in the generation of a distant coacervate layer. The outcome of this process results into two phases; out of two one is super concentrated and other contains a very low concentration of biopolymer.

Complex coacervation is considered to have no definite types and is an expensive method for encapsulation of food ingredients however the potential benefits it offers especially to high value, labile functional ingredients (Bungenberg and Kruyt 1929). Yuan et al. (Yuan et al. 2017) investigated the encapsulation efficiency and oxidative stability of two different microencapsulated systems prepared via the coacervation process. Astaxanthin, a food colorant which is sensitive to thermal treatment. Due to this, it was encapsulated in gelatin/cashew gum fabricated using complex coacervation which not only increases stability, but the encapsulation was found to be 59.9 ± 0.01%. This improves the coloring capacity of astaxanthin (Gomez et al. 2016). Mendanha et al. (Mendanha et al. 2009) microencapsulated casein hydroxylate via complex coacervation using a fixed ratio of soy protein isolate and pectin which results in controlled release of the core material.

Surfactants are also one of the materials used in the coacervation process other than polymers. With advancement in surfactant chemistry, surfactants with novel structures have been developed and found to be efficient in coacervation process. Among the various developed surfactants, gemini surfactants are of special interest due to its extremely low concentration and large concentration range for formation of coacervates. Menger et al. (Menger et al. 2000) explored the gemini surfactants of intermediate size resulting in sponge framework and coalesce into coacervate. So, coacervation is an interesting process for encapsulation and have great potential in this field for research and development.

3.3 Advantages of Coacervation

Encapsulation by coacervation method is increasingly being utilized in the food industry due to high encapsulation efficiency and mild processing conditions which is essential for sensitive food ingredient (De et al. 2004). The water-insoluble ingredients need to be addressed for controlled release profile which can be achieved via encapsulation through coacervation process. The coacervation process not only increases the loading efficiency but also increases the shelf live of the food ingredient if sensitive to the environment. It protects the encapsulated ingredient against nutritional losses and interacting with harmful environmental factors like heat, moisture, air, and light. The process can modify the physical property of the original ingredient like converting the liquids one to dry solids for easy handling.

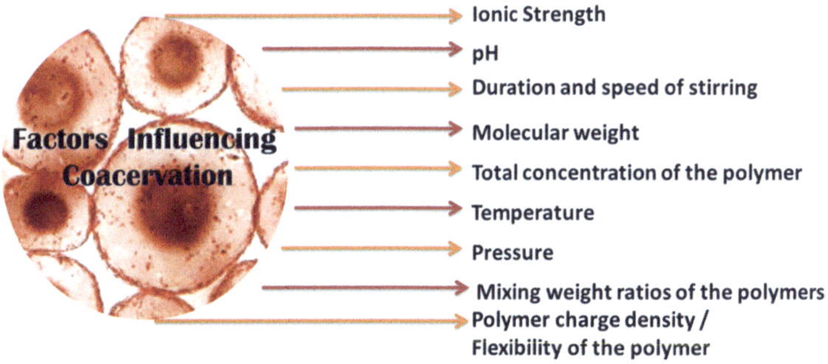

Scheme 2 Factors affecting coacervation process

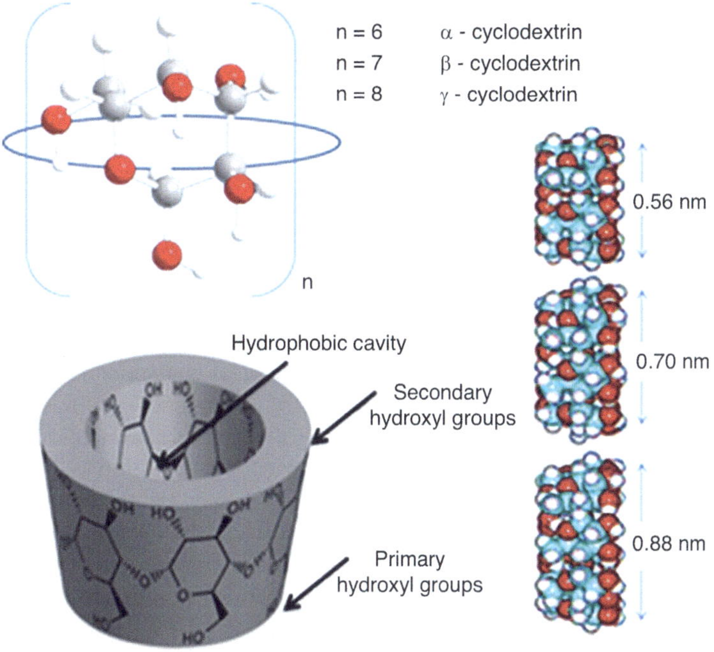

Fig. 7 Classification of cyclodextrins

3.4 *Factors Influencing Coacervation*

Coacervation depends on the mixing ratio of the two polymers because the electrostatic contact depends on the mixing amount or the active sites available on the biopolymers. Under high concentration of salt, the macromolecule is having low charge density which results in suppression of complex coacervation (Scheme 2). Low molecular weight polymer macromolecules also result in fewer chances of coacervation due to ion pairing. The interaction of biopolymers is electrostatic in nature and coacervation is strongly dependent on the molecular charge density, pH and ionic strength (Weinbreck et al. 2003).

4 Cyclodextrin

4.1 What Are Cyclodextrins?

CDs (Fig. 7) are tasteless, odorless and non-digestible cyclic oligosaccharides comprising of glucose units. 1,4 glycosidic bonds link these glucose units together giving them a truncated cone shape with a hydrophobic interior and hydrophilic exterior (Su Ha and He 2008). The hydrophobic cavity has the potential to encapsulate a large number of hydrophobic drugs, and bioactive, without changing their chemical nature. Unmodified or native cyclodextrins have been extensively used in the food and pharmaceutical industry owing to their non-carcinogenic and non-caloric characteristics (Szejtli and Szente 2005). They are enzyme modified starch derivatives produced industrially for versatile application in food, cosmetics, environmental, agricultural, pharmaceuticals industry.

They play a key role in solubilizing, stabilizing, encapsulating and protecting the bioactive components. These macrocyclic oligosaccharides were first discovered by Villiers in 1891 and since their discovery, their application has been many-fold (Lacovino et al. 2017).

CDs have been widely used either for stabilization of flavors and fragrances or for the elimination of undesired tastes, microbiological contaminations, and other undesired compounds in food, cosmetic industry. Over time, CDs has become a boom to the food industry. They are carriers for molecular encapsulation of flavors and other sensitive ingredients (Fenvvesi et al. 2016). They are used as dietary fibers as they have a proven potential for controlling the body weight by reducing the digestion of carbohydrates and lipids. They are also instrumental in improving the blood lipid profile. They are prebiotics in nature and are approved by food authority as carriers for food supplements and nutraceuticals. CDs have been extensively exploited in myriads of fields. α-, β-, and γ-CDs with 6,7 and 8 glucose units respectively, are the most excepted members of CDs family. CDs can form inclusion complexes with molecules of low hydrophobicity and proper geometrical size (Ryzhakov et al. 2016). To further improve the hosting features of these native CDs, they are chemically modified to enhance their solubility and reduce toxicity and improve inclusion capability. Due to the presence of a large number of free reactive hydroxyl on the exterior of native CDs, their functionality can be amplified by many folds through chemical modification. The chemical modification of CDs has made them a solid market product for encapsulation of food materials and neutraceuticals. There are around 30 marketed pharmaceuticals products based on CDs inclusion to improve the physical and chemical properties, or to enhance the bioavailability of poorly soluble drugs.

4.2 Why some Food Needs Cyclodextrin Encapsulation?

Food substances (natural or synthetic) are an integral part of human nutrition but the direct use of these raw materials is hindered due to following reasons (Astray et al. 2009, Fenyvesi et al. 2016):

- *Variable composition:* The aroma/ flavor content of these species can be preserved by forming complexes with CDs and it enhances the reproducibility of the marketed product.
- *Contaminations:* Certain natural products may be contaminated by microbes or parasitic infections and hence their direct utilization is hindered. CDs act as preservatives for such food products and prevent them from such infestations.
- *Storage:* The aroma content of the flavored substances reduces on storage. These natural substances also undergo some undesirable alterations in presence of heat, light, and moisture. Thus, CDs act as protective agents and help store their original properties over a considerable long period of time.
- *Shelf life:* The shelf life of food extracts is generally limited due to their extractions by steam distillation method. In order to enhance their shelf life, these components are converted to solid powders like product by encapsulating them with starch derivatives.

The inclusion complexes formed by CDs makes the use of sensitive ingredients (hydroscopic gels, oils sensitive to light etc.) more practical and renders them marketable. These powdered products offer the food manufacturers a clean and standardized form of variable and unreliable ingredients.

4.3 Methods of Preparation of Cyclodextrin Complexes

There are many innovative methods reported in the literature for the preparation of an inclusion complex using CDs. But concerted efforts are being made to find an effortless and proficient method which would make the process more cost-effective. A number of complexation studies are available in the literature (Kaur et al. 2015), most common among them is the simple physical stirring using the magnetic stirrer. Although this standard procedure is easy, it needs long hours for the complexation process to complete. To reduce the time, the microwave energy is being used. This fabrication process is quick and facile and the product is obtained within a few minutes, but it suffers on economic grounds. The commercial microwaves are expensive and involve high maintenance cost inflating the overall price of the market product. Lately, the cavitation technique is being explored in this area which is showing promising results (Kaur et al. 2015). It neither requires laborious, endless stirring nor is the use of any toxic solvents required. Research is going on in this field to scale up the product using sonication. Other obsolete techniques involved are grinding, agitation and use of mortar and pestle.

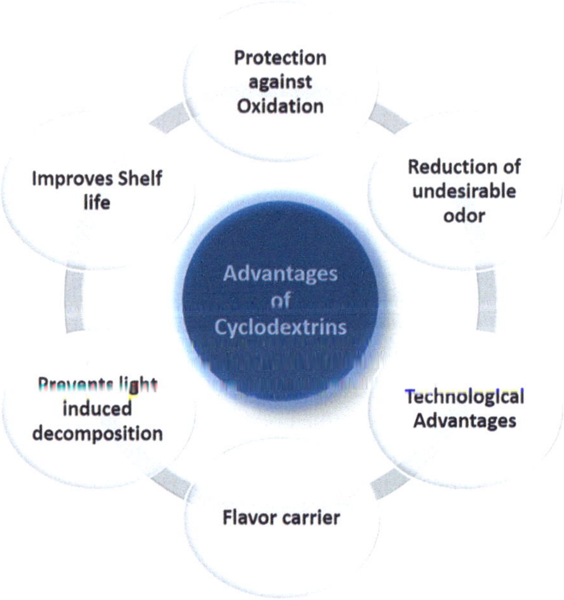

Scheme 3 Advantages of cyclodextrin

4.4 Advantages of Cyclodextrin (Scheme 3)

4.4.1 Applications in Food Industry

CDs are widely used in the food industry as hosts for the encapsulation of flavors. Flavors play a very vital role in customer satisfaction which further has a direct impact on the marketing and consumption of food. But these ingredients being very fragile gets degraded during processing and storage, therefore, the formation of flavor/CD inclusion complex offers great potential for the protection of these labile materials. Complexation with CDs is preferred over the other existing encapsulation practices like spray drying, fluidized bed coating, coacervation, co-crystallization etc. due to its ability to protect the guest throughout the food processing operation withstanding the freezing, thawing and thermal processing.

Besides flavors, CDs are also capable of protecting the lipophilic food materials sensitive to photodegradation and sensitive to oxygen. Numerous enzymes catalyzed browning reactions takes place in the production of juices due to the mechanical damage. The compounds are treated with CDs to remove polyphenol-oxidase by complexation (Szejtli 2004).

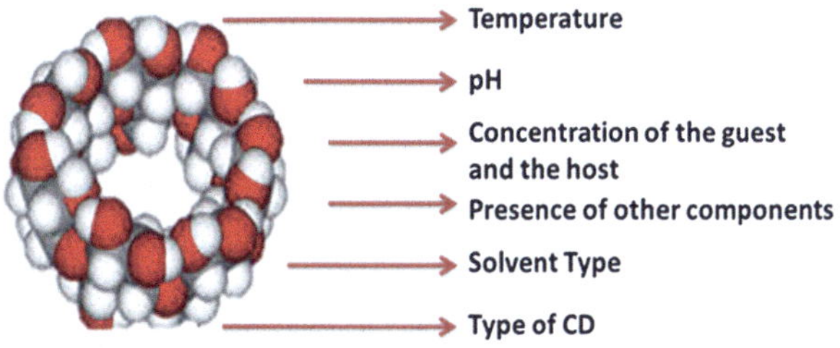

Scheme 4 Factors influencing cyclodextrin encapsulation

4.4.2 Application in Pharmaceutical Industry (Davis and Brewster 2004)

4.4.2.1 Factors Influencing Cyclodextrin Encapsulation

Only the surface of the CD is available for complexation in the crystalline form, but the CDs become more accessible in the solution form thereby improving the chances of the formation of an inclusion complex. Water, being the universal solvent, is used in most of the reactions, the amount of water directs the rate of complex formation (Scheme 4). Since many guest molecules are hydrophobic in nature, the nature of the solvent also affects the complexation process. Under very dilute conditions, i.e. excess solvent, the distance between the CD and the guest molecule increases so much that it curtails the complexations. They are unable to get in contact with each other and hence the complexation becomes very slow or impossible to happen. Another prominent factor affecting the CD-guest formation is -Temperature.

With the rise in temperature, the molecular vibrations increase which could lead to destabilization of the host-guest complex. To make their industrial use more viable, enormous research is being done in this field in recent years (Yang et al. 2014).

4.5 Use and Regulatory of Cyclodextrins

Presently there is an annual increase of around 20–30% in the use of CDs and it can be attributed to its widespread utilization in pharmaceutical, food and chemical industries. CDs and their modified products are used extensively in drugs as complexing agents, solubilizing agents, diluting purposes or as additives to improve the physical and chemical properties of the sensitive bioactive. CDs also hold an important place in the catalytic industry. While it can enhance the selectivity of reactions it can also be used for the separation of industrial products. The food industry is indebted to CDs for the stabilization of flavors and fragrances; elimination and reduction of undesirable taste and odor. Since most of the bioactive and healthy food components have unpleasant sensory characteristics, CDs form inclusion complexes with healthy bioactives and make the food product more consumer-friendly. It bridges the gap between health and commercialization in the food industry.

Despite tremendous uses of CDs and their derivatives, the regulatory agencies keep a austere eye on the safety quotient of commercial use of CDs. The regulatory status of CDs is variable in different countries. In the United States of America, α,β and γ CDs have been included in the "generally regarded as safe" list and can be commercialized while the Japanese consider the CDs as Natural Products and limits its industrial use in food and pharmaceutical sector based on its concentration and purity. For Australia and New Zealand, CDs are classified as Novel Food from 2004 and 2003, respectively (Cravotto et al. 2006). The U.S. Environmental Protection Agency (EPA), has put an end to the need of requirement to evince a maximum permissible level for residues of α,β and γ CDs in various food commodities.

5 A Brief Overview of the Literature

Bioactive Ingredient	Structure	Wall material
Inclusion complex		
Curcumin (turmeric) (Li et al. 2018) 	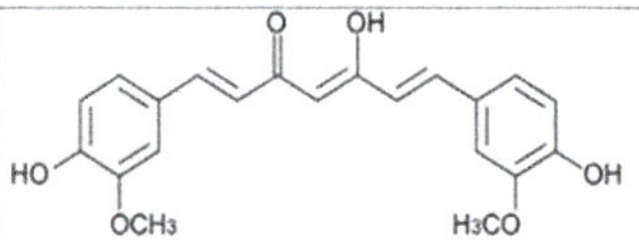	α-, β-, γ-CD, hydroxypropyl (HP)-β-CD, and dimethyl (DM)-β-CD
Vitamin A (Lin et al. 2000) 	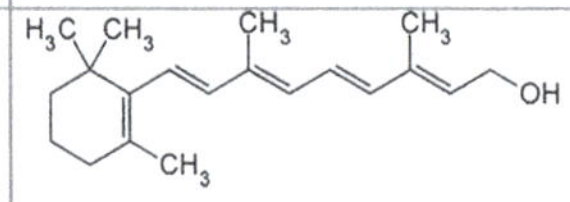	β-CD and hydroxyl propyl β-CD
Vitamin D (Kong et al. 2018) 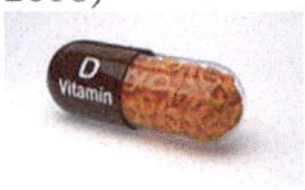	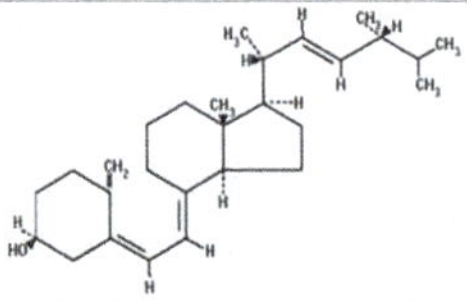	Dextran nanoparticles
DHA (L) (Zimet and Livney 2009) 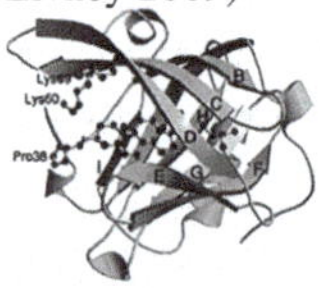	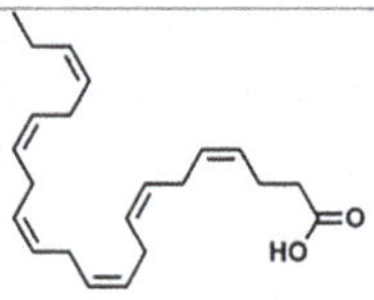	β-lactoglobulin and low methoxyl pectin
Linoleic acid (L) (Hadarunga et al. 2006) 	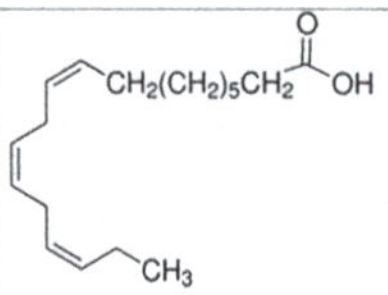	α- and β-cyclodextrin

Bioactive Ingredient	Structure	Wall material
Resveratrol (Kaur et al. 2015), 	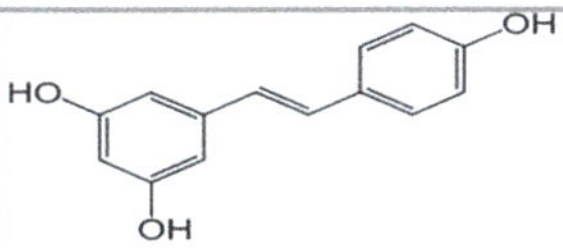	(HP)-β-CD
Benzyl Isothiocyanate (Uppal et al. 2017) 	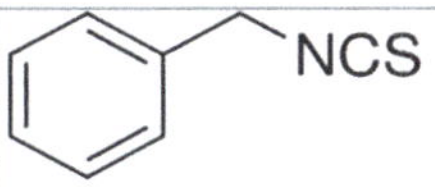	β-,CD, and hydroxypropyl (HP)-β-CD,
β-carotene (Kong et al. 2018) 	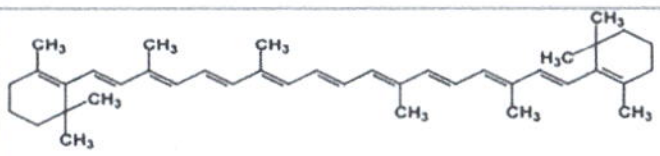	Amylose-surfactant and amylose/ starch-ascorbyl palmitate (AscP) inclusion complexes
Vanillin (Karathanos et al. 2007) 	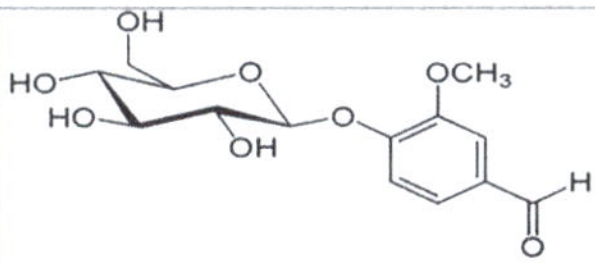	β-cyclodextrin
Astaxanthin (Yuan et al. 2008) 	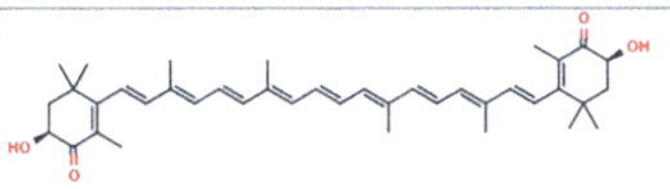	hydroxypropyl-β-cyclodextrin
Trans-ferulic acid (Wang et al. 2011) 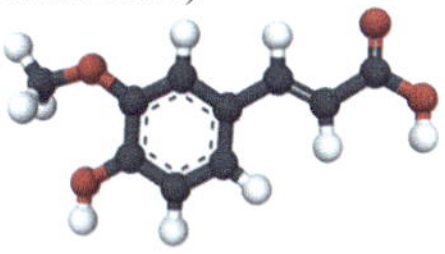	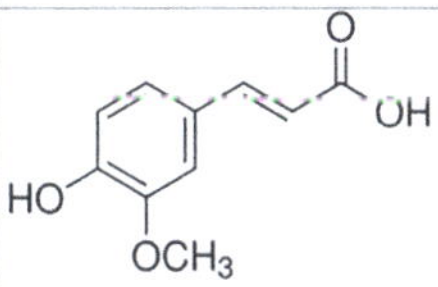	hydroxypropyl-β-cyclodextrin
Rutin (Nguyen et al. 2013) 	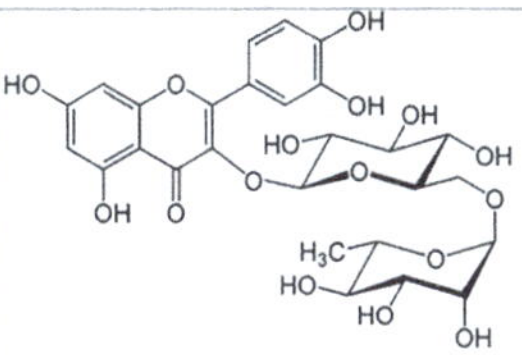	β-cyclodextrin (β-CD), (2-hydroxypropyl)-α-cyclodextrin (HP-α-CD), (2-hydroxypropyl)-β-cyclodextrin (HP-β-CD) and (2-hydroxypropyl)-γ-cyclodextrin (HP-γ-CD)

Bioactive Ingredient	Structure	Wall material
Paeonol (Tsai et al. 2010) 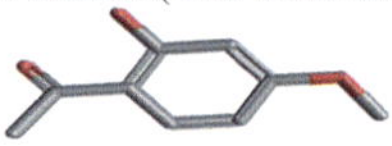	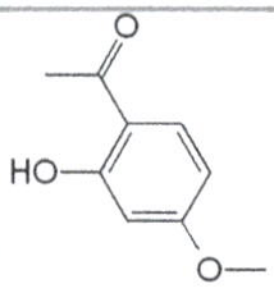	β-cyclodextrin
Coacervation		
Vitamin A (Junyaprasert et al. 2001) 	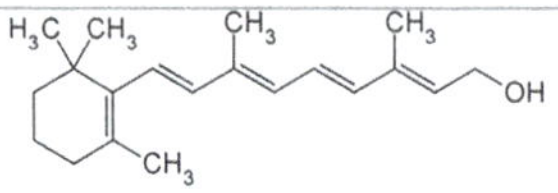	Gelatin and gum Arabic
Folic acid (vitamin B9) (Chapeau et al. 2017) 	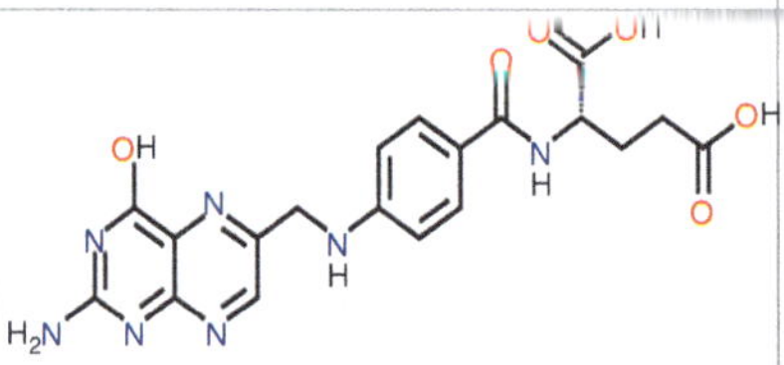	Lactoferrin and b-lactoglobulin co-assembly
Capsaicin (L) (Wang et al. 2008) 	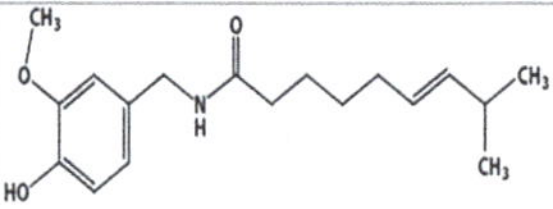	Gelatin, maltodextrin and Tannins
Vitamin B_{12} (Chalasani et al. 2007) 	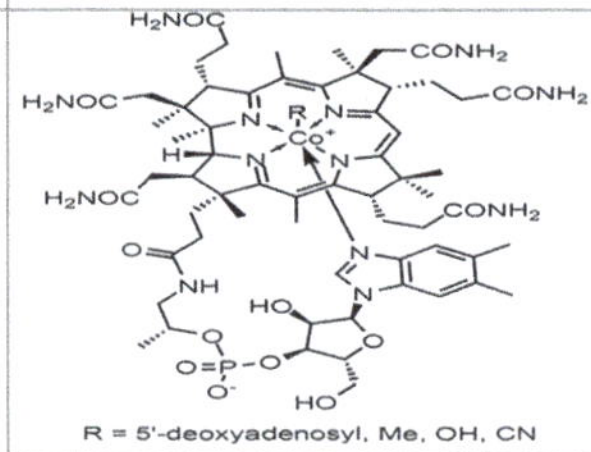	Dextran nanoparticles
Folic acid (vitamin B9) (Penalva et al. 2015) 	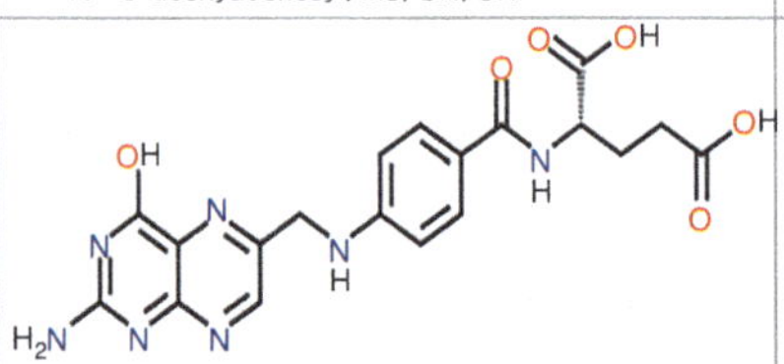	Casein nanoparticles
Capsaicin (L) (Jingcheng et al. 2010) 	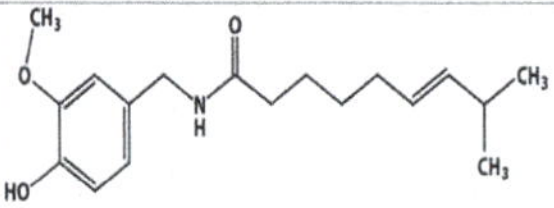	Gelatin, maltodextrin and Tannins

Bioactive Ingredient	Structure	Wall material
Vitamin B_{12} (Wang et al. 2011) 	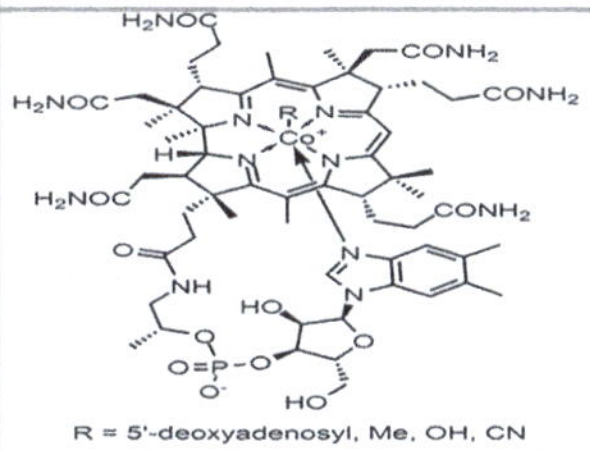	Dextran nanoparticles
Folic acid (vitamin B9) (Penalya et al. 2005) 	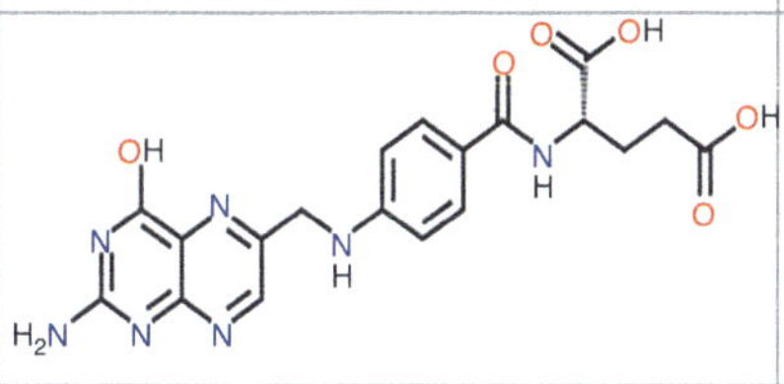	Casein nanoparticles
Capsaicin (L) (Xing et al. 2005) 	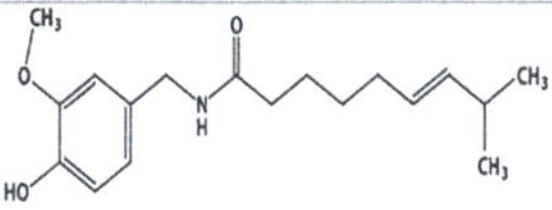	Gelatin, acacia, and hydrolyzable tannins
Capsaicin (L) (Xing et al. 2004) 	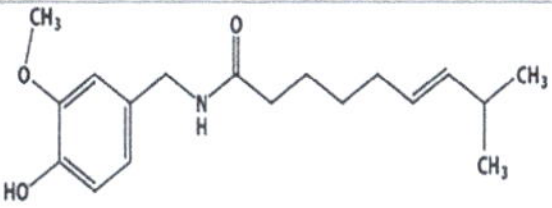	Gelatin, acacia, and tannins

6 Conclusion and Future Scope

In upcoming years, nanoencapsulation is emerging as a beneficial method for the entrapment of environmentally sensitive compounds. By entrapment of different combinations of compounds, the synergic effect can be achieved that enriches the value of the food. From the literature, it is evident that the utilization of encapsulated compounds instead of free one can increase the stability and bioavailability of the compounds which increases their utilization for the administration process. Techniques like coacervation and host-guest chemistry are enduring methods for the efficient entrapment of the ingredients. Basic understanding of supramolecular chemistry and non-covalent interactions can be manipulated to create desirable micro and nanostructures which are widely applied in various fields of bio-medical. With delicate designing and work plan, these systems generate formulations that have many advantages like prolonged circulation, unique size and shape characteristics for passive targeting as well as specific cellular trafficking pathways. Coacervation is one of the emerging nanoencapsulation technique. It is subdivided into two forms: Simple and complex coacervation. Coacervation process deals with proteins, surfactants, and polysaccharides which results into the coacervate formation. But the gemini surfactants and globular proteins are new into the world of coacervation. They are rapidly developing in this field due to their unusual behaviour which creates curiosity into the researchers to explore this field. In order to apply these gemini surfactants into the formulation process, the in-depth understanding of the behaviour of these surfactants is very important. The coacervation process involving globular proteins is an interesting topic to explore as the interdependent specificities in the heteroprotein process can create many new systems including gels, high concentration systems. The urge to find healthier food options and to maintain the taste, odour and health benefits of fragile food components the CDs encapsulation has played a vital role. CDs have proven their niche in the development of good food and its healthy components. When administered orally, CDs are non-toxic as they do not get absorbed in the gastrointestinal tract and are also unsusceptible to α-amylase (Valle and Martin 2004). Thus, the augmentation in the society for healthy food demands is the thrust behind the rapid commercialization of new, smart food using techniques like encapsulation with CDs and coacervation.

References

Ambinder PS, Deladino L, Navarro AS, Amalvy JI, Martino MN (2011) Yerba mate extract encapsulation with alginate and chitosan systems: interactions between active compound encapsulation polymers. JEAS 1:80–87. https://doi.org/10.4236/jeas.2011.14011

Astray G, Gonzalez-Barreiro C, Mejuto JC, Rial-Otero R, Simal-Gándara J (2009) A review on the use of cyclodextrins in foods. Food Hydrocoll 23(7):1631–1640. https://doi.org/10.1016/j.foodhyd.2009.01.001

Augustin MA, Sanguansri P (2009) Nanostructured materials in the food industry. Adv Food Nutr Res 58(4):183–213. https://doi.org/10.1016/S1043-4526(09)58005-9

Chalasani KB, Russell-Jones GJ, Jain AK, Diwan PV, Jain SK (2007) Effective oral delivery of insulin in animal models using vitamin B12-coated dextran nanoparticles. J Control Release 122(2):141–150. https://doi.org/10.1016/j.jconrel.2007.05.019

Chapeau AL, Hamon P, Rousseau F, Croguennec T, Poncelet D, Bouhallab S (2017) Scale-up production of vitamin loaded heteroprotein coacervates and their protective property. J Food Eng 206:67–76. https://doi.org/10.1016/j.jfoodeng.2017.03.005

Chau CF, Wu SH, Yen GC (2007) The development of regulations for food nanotechnology. Trends Food Sci Technol 18(5):269–280. https://doi.org/10.1016/j.tifs.2007.01.007

Cheong JN, Tan CP, Yaakob B, Che M, Misran M (2008) α-Tocopherol nanodispersions: preparation, characterization and stability evaluation. Int J Food Eng 89(2):204–209

Cravotto G, Binello A, Baranelli E, Carraro P, Trotta F (2006) Cyclodextrins as food additives and in food processing. Curr Nutr Food Sci 2:343–350. https://doi.org/10.2174/157340106778699485

Davis ME, Brewster ME (2004) Cyclodextrin-based pharmaceutics: past, present and future. Nat Rev Drug Discov 3:1023–1035. https://doi.org/10.1038/nrd1576

Desai KGH, Park HJ (2005) Recent developments in microencapsulation of food ingredients. Dry Technol 23:1361–1394. https://doi.org/10.1081/DRT-200063478

Fenyvesi E, Vikmon M, Szente L (2016) Cyclodextrins in food technology and human nutrition: benefits and limitations. Crit Rev Food Sci Nutr 56(12):1981–2004. https://doi.org/10.1080/10408398.2013.809513

Feynman RP (1961) There's plenty of room at the bottom. In: Gilbert HD (ed) Miniaturization. Reinhold, New York

Gomez EJ, Comunian TA, Montero P, Ferro FR, Favaro-Trindade CS (2016) Encapsulation of an astaxanthin-containing lipid extract from shrimp waste by complex coacervation using a novel gelatin–cashew gum complex. Food Hydrocoll 61:155–162. https://doi.org/10.1016/j.foodhyd.2016.05.005

Gouin S (2004) Microencapsulation: industrial appraisal of existing technologies and trends. Trends Food Sci Technol 15(7–8):330–347. https://doi.org/10.1016/j.tifs.2003.10.005

Jincheng W, Xiaoyu Z, Sihao C (2010) Preparation and properties of nanoencapsulated capsaicin by complex coacervation method. Chem Eng Commun 197:919–933. https://doi.org/10.1080/00986440903249700

Junyaprasert VB, Mitrevej A, Sinchaipanid N, Boonme P, Wurster DE (2001) Drug Dev Ind Pharm 27(6):561–566. https://doi.org/10.1081/DDC-100105181

Karathanos VT, Mourtzinos I, Yannakopoulou K, Andrikopoulos NK (2007) Study of the solubility, antioxidant activity and structure of inclusion complex of vanillin with β-cyclodextrin. Food Chem 101:652–658. https://doi.org/10.1016/j.foodchem.2006.01.053

Kaur K, Uppal S, Kaur R, Agarwal J, Mehta SK (2015) Energy efficient, facile and cost-effective methodology for formation of an inclusion complex of resveratrol with hp-β-CD. New J Chem 39:8855. https://doi.org/10.1039/C5NJ00924C

Kong L, Bhosale R, Ziegler GR (2018) Encapsulation and stabilization of β-carotene by amylose inclusion complexes. Food Res Int 105:446–452

Lacovino R, Caso JV, Donato CD, Malgieri G, Palmieri M, Russo L, Isernia C (2017) Curr Org Chem:21

Lakkis JM (2007) Encapsulation and controlled release technologies in food systems. Blackwell, Iowa

Leong TSH, Wooster TJ, Kentish SE, Ashokkumar M (2009) Minimising oil droplet size using ultrasonic emulsification. Ultrason Sonochem 16(6):721–727. https://doi.org/10.1016/j.ultsonch.2009.02.008

Li X, Uehara S, Sawangrat K, Morishita M, Kusamori K, Katsumi H, Sakane T, Yamamoto A (2018) Improvement of intestinal absorption of curcumin by cyclodextrins and the mechanisms underlying absorption enhancement. Int J Pharm 535:340–349. https://doi.org/10.1016/j.ijpharm.2017.11.032

Liang R, Shoemaker CF, Yang X, Zhong F, Huang Q (2013) Stability and bioaccessibility of β-carotene in nanoemulsions stabilized by modified starches. J Agric Food Chem 61(6):1249–1257. https://doi.org/10.1021/jf303967f

Lin HS, Chean CS, Ng YY, Chan SY, Ho PC (2000) 2-Hydroxypropyl-β-cyclodextrin increases aqueous solubility and photostability of all-trans-retinoic acid. J Clin Pharm Ther 25:265–269. https://doi.org/10.1046/j.1365-2710.2000.00285.x

Liu J, Shim YY, Wang Y, Reaney MJT (2015) Intermolecular interaction and complex coacervation between bovine serum albumin and gum from whole flaxseed (Linum usitatissimum L.). Food Hydrocoll 49:95–103. https://doi.org/10.1016/j.foodhyd.2015.02.035

Lv Y, Yang F, Li X, Zhang X, Abbas S (2014) Formation of heat-resistant nanocapsules of jasmine essential oil via gelatin/gum Arabic based complex coacervation. Food Hydrocoll 35:305–314. https://doi.org/10.1016/j.foodhyd.2013.06.003

Mangolim CS, Moriwaki C, Nogueria AC, Sato F, Bacsso ML, Neto AM, Matioli G (2014) Curcumin–β-cyclodextrin inclusion complex: stability, solubility, characterization by FT-IR, FT-Raman, X-ray diffraction and photoacoustic spectroscopy, and food application. Food Chem 153:361–370. https://doi.org/10.1016/j.foodchem.2013.12.067

McClements DJ, Decker EA, Park Y, Weiss J (2009) Structural design principles for delivery of bioactive components in nutraceuticals and functional foods. Crit Rev Food Sci Nutr 49(6):577–606. https://doi.org/10.1080/10408390902841529

Mendanha DV, Ortiz SEM, Favaro-Trindade CS, Mauri A, Monterrey-Quintero ES, Thomazini M (2009) Microencapsulation of casein hydrosylate by complex coacervation with SPI/pectin. Food Res Int 42:1099–1104. https://doi.org/10.1016/j.foodres.2009.05.007

Menger FM, Peresypkin AV, Caran KL, Apkarian RP (2000) A sponge morphology in an elementary coacervate. Langmuir 16(24):9113–9116. https://doi.org/10.1021/la0010626

Mishra B, Patel BB, Tiwari S (2010) Colloidal nanocarriers: a review on formulation technology, types, and applications toward targeted drug delivery. Nanomedicine 6:9–24. https://doi.org/10.1016/j.nano.2009.04.008

Mun S, Kim YR, Shin M, McClements DJ (2015) Control of lipid digestion and nutraceutical bioaccessibility using starch-based filled hydrogels: influence of starch and surfactant type. Food Hydrocoll 44:380–389. https://doi.org/10.1016/j.foodhyd.2014.10.013

Nguyen TA, Liu B, Zhao J, Thomas DS, Hook JM (2013) An investigation into the supramolecular structure, solubility, stability and antioxidant activity of rutin/cyclodextrin inclusion complex. Food Chem 136:186–192. https://doi.org/10.1016/j.foodchem.2012.07.104

Norton IT, Frith WJ (2001) Microstructure design in mixed biopolymer composites. Food Hydrocoll 15:543–553. https://doi.org/10.1016/S0268-005X(01)00062-5

Penalva R, Esparza I, Agüeros M, Navarro CJG, Herrero CG, Irache JM (2015) Casein nanoparticles as carriers for the oral delivery of folic acid. Food Hydrocoll 44:399–406. https://doi.org/10.1016/j.foodhyd.2014.10.004

Risch JS (1995) Encapsulation and controlled release of food ingredients. ACS Symp Ser 590:2–7. https://doi.org/10.1021/bk-1995-0590.ch001

Ryzhakov A, Do Thi T, Stappaerts J, Bertoletti L, Kimpe K, Rodrigues Sà Couto A, Saokham P, Van den Mooter G, Augustijns P, Somsen GW, Kurkov S, Inghelbrecht S, Arien A, Jimidar MI, Schrijnemakers K, Loftsson T (2016) Self-assembly of cyclodextrins and their complexes in aqueous solutions. J Pharm Sci:2556–2569. https://doi.org/10.1016/j.xphs.2016.01.019

Sahoo SK, Parveen S, Panda JJ (2007) The present and future of nanotechnology in human health care. Nanomedicine: NBM 3:20–31. https://doi.org/10.1016/j.nano.2006.11.008

Sanguansri P, Augustin MA (2006) Nanoscale materials development— a food industry perspective. Trends Food Sci Technol 17(10):547–556. https://doi.org/10.1016/j.tifs.2006.04.010

Shahidi F, Han XQ (1993) Encapsulation of food ingredients. Crit Rev Food Sci Nutr 33:501–547. https://doi.org/10.1080/10408399309527645

Shutava TG, Balkundi SS, Lvou YM (2009) (−)-Epigallocatechin gallate/gelatin layer-by-layer assembled films and microcapsules. J Colloid Interface Sci 330(36):276–283. https://doi.org/10.1016/j.jcis.2008.10.082

Silvestre C, Duraccio D, Sossio C (2011) Food packaging based on polymer nanomaterials. Prog Polym Sci 36:1766–1782. https://doi.org/10.1016/j.progpolymsci.2011.02.003

Singh S, Demco DE, Rahimi K, Fechete R, Rodriguez-Cabello JC, Möller M (2016) Coacervation of elastin-like recombinant microgels. Macromol Rapid Commun 37:181–186. https://doi.org/10.1002/marc.201500457

Stella VJ, He Q (2008) Cyclodextrins. Toxicol Pathol 36:30–42. https://doi.org/10.1177/0192623307310945

Szejtli J (2004) Past, present and future of cyclodextrin research. Pure Appl Chem 76:1825–1845. https://doi.org/10.1351/pac200476101825

Szejtli J, Szente L (2005) Elimination of bitter, disgusting tastes of drugs and foods by cyclodextrins. Eur J Pharm Biopharm 61:115–125. https://doi.org/10.1016/j.ejpb.2005.05.006

Taylor J, Taylor JRN (2010) Patent Cooperation Treaty Application 2010. WO 2010/041203 A1 Process for Producing Protein Microparticles

Tsai Y, Tsai HH, Wu CP, Tsai FJ (2010) Food Chem 120:837–841

Uppal S, Kaur K, Kumar R, Kahlon NK, Singh R, Mehta SK (2017) Encompassment of Benzyl Isothiocyanate in cyclodextrin using ultrasonication methodology to enhance its stability for biological applications. Ultrason Sonochem 39:25–33. https://doi.org/10.1016/j.ultsonch.2017.04.007

Valle D, Martin EM (2004) Cyclodextrins and their uses: a review. Process Biochem 39:1033–1046. https://doi.org/10.1016/S0032-9592(03)00258-9

Verma S, Domb AJ, Kumar N (2011) Nanomaterials for regenerative medicine. Nanomedicine (Lond) 6:157–181. https://doi.org/10.2217/nnm.10.157

Wang JC, Chen SH, Xu C (2008) Synthesis and properties research on the nanocapsulated capsaicin by simple coacervation method. J Dispers Sci Technol 29:687–695. https://doi.org/10.1080/01932690701756651

Wang J, Cao Y, Sun B, Wang C (2011) Characterisation of inclusion complex of trans-ferulic acid and hydroxypropyl-β-cyclodextrin. Food Chem 124:1069–1075. https://doi.org/10.1016/j.foodchem.2010.07.080

Weinbreck F, de Vries R, Schrooyen P, de Kruif CG (2003) Complex coacervation of whey proteins and gum Arabic. Biomacromolecules 4:293–303. https://doi.org/10.1021/bm025667n

Xing F, Cheng G, Yang B, Ma L (2004) Microencapsulation of capsaicin by the complex coacervation of gelatin, acacia and tannin. J Appl Polym Sci 91:2669–2675. https://doi.org/10.1002/app.13449

Xing F, Cheng G, Yi K, Ma L (2005) Nanoencapsulation of capsaicin by complex coacervation of gelatin, acacia, and tannins. J Appl Polym Sci 96(6):2225–2229. https://doi.org/10.1002/app.21698

Yaktine A, Pray L (2009) Nanotechnology in food products: workshop summary. National Academies Press Washington DC doi: https://doi.org/10.17226/12633

Yang H, Yuan B, Zhang X (2014) Supramolecular chemistry at interfaces: host–guest interactions for fabricating multifunctional biointerfaces. Acc Chem Res 47(7):2106–2115. https://doi.org/10.1021/ar500105t

Yao M, McClements DJ, Xiao H (2015) Improving oral bioavailability of nutraceuticals by engineered nanoparticle-based delivery systems. Curr Opin Food Sci 2:14–19. https://doi.org/10.1016/j.cofs.2014.12.005

Yuan C, Jin Z, Zhuang XH, Shen W (2008) Preparation and stability of the inclusion complex of astaxanthin with hydroxypropyl-β-cyclodextrin. Food Chem 109:264–268. https://doi.org/10.1016/j.foodchem.2007.07.051

Yuan Y, Kong ZY, Sun YE, Zeng QZ, Yang XQ (2017) Complex coacervation of soy protein with chitosan: constructing antioxidant microcapsule for algal oil delivery. LWT Food Sci Technol 75:171–179. https://doi.org/10.1016/j.lwt.2016.08.045

Zimet P, Livney YD (2009) Beta-lactoglobulin and its nanocomplexes with pectin as vehicles for ω-3 polyunsaturated fatty acids. Food Hydrocoll 23:1120–1126. https://doi.org/10.1016/j.foodhyd.2008.10.008

Principles and Potential Applications of Cavitation Technology for Nano-Foods

Liang Ee Low, See Kiat Wong, Chien Lye Chew, and Siah Ying Tang

Abbreviations

US	Ultrasound
LPU	Low Power Ultrasound
HPU	High Power Ultrasound
MS	Manosonication
TS	Thermosonication
MTS	Manothermosonication
HIFU	High Intensity Focused Ultrasound
HUTR	High-intensity Ultrasonic Tubular Reactor
HHP	High Hydrostatic Pressure
SAEW	Static Acidic Electrolyzed Water
USC	Ultrasound Cooling

L. E. Low
Institute of Pharmaceutics, College of Pharmaceutical Sciences, Zhejiang University, Hangzhou, Zhejiang, China

C. L. Chew
Chemical Engineering Discipline, School of Engineering, Monash University Malaysia, Bandar Sunway, Selangor, Malaysia

Processing Technology, Sime Darby Research, R&D Centre – Carey Island, Pulau Carey, Selangor, Malaysia

S. K. Wong
Chemical Engineering Discipline, School of Engineering, Monash University Malaysia, Bandar Sunway, Selangor, Malaysia

S. Y. Tang (✉)
Chemical Engineering Discipline, School of Engineering, Monash University Malaysia, Bandar Sunway, Selangor, Malaysia

Advanced Engineering Platform, Monash University Malaysia, Bandar Sunway, Selangor, Malaysia
e-mail: patrick.tang@monash.edu

U. Hebbar et al. (eds.), *Nano-food Engineering*, Food Engineering Series,
https://doi.org/10.1007/978-3-030-44552-2_5

1 Introduction

Ultrasound (US) cavitation has attracted much attention recently in the scientific community as green and energy saving approach for nanofood preparation. It continue to serve as one of the rapidly emerging methods that promote mild processing conditions while able to improve the food quality (Majid et al. 2015; Manickam et al. 2014). US is a sound wave above the threshold of human hearing range with frequency of 20 kHz and above up to few hundred MHz (Avison 1989). Generally, it can be categorized two operating modes, namely the low power ultrasound (LPU) and high power ultrasound (HPU) mode. LPU have frequency above mega-hertz ranges while low power modes generally have frequencies ranging from 20–500 kHz (Mason et al. 2011). It generally operates as analytical techniques to obtain various information regarding the physicochemical properties of nanofood including sugar content, texture hardness, ripeness and others (Soria and Villamiel 2010). On the other hand, the applications of HPU facilitates turbulent mixing and secondary effects attributed to acoustic cavitation (the formation, growth, and collapse of bubbles) (Suslick 1990; Suslick and Price 1999), which creates short burst of extreme pressure ($\approx$ 100 MPa), temperature ($\approx$ 5000 °C) and free radicals (•H & •OH) as well as rapid heating and cooling rates above 10^{10} K/s during the implosion of the cavitation bubbles (Suslick 1990). This not only promotes the changes in physical and chemical properties of food (Piyasena et al. 2003), but also capable of reducing the sizes of food to the nanometer ranges. Additionally, the free radicals produced from US cavitation also found to be able to deactivate microorganisms activities that will caused food decomposition, which thus led to a reduced shelf life of food. Thus, HPU mode is considered to serve a more important role for the nanofood production and preservation as regarded by its vigorous cavitation phenomenon due to high dispersing power. To date, the US cavitation technology has found its application in dairy processing, emulsion-based food products, commercial cooking oils, food gels, etc. (Majid et al. 2015).

Throughout this chapter, the basic principle of US cavitation and its potential applications in emulsification, spray drying and pasteurization will be described. On the other hand, some other potential applications of the US cavitation will also be mentioned briefly. The outline of this chapter can be seen in Fig. 1. It is important to remark the importance of this technique because US cavitation technology is a green and mild processing method that hold promising potential to effectively preserve the nutrient, bioactive and physical characteristics of a food.

2 Nanotechnology and Nanofood

Nanotechnology has been receiving increasing attentions in the food industry in recent years for food engineering and processing. According to world nanoforum, a food is defined as nanofood if nanotechnology techniques or tools are used during

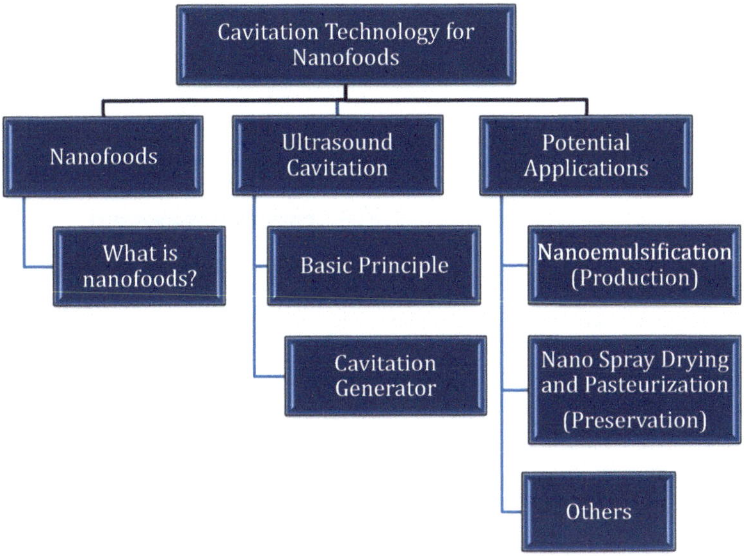

Fig. 1 Chart showing the outline of the chapter

cultivation, production, processing, or packaging of the food. Literature survey has revealed that the total market value of nanofood had reach US$5.8 billion in 2012, and continued to increase in recent years, offering heavy profit to economy (Chellaram et al. 2014). Food nanotechnology has its history since the introduction of Pasteurization process to kill the spoilage bacteria and improve the quality of foods. Soon after, work done by Watson and Crick (1953) on the model of DNA structure that is about 2.5 nm further opened the gateway of applications in biomedical, agriculture and food production fields.

The implication of nanotechnology in food industry has leads to several advantages in the food production and packaging processes. This includes the better shelf life and nutrient delivery of the produced nanofoods as well as food packaging that are more perseverance, consumer and environmental friendly (Chellaram et al. 2014). Nanofood was readily available in human everyday life, even before the modern days. Some examples of traditional nanofood products include milk, butter, ice-cream, yogurt, etc., that consist of the dispersion of nano-sized oil droplets in aqueous suspensions (Martirosyan and Schneider 2014; Pray and Yaktine 2009; Sekhon 2010). Figure 2 show the example of the nanofood with nano-sized droplets (milk). With recent advancement on the development of nanotechnology and nanomaterials, many other approaches have been performed for development and improvement of foods with enhanced health benefits or better shelf life. For example, recent reports have shown that the employment of nutrient encapsulated nanostructured capsule/particles as food supplement improved the nutrient absorption in human body due to the nanomaterial's high surface area to volume ratio that leads

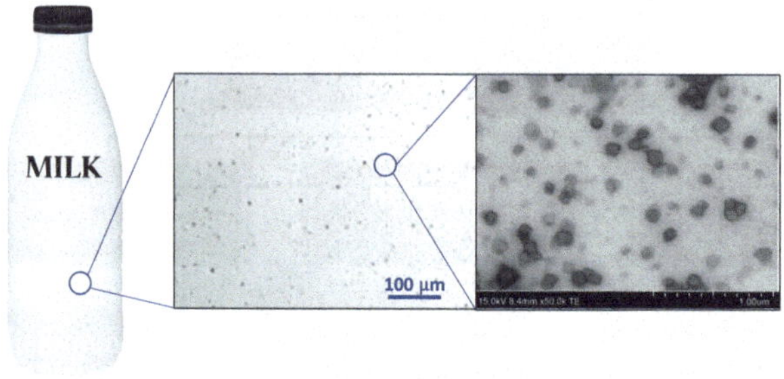

Fig. 2 Example of traditional nanofood (milk) and its optical microscopic form

to effective transport of nutrient as compared to common supplements (Alfadul and Elneshwy 2010; Chaudhry et al. 2008; Mishra et al. 2017; Pathakoti et al. 2017). Besides that, literature also showed that the used of nano-sized self-assembled structural lipids successfully inhibited the transportation of cholesterol from the digestive system into the bloodstream (Dingman 2008).

Although nanofood offers exciting possibilities in the food industry and products, it should be noticed that the nanofood, including the nanomaterials for the packaging are difficult to be produced without heavy consumptions of time, energy and sometimes, extreme treatment conditions (Majid et al. 2015). More importantly, from recent years food industry has demand for minimal processing of food under critical conditions since these may lower their nutrient level and bioavailability (Czechowska-Biskup et al. 2005). Thus, it is advisable for the food industry to employ newer processing method with mild treatment conditions in order to retain the nutrient, bioactive and physical characteristics of a food.

3 Ultrasound Cavitation

3.1 Principle of Ultrasonic Cavitation and its Development for Nanofood

The utilization of ultrasound (US) began in circa 1700, when a physiologist named Lazzaro Spallanzani first studied ultrasonic physics by deducing that bats used US to navigate by echolocation (Tsung 2011). From there, research and innovations started to follow one after another. For example, the invention of first transducer in 1915 by physicist Paul Langevin after Titanic sunk in 1912 to detect icebergs and submarines, the application of US on medical diagnosis by Karl Dussick in 1942, and more (Tsung 2011). In the twenty-first century, there had been great advances in the ultrasonic field and many equipment/instrument were invented to be equipped with ultrasonic capabilities that is differentiable in two categories; low power ultrasound (LPU) with high-frequency, and high power ultrasound (HPU) with low-frequency.

HPU utilizes US ranging from 20 to 100 kHz to generate cavitation bubbles that served as one of the important alternative for a variety of applications such as cleaning, food processing, chemical synthesis and more (Nithila et al. 2014). In acoustic cavitation, the bubble inside a liquid is continuously getting compressed and expanded through the alternating positive and negative pressures produced from the acoustic energy input. The bubbles grow along with the oscillating sound wave until a big void (big bubbles) is created in the liquid medium, where the big void then imploded and caused a short bursting release of thermal, mechanical and chemical effects (see Fig. 3). Thermal effects compose of the generation of heat (up to 5000 °C) with rapid cooling rates of 10^{10} K/s (Suslick 1990; Suslick and Price 1999). The mechanical effects include an extremely high pressure (100 MPa) as well as shockwaves and microjets (Suslick and Price 1999; Yusaf and Al-juboori 2014). Besides, the collapse of the cavitation bubbles also exerted free radicals (•OH and •H) as chemical effects from US cavitation.

The aids of US cavitation in generating nanofoods can be categorized into mechanical, chemical and thermal effects from the collapse of cavitation bubbles (Fig. 4). For instance, the high pressure shear force and microjet produced mechanical effects of the acoustic cavitation can results in the delivery of high energy for production and processing of food, thereby reducing the sizes of the ultrasonically treated foods to nano-sizes range. An example of such employment is the ultrasonic preparation of nanoemulsions (Manickam et al. 2014). Nanoemulsions can take various forms in our daily life as food including milk, cheese, emulsion-based supplements, ice-cream and mayonnaise (Martirosyan and Schneider 2014; Pray and Yaktine 2009; Sekhon 2010). The process of US-induced nanoemulsification utilized the high-pressure microjets and energy exerted from US cavitation to first disperse one liquid into another immiscible liquids to create a coarse emulsion droplet, and followed by breaking it down to finer size ranges. Recent literature has demonstrated that US is, in fact a relatively greener and more energy efficient

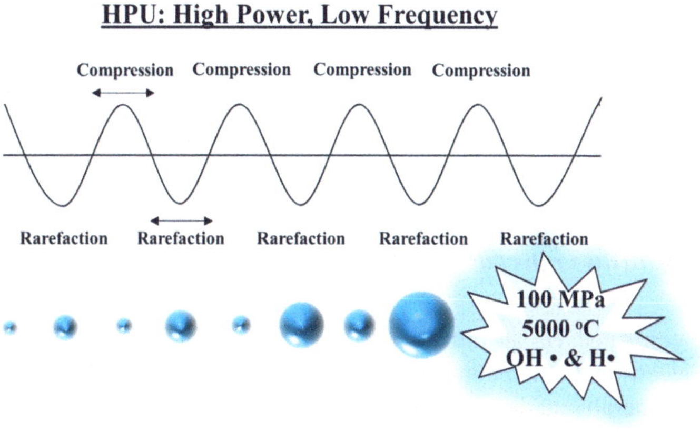

Fig. 3 Theoretical mechanisms and outcome of the growth and collapse of a cavitation bubble when it is subject to ultrasound field

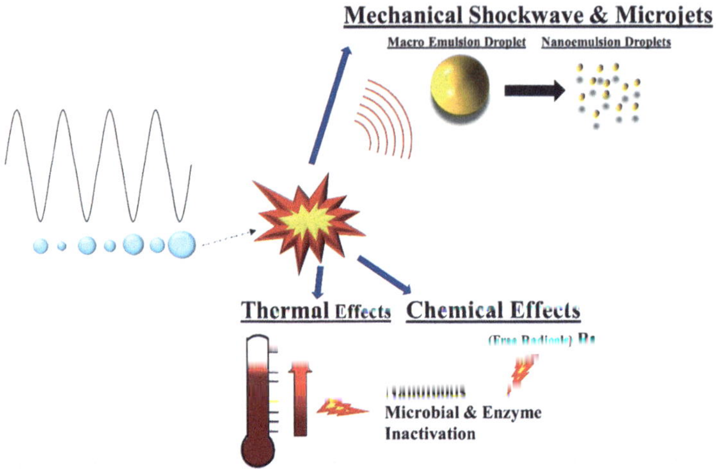

Fig. 4 Possible mechanical, chemical and thermal effects on nanofoods based on ultrasound sonochemistry

techniques for the preparation of nanoemulsions as compared to conventional methods (Manickam et al. 2014). Besides that, the radicals and thermal heat produced from US cavitation can also assist in the inactivation of microorganism and enzyme in food products to improve the food quality and shelf life. Microorganisms such as bacteria and fungi tend to secrete enzymes to decompose the lipid and proteins in the food product, which then accelerates the food decaying process and caused undesirable food spoilage. The use of ultrasonically induced radicals was found to successfully disrupted the microbes and leads to improved food quality and preservation (Alvarez et al. 2003; Condon-Abanto et al. 2016; Evelyn and Silva 2015; Gabriel 2015; Moody et al. 2014; Pagan et al. 1999). Upon disruption of microbe cells by the high pressure and temperature generated through US irradiation, the hydroxyl radicals produced oxidized the microorganisms and thus deactivate its enzyme secretion, thereby preventing the unwanted damage from the microorganisms. In overall, US cavitation technology has showed tremendous potential in the preparation and preservation of nanofoods. The mentioned potential applications will be further discussed in the later sections of this chapter.

3.2 *Ultrasound Generation*

US can be generated via an ultrasonic transducer that utilize the high-frequency oscillating electric field and converted it to mechanical vibrations of the same frequency. It is generally made up of either the piezoelectric or magnetostrictive materials that expand or contract when they are placed in electromagnetic fields (Suslick 1989), where US is produced as a result of the expansion and contraction of the material. Due to the simplicity of US generation and the energy concentration effect of acoustic cavitation, US has been widely used to intensify and to enhance the process of chemical reactions, bioactives encapsulation and extraction (Tang et al. 2012). The energy efficient property of ultrasound has also led to the increased interest in extensive research on the use of US for nanofoods production and preservation. As an example, the nanoemulsion is one of the most well documented nanofood systems producible by US cavitation. In emulsification process, US is generally emitted either by cylindrical sonotrodes immersed in the liquid (Fig. 5a). However, one vital drawback of such US process is the low processing volume and high chances for cross contamination due to the direct contact of the ultrasonic transducer and the liquid sample. This greatly limited practicality of US equipment for large-scale preparation of nanoemulsions. To tackle this issue, researchers tend to utilize the ultrasonic bath system for several US processes (Fig. 5b). Although the ultrasonic bath system can avoid the cross contamination of samples during the sonication, the US intensity implied by the US bath is, unfortunately, too weak for an effective US irradiation as compared to the horn system. As an alternative, a newly invented high-intensity ultrasonic tubular reactor (HUTR) revealed promising capability in preparing highly stable Pickering emulsions (Low et al. 2019). The outcome demonstrated the possible employment of the HUTR as a replacement for both the US horn and bath systems. However, further researches will be needful to evaluate the potential of the HUTR for nanofood engineering. Overall, the aforementioned issue is expected to be addressed with the future advances in the ultrasonic technology.

4 Application of Cavitation in Nanoemulsification

4.1 Cavitation-Induced Nanoemulsification

The utilization of cavitation technology for nanoemulsion preparation has been extensively studied over the years (Manickam et al. 2014). Nanoemulsions are formed through the mixing of two immiscible liquids with the presence of surface-active

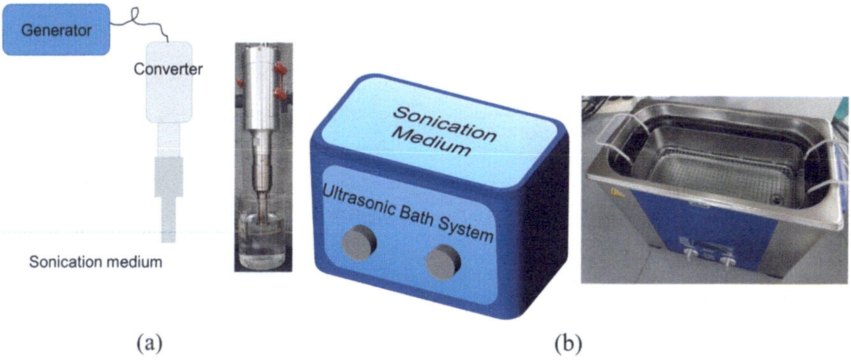

Fig. 5 US generated by the (**a**) horn system and (**b**) bath system

agent and forms homogeneous dispersion (emulsification). The droplet size of the nanoemulsion is ranging from 20 nm to 500 nm (nanoscale) with high stability against separation due to its nanosized droplets having lower attraction force as compared to normal macro−/micro- emulsion (Bilbao-Sáinz et al. 2010; Silva et al. 2011). In most cases the mixing is performed between water and oil, where the oil phase are normally vegetable oil, essential oil with loaded vitamins, antioxidants, and bioactive agents, whereas the water phase includes water, gums, emulsifiers, sweeteners, salts, preservatives, and protein (Chanamai and McClements 2000). The mixing process required external energy through either mechanical agitation (high shear) to break the oil and water into smaller droplets with the presence of air bubbles in between of both phases, or with US cavitation that involved the formation, grows, and collapses of cavity bubbles, where the collapse of cavitation bubbles released high energy, forming asymmetrical microjets between the interface of two immiscible liquids which then aiding the emulsification process to form emulsion (Thompson and Doraiswamy 1999). Li and Fogler (1978a, b) mentioned that the emulsification mechanism implied by US contained two stages: (1) the generation of coarse droplets by the acoustically produced interfacial waves, which erupted the oil phase into aqueous phase in the form of droplets. (2) The coarse droplets then imploded violently and causes localized intense turbulence and shear forces that produce microjets to further breaking the primarily generated oil droplets into finer ones (see Fig. 6).

Between the two mentioned methods, US cavitation technology is preferred for nanofoods processing due to its low power supply and ability to promotes mild processing conditions (Majid et al. 2015; Manickam et al. 2014). Abismail et al. (1999) reported that emulsification process with US provides better results as compared to mechanical mean in term of surfactant requirement and emulsion stability. The mixture of oils and water under US treatment formed smaller droplets and bigger dispersed phase volume, which provide better stability for the emulsion (Gaikwad and Pandit 2008). Generally, there are three parameters in the formation of nanoemulsion with US namely frequency, power output and exposure time (Cardoso-Ugarte et al. 2016). Abbas et al. (2013) reported that most of the nanoemulsion is produced with low frequency of US from 20 kHz to 24 kHz range due to the fact that lower frequency US leads to a more vigorous cavitation. Moreover,

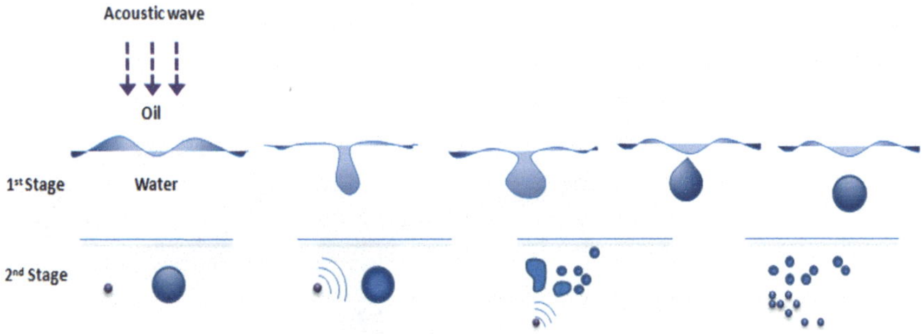

Fig. 6 US emulsification: droplet formation and break-up (adapted from Tang et al. 2012)

longer exposure time of the emulsion to US reduces the droplet size and forms a more stable emulsion (Leong et al. 2009). It was reported that nanoemulsion is an important food ingredient due to the diffusion and migration on the materials content in the food matrix. The advantage of nanoemulsions is their ability to improve the bioavailability of the encapsulated compounds because these have a high surface-volume ratio in the system (Dickinson 2003). Goindi and co-workers (2016) categories the application of nanoemulsion as food additive in 3 functions, namely the carrier of antimicrobial agents, antioxidant agents and nutraceutical ingredient.

4.2 *Preparation of Essential Oil-Based Nanoemulsions*

Essential oil is known as a natural preservative with antimicrobial properties in food products (Goindi et al. 2016). One of the limitations of essential oil is its hydrophobic characteristic that retarded its antimicrobial activity in many food products (Sugumar et al. 2015). As such, various studies were conducted on the application of essential oil in the form of nanoemulsion as natural antimicrobials and preservatives in food industry without compromising the food quality (Donsì and Ferrari 2016). For instance, Bhargava et al. (2015) reported that oregano oil nanoemulsions can effectively inhibit the foodborne bacteria such as Listeria monocytogenes, Salmonella typhimurium, Escherichia coli. A study was conducted on the orange essential oil in the nanoemulsion form to study its antimicrobial activity against the yeast known as Saccharomyces cerevisiae that responsible for the apple juice spoilage. The results showed that the presence of essential oil in nanoemulsion exhibit complete antimicrobial activity as compared to the control sample (Fig. 7). Another evaluation was conducted on the effect of lemon grass oil in nanoemulsions form against Salmonella typhimurium and Escherichia coli (Kim et al. 2013).

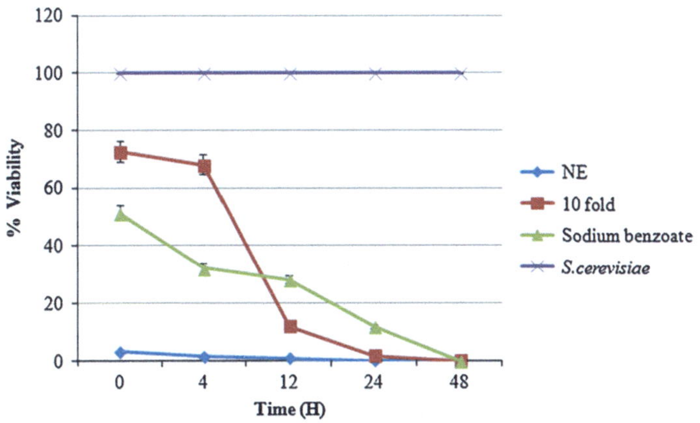

Fig. 7 Time kill analysis curve of orange oil nanoemulsion in real food system-apple juice (adapted from Sugumar et al. 2015)

By encapsulating the lemon grass oil nanoemulsions in plums, the findings revealed that the lemon grass oil-containing plums have longer shelf life and better stability as compared to the control plums due to the antimicrobial activity of the encapsulated nanoemulsions (Fig. 8a and b). Another trial conducted on the bread packed in films that contain nanoemulsions shows high antimicrobial activity with longer life span. The nanoemulsion produced from essential oils of clove bud and oregano was assimilated into edible films to form a protective layer against microorganism and hence improve the shelf life of the food products (Otoni et al. 2014).

4.3 Nanoencapsulation of Food/Nutraceutical Ingredients in Nanoemulsions

US irradiation promotes the nanoencapsulation of food additive and supplements in nanoemulsions to favor products with increased bioactive content (Chaudhry et al. 2008). Nanoencapsulation is a method of nanocarrier system where additives such as antioxidant, color and flavors are loaded into a food carrier such as nanoemulsions (Ezhilarasi et al. 2013). In general, food products contain flavors and colorants that are prone to oxidation that will degrade the food quality (Goindi et al. 2016). As such, nanoemulsification is introduced to encapsulate the functional food ingredient to preserve and improve its stability. For example, in hazelnut paste industry, nanoemulsion is used to improve the phenolic efficiency against lipid oxidation to enhance the product's antioxidant activity (Spigno et al. 2013). In another study by Yang et al. (2011), oil-in-water nanoemulsions is found to decrease the formation of

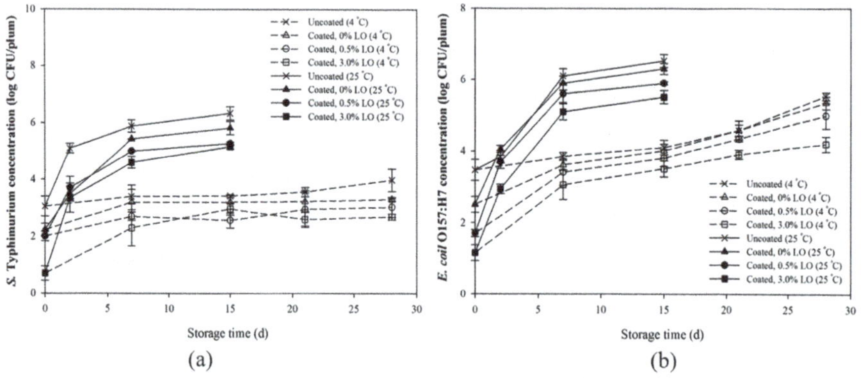

Fig. 8 Effect of lemon grass oil (LO) in nanoemulsions form against the growth of (**a**) Salmonella typhimurium and (**b**) Escherichia coli on plums during storage at 4 and 25 °C at different LO concentration (adapted from Kim et al. 2013)

undesired compounds while stabilizing the citral chemically. Their extended study on the citral nanoemulsion further demonstrated that assimilating the citral with antioxidants compounds such as beta-carotene resulted in enhanced stability of the citral against degradation and oxidation (Fig. 9).

Nutraceuticals ingredient is the derivative of food products that contain additional nutritional value that is beneficial to the consumer in term of health. Polyunsaturated fatty acids (PUFA) that is rich of omega-3 is known for benefits in preventing various diseases such as cardiovascular diseases and cancer. However, its application is limited due to its high oxidative degradation rate as regarded by the high numbers of weak double and triple bonds (Kris-Etherton et al. 2009). Fortunately, the encapsulation of the omega-3 rich PUFAs with nanoemulsions showed feasibility in its protection against oxidation. In Esquerdo and others' (2015) study on fish oil based omega-3 PUFAs encapsulated emulsions, the oxidation rate of PUFAs was reported to be significantly enhanced due to the shielding effects of emulsion that limited the contact of omega-3 PUFAs with foreign compound. Besides that omega-3, Resveratrol is another interesting natural polyphenol that possessed various health benefits such as anti-inflammatory, anti-carcinogenic, anti-obesity, protection to heart and nervous system. High amount of resveratrol are detected in various fruits and berries such as blueberry, grape and raspberry (Neves et al. 2013). However, the resveratrol is sensitive to ultraviolet light and prone to degradation and isomerization when it was exposed. As such, studies were conducted to encapsulate the resveratrol in the form of nanoemulsion to shelter it against ultraviolet light and the results are positive (Fig. 10). To date, nanoemulsion is known as promising nano-vehicles in the pharmaceutical industry. Therefore, it showed high potential in food industry as food supplements and other nanofoods with increased nutrient contents.

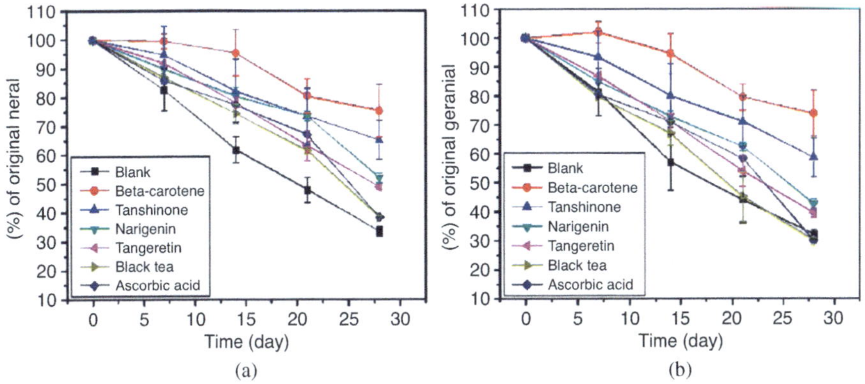

Fig. 9 Degradation of (**a**) neral and (**b**) geranial in emulsions with and without different antioxidants stored at 25 oC (adapted from Yang et al. 2011)

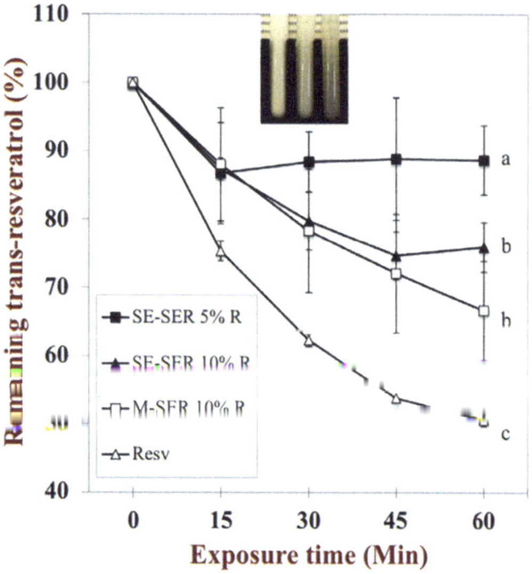

Fig. 10 Percentage of remaining trans-resveratrol after direct exposure to UV-light Different letters are significantly different ($P < 0.05$). From left to right, the photograph shows SE-SER 5% R, SE-SER 10% R and M-SER 10% R (adapted from Davidov-Pardo and McClements 2015)

5 Application of Cavitation in Nanofood Preservations

5.1 Spray Drying

Drying has been widely adopted since prehistorical times as a process for food preservation to ensure good quality by avoiding physiological changes during storage, to extend its shelf life and for the ease of transportation (Huang et al. 2017). Drying involves the removal of volatile liquid, which often known as moisture or water content from the food products via thermal treatment, pressurize condition or both. The preservation of food can be achieved when the water content is reduced to a level where it inhibits the growth of microorganisms and hinders the potential chemical reactions such as enzymatic reaction or lipid oxidation that will contribute to the deterioration of food products (Anandharamakrishnan 2017). Freeze drying, drum drying and spray drying are some of the most commonly used approaches in the formation of powder from liquid, emulsions or suspensions. In which, spray drying has attracted a lot of attention specifically in the drying of dairy products (Schuck et al. 2016).

Spray drying is a continuous process that produced dried powders from emulsion by evaporating the volatile liquid. This process is achieved by mixing a hot, flowing gas (usually 150 °C – 250 °C) with the emulsion which atomized into fine droplets (10–150 μm) within the drying chamber, causing the solvent to evaporate and form powder. The drying kinetics are influenced by the air-liquid interface area and the process typically occurs within a few seconds. This is the main benefits when compared with freeze drying which might take up to hours or days. In addition, spray drying also signifies lower specific energy cost as well as higher productivity which makes it an efficient dryer for continuous processing operations. (Whitby et al.

2017; Huang et al. 2017) Over the years, dairy products has served as one of the top quality and multifunctional protein source that is preferred in high demand by consumers. This has put researchers to work on new and innovative methods to increase the productivity of dairy products while conserving its nutritional value during the food processing. Several studies have shown that the extended storage time and temperature have adverse effects on the solubility of milk protein concentrate powders due to the physiochemical interactions between proteins, salts and sugars (Anema et al. 2006; Havea 2006). It is therefore essential in maintaining and modifying the stability and functional properties of milk proteins for the desired end-product characteristics such as the appearance, viscosity, taste and dietary requirements. For instance, protein modification can be done to decrease its viscosity during spray drying to greatly increase the solid levels, thus reducing overall energy cost and the operating cost (Schuck et al. 2016).

Besides that, the incorporation of cavitation technology with spray drying has emerged as one of the potential application in protein modification. The cavitation process generates millions of cavities (bubbles) under high pressure and when the pressure recovers, the cavities collapsed, resulting in a powerful, energy shockwave to the surrounding liquid. The exerted shockwave forces the surrounding liquid to mix, forming emulsion and generates high temperature to facilitate optimal heating via heat and mass transfer. In brief, the US cavitation generates different variations of pressure waves at liquid/gas interface and causes alternative compressions and expansions (sponge effects) that causes the liquid to flow out of the food and increases the evaporation rate during the food drying process. Moreover, the US waves also reduced the boundary layer between the liquid and hot air interface and thus increased the total mass transfer to achieve higher diffusion rate, which is essential as it determines the required process temperature, flow rate and pressure to accomplish better drying process. A brief illustration of the cavitation aided spray drying process is shown in Fig. 11. In fact, it has been proven that this technology is able to enhance the drying efficiency typically during spray drying that promote

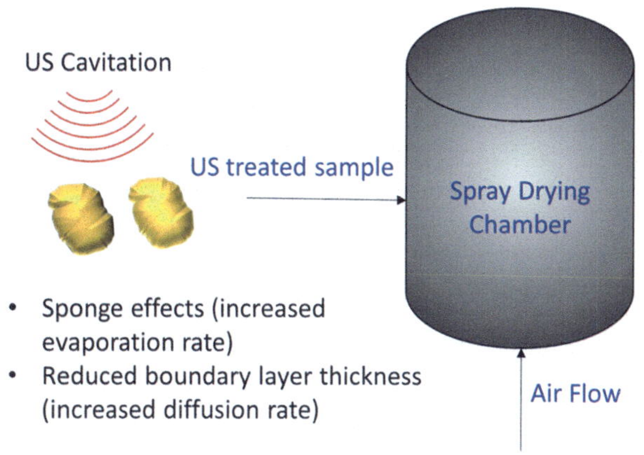

Fig. 11 Brief illustration on the US cavitation effects on nanofoods prior to spray drying process

emulsification, increase powder hydration and functionalization as well as enhancing the encapsulation of bioactive compounds (Li et al. 2018). One study has shown that the viscosity of whey protein concentrate 80 has been reduced by 20% which increased the solid levels of more than 15% when incorporating cavitation technology prior to spray drying (Oestergaard 2016).

Furthermore, de Barros Fernandes and his co-workers have demonstrated that the encapsulation of ginger essential oil using blends of gum Arabic, maltodextrin and inulin as wall materials assisted by cavitation technology with spray drying displayed better characteristics for wettability, encapsulation efficiency and hygroscopicity (de Barros Fernandes et al. 2016). Li and her co-workers have successfully reduced the viscosity of milk protein by 20% and 56% upon the application of cavitation rotor speed of 35 Hz and 50 Hz prior to spray drying (Li et al. 2018). Zisu and his co-workers (2010) have demonstrated that gels formed after batch sonication were denser and firmer as compared with unsonicated powders (see Fig. 12). Moreover, the viscosity and particle size of whey proteins retentate were lowered by more than 50% when sonicated with 50–100% amplitude as shown in Fig. 13. The

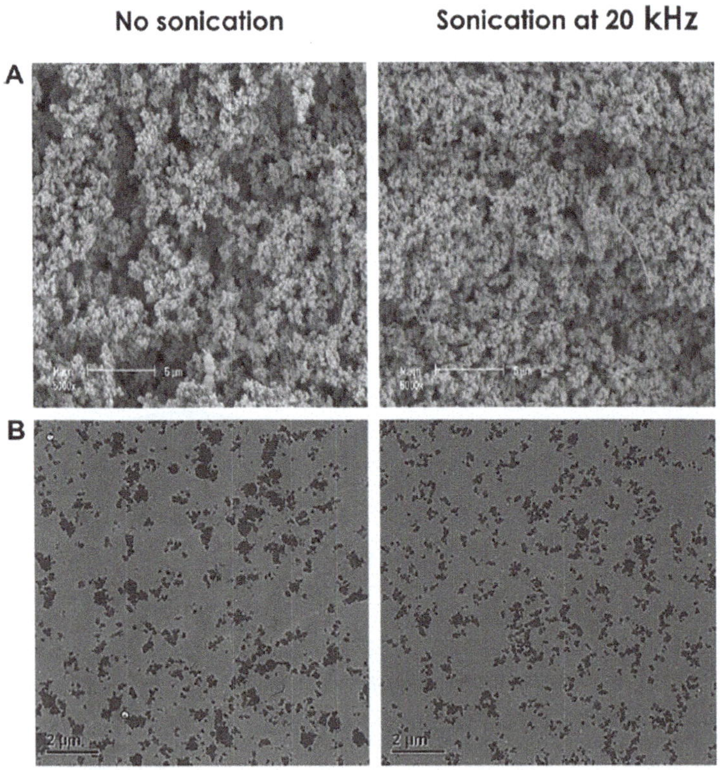

Fig. 12 Protein networks of whey protein retentate by (**a**) scanning electron microscopy (**b**) transmission electron microscopy with and without sonication at 20 kHz (Adapted from Zisu et al. 2010)

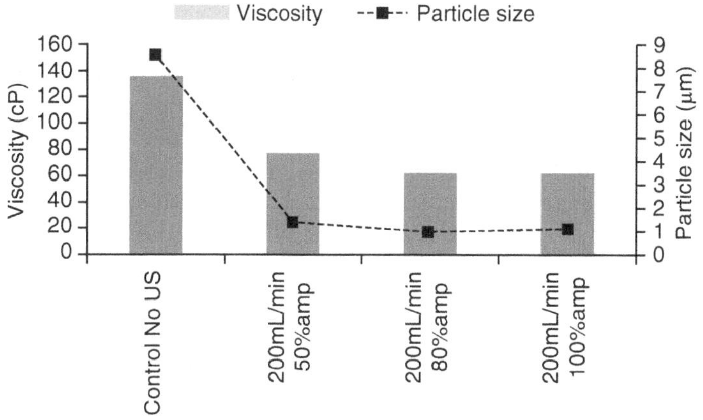

Fig. 13 Viscosity and particle size of whey protein retentate with and without sonication at 20 kHz before spray drying (Adapted from Zisu et al. 2010)

addition of cavitation approach prior to spray drying could improves drying efficiency and reduces the overall energy required.

By incorporating the cavitation technology into food drying process, it was found that the high-intensity air-borne US enhanced the food drying process by increasing the evaporation rate through either disturbing the liquid/gas interface via acoustic pressure waves or through reducing the boundary layer of the interfaces to ease the mass transfer rate. It is no doubt that the hot air present during spray drying could lead to the degradation of food quality especially when it is used in a long period of time. Therefore, the incorporation of cavitation technology can be one appropriate approach when dealing with heat sensitive materials (Yao 2016). The parameters of US cavitation assisted spray drying process are important in modifying and tailoring the dairy products according to the needs of end-products to ensure enhanced processing efficacy and to maintain the overall food quality. The incorporation of cavitation technology with spray drying showed numerous advantages over other types of drying methods and this mutual method can be further explored to develop additional useful approaches in handling other types of food and in a larger scale.

5.2 *Pasteurisation and Sterilisation*

Nanofood has become the new trend in food industry after the advancement of nanotechnology in other fields such as biomedical, information technology and manufacturing industries. It is related to the application, production and processing of materials with sizes of less than 1000 nm. At the size of less than 1 micron, the properties and characteristics of food materials would be different, and this provides new understanding about the properties of the food materials. Thus, better functional properties of new food system can be developed to improve the well-being and health of individuals. In addition, nanofood can also be produced by using various nanotechnology such as US emulsification, dry milling and high-pressure homogenization (Sanguansri and Augustin 2006). In this section, the potential application of US cavitation technology in pasteurization and sterilization of nanofood products will be addressed.

The traditional thermal processing method has been used to kill microorganism and to inactivate enzyme effectively. However, the nutrients lost, formation of undesirable flavours, colours and the decreased of overall food product quality were inevitable. Its effectiveness is highly dependent on the processing temperature and time, while these two factors are strongly associated with the drawbacks as mentioned above (Chemat et al. 2011). Therefore, a new processing method for pasteurization and sterilization of nanofood products has led to the development of US treatment due to the increased in demand by consumers in the recent years to replace the conventional thermal processing method.

Food deterioration is associated with the activity of microorganism and the action of enzyme in a food product. Microorganisms such as bacteria and fungi consumed nutrients by secreting enzymes including lipases and proteases to breakdown the lipid and proteins found within the food product. This renders the food decaying process, resulting in food spoilage that is undesirable for consumption. To date, the US treatment alone has been widely used by biologist or microbiologist for cell lysis and denaturation of enzyme but its applications are still new in food processing. By using sufficient high intensity US, the cell membranes can be disrupted through thinning of cell membranes by localized generated heat energy and the formation of cavitation bubbles (Butz and Tauscher 2002). The formation of cavitation bubbles undergoes irregular oscillations and induces implosion which generate high temperature and pressure that would disrupt the cellular structures and kill the cells. The mechanical shear force generated during the bubbles implosion as well has contributed sufficient physical energy to disrupt the cell membrane (Butz and Tauscher 2002). Once the cell membrane has been disrupted, water can be easily influx through osmotic pressure and caused swelling, which eventually lead to cell lysis. (see Fig. 14).

The physical effects of cavitation have been widely studied while the chemical effects of cavitation are less explored until recent years. The formation of free radicals during acoustic cavitation is also responsible for inactivation of microorganism and enzyme. Hydroxyl radicals (•OH) are produced during sonolysis of water molecules ($H_2O \rightarrow OH^- + H^+$) and these radicals are highly reactive against living cells which could cause oxidative stress (see Fig. 15). The cavitation intensity can be measured as the rate of peroxide generation ($H_2O + OH^- + H^+ \rightarrow H_2O_2 + H_2$) and this is one of the significant inactivation mechanism (O'Donnell et al. 2010). It is known that the concentration of hydroxyl radicals plays an important role in the activity of antioxidants in food while the degree of free radicals formation can be controlled via adjusting of the frequency of US. It was reported that by using lower

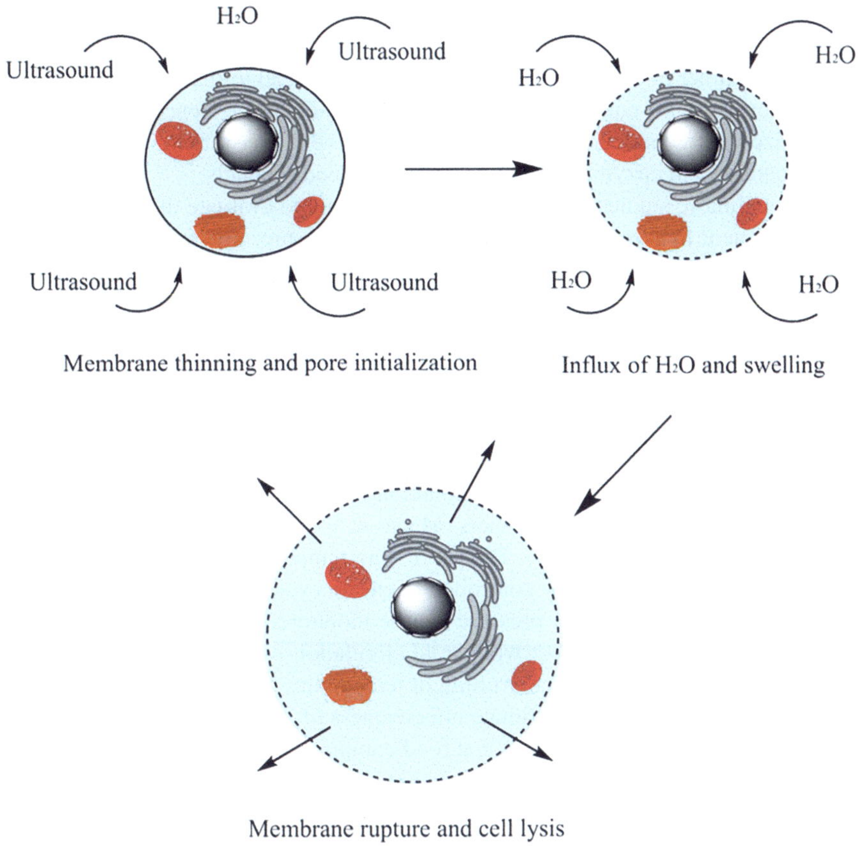

Fig. 14 Schematic diagram of cell lysis induced by ultrasound treatment

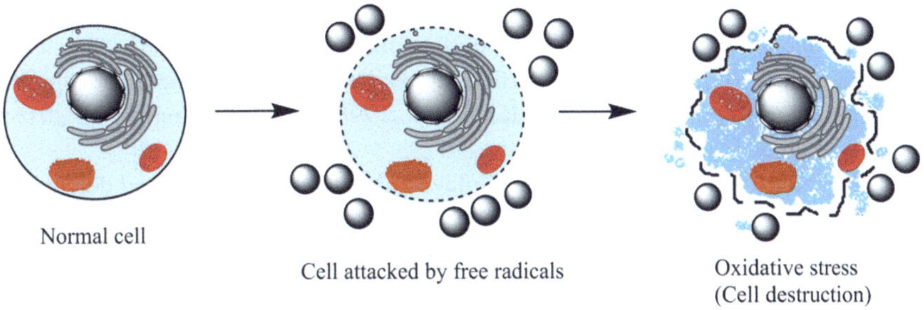

Fig. 15 Schematic diagram of cell destruction induced by free radicals

frequency of US treatment, the amounts of free radicals produced are lesser compared to high frequency US. This is due to the higher localized generated heat and the higher number of active cavitation bubble that can lead to the formation of free radicals during high frequency US (Soria and Villamiel 2010). Hence, it is essential to optimize the frequency and temperature to achieve higher rate in inactivation of microorganism and enzyme to better preserve the food products.

More and more studies have been conducted lately to evaluate the effectiveness of US treatment in combination of both heat and pressure. The critical processing factors in inactivation of microorganism and enzyme are the nature of ultrasonic waves, the processing time and temperature, the volume and composition of food product, as well as the chemical environment (pH). (Chemat et al. 2011) It is therefore important to analyse and optimise the time and temperature for US treatment to achieve greater food quality and to reduce the resultant damage. Application involving US with heat or pressure or also known as manosonication, MS (combination of US and pressure), thermosonication, TS (combination of US with heat) and manothermosonication, MTS (combination of US with pressure and heat) have been investigated by several researchers (Manas et al. 2000; Pagan et al. 1999; Raso et al. 1998; Ordonez et al. 1984). The processing temperature and pressure during US treatment must be precisely controlled to ensure even distribution on the resultant product. For instance, the pressure can be monitored by manometer that was placed in the treatment chamber while cold circulation water coil can be installed around the chamber for better controlling of temperature. Despite the regulation of temperature and pressure, the ultimate effectiveness of US treatment is still largely relying on the microorganism types and food composition.

The use of US alone in food industry is not so promising at the current moment but the use of US with pressure and heat has demonstrated synergistic effects in the inactivation of microorganism especially in bacteria. These combined US technologies (MS, TS, and MTS) required less energy intake, more cost-effective and more environmental friendly as compared with conventional thermal processing. Also, the overall quality of food product can be retained. Some example of successfully inhibited bacteria done by various researchers are *Listeria* monocytogenes, rod-shaped and gram-positive bacteria (Pagan et al. 1999; Gabriel 2015), *Escherichia coli*, rod-shaped and gram-negative bacteria (Brayman et al. 2017; Moody et al. 2014), *Salmonella enterica*, rod-shaped, flagellated and gram-negative bacteria (Alvarez et al. 2003; Luo and Oh 2016) and *Bacillus spp.*, rod-shaped, endospore-forming and gram-positive bacteria (Raso et al. 1998; Condon-Abanto et al. 2016; Jambrak et al. 2018).

Besides microorganism, enzyme also played an important role in food deterioration. The activity of enzyme will need to be reduced or inactivated to prevent defects on food product properties such as flavour, nutritional value, appearance and colour. Heat is the most common energy or process to silent the protein activities but this treatment always come with some drawbacks and it is not effective against high heat resistant protein. Thus, the alternative method for enzyme inactivation has led researcher to focus on US wave treatment in combination with heat or pressure, to improve the effectiveness in denaturing the proteins while greatly reducing the harmful effects to the food products. Generally, the inactivation of enzyme is achieved by depolymerizing the polymeric globular protein into polypeptide chains. Followed by the loss of molecular function, the prolong US treatment can further hydrolysed the polypeptide chains to break down into smaller subunit. (Chemat et al. 2011). The effectiveness of US treatment in inactivation of enzyme can be quantified by calculating the activation energy (Ea) from the Arrhenius equation as shown below:

$$k = A.\exp\left(-Ea / RT\right) \tag{1}$$

k is the rate of reaction and it can be used to calculate the total Ea by plotting log k against the inverse of the absolute temperature (T). To put in simple term, the activation energy is the required energy or the threshold for a reaction to proceed (Mason et al. 2003).

It is important to determine the factors that can affect the effectiveness on inactivation of enzyme using US treatment together with the heat or pressure. These factors included the rate of US waves, the temperature and time for the process as well as the nature of the enzyme. Enzymes are macromolecules biocatalyst which made up of different types of proteins in a 3-dimensional structure conforming by forces such as hydrogen bonding, van der Waals forces of attraction, electron static force and steric hindrance. All these mentioned molecular forces must be broken down or altered to decrease the enzyme activity and hence inactivate them (Chemat et al. 2011). Generally, environment factors such as temperature and pH have been widely used for this purpose as the enzyme's activity decreases remarkably outside its optimum temperature and pH. For instance, heat treatment is the most traditional way in enzyme inactivation but these US technologies (MS, TS, and MTS) have replaced the conventional heat treatment to counter the variety of enzymes available, different amino acid sequences and different molecular structures (O'Donnell et al. 2010). Studies were conducted using US treatment in inactivation of several enzymes such as peroxidases (POD) (Cao et al. 2018; Silva et al. 2015), polyphenol oxidase (PPO) (Cao et al. 2018; Silva et al. 2015; Zhou et al. 2016), lipoxygenase (LOX) (Lopez et al. 1994) and pectinmethylesterase (PME) (Raviyan et al. 2005).

Conventional thermal processing method may not be adequate in processing many types of fruits as it often results in the loss of flavour, appearance, and nutritional value of the products. In this regard, the emerge of US approach in nanofood technology has shown numerous benefits in preserving the quality food and at the same time, inactivate harmful microorganism and enzymes that may contribute towards food deterioration. The combination technique of US with both thermal and pressure treatment has been even proven to be having synergistic effects. It was found to be very effective against various types of microorganisms and several enzymes that are commonly found in nanofood product such as dairy products. Zhang and his co-workers have demonstrated by using low intensity ultrasound combined with heat (LIUH) pretreatment was efficient in deactivation of enzymes (POD, AAO and PPO) in daylilies compared to thermal pretreatment alone (see Fig . 16).

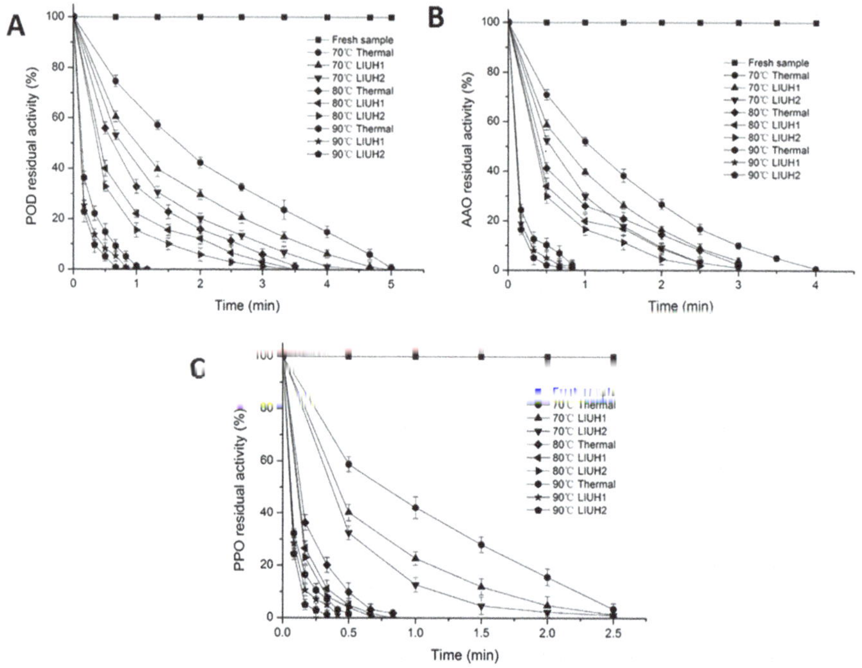

Fig. 16 Effects of thermal and LIUH on activities of (**A**) POD, (**B**) AAO, and (**C**) PPO from daylilies. LIUH: 0.2 W/cm^2; LIUH: 0.4 W/cm^2 (Adapted from Zhang et al. 2017)

In order to achieve even greater inactivation rate and to reduce the harmful effects on the food products, the US amplitude, temperature and pressure can be optimized further and studies are currently on-going to try to apply this technique on many others food product to extend their shelf-life while preserving the food quality at its best. No doubt all the treatment parameters can be optimized to obtain better outcome in the inactivation of microorganism and enzyme but the effectiveness are still largely dependent on the types and properties of the microorganism and enzyme.

6 Other Applications

Besides the application of US on emulsion, pasteurization and spray drying, there are others potential application in nanofood production. One of the potential application is the extraction of vegetable oil and phytonutrient from plants, where they can be further processed to form lipid nanoemulsions that aids in increasing the flavor, bioactive delivery efficiencies, shelf life, etc. (Juliano et al 2013; Medina-Torres et al 2017). US cavitation application, particularly the HPU produces micro-bubbles along with the high shear forces from cavitation. When the micro-bubbles collapse, it releases high amount of localized energy that cause water molecules to form free radicals. This phenomenon increases the chemical reaction when contacted to the matter and enhanced the mass transfer rate (Takahashi et al. 2007). In oil extraction, the application of US assists in the disruption of tissue structures such as membrane that encapsulate the oil, which allows the increment in mass transfer rate of trapped oil to the surface and thus increasing oil extraction rate (Amirante and Paduano 2016; Clodoveo et al. 2014; Vilkhu et al. 2008). The mechanism of US cavitation on the cell disruption is shown in Fig. 17.

In fact, recent study of the application of US in olive oil extraction process had shown that US-assisted extraction improvised the extraction efficiency and the production throughput due the mechanical and thermal effect from US (Jiménez et al. 2007; Clodoveo et al. 2017; Cravotto et al. 2008). It should be noted that the employment of US application not only improved the oil yield during extraction by 1%, but also resulted in oil that contained higher content of antioxidant such as tocopherols and tocotrienols (Jiménez et al. 2007; Clodoveo et al. 2013; Hashemia et al. 2015). As an example, Amirante et al. (2017) reported that the oil extraction with US application increases the oil yield and polyphenol by 5% and 12% respectively. Polyphenol is a type of antioxidant where it delays the oxidation process, which thus increases the shelf life of the extracted virgin oil (Clodoveo et al. 2013). With the increase in the extraction of the nanonutrients such as vitamin and antioxidant in oil, one can envision that the use of cavitation technologies in oil extraction can greatly improve the shelf life and quality of the virgin oil. In another example on chia seed's oil, De Mello and co-workers (2017) reported a higher oil yield without alteration of the oil fatty acid composition upon utilizing US for extraction process as compared to conventional solvent extraction method. Besides that, Benito-Román et al. (2013) reported that application of US boosted the yield of β-glucans extraction from barley with shorter processing time and energy utilization. The US assisted extraction took just three minutes and consumed 170 kj/L of energy as compared to traditional extraction that took three hours under heat at 55 °C with stirring at 1000 rpm (Benito-Román et al. 2013). Research was also conducted on soybean where the slurry was treated with US with different temperature and duration to improve the protein yield extraction (Fahmi et al. 2014). The work revealed that the protein content extraction rate was enhanced by 6.3% under US treatment with the frequency of 35 kHz (Fahmi et al. 2014). From the above examples, it is clearly shown that the cavitation application able to enhance extraction yield of nanonutrient compounds and oil from plant. In addition, US assisted extraction

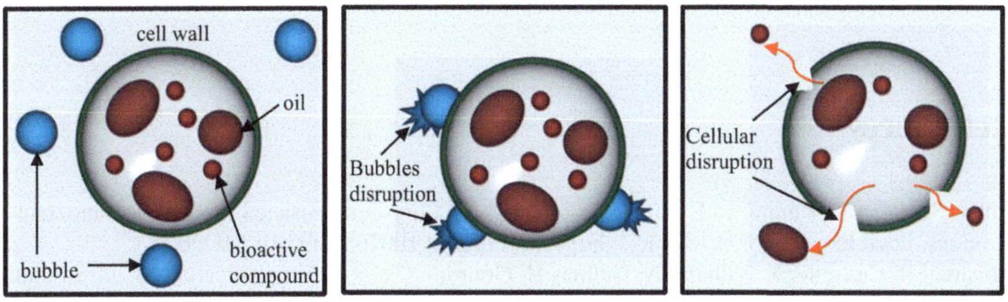

Fig. 17 Mechanism of US cavitation in cell disruption during the oil and bioactive compounds extractions

treatment in generally reduces extraction time and consume less energy. This showed high potential in producing down-stream food products such as nanoemulsions that can be utilized to improve the bioavailability stability of the extracted oil and nutrients.

7 Conclusion

In summary, ultrasonic cavitation technology plays a prominent role in food-related applications. The high pressure shear of ultrasonication can functioned well in breaking the cell wall of plant, thus improving the extraction of the lipid and other nutrients in the plant. Besides that, US irradiation with high power also supplements the formation and nanoencapsulation of nutritious lipid nanoemulsion. The main mechanisms should be attributed to the high pressure sound waves that first disperse the plant lipids into another immiscible aqueous liquids to create a big oil droplet, and then further breaking it down to smaller size. This mechanism is also commonly known as emulsification. Besides emulsification, the high pressure jets from US cavitation as well as the continuous compression and extension of the sound waves can also aids in reducing the boundary thickness of food products to improve the water evaporation rate in the food, which thus generate a greener route for spray drying of nanofood. The reactive radicals and heat produced through US cavitation aided well in the preservation of food by deactivating the microorganisms and enzyme activities via cell lysis and enzyme depolymerization (pasteurization). Combining the US cavitation with heat and pressure showed superior effectiveness in pasteurization process as compared to conventional methods. However, in depth analysis and optimization of processing parameters are required for each microorganism types and food composition to ensure the maximum preservation effects as well as minimum destructive damage to the foods. Besides the mentioned applications, US cavitation also found its role in enhancing the extraction yield of oil and nutrient from various sources, where the product can be further processed using cavitation technology to further processed into nanofoods. Although US cavitation has been extensively studied for nanofoods processing, nevertheless, the future of US in the development of nanofood deserves substantial attention since there are still tremendous possibility for the US cavitation in applications involving the quality control of nanofoods.

References

Abbas S, Hayat K, Karangwa E (2013) An overview of ultrasound-assisted food grade nanoemulsions. Food Eng Rev 5(3):139–157. https://doi.org/10.1007/s12393-013-9066-3

Abismail B, Canselier J, Wilhelm A, Delmas H, Gourdon C (1999) Emulsification by ultrasound: Drop size distribution and stability. Ultrason Sonochem 6(1–2):75–83. https://doi.org/10.1016/S1350-4177(98)00027-3

Alfadul SM, Elneshwy AA (2010) Use of nanotechnology in food processing, packaging, and safety-review. Afr J Food Agric Nutr Dev 10(6):2719–2739. https://doi.org/10.4314/ajfand.v10i6.58068

Alvarez I, Manas P, Sala FJ, Condon S (2003) Inactivation of Salmonella enterica serovar enteritidis by ultrasonic waves under pressure at different water activities. Appl Environ Microbiol 69(1):668–672. https://doi.org/10.1128/AEM.69.1.668-672.2003

Amirante R, Paduano A (2016) Ultrasound in Olive Oil Extraction. In: Boskou D, Clodoveo ML, editors. Prod. from Olive Tree. 2016th ed., InTech; . https://doi.org/10.5772/64765

Amirante R, Distaso E, Tamburrano P, Paduano A, Pettinicchio D, Clodoveo ML (2017) Acoustic cavitation by means ultrasounds in the extra virgin olive oil extraction process. Energy Procedia 126:82–90. https://doi.org/10.1016/j.egypro.2017.08.065

Anandharamakrishnan C (2017) Handbook of drying for dairy products. Wiley, Chichester/Hoboken

Anema SG, Pinder DN, Hunter RJ, Hemar Y (2006) Effects of storage temperature on the solubility of milk protein concentrate (MPC85). Food Hydrocoll 20:386–393. https://doi.org/10.1016/j.foodhyd.2005.03.015

Avison JH (1989) Sound waves. In: Avison JH (ed) The world of physics, 2nd edn. Thomas Nelson & Sons Ltd, Cheltenham, pp 450–470

Benito-Román Ó, Alonso E, Cocero MJ (2013) Ultrasound-assisted extraction of β-glucans from barley. LWT–Food Science and Technology 50(1):57–63. https://doi.org/10.1016/j.lwt.2012.07.006

Bhargava K, Conti DS, da Rocha SRP, Zhang Y (2015) Application of an oregano oil nanoemulsion to the control of foodborne bacteria on fresh lettuce. Food Microbiol 47:69–73. https://doi.org/10.1016/j.fm.2014.11.007

Bilbao-Sáinz C, Avena-Bustillos RJ, Wood D, Williams T, McHugh T (2010) Nanoemulsions prepared by a low-energy emulsification method applied to edible films. J Agric Food Chem 58(22):11932–11938. https://doi.org/10.1021/jf102341r

Brayman AA, MacConaghy BE, Wang Y-N, Chan KT, Monsky WL, McCleny AJ, Matula TJ (2017) Inactivation of planktonic Escherichia coli by focused 2-MHz ultrasound. Ultrasound Med Biol 43(7):1476–1485. https://doi.org/10.1016/j.ultrasmedbio.2017.03.009

Butz P, Tauscher B (2002) Emerging technologies: chemical aspects. Food Res Int 35(2–3):279–284. https://doi.org/10.1016/S0963-9969(01)00197-1

Cao X, Cai C, Wang Y, Zheng X (2018) The inactivation kinetics of polyphenol oxidase and peroxidase in bayberry juice during thermal and ultrasound treatments. Innov Food Sci Emerg Technol 45:169–178. https://doi.org/10.1016/j.ifset.2017.09.018

Cardoso-Ugarte GA, López-Malo A, Jiménez-Munguía MT (2016) Application of nanoemulsion technology for encapsulation and release of lipophilic bioactive compounds in food. In: Grumezescu AM (ed) Emulsions, vol 3. Elsevier, USA, pp 227–255. https://doi.org/10.1016/B978-0-12-804306-6.00007-6

Chanamai R, McClements JD (2000) Dependence of creaming and rheology of monodisperse oil-in-water emulsions on droplet size and concentration. Colloids Surf A Physicochem Eng Asp 172:79–86. https://doi.org/10.1016/S0927-7757(00)00551-3

Chaudhry Q, Scotter M, Blackburn J, Bryony R, Boxall A, Castle L, Aitken R, Watkins R (2008) Applications and implications of nanotechnologies for the food sector. Food Addit. Contam. Part A Chem. Anal. Control Expo. Risk Assess 25(3):241–258. https://doi.org/10.1080/02652030701744538

Chellaram C, Murugaboopathi G, John AA, Sivakumar R, Ganesan S, Krithika S, Priya G (2014) Significance of nanotechnology in food industry. APCBEE Procedia 8:109–113. https://doi.org/10.1016/j.apcbee.2014.03.010

Chemat F, Huma Z, Khan M (2011) Application of ultrasound in food technology: processing, preservation and extraction. Ultrasonic Sonochemistry 18:813–835. https://doi.org/10.1016/j.ultsonch.2010.11.023

Clodoveo ML, Durante V, La Notte D (2013) Working towards the development of innovative ultrasound equipment for the extraction of virgin olive oil. Ultrason Sonochem 20:1261–1270. https://doi.org/10.1016/j.ultsonch.2013.02.001

Clodoveo ML, Hbaieb RH, Kotti F, Mugnozza GS, Gargouri M (2014) Mechanical strategies to increase nutritional and sensory quality of virgin olive oil by modulating the endogenous enzyme activities. Compr Rev Food Sci Food Saf 13:135–154. https://doi.org/10.1111/1541-4337.12054

Clodoveo ML, Moramarco V, Paduano A, Sacchi R, Di Palmo T, Crupi P, Corbo F, Pesce V, Distaso E, Tamburrano P, Amirante R (2017) Engineering design and prototype development of a fullscale ultrasound system for virgin olive oil by means of numerical and experimental analysis. Ultrason Sonochem 37:169–181. https://doi.org/10.1016/j.ultsonch.2017.01.004

Condon-Abanto S, Arroyo C, Alvarez I, Condon S, Lyng JG (2016) Application of ultrasound in combination with heat and pressure for the inactivation of spore forming bacteria isolated from edible crab (cancer pagurus). Int J Food Microbiol 223:9–16. https://doi.org/10.1016/j.ijfoodmicro.2016.02.001

Cravotto G, Boffa L, Mantegna S, Perego P, Avogadro M, Cintas P (2008) Improved extraction of vegetable oils under high-intensity ultrasound and/or microwaves. Ultrason Sonochem 15:898–902. https://doi.org/10.1016/j.ultsonch.2007.10.009

Czechowska-Biskup R, Rokita B, Lotfy S, Ulanski P, Rosiak JM (2005) Degradation of chitosan and starch by 360-kHz ultrasound. Carbohydr Polym 60:175–184. https://doi.org/10.1016/j.carbpol.2004.12.001

Davidov-Pardo G, McClements DJ (2015) Nutraceutical delivery systems: resveratrol encapsulation in grape seed oil nanoemulsions formed by spontaneous emulsification. Food Chem 167:205–212. https://doi.org/10.1016/j.foodchem.2014.06.082

de Barros Fernandes RV, Borges SV, Silva EK, da Silva YF, de Souza HJB, do Carmo EL, de Oliveira CR, Yoshida MI, Botrel DA (2016) Study of ultrasound-assisted emulsions on microencapsulation of ginger essential oil by spray drying. Ind Crop Prod 94:413–423. https://doi.org/10.1016/j.indcrop.2016.09.010

De Mello BT, Santos Garcia VA, Silva C (2017) Ultrasound-Assisted Extraction of Oil from Chia (Salvia hispânica L.) Seeds: Optimization Extraction and Fatty Acid Profile. J. Food Process Eng. 40(1):e12298. https://doi.org/10.1111/jfpe.12298

Dickinson E (2003) Hydrocolloids at interfaces and the influence on the properties of dispersed systems. Food Hydrocoll 17(1):25–39. https://doi.org/10.1016/S0268-005X(01)00120-5

Dingman J (2008) Nanotechnology: its impact on food safety. J Environ Health 70(6):47–50

Donsì F, Ferrari G (2016) Essential oil nanoemulsions as antimicrobial agents in food. J Biotechnol 233:106–120. https://doi.org/10.1016/j.jbiotec.2016.07.005

Esquerdo VM, Dotto GL, Pinto LAA (2015) Preparation of nanoemulsions containing unsaturated fatty acid concentrate-chitosan capsules. J Colloid Interface Sci 445:137–142. https://doi.org/10.1016/j.jcis.2014.12.094

Evelyn, Silva FVM (2015) Inactivation of Byssochlamys nivea ascospores in strawberry puree by high pressure, power ultrasound and thermal processing. Int J Food Microbiol 214:129–136. https://doi.org/10.1016/j.ijfoodmicro.2015.07.031

Ezhilarasi PN, Karthik P, Chhanwal N, Anandharamakrishnan C (2013) Nanoencapsulation techniques for food bioactive components: a review. Food Bioprocess Technol 6(3):628–647. https://doi.org/10.1007/s11947-012-0944-0

Fahmi R, Khodaiyan F, Pourahmad R, Emam-Djomeh Z (2014) Effect of ultrasound assisted extraction upon the Genistin and Daidzin contents of resultant soymilk. J Food Sci Technol 51(10):2857–2861. https://doi.org/10.1007/s13197-012-0744-6

Gabriel AA (2015) Inactivation of Listeria monocytogenes in milk by multifrequency power ultrasound. J. Food Process. Preserv 39:846–853. https://doi.org/10.1111/jfpp.12295

Gaikwad SG, Pandit AB (2008) Ultrasound emulsification: effect of ultrasonic and physicochemical properties on dispersed phase volume and droplet size. Ultrason Sonochem 15(4):554–563. https://doi.org/10.1016/j.ultsonch.2007.06.011

Goindi S†, Kaur A, Kaur R, Kalra A, Chauhan P (2016) Nanoemulsions: An Emerging Technology In The Food Industry. In: Grumezescu AM (ed) Emulsions, 1st edn. Academic, Amsterdam, pp 651–688. https://doi.org/10.1016/B978-0-12-804306-6.00019-2

Hashemia SMB, Michiels J, Yousefabad SHA, Hosseini M (2015) Kolkhoung (Pistacia khinjuk) kernel oil quality is affected by different parameters in pulsed ultrasound-assisted solvent extraction. Ind Crop Prod 70:28–33. https://doi.org/10.1016/j.indcrop.2015.03.023

Havea P (2006) Protein interactions in milk protein concentrate powders. Int Dairy J 16:415–422. https://doi.org/10.1016/j.idairyj.2005.06.005

Huang S, Vignolles ML, Chen XD, Loir YL, Jan G, Schuck P, Jeantet R (2017) Spray drying of probiotics and other food-grade bacteria: A review. Trends Food Sci Technol 63:1–17. https://doi.org/10.1016/j.tifs.2017.02.007

Jambrak AR, Simunek M, Evacic S, Markoc K, Smoljanic G, Frece J (2018) Influence of high power ultrasound on selected moulds, yeasts and alicyclobacillus acidoterrestris in apple, cranberry and blueberry juice and nectar. Ultrasonics 83:3–17. https://doi.org/10.1016/j.ultras.2017.02.011

Jiménez A, Beltrán G, Uceda M (2007) High-power ultrasound in olive paste pretreatment. Effect on process yield and virgin olive oil characteristics. Ultrason Sonochem 14:725–731. https://doi.org/10.1016/j.ultsonch.2006.12.006

Juliano P, Swiergon P, Lee KH, Gee PT, Clarke PT, Augustin MA (2013) Effects of pilot plant-scale ultrasound on palm oil separation and oil quality. J Am Oil Chem Soc 90:1253–1260. http://doi.org/10.1007/s11746-013-2259-3

Kim IH, Lee H, Kim JE, Song KB, Lee YS, Chung DS, Min SC (2013) Plum coatings of lemongrass oil-incorporating carnauba wax-based nanoemulsion. J Food Sci 78:E1551–E1559. https://doi.org/10.1111/1750-3841.12244

Kris-Etherton PM, Grieger JA, Etherton TD (2009) Dietary reference intakes for DHA and EPA. Prostaglandins Leukot Essent Fatty Acids 81(2–3):99–104. https://doi.org/10.1016/j.plefa.2009.05.011

Leong TSH, Wooster TJ, Kentish SE, Ashokkumar M (2009) Minimising oil droplet size using ultrasonic emulsification. Ultrason Sonochem 16(6):721–727. https://doi.org/10.1016/j.ultsonch.2009.02.008

Li MK, Fogler HS (1978a) Acoustic emulsification. Part 1. The instability of the oil-water interface to form the initial droplets. J Fluid Mech 88(3):499–511. https://doi.org/10.1017/S0022112078002232

Li MK, Fogler HS (1978b) Acoustic emulsification. Part 2: Break-up of the larger primary oil droplets in a water medium. J Fluid Mech 88(3):513–528. https://doi.org/10.1017/S0022112078002244

Li K, Woo MW, Patel H, Metzger L, Selomulya C (2018) Improvement of rheological and functional properties of milk protein concentrate by hydrodynamic cavitation. J Food Eng 221:106–113. https://doi.org/10.1016/j.jfoodeng.2017.10.005

Lopez P, Sala FJ, de la Fuente JL, Condon S, Raso J, Burgos J (1994) Inactivation of peroxidase, lipoxygenase, and polyphenol oxidase by manothermosonication. J Agric Food Chem 42(2):252–256. https://doi.org/10.1021/jf00038a005

Luo K, Oh D-H (2016) Inactivation kinetics of Listeria monocytogenes and Salmonella enterica serovar Typhimurium on fresh-cut bell pepper treated with slightly acidic electrolyzed water combined with ultrasound and mild heat. Food Microbiol 53(Part B):165–171. http://doi.org/10.1016/j.fm.2015.09.014

Low LE, Wong SK, Tang SY, Chew CL, de Silva HA, Lee JMV, Hoo CH, Kenrick K (2019) Production of highly uniform Pickering emulsions by novel high-intensity ultrasonic tubular reactor (HUTR). Ultrason Sonochem 54:121–128. http://doi.org/10.1016/j.ultsonch.2019.02.008

Majid I, Nayik GA, Nanda V (2015) Ultrasonication and food technology: A review. Cogent Food Agric 1(1071022):1–11. https://doi.org/10.1080/23311932.2015.1071022

Manas P, Pagan R, Raso J, Sala FJ, Condon S (2000) Inactivation of Salmonella typhimurium, and Salmonella Senftenberg by ultrasonic waves under pressure. J Food Prot 63(4):451–456. PMID:10772209

Manickam S, Tang SY, Tan KW (2014) Cavitation technology – a greener processing technique for the generation of pharmaceutical nanoemulsions. Ultrason Sonochem 21:2069–2083. https://doi.org/10.1016/j.ultsonch.2014.03.025

Martirosyan A, Schneider Y (2014) Engineered nanomaterials in food: Implications for food safety and consumer health. Int J Environ Res Public Health 11:5720–5750. https://doi.org/10.3990/ijerph110605720

Mason TJ, Paniwnyk L, Chemat F. Ultrasound as a preservation technology. In: Zeuthen P, Bøgh-Sørensen L, editor. Food Preservation Techniques. Boca Raton, CRC Press, 2003. p. 303–337

Mason TJ, Chemat F, Vinatoru M (2011) The extraction of natural products using ultrasound or microwaves. Curr Org Chem 15:237–247. https://doi.org/10.2174/138527211793979871

Medina-Torres N, Ayora-Talavera T, Espinosa-Andrews H, Sánchez-Contreras A, Pacheco N (2017) Ultrasound assisted extraction for the recovery of phenolic compounds from vegetable sources. Agronomy 7:47. http://doi.org/10.3390/agronomy7030047

Mishra P, Singh A, Mishra RK, Ong S, Prasad SM (2017) Application of nanotechnology to enhance the nutrient quality of food crops and agricultural production. In: Ahmad P, Tripathi DK, Sharma S, Chauhan D, Dubey NK (eds) Nanomaterials in plants, algae, and microorganisms, 1st edn. Elsevier, USA, pp 143–154. https://doi.org/10.1016/B978-0-12-811487-2.00020-7

Moody A, Marx G, Swanson BG, Aguirre DB (2014) A comprehensive study on the inactivation of Escherichia coli under nonthermal technologies: high hydrostatic pressure, pulsed electric fields and ultrasound. Food Control 37:305–314. https://doi.org/10.1016/j.foodcont.2013.09.052

Neves AR, Lucio M, Martins S, Lima JLC, Reis S (2013) Novel resveratrol nanodelivery systems based on lipid nanoparticles to enhance its oral bioavailability. Int J Nanomedicine 8:177–187. https://doi.org/10.2147/IJN.S37840

Nithila SD, Anandkumar B, Vanithakumari SC, George RP, Mudali UK, Dayal RK (2014) Studies to control biofilm formation by coupling ultrasonication of natural waters and anodization of titanium. Ultrason Sonochem 21:189–199. https://doi.org/10.1016/j.ultsonch.2013.06.010

O'Donnell CP, Tiwari BK, Bourke P, Cullen PJ (2010) Effect of ultrasonic processing on food enzymes of industrial importance. Trends Food Sci Technol 21:358–367. https://doi.org/10.1016/j.tifs.2010.04.007

Oestergaard B (2016) Hydrodynamic cavitation helps dairy powder processing. SPX Flow Technology, USA

Ordonez JA, Sanz B, Hernandez PE, Lopez-Lorenzo P (1984) A note on the effect of combined ultrasonic and heat treatments on the survival of thermoduric streptococci. J Appl Bacteriol 54:175–177. https://doi.org/10.1111/j.1365-2672.1984.tb04711.x

Otoni CG, Pontes SFO, Medeiros EAA, Soares NDFF (2014) Edible films from methylcellulose and nanoemulsions of clove bud (Syzygium aromaticum) and oregano (Origanum vulgare) essential oils as shelf life extenders for sliced bread. J Agric Food Chem 62:5214–5219. https://doi.org/10.1021/jf501055f

Pagan R, Manas P, Alvarez I, Condon S (1999) Resistance of Listeria monocytogenes to ultrasonic waves under pressure at sublethal (manosonication) and lethal (manothermosonication) temperatures. J.Food Microbiol 16:139–148. https://doi.org/10.1006/fmic.1998.0231

Pathakoti K, Manubolu M, Hwang H (2017) Nanostructures: Current uses and future applications in food science. J Food Drug Anal 25:245–253. https://doi.org/10.1016/j.jfda.2017.02.004

Piyasena P, Mohareb E, McKellar RC (2003) Inactivation of microbes using ultrasound: A review. Int J Food Microbiol 87(3):207–216. https://doi.org/10.1016/S0168-1605(03)00075-8

Pray L, Yaktine A (2009) Nanotechnology in Food Products: Workshop Summary. Institute of Medicine of the National Academics, Washington

Raso J, Palo A, Pagan R, Condon S (1998) In activation of Bacillus subtilis spores by combining ultrasonic waves under pressure and mild heat treatment. J Appl Microbiol 85:849–854. https://doi.org/10.1046/j.1365-2672.1998.00593.x

Raviyan P, Zhang Z, Feng H (2005) Ultrasonication for tomato pectionmethylesterase inactivation: effect of cavitation intensity and temperature on inactivation. J Food Eng 70(2):189–196. https://doi.org/10.1016/j.jfoodeng.2004.09.028

Sanguansri P, Augustin MA (2006) Nanoscale materials development a food industry perspective. Trends Food Sci Technol 17:547–556. https://doi.org/10.1016/j.tifs.2006.04.010
Schuck P, Jeantet R, Bhandari B, Chen XD, Perrone IT, Carvalho AF (2016) Recent advances in spray drying relevant to the dairy industry: A comprehensive critical review. Dry Technol 34(15):1773–1790. https://doi.org/10.1080/07373937.2016.1233114
Sekhon BS (2010) Food nanotechnology – an overview. Nanotechnol Sci Appl 2010(3):1–15. https://doi.org/10.2147/nsa.s8677
Silva H, Cerqueira M, Vicente A (2011) Nanoemulsions for food applications: Development and characterization. Food Bioprocess Technol 5(3):854–867. https://doi.org/10.1007/s11947-011-0683-7
Silva LCA, Almeida PS, Rodrigues S, Fernandes FAN (2015) Inactivation of Polyphenoloxidase and peroxidase in apple cubes and in apple juice subjected to high intensity power ultrasound processing. J Food Process Preserv 39(6):1745–4549. https://doi.org/10.1111/jfpp.12451
Soria AC, Villamiel M (2010) Effect of ultrasound on the technological properties and bioactivity of food: A review. Trends Food Sci Technol 21(7):323–331. https://doi.org/10.1016/j.tifs.2010.04.003
Spigno G, Donsì F, Amendola D, Sessa M, Ferrari G, De Faveri DM (2013) Nanoencapsulation systems to improve solubility and antioxidant efficiency of a grape marc extract into hazelnut paste. J Food Eng 114(2):207–214. https://doi.org/10.1016/j.jfoodeng.2012.08.014
Sugumar S, Singh S, Mukherjee A, Chandrasekaran N (2015) Nanoemulsion of orange oil with non ionic surfactant produced emulsion using ultrasonication technique: Evaluating against food spoilage yeast. Appl Nanosci 6:113–120. https://doi.org/10.1007/s13204-015-0412-z
Suslick KS (1989) The chemical effects of ultrasound. Sci Am 260(2):80–87
Suslick KS (1990) Sonochemistry. Science 247(4949):1439–1445. https://doi.org/10.1126/science.247.4949.1439
Suslick KS, Price GJ (1999) Applications of ultrasound to materials chemistry. Annu Rev Mater Sci 29(1):295–326. https://doi.org/10.1146/annurev.matsci.29.1.295
Takahashi M, Chiba K, Li P (2007) Free-radical generation from collapsing microbubbles in the absence of a dynamic stimulus. J Phys Chem B 111(6):1343–1347. https://doi.org/10.1021/jp0669254
Tang SY, Tan KW, Manickam S (2012) Ultrasound cavitation as a green processing technique in the design and manufacture of pharmaceutical Nanoemulsions in drug delivery system. In: Sanghi R, Singh V (eds) Green chemistry and environmental remediation, 1st edn. Scrivener Publishing, USA, pp 153–208. https://doi.org/10.1016/B978-0-12-804306-6.00019-2
Thompson LH, Doraiswamy LK (1999) Sonochemistry: Science and engineering. Ind Eng Chem Res 38(4):1215–1249. https://doi.org/10.1021/ie9804172
Tsung J (2011) History of Ultrasound and Technological Advances. Available: http://www.wcume.org/wp-content/uploads/2011/05/Tsung.pdf. Last accessed 26th April 2018
Vilkhu K, Mawson R, Simons L, Bates D (2008) Applications and opportunities for ultrasound assisted extraction in the food industry - a review. Innovative Food Sci Emerg Technol 9(2):161–169. https://doi.org/10.1016/j.ifset.2007.04.014
Watson JD, Crick FH (1953) Genetical implications of the structure of deoxyribonucleric acid. Nature 171(4361):964–967. https://doi.org/10.1016/j.apcbee.2014.03.010
Whitby CP, Scarborough H, Ngothai Y (2017) Drying oil-in-water Pickering emulsions to make redispersible powders. Adv Powder Technol 28:2940–2946. https://doi.org/10.1016/j.apt.2017.09.001
Yang X, Tian H, Ho CT, Huang Q (2011) Inhibition of citral degradation by oilin-water nanoemulsions combined with antioxidants. J Agric Food Chem 59:6113–6119. https://doi.org/10.1021/jf2012375
Yao Y (2016) Enhancement of mass transfer by ultrasound: application to adsorbent regeneration and food drying/dehydration. Ultrason Sonochem 31:512–531. https://doi.org/10.1016/j.ultsonch.2016.01.039
Yusaf T, Al-Juboori RA (2014) Alternative methods of microorganism disruption for agricultural applications. Appl Energy 114:909–923. https://doi.org/10.1016/j.apenergy.2013.08.085

Zhang Z, Niu L, Li D, Liu C, Ma R, Song J, Zhao J (2017) Low intensity ultrasound as a pretreatment to drying of daylilies: Impact on enzyme inactivation, color changes and nutrition quality parameters. Ultrason Sonochem 36:50–58. https://doi.org/10.1016/j.ultsonch.2016.11.007
Zhou L, Liu W, Xiong Z, Zou L, Liu J, Zhong J, Chen J (2016) Effect of ultrasound combined with malic acid on the activity and conformation of mushroom (Agaricus bisporus) polyphenoloxidase. Enzym Microb Technol 90:61–68. https://doi.org/10.1016/j.enzmictec.2016.05.001
Zisu B, Bhaskaracharya R, Kentish S, Ashokkumar M (2010) Ultrasonic processing of dairy systems in large scale reactors. Ultrason Sonochem 17:1075–1081. https://doi.org/10.1016/j.ultsonch.2009.10.014

Microfluidization in Nano-Food Engineering

Ilse Monroy-Rodríguez, Amor Monroy-Villagrana, Maribel Cornejo-Mazón, Santiago García-Pinilla, Humberto Hernández-Sánchez, and Gustavo F. Gutiérrez-López

1 Introduction

A wide variety of nano-food related products have been designed in the last few years which have been prepared by using relatively new techniques and operations. One of them is micro-channeling of fluids by using high pressure homogenizers or microfluidizers. Their design is based on the collision of fluids at very high velocities being conducted through micro-channels into interaction chambers (IX). Design and operation of these pieces of equipment are complex issues and many aspects of them are still under investigation given the presence of turbulent flows that change direction and collide in the IX giving place to local differences of momentum, mass and heat transfer. Also, there is a wide variety of nano-foods and nano-ingredients produced by using this technique that have proved to have advantages over traditionally preparations with the same ingredients. In this chapter a revision of engineering principles, design considerations and a description of selected nano-foods produced by microfluidization as well as an example of effects of microfluidization (MF) on functional properties of a microparticulated protein are presented.

I. Monroy-Rodríguez · A. Monroy-Villagrana · S. García-Pinilla · H. Hernández-Sánchez
G. F. Gutiérrez-López (✉)
Departamento de Ingeniería Bioquímica, Escuela Nacional de Ciencias Biológicas, Instituto Politécnico Nacional, Carpio y Plan de Ayala S/N, Ciudad de México 11340, Mexico

M. Cornejo-Mazón
Departamento de Biofísica, Instituto Politécnico Nacional Escuela Nacional de Ciencias Biológicas, Carpio y Plan de Ayala S/N, Ciudad de México 11340, Mexico

U. Hebbar et al. (eds.), *Nano-food Engineering*, Food Engineering Series,
https://doi.org/10.1007/978-3-030-44552-2_6

2 Design and Operating Principles of Microfluidization

Microfluidization comprises the transfer of mechanical energy to the particles from a high-pressure/velocity fluid (Kasaai et al. 2003). It is a form of homogenization that uses micro-channel Interaction Chambers (IX) to promote cavitation, with high shear and impact.

The microfluidizer contains an intensifier pump that generates high pressures, which are used to force the product through an IX formed by micro channels that will divide the product into streams which are further forced to collide with one another so giving place to size reduction, dispersion, or formation of emulsions (Kasaai et al. 2003). The pneumatically pump is capable of rising the inlet compressed air (150–600 kPa) up to about 150 MPa (Microfluidics 2008; Jafari et al. 2007b). (Jafari et al. 2006). Figure 1 shows the equipment M-110Y produced by Microfluidics Corp. The chamber is essentially a continuous microreactor that uses a turbulent mixer.

There are mainly, two types of IX: "Y" and "Z." which may have a variety of sizes, ranging from the ones that are used in laboratory to those applied in large-scale production. These chambers can be used in combination with an auxiliary processing module (Microfluidics 2008). A number of works have shown that the homogenization by MF has advantages when compared to other types of conventional homogenization systems. The particle size distribution produced by a microfluidizer appears narrower and smaller than that obtained in systems produced by traditional methods of homogenization (Jafari et al. 2006, 2007a, b; Perrier-Cornet et al. 2005 and Wooster et al. 2008). Even more, a number of cycles or passes through the equipment may be used as to obtain the desired product and the characteristics of the final product will depend specially on the applied pressure, number cycles or passes used, and the type of the product. In the case of emulsions, it has been reported that MF at very high pressures and longer emulsification times and cycles, produces to overprocessing of the product which consist in recoalescence of the droplets and an increment of the droplet sizes (Jafari et al. 2006, 2007a, b and

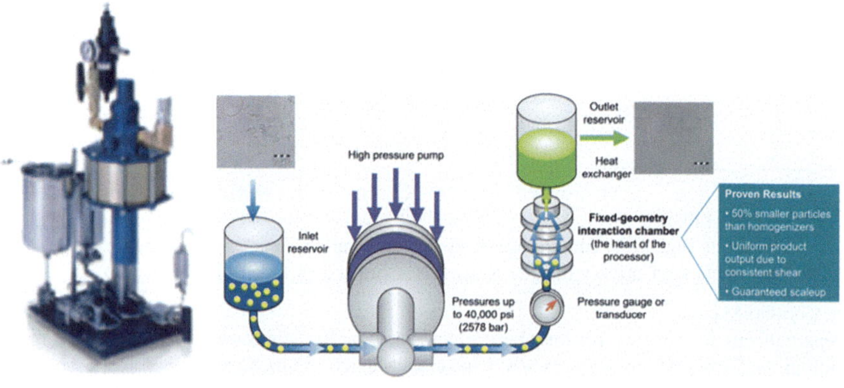

Fig. 1 M-110Y Microfluidizer® M-110Y and schematics of the MF process (Microfluidics Corp) Microfluidics International Corporation (2014)

Olson et al. 2004). Nowadays, microfluidizers are designed aiming at preventing cavitation since although it is a strong force that may break particles, it may damage the equipment. Moreover, the relationship between the turbulent mixer, the localized energy dissipation device, and a fixed geometry are the responsible parts which generate a uniform pressure profile that creates a precise and repeatable size distribution in the nanosystem. This distribution depends on (Lee and Norton 2013 and Qian and McClements 2011):

1. Hydrodynamic conditions within the break-up zone, the energy dissipation, the viscosity of the two phases.
2. Residence time in the break-up zone, type of oil, emulsifier, and concentrations.
3. Physicochemical properties of each phase such as interfacial tension.

The size of the channels at large-scale operations is very similar to those at laboratory scale and a wide variety of micro-channel sizes are available as to produce a range of shears and modes of collision in such a way that, for scale-up procedures, the use of multiple microchannels in parallel is recommended to attain similar shears at the production scale similar to those reached at the laboratory scale. The chambers are shown in Fig. 2.

3 MF Hydrodynamics: Mixing in Microchannels and Residence Time Distribution (RTD)

The shear forces caused by differences in local pressure are different among the rheology of the system, in laminar flows the deformation is performed only by shear forces, while in turbulent flow the inertial forces also influence the break-up, which is executed by three mechanisms of rupture (Stang et al. 2001):

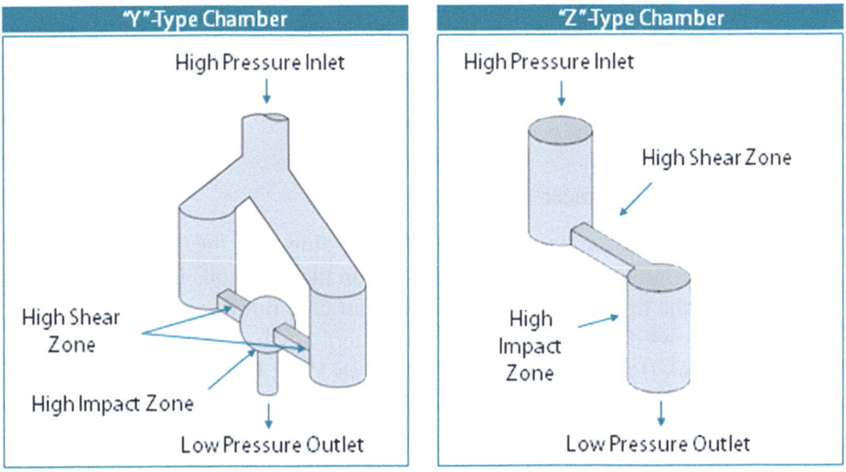

Fig. 2 Microfluidizer IX "Z" and "Y" (Microfluidics Corp)

1. Rupture due to shear forces in laminar flow
2. Rupture due to shear forces in turbulent flow
3. Inertial forces in turbulent flow

Knowing the hydrodynamics within the microfluidizer through the measurement of residence time distribution (RTD), inside and outside the IX, is a useful tool since on the one hand it offers the possibility of characterizing the flow within the equipment and correlate them with the products obtained, this because the turbulence associated with the shocks, caused by the different types of flow cause different RTD influencing the characteristics of the product and will depend on the type of IX used and the pressure (Levenspiel 1999; Gutierrez et al. 2010 and Monroy-Villagrana et al. 2015).

In the case of emulsions, liposome formation, homogenization of dairy products and coencapsulation, this time, within the IX, must be sufficient to allow the breaking of the drops and the absorption of the emulsifiers to be carried out in order to stabilize the new interfaces formed and generate stable systems, with a homogeneous particle size distribution. Regarding other types of systems, it is important to know the desirable final characteristics of the products to be obtained in order to optimize the flow and mixing conditions.

3.1 E Curve. Distribution of Fluid Age

To get the RTD is necessary to determine the RT of each portion of the fluid stream, also called C curve (Levenspiel 1999). This curve corresponds to the curve of age distribution at the output when a pulse of tracer substance (i.e., dye) is introduced into the fluid stream. The average residence time (RT) is calculated by Eq. (Levenspiel 1999).

$$RT = \frac{\int_0^\infty tC\,dt}{\int_0^\infty C\,dt} = \frac{\sum_i^n t_i C_i \Delta t_i}{\sum_i^n C_i \Delta t_i} \tag{1}$$

Where

RT = average residence time
t = time
C = concentration of the tracer

There are three different types of ideal flow: the plug flow, the completely stirred tank, and the intermittent stream which depend on the variations in the distribution or dispersion of the fluid within the device (in our case throughout the microchannels). The real flows deviate from ideal flow patterns such as piston flow and completely stirred tank (Pinheiro-Torres and Oliveirah 1998; Gutierrez et al. 2010 and Levenspiel 1999) and, in general, the flow patterns are calculated from dispersion models (Pinheiro-Torres and Oliveirah 1998; Levenspiel 1999 and Gutierrez et al. 2010). This model consider the piston flow of a fluid with some degree of back mixing. Deviations from ideal plug flow arise in real systems when different liquid flows and associated turbulences are present and the dispersion number (D/vL) characterizes such deviations.

$$\frac{D}{\nu L} \rightarrow 0 \text{ Small dispersion, tends to plug flow,}$$

$$\frac{D}{\nu L} \rightarrow \infty \text{ Great dispersions, which tend to completely stirred tank flow.}$$

To calculate the dispersion in the equipment it is necessary to evaluate the variance of the RT of the individual samples and the dimensionless variances from the average RT (Eq. 1) and dimensionless variances, Eq. (2) as follows:

$$\sigma^2 = \int_{\infty}^{0} (t - RT) * E(t) dt \tag{2}$$

$$\sigma_\theta^2 = \sigma^2 / RT^2 \tag{3}$$

From the calculated values of the dimensionless variance, the dispersion module is calculated by means of Eq. (4) for closed containers. Once obtained the dispersion module (*D*/*vL*) an E_θ curve (dimensionless) as a function of the dimensionless time may be obtained by considering the corresponding dispersion module.

$$\sigma_\theta^2 = 2\frac{D}{\nu L} - 2\left[\frac{D}{\nu L}\right]^2 \left(1 - e^{-\frac{\nu L}{D}}\right) \tag{4}$$

3.2 *Energy Density and Pressure Drop During Mixing in a Microfluidizer*

The analysis of the RTD curve must consider the fluid path before reaching the IX, the type of flow in terms of the Re number, the amount of dissipated energy and the pressure drop through the different parts of the equipment for finally evaluating the energy per volume applied or energy density in the IX that is the site of maximum shear. Since shear, turbulence, and impact occur simultaneously in the IX, a considerable amount of the mechanical energy applied is dissipated as heat (Skurtys and Aguilera 2008; Stang et al. 2001).

Monroy-Villagrana et al. (2015) performed an experiment to evaluate the hydrodynamic characterization on the formation of α-tocopherol emulsions using MD and gum Arabic (GA) as emulsifying agents (1:1, respectively) changing: oil-phase concentrations (17–28%), pressure (69–103 MPa), 1 cycle of MF and comparing the behavior of the emulsion with a red-dye tracer in water (0.05% w/v red 40 dye). Authors found that the pressure drop primarily occurs in the IX. The flow in the pipes before reaching the IX is laminar and in the IX is turbulent, emulsification by MF is primarily caused by collision of liquid streams at high shears, applied pressure and effective viscosities which control type of flow regime and particle-size reduction. When the aim of the MF is to stabilize surface-active biopolymers, small particle sizes are obtained by increasing the amount of the oil fraction and increasing the residence time in the IX and by applying the lowest possible pressure (69 MPa).

4 Applications of MF in Nano-Food Engineering

As a homogenization technique, MF can change the properties of the food depending on its composition and some practical applications of MF will be discussed in this section. This technology has been investigated as an alternative method for homogenization, dispersion, emulsification and as a particle size reduction method (Sinaga et al. 2018). One of the main consequences of treating fluids by MF is the possibility of having a more homogenous particle size distribution, changes on particle charge, increased stability of colloidal systems and changes on the microstructure that may favor the properties of the products. It is important to analyze the conditions of MF treatments depending on the type of food-related fluids to obtain the minimum particle size, maximum stability and functionality which will strongly depend on the type of the IX, operating conditions, sample composition, interfacial tension, and rheological properties (Maherani et al. 2018). As a consequence of the many variables involved in MF and given the vast number of interactions amongst them, heuristic approaches to the design of operating conditions are frequently applied and may be part of an artificial intelligence algorithm to obtain rational and systematic designed processes and products.

4.1 MF as a Pretreatment for Encapsulation

Encapsulation is a technique widely used in the food areas with the purpose of preserving bioactive compounds. The process of homogenization prior to the formation of solid encapsulates is relevant for determining their particle size, structure, functionality and stability. The advantage of nanocapsules (as compared to micro and macrocapsules) is that they provide more surface area and have the potential to increase solubility, enhance bioavailability, functionality, and to improve time-controlled release (Mozafari et al. 2008). Some of the types of bioactive compounds that have been analyzed mainly for the formation of encapsulates by MF are microbial compounds, since it has been observed that MF does not affect their chemical composition.

García-Toledo et al. (2018) produced nanoencapsulates of bacteriocin from *Pediococcus acidilactici* with phosphatidylcholine from soybean using a rotor-stator homogenizer (Ultra-Turrax® IKA® T-25) and a microfluidizer (Microfluidics M-110)and their results did not show statistical difference when applying 1 and 2 MF cycles in respect to the particle size and polydispersity index (PDI). In addition, particle size distribution (PSD) was unimodal in most cases when using 2 MF cycles, with a narrow PDI (0.4). Besides, these authors did not observe overprocessing under the homogenization conditions used. They concluded that by applying one MF cycle, highly stable encapsulates were generated since there is a particle size reduction limit in which interaction forces could decrease, giving place to flocculation and coalescence (García-Toledo et al. 2018).

Moreover, Arredondo-Ochoa et al. (2017) compared different homogenization techniques such as ultrasound and microluidization for the formation of beeswax and starch nanoemulsions (wall materials) of different antimicrobial agents. Their best products were obtained by using MF with a particle size of 77.6 ± 6.2 nm, PDI of 0.4 ± 0.0 leading to a good stability (ζ-potential values of −36 ± 4 mV) and translucency as measured by a whiteness index (WI) of 31.8 ± 0.8. Nanoemulsions showed a Newtonian behavior, and good wettability properties. Additionally, they observed that antimicrobial activity was not affected by MF.

Maherani et al. (2018) optimized the MF conditions to prepare an emulsion with antioxidant and antimicrobial properties, they obtained a reduction of particle size from 170 to 70 nm and viscosity by increasing the MF pressure and cycles. No significant changes on particle size were observed at pressures higher than 172.37 MPa.

On the other hand, the encapsulation of antioxidant compounds and vitamins has been reported by using MF (Monroy-Villagrana et al. 2014; Granillo-Guerrero et al. 2017 and Villalobos-Castillejos et al. 2018). Bamba et al. (2018) examined the influence of homogenization on the coencapsulation of phenolics and anthocyanins when preparing a double emulsion stabilized with a whey protein isolate. They observed that overprocessing arises at 200 MPa (particle size increased from 327 ± 5.46 nm when it was processed at 100 MPa up to 1580 ± 100 nm at 200 MPa) affecting encapsulation efficiency which could be observed by an increase of zeta potential values (from −44.53 ± 0.47 to −31.43 ± 0.50 mV). The droplet coalescence is critical since may cause the loss of encapsulated compounds and a deterioration of emulsion properties.

Also, Wang et al. (2017) found no adverse effects of homogenization pressure in the range 5 to 65 MPa on the oil droplet size in $W^1/O/W^2$ emulsions. On the other hand, it is important to consider the influence of other homogenizations systems such as different nozzles in the formation of encapsulates by spray drying since this process could modify the characteristics of the original emulsions (Villalobos-Espinosa et al. 2019). Detailed pre-treatments, MF conditions as well type of microfluidizer used in the above described works are shown in Table 1.

Table 1 Pre-treatment conditions for MF and type of microfluidizer used in Section 1.4.1

MF treatment	Sample	Reference
Pre-homogenization: ULTRA-TURRAX® 13,500 rpm for 3 min. Microfluidizer (Microfluidics M-110) 50 or 100 MPa pressure for 2 cycles	Bacteriocin (Pediocin) encapsulation with phosphatidylcholine	García-Toledo et al. (2018)
Pre-homogenization: ULTRA-TURRAX® 21,500 rpm for 3 min. Microfluidizer (M110P Nano DeBEE, Easton, MA, USA) 150 MPa pressure for 3 cycles	Beeswax, starch, emulsifyers and antimicrobial agents	Arredondo-Ochoa et al. (2017)
Pre-homogenization: ULTRA-TURRAX® 10,000 rpm 40s, Microfluidizer (electric-hydraulic M-110P) equipped with a diamond IX for emulsions downstream (Microfluidics international Corp., Newton, MA, USA). Optimization was performed at pressures of 103.42, 137.9 and 172.37 MPa in 1–2 cycles (at 25 °C)	The oil-in-water (O/W) microemulsion contains plant oil extracts, BIOSECUR F440D® with ratio of 1:3 (W/W), 1% (W/W) lecithin and sucrose monopalmitate with different ratio (45:55%, 36:64%) as emulsifier	Maherani, et al. (2018)
Pre-homogenization: 10,000 rpm for 10 min (W1/O) and 6000 rpm for 15 min (W1/O/W2). The effect of MF was assessed by varying the microfluidizer pressure from 50 to 200 MPa with one cycle through the microfluidizer (Microfluidics M-110P)	Coencapsulation of polyphenols and anthocyanins from blueberry pomace by double emulsion stabilized by whey proteins	Bamba et al. (2018)

4.2 *MF as Nanoemulsifying Technique*

Nanoemulsions refer to emulsion systems (simple or multiple) in which the size of the dispersed droplets range from 50 to 200 nm in diameter (transparent emulsions) or up to 500 nm resulting in a milky appearance (Jafari et al. 2006). Nanoscale droplets can be obtained by high pressure/shear homogenization methods such as MF.

As in encapsulation, the formation of nanoemulsions have the objective of stabilizing and preserving bioactive compounds present in the oil phase (García-Toledo et al. 2018). Nanoemulsions as a conservation technique has advantages such as greater stability, lower susceptibility to creaming and sedimentation levels than microemulsions, since the Brownian motion prevails over gravitational forces (Jafari et al. 2017). Besides, the droplet size also influences the appearance, bioavailability, bioaccessibility, texture and stability of the final product (Qian and McClements 2011; Salvia-Trujillo et al. 2013 and Tang and Liu 2013). In this respect, MF processing, consistently generate significantly higher shear than other methods and more uniform particle size reductions (Sadeghpour et al. 2015).

Although the concentration and type of stabilizer is a determining factor in the stability of emulsions, the MF conditions change depending on the nanoemulsion's formulation. Santos et al. (2018) used a new type of xanthan gum (advanced performance xanthan gum) in combination with egg protein and sunflower oil and their results indicated that pressures above 68.95 MPa provoked an increase of flocculation. However, emulsions prepared by using MF were more stable than those untreated.

On the other hand, rheology of the products may change aiding to obtain products with modified texture depending on the nature of the emulsifier used since MF could give place to the formation of gel-like colloids. This has been reported in whey protein isolate (Zong-Cai Tu et al. 2017) gliadin particles (Liu et al. 2017) and soy protein isolate (Liu and Tang. 2013) with the advantage of replacing treatments which use aggressive chemical and thermal treatments as well the incorporation of food additives.

Detailed pre-treatments, MF conditions as well type of microfluidizer used in the above described works are shown in Table 2.

4.3 *MF as Liposome Formation Technique*

A liposome is a spherical vesicle (small bubble) with a membrane composed of a double layer of phospholipids, consisting of water-soluble and liposoluble phases. The traditional liposome preparation method involves the utilization of detergents or organic solvents, which are either undesirable or impermissible components in

Table 2 Pre-treatment conditions for MF and type of microfluidizer used in Section 1.4.2

MF treatment	Sample	Reference
Pre-homogenization: 8000 rpm for 3 min. Microfluidizer M110P (chamber Y) at pressures 34.4, 68.9, 103.4 and 137.89 MPa.	O/W emlusions, sunflower oil, advanced performance xanthan gum and egg protein.	Santos et al. (2018)
MF using an air-driven high-pressure microfluidizer (Microfluidics PureNano, Newton, MA, USA) with a 75 μm chamber (F20Y) operated at a fixed pressure of 90 MPa for three passes.	Oil-in-water nanoemulsions stabilized by whey protein isolate (WPI).	Zong-Cai Tu et al. (2017)
Pre-homogenization: ULTRA-TURRAX® 23,000 rpm for 2 min. Microfluidizer (M110EH model, Microfluidics International Corporation, Newton, MA) for one pass at a pressure level of 40, 80, and 120 MPa, Respectively at 25 °C	Algal oil (50%, v/v) and gliadin particle solutions (50%, v/v).	Liu et al. (2017)
Pre-homogenization: ULTRA-TURRAX® T50 (IKA, Shanghai, China). 2000 rpm for 120 s. Microfluidizer M110P at different pressures (10, 35, 60 and 70 MPa) for one pass.	Biopolymers (Rhamsan and/ or welan gum) surfactant, AMIDET®N and thyme oil.	Trujillo-Cayado, et al. (2019)

food products. The utilization of high-pressure homogenizers, such as microfluidizers, to fabricate liposomes overcomes many of the current processing challenges (Peng et al. 2018)

Recent research has confirmed the viability to produce liposomes by MF such as curcumin-loaded liposomes formed from sunflower lecithin (Peng et al. 2018) and alkaloid-catechin-cocoa extract loaded liposomes (Toro-Uribe et al. 2018). In both studies, the MF treatment produced a decrease in particle size of the liposomes and therefore an improved stability.

Detailed pre-treatments, MF conditions as well type of microfluidizer used in the above described works are shown in Table 3.

4.4 MF in Dairy Processing

Homogenization in milk processing is applied to produce more consistent raw milk through the breaking down of the fat globules avoiding creaming (López-Fandino 2006). Another research, suggested that high pressure homogenization can also be used to produce particular changes in the molecular structure of proteins, that might modify milk constituents and properties of final products (Garcia-Risco et al. 2002 and López-Fandino 2006).

Bucci et al. (2018) analyzed the changes in the physical, chemical and microbial properties in milk homogenized by MF. They observed a pressure-dependent increase in outlet temperature, inactivation of alkaline phosphatase in raw and thermized samples at 125 and 170 MPa, observing that the microfluidized milk was a stable emulsion of fat droplets ranging from 0.390 to 0.501 μm, compared with 7.921 μm (control) and 4.127 (homogenized control) μm. No chemical composition changes in these last treated products were found, but the t protein content was reduced to approximately 3.1–2.2% as detected by infrared spectroscopy. Microbiological results indicated a decrease in mesophilic aerobic and psychrophilic milk microflora with increasing temperature and pressure, suggesting that MF may eliminate bacterial load. Tthe viscosity increased up to 0.7 Pa•s in microfluidized milk in comparison with the control (0.3 Pa•s). Addittttionally, Hardham

Table 3 Pre-treatment conditions for MF and type of microfluidizer used in Section 1.4.3

MF treatment	Sample	Reference
Microfluidizer (M-110P, Microfluidics Corporation, Newton, MA, USA) pressure of 117.21 MPa for 3 cycles.	Curcumin-Loaded Liposomes Formed from Sunflower Lecithin	Peng et al. (2018)
Pre-homogenization: high shear blender (T10 ULTRA-TURRAX®, IKA, Staufen, Germany) at 20,000 rpm for 2 min at 25 °C. Microfluidizer (110 T, Microfluidics, Newton, MA, USA) at 137.9 MPa for several passages (up to 8 cycles) in an ice bath (4 °C)	alkaloid-catechin-, and cocoa extract-loaded liposomes	Toro-Uribe et al. (2018)

et al. (2000) reported a decrement of fat separation on microfluidized milk during storage.

The properties of milk are likely to be dependent on the casein micelle size and various processing technologies produce a particular change in the average size of casein micelles. Sinaga et al. 2018 analyzed the changes in casein micelle size by subjecting milk to MF. Their results also showed that microfluidized milk had smaller particle size (3%) with treatments at pressures up to 126 MPa. No further reduction in particle size was observed after increasing the microfluidized cycles up to a maximum of 6. Although the average casein micelle sizes were similar amongst MF treatments, elevating the processing pressure resulted in narrower size distributions. In contrast, increasing the number of cycles had little effect on casein micelle size distribution.

Another application of MF in diary processing are, the production of non-fat or low-fat ice creams that usually had a slower meltdown without affecting sensory properties. Olson et al. (2003) changed mozzarella cheese microstructure by reducing fat globes size (Tunick et al. 2000). On the other hand, by increasing of interconnectivity in the protein networks with embedded fat globules in yogurt did not affect water retention while improving the thickness, creaminess, cohesiveness and viscosity of this product (Ciron et al. 2010; Ciron et al. 2011 and Penna et al. 2014).

4.5 MF as a Structural Modifier in Starch

Starch is one of the polysaccharides with higher occurrence in nature (Sweedman et al. 2013). In the food industry, the starch has been used in different applications, such as texturizer, as additive to increase yields and for the stabilization of emulsions (Jobling 2004 and Leal-Castañeda et al. 2018).

As discussed above, MF could provoke changes on the structure of products and therefore in their functional properties and in this regard, some studies have analyzed the structural changes caused in starch by MF. Wang et al. (2018) evaluated the structural characteristics of rice amylose treated by MF, observing an increment of hollows and grooves on the surface of the treated rice amylose in comparison with the untreated starch. Also, the particle size decreased and became uniform with an increase in MF pressure. MF treatments above 100 MPa were sufficient to cause the complete destruction of the crystalline structure of rice amylose. In other cases, it has been observed that MF treatment increased the starch particle size due to its partial gelatinization as a consequence of the high pressures and temperatures reached (Che et al. 2007 and Wang et al. 2018). The same behavior was observed in cassava starch granules (Kasemwong et al. 2011). Błaszczak et al. (2005) claimed that MF treatments could lead to the filling of inner parts of the granules with a gel-like network. Augustin et al. (2008) found that MF could improve the functionality of resistant starch by increasing water retention and viscosity.

Other researchers suggested that reduced particle size and starch degradation as a consequence of MF's treatments could decrease solubility and apparent viscosity of starch pastes (Wang et al. 2012; Tu et al. 2013 and Duan et al. 2016).

Due to the changes provoked in the structure of the starch by MF, some practical considerations could be applied to facilitate the modification of the starch by chemical methods such as those reposted by Li et al. (2018) who proved that MF was useful as a pretreatment during OSA modification of rice starch due to the changes of starch structure induced by MF. Their results showed that the surface of starch particles became rougher and was destroyed with a decreasing degree of crystallinity caused by the MF treatments. They also pointed out that the degree of substitution increased with MF pressures. Resultant starch showed a significantly higher peak viscosity, emulsifying activity, and emulsion stability but lower pasting temperature as compared with the native rice starch.

MF conditions as well type of microfluidizer used in the above described works are shown in Table 4.

4.6 *MF for Improving the Extraction of Bioactive Compounds*

Due to the high shear forces and pressures applied in MF, this process can cause the rupture of cells and particles and therefore has the potential to improve the extraction yield of bioactive compounds from food and food waste including those which sizes are in the range of a few nm.

Xing et al. (2018) used MF as a pretreatment for sulforaphane extraction from broccoli (*Brassica oleracea*) seeds, by using a Microfluidizer (M110-Y, Microfluidics Corp. USA) equipped with a Z type chamber (H10Z, 100 μm),as they could reduce the particle size 2–10 times of the broccoli seeds and obtained a unimodal distribution by increasing the MF pressure and number of passes. The highest sulforaphane content was obtained at 34.47 MPa and 5 passes, achieving three times the yield with respect to the control sample. Also, structural changes were confirmed by means of SEM images. From this research it could be assumed that there were various relationships between the decrease in particle size and the extraction yield.

Table 4 MF and type of microfluidizer used in Section 1.4.5

MF treatment	Sample	Reference
Pre-treatment: Colloid milling (JMS-80, Langfang Tongyong Machinery Manufacturing Co.,Ltd., Henan, China) and homogenization (GYB60-6S, Donghua Co.,Shanghai, China) were used to mix the suspension above sequentially before MF. MF was performed using an M-110EH-30 microfluidizer (Microfluidics Co., Newton, MA, USA) at 60, 100, 140, and 180 MPa, respectively.	Rice amylose suspension	Wang et al. (2018)
Microfluidizer (M-110EH, Microfluidizer, USA) at pressure of 0, 60, 120, or 180 MPa for 2 cycles.	Rice starch	Li et al. (2018)

Prasad et al. (2009) also found that ultrasonication and high-pressure extractions had a potential to achieve high extraction efficiency from plant tissues.

4.7 MF as a Protein Structural Modifier

In food technology, proteins have different useful functional properties such as stabilizing agents for colloidal systems and some research has focused on determining MF conditions to cause changes in the structure and functionality of different types of proteins.

MF could be considered as a milling technique to modify corn gluten meal (a sub product rich in zein). Ozturk and Mert 2018 evaluated the potential of MF as a value-adding process to this product and their results showed that pH modifications and MF caused deformation within the protein structure such as the breakages of bonds, unfolding of hydrophobic amino acids and hydrolysis. Due to these changes, water holding ability was improved.

Sha et al. (2018) treated fish gelatin by applying MF and their results showed improvement on its emulsifying and foaming properties, the microstructure was analyzed by SEM and they could see endowed fish gelatin with many rough porous but uncracked. On the other hand, McCarthy et al. (2016) suggested that MF treatment in pea protein isolate could be used to produce cold-set-gels since MF created more uniform size distributions, which can contribute to increase the viscosity of the product (Qian and McClements 2011 and Román et al. 2015).

Oliete et al. (2019) studied the effect of MF as a pretreatment in the obtention of spray dried pea globulin aggregates and their results showed a decrease in aggregates size and surface hydrophobicity due to protein re-arrangements that caused higher solubility of the aggregates.

It is assumed that MF could affect the antigenicity of proteins because some epitopes could result exposed during this process. Recently, Chen et al. (2018) evaluated the antigenicity changes after MF and in vitro digestion in beta-lactoglobulin (β-LG) and their results showed that antigenicity declined from 13.41 to 12.27 and 7.19 lg mL^{-1} at 0.1, 80 and 160 MPa of MF pressure respectively. Also, the reduction of antigenicity has also been evaluated by combining the β-LG with oleic acid by MF (Zhong et al. 2019). The effect of MF pressure on β-LG particle size indicated that the average particle size decreased as the pressure increased to 80 MPa, while it increased at pressure of 160 MPa possible due to overprocessing effects (Zhong et al. 2012). These changes were observed by the values of PSD and atomic force microscope images.

Although MF promotes changes in the functional and structural properties of proteins, some studies suggested that the primary structure of some proteins kept stable and was not affected by physical processing such as HHP and MF (Liu et al. 2010 and Meng et al. 2017).

Detailed pre-treatments, MF conditions as well type of microfluidizer used in the above described works are shown in Table 5.

Table 5 Pre-treatment conditions for MF and type of microfluidizer used in Section 1.4.7

MF treatment	Sample	Reference
Microfluidizer (M-110EH-30, Microfluidics, Newton, MA, USA) at pressures of 0.1, 80 and 160 MPa for 3 cycles.	β-lactoglobulin	Chen et al. (2018)
Microfluidizer M-110Y, Microfluidics, USA) equipped with two chambers: the first chamber having 200 μm size and 50 MPa pressure and the second chamber having 100 μm size and 125 MPa pressure. Samples were MF three times.	Corn gluten meal	Ozturk and Mert (2018)
Microfluidizer (M-100EH-30 Microfluidics Co., Newton) at pressures of 40, 80, 120, and 160 MPa for 5 cycles, respectively at 25 °C.	Fish gelatin	Sha et al. (2018)
Pre-homogenization: ULTRA-TURRAX® (IKA, Staufen, Germany) 10 000 rpm for 1 min. Microfluidizer (M110-EH) equipped with a 75 μm Y-type ceramic IX (MicrofluidicsInternational Corp., Newton, MA, USA) at 50 MPa for one cycle.	Pea protein	McCarthy et al. (2016)
Microfluidizer (LM10 Microfluidics, Newton, MA, USA) fitted with a Z-type chamber (G10Z) at 130 MPa for 3 cycles.	Pea protein isolated	Oliete et al. (2019)

5 A Case Study on Evaluation of MF Effect on Soy Protein Isolate Foaming Properties

Aiming at illustrating the effects of MF on the functionality of polymers, in this section a case-study on the effects over soy protein isolate (SPI) foaming capacity is presented. It has been reported that, protein microparticles (with particle sizes between 0.1–10 μm) are obtained from protein aggregates (by pH changes and thermal treatments) after MF. The applications of microparticles have recently been evaluated aiming at be added to food preparations as a fat substitute (Ipsen 2016).

In our case, the protein was treated by MF to achieve improved foaming properties by modifying its structure, since the reduction in particle size due to MF changes could be expected in the stabilization of this dispersed systems. On the other hand, the structure of the protein treated by high pressure homogenization could promote the exposure of hydrophobic and hydrophilic groups thus modifying its solubility.

5.1 Design of Experiments and Preparation of Samples

A Box-Behnken experimental design with three numeric variables were chosen; ionic strength (IS) (100, 300 and 500 mM NaCl), pH (3, 5 and 7) and pressure 12,000, 14,000 and 16,000 psi (82.73, 96.53 and 101.32 MPa) and choosing as categorical variable the number of MF cycles (1 and 3).

Suspensions of a commercial isolate of 6% (w / v) soy protein were prepared and NaCl was added to control ionic strength of the media. Subsequently, the pH was adjusted with 2 M solutions of HCl and NaOH and 0.02% of sodium azide was added to prevent microbial growth. Finally, samples were subjected to thermal treatment (93 °C for 15 min) before MF as reported by Liu and Tang (2013).

5.2 Particle Size (PS) and Zeta Potential (ζ)

The PS and (ζ) were measured in the Zeta Sizer (Malvern Nano, S USA) as reported by Monroy-Villagrana et al. (2014).

5.3 Evaluation of Foaming Capacity (FC) and Foam Stability (FS)

To determine the capacity of foaming, the formation, drainage and foam collapse described below were evaluated (Morales et al. 2015):

15 ml of the protein suspensions were taken and subjected to shaking in a graduated tube for 1.5 min at 15,500 rpm with an ULTRA-TURRAX® homogenizer (T18 IKA USA) at room temperature. The foaming capacity was calculated with Eq. 5.

$$CF(\%) = \left[\frac{(B - A)}{A}\right] x100 \tag{5}$$

Where A and B are the volumes before and after stirring respectively.

The stability of the foam was evaluated by using Eq. 6.

$$FS(\%) = \left[\frac{(V30 - B)}{(A - 100)}\right] x100 \tag{6}$$

Where V_{30} is the volume after 30 min, A and B are the volumes before and after stirring respectively.

5.4 Particle Size

Nicolai et al. 2011 and Ryan et al. 2013 reported that size of protein aggregates was influence by the concentration of the protein, heat treatments, type of protein, ionic strength and pH of the media. The effect of concentration of the protein in the media, heat treatments, type of protein, ionic strength and pH variables on particle sizes of whey protein aggregates was reported by Nicolai et al. 2011 and Ryan et al. 2013

The particle sizes (PS) of microparticulated protein were in the range of 647 nm to 10.07 μm which resulted in the smallest protein particle sizes (0.679–1.32 μm) obtained at a pH value of 7 at 1 MF cycle, with 3 cyles of MF the range of particle size was 0.627–0.866 μm. It was observed that when the IS increased, the particle size value decreased. On the other hand, at a pH of 3 and 5 the particle size increased

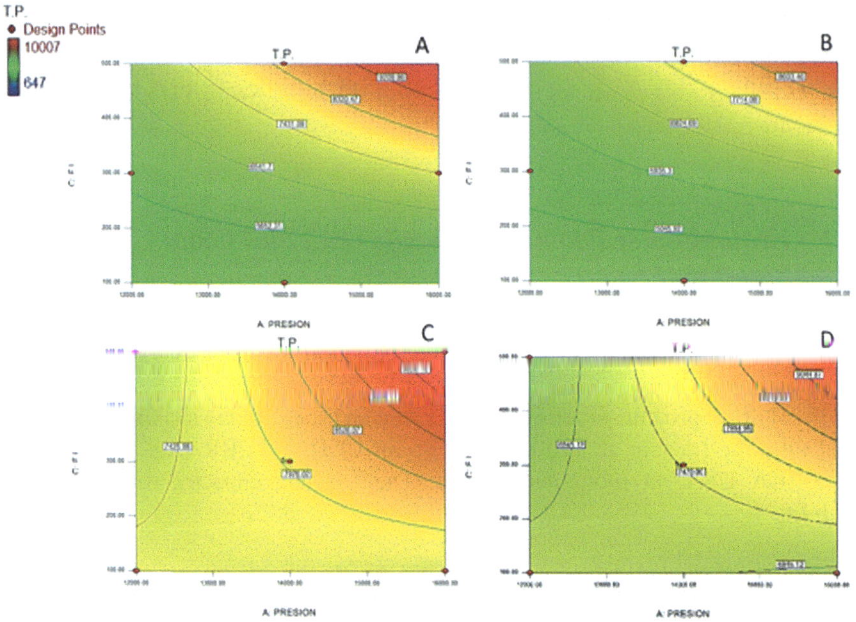

Fig. 3 PS contour graphs (y-axis; ionic strength, x-axis; pressure). (**A**) pH 3, 1 cycle (**B**) pH 3, 3 cycles (**C**) pH 5, 1 cycle (**D**) pH 5, 3 cycles

considerably in values between 5 and 10 μm and as the number of cycles increased, the PS decreased. In these pH values, the size of particles increased by increasing the value of IS as shown in Fig. 3 (in the red area the largest particles are found).

The size of the protein aggregates seems to be affected by the solubility decrement, since when it is less soluble (pH 3–5), the PS was higher, whereas at pH 7, the PS considerably decreased. Achouri et al. (2005) analyzed the solubility in SPI at different pH values, the solubility was lower in a pH range of 5 to 8, the isoelectric point of SPI proteins is between 3 and 4.5. In addition, the heat treatment is also a key factor for the formation of the aggregates since, at the temperature reached 93 ° C during the first 15 minutes it was expected that the solubility of the protein decreased due to the exposure of the hydrophobic groups (Guo et al. 2015). During MF treatments this effect was also expected due to the high pressures reached this process.

5.5 *Zeta Potential (ζ)*

The ζ is indicative of the electric charge of the particles in the media. When the charged particles approach each other at high absolute values of $\zeta \geq |30|$ mV, they repel to each other, promoting stability of the dispersed particles (González et al. 2007; Luo and Gu 2007; Kaszuba et al. 2008 Cano-Sarmiento et al. 2018).

The ζ results for our samples were between −22.56 and − 35.15 mV. At pH 7 at lower pressure (82.73 MPa) the value of this parameter was lower (−35.15 ± 0.1 mV) compared to the microparticles at pH 3 subjected at higher MF pressure (101.32 MPa) the ζ was −22.56 ± 0.51 mV. Liu and Tang (2013) observed an increment of ζ (from −25 to -10 mV) by increasing the value of IS in emulsions stabilized with soybean protein subjected to MF, in our case we observed that MF pressure and pH had more impact on ζ.

5.6 *Foaming Capacity*

The FC obtained values (150–190%) are considered to be high under conditions of pH 7 and an IS of 500 mM. It has been reported that the conditions of the heat treatment in SPI reached improved the FC, on the other hand, if the heating time is higher to 30 min the FC could decrease (Guo et al. 2015) These results could be explained due to the decrement of solubility of the protein which might cause the decrement of the surface tension of the system, favoring the entry of air into the media, resulting in a greater volume of the foam, improving the FC. In Fig. 4, the best FC conditions can be observed in graphs C and F at higher IS (500 mM) in 1 and 3 cycles of MF.

5.7 *Foaming Stability*

The most stable foams were obtained at pH 3 and 5 at a pressure lower than 96.53 MPa with one MF cycle and IS between 100–300 mM as shown in Fig. 5, in graphs A and B.

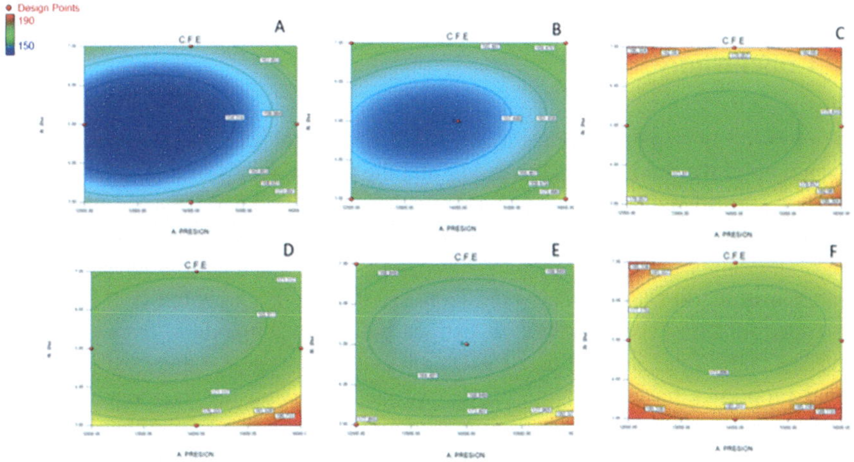

Fig. 4 FC contour graphs (y-axis; pH, x-axis; pressure). (**A**) IS 100 mM, 1 cycle (**B**) IS 300 mM, 1 cycle (**C**) IS 500 mM, 1 cycle (**D**) IS 100 mM, 3 cycles (**E**) IS 300 mM, 3 cycles (**F**) IS 500 mM, 3 cycles

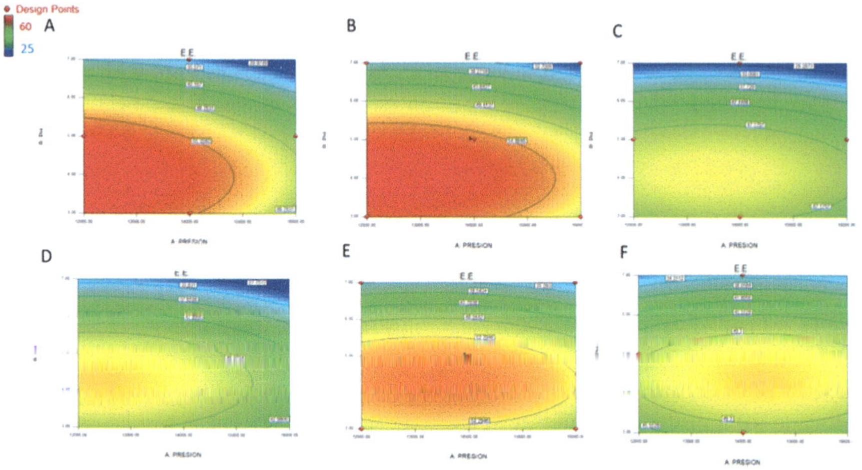

Fig. 5 FS contour graphs (y-axis; pH, x-axis; pressure). (**A**) IS 100 mM, 1 cycle (**B**) IS 300 mM, 1 cycle (**C**) IS 500 mM, 1 cycle (**D**) 100 mM, 3 cycles (**E**) IS 300 mM, 3 cycles (**F**) IS 500 mM, 3 cycles

Zúñiga et al. (2011) concluded that smaller and less dense aggregates could better stabilize the foam structure due to the formation of a continuous network of a thin layer in whey protein. On the other hand, Guo et al. (2015) observed that FS is high when the protein aggregates are larger in SPI, in our case we observed a similar behavior. As a result of structural changes in the protein aggregates, their surface properties are modified in the same way, causing changes in their emulsifying and foaming capacity (Palazolo et al. 2011; Moro et al. 2011 and Ryan et al. 2013).

6 Conclusions

MF engineering is complex since many interacting variables are involved in microchanneling of the fluids subjected to MF and the fundamentals of high-speed fluids collision are still being investigated and represent a challenge for process engineers. However MF is widely applied in the food area and a variety of nano-foods and nano-ingredients have been prepared by using it. As for example, encapsulates, nanoemulsions, liposomes, dairy products are commonly prepared by MF. Also, this operation has been applied to extract bioactive compounds from complex matrixes and, in the characteristics of starch and protein preparations which functional properties have been improved improved by using this technique. Future studies must include overprocessing considerations as well as the effect of nanosizing in structure-function relationships.

References

Arredondo-Ochoa T, García-Almendárez BE, Escamilla-García M, Martín-Belloso O, Rossi-Márquez G, Medina-Torres L, Regalado-González C (2017) Physicochemical and antimicrobial characterization of beeswax – starch food-grade Nanoemulsions incorporating natural antimicrobials. Int J Mol Sci 2017(18):2712. https://doi.org/10.3390/ijms18122712

Augustin MA, Sanguansri P, Htoon A (2008) Functional performance of a resistant starch ingredient modified using a microfluidiser. Innovative Food Sci Emerg Technol 9(2):224–231. https://doi.org/10.1016/j.ifset.2007.11.003

Bamba BSB, Shi J, Tranchant CC, Jun Xue S, Forney CF, Lim LT, Xu W, Xu G (2018) Coencapsulation of polyphenols and anthocyanins from blueberry Pomace by double emulsion stabilized by whey proteins: effect of homogenization parameters. Molecules 2018(23):2525. https://doi.org/10.3390/molecules23102525

Błaszczak W, Fornal J, Valverde S, Garrido L (2005) Pressure induced changes in the structure of corn starches with different amylose content. Carbohydr Polym 61(2):132–140. https://doi.org/10.1016/j.carbpol.2005.04.005

Bucci AJ, Van Hekken DL, Tunick MH, Renye JA, Tomasula PM (2018) The effects of microfluidization on the physical, microbial, chemical, and coagulation properties of milk. J Dairy Sci 101:1–12. https://doi.org/10.3168/jds.2017-13907

Cano-Sarmiento C, Téllez-Medina DI, Viveros-Contreras R, Cornejo-Mazón M, Figueroa-Hernández CY, García-Armenta E, Alamilla-Beltrán L, García HS, Gutiérrez-López GF (2018) Zeta potential of food matrices. Food Eng Rev. https://doi.org/10.1007/s12393-018-9176-z

Che L, Dong L, Wang L, Özkan N, Chen XD, Mao Z (2007) Effect of high-pressure homogenization on the structure of cassava starch. Int J Food Prop 10(4):911–922. https://doi.org/10.1080/10942910701223315

Chen H, Hong Q, Zhong J (2018) The enhancement of gastrointestinal digestibility of b -LG by dynamic high-pressure microfluidization to reduce its antigenicity. Int. J. Food Sci. Technol.:1–7. https://doi.org/10.1111/ijfs.14044

Ciron CIE, Gee VL, Kelly AL, Auty MAE (2010) Comparison of the effects of high-pressure microfluidization and conventional homogenization of milk on particle size, water retention and texture of non-fat and low-fat yoghurts. Int Dairy J 20(5):314–320. https://doi.org/10.1016/j.idairyj.2009.11.018

Ciron CIE, Gee VL, Kelly AL, Auty MAE (2011) Effect of microfluidization of heat-treated milk on rheology and sensory properties of reduced fat yoghurt. Food Hydrocoll 25(6):1470–1476. https://doi.org/10.1016/j.foodhyd.2011.02.012

Duan D, Tu Z, Wang H, Sha X, Zhu X (2016) Physicochemical and rheological properties of modified rice amylose by dynamic high- pressure microfluidization. Int J Food Prop 20(4):734–744. https://doi.org/10.1080/10942912.2016.1178283

Garcia-Risco MR, Ramos M, Lopez-Fandino R (2002) Modifications in milk proteins induced by heat treatment and homogenization and their influence on susceptibility to proteolysis. Int Dairy J 12(8):679–688. https://doi.org/10.1016/S0958-6946(02)00060-2

García-Toledo JA, Torrestiana-Sánchez B, Martínez-Sánchez CE, Tejero-Andrade JM, García-Bórquez A, Mendoza-García PG (2018) Nanoencapsulation of a bacteriocin from pediococcus acidilactici ITV26 by microfluidization. Food Bioprocess Technol 12:88–97. https://doi.org/10.1007/s11947-018-2184-4

Granillo-Guerrero VG, Villalobos-Espinosa JC, Alamilla-Beltrán L, Téllez-Medina DI, Hernández-Sánchez H, Dorantes-Álvarez L, Gutiérrez-López GF (2017) Optimization of the formulation of emulsions prepared with a mixture of vitamins d and e by means of an experimental design simplex centroid and analysis of colocalization of its components. Rev Mex Ing Quim 16(3):861–872

Guo F, Xiong YL, Qin F, Jian H, Huang X, Chen J (2015) Surface properties of heat-induced soluble soy protein aggregates of different molecular masses. J Food Sci 80(2):C279–C287. https://doi.org/10.1111/1750-3841.12761

Gutierrez CGCC, Dias EFTS, Gut JAW (2010) Residence time distribution in holding tubes using generalized convection model and numerical convolution for non-ideal tracer detection. J Food Eng 98(2):248–256. https://doi.org/10.1016/j.jfoodeng.2010.01.004

Hardham JF, Imison BW, French HM (2000) Effect of homogenisation and microfluidisation on the extent of fat separation during storage of UHT milk. Aust J Dairy Technol 55(1):16–22

Ipsen R (2016) Microparticulated whey proteins for improving dairy product texture. Int Dairy J 67:73–79. https://doi.org/10.1016/j.idairyj.2016.08.009

Jafari SM, He Y, Bhandari B (2006) Nano-emulsion production by sonication and microfluidization: a comparison. Int J Food Prop 2006(9):475–485. https://doi.org/10.1080/10942910600596464

Jafari SM, He Y, Bhandari B (2007a) Optimization of nano-emulsions production by microfluidization. Eur Food Res Technol 225(5–6):733–741. https://doi.org/10.1007/s00217-006-0476-9

Jafari SM, He YH, Bhandari B (2007b) Production of sub-micron emulsions by ultrasound and microfluidization techniques. J Food Eng 82(4):478–488. https://doi.org/10.1016/j.jfoodeng.2007.03.007

Jafari SM, Paximada P, Mandala I, Assadpour E, Mehrnia MA (2017) Encapsulation by nanoemulsions. In: Nanoencapsulation technologies for the food and Nutraceutical industries. Academic, New York, pp 36–73

Jobling S (2004) Improving starch for food and industrial applications. Curr Opin Plant Biol 7:210–218. https://doi.org/10.1016/j.pbi.2003.12.001

Kasaai MR, Charlet G, Paquin P, Arul J (2003) Fragmentation of chitosan by microfluidization process. Innov Food Sci Emerg 4(4):403–413. https://doi.org/10.1016/S1466-8564(03)00047-X

Kasemwong K, Ruktanonchai UR, Srinuanchai W, Itthisoponkul T, Sriroth K (2011) Effect of high-pressure microfluidization on the structure of cassava starch granule. Starch-Stärke 63(3):160–170. https://doi.org/10.1002/star.201000123

Leal-Castañeda EJ, García-Tejeda Y, Hernández-Sánchez H, Alamilla-Beltrán L, Téllez-Medina DI, Calderón-Domínguez GHS, Gutiérrez-López GF (2018) Pickering emulsions stabilized with native and lauroylated amaranth starch. Food Hydrocoll 80:177–185. https://doi.org/10.1016/j.foodhyd.2018.01.043

Lee L, Norton IT (2013) Comparing droplet breakup for a high-pressure valve homogeniser and a microfluidizer for the potential production of food-grade nanoemulsions. J Food Eng 114(2):158–163

Levenspiel O (1999) Ingeniería de las Reacciones Químicas. Reverte 9:277–347

Li YT, Wang RS, Liang RH, Chen J, He XH, Chen RY, Liu W, Liu CM (2018) Dynamic high-pressure microfluidization assisting octenyl succinic anhydride modification of rice starch. Carbohydr. Polym. S0144-8617(18):30367–30369. https://doi.org/10.1016/j.carbpol.2018.03.103

Liu F, Tang C (2013) Soy protein nanoparticle aggregates as Pickering stabilizers for oil- in-water emulsions. J Agric Food Chem 2013(61):8888–8898. https://doi.org/10.1021/jf401859y

Liu W, Zhang ZQ, Liu CM, Xie MY, Tu ZC, Liu JH, Liang RH (2010) The effect of dynamic high-pressure microfluidization on the activity, stability and con- formation of trypsin. Food Chem 123:616–621. https://doi.org/10.1016/j.foodchem.2010.04.079

Liu X, Liu Y, Guo J, Yin S, Yang X (2017) Microfluidization initiated cross-linking of gliadin particles for structured algal oil emulsions. Food Hydrocoll 73:153–161. https://doi.org/10.1016/j.foodhyd.2017.07.001

López-Fandino R (2006) Functional improvement of milk whey proteins induced by high hydrostatic pressure. Crit Rev Food Sci Nutr 46(4):351–363. https://doi.org/10.1080/10408690590957278

Maherani B, Ali M, Stephane K, Monique S (2018) Design of biosystems to provide healthy and safe food. Part A : effect of emulsifier and preparation technique on physicochemical, antioxidant and antimicrobial properties. Eur Food Res Technol. https://doi.org/10.1007/s00217-018-3108-2

McCarthy NA, Kennedy D, Hogan SA, Kelly PM, Thapa K, Murphy KM, Fenelon MA (2016) Emulsification properties of pea protein isolate using homogenization, microfluidization and ultrasonication. Food Res. Int. 89(1):415–421. https://doi.org/10.1016/j.foodres.2016.07.024

Meng X, Bai Y, Gao J, Li X, Chen H (2017) Effects of high hydrostatic pressure on the structure and potential allergenicity of the major allergen bovine b-lactoglobulin. Food Chem 219:290–296. https://doi.org/10.1016/j.foodchem.2016.09.153

Microfluidics International Corporation (2008) Manual of 110 Y microfluidizer. Microfluidics Corp, Newton

Microfluidics International Corporation (2014) Digitally controlled lab unit for small sample material processing. http://www.microfluidicscorp.com/images/stories/pdf/lm10.pdf. Accessed 10 Oct 2014

Monroy-Villagrana A, Alamilla-Beltrán L, Hernández-Sánchez H, Gutiérrez-López GF (2015) Hydrodynamic characterization of the formation of alpha-tocopherol nanoemulsions in a microfluidizer. In: Food Nanoscience and nanotechnology. Springer International Publishing, pp 163–175

Monroy-Villagrana A, Cano-Sarmiento C, Alamilla-Beltrán L, Hernández-Sánchez H, Gutiérrez-López GF (2014) Coupled Taguchi-RSM optimization of the conditions to emulsify α-tocopherol in an Arabic gum-maltodextrin matrix by microfluidization. Rev Mex Ing Quim 13(3):679–688

Morales R, Martínez KD, Ruiz-Henestrosa VMP, Pilosof AMR (2015) Ultrasonics Sonochemistry modification of foaming properties of soy protein isolate by high ultrasound intensity: particle size effect. Ultrason Sonochem 26:48–55. https://doi.org/10.1016/j.ultsonch.2015.01.011

Moro A, Báez GD, Busti PA, Ballerini GA, Delorenzi NJ (2011) Effects of heat-treated β-lactoglobulin and its aggregates on foaming properties. Food Hydrocoll 25:1009–1015. https://doi.org/10.1016/j.foodhyd.2010.09.021

Mozafari MR, Johnson C, Hatziantoniou S, Demetzos C (2008) Nanoliposomes and their applications in food nanotechnology. J Liposome Res 18(4):309–327. https://doi.org/10.1080/08982100802465941

Nicolai T, Britten M, Schmitt C (2011) β-Lactoglobulin and WPI aggregates: formation, structure and applications. Food Hydrocolloids 25:1945–1962. https://doi.org/10.1016/j.foodhyd.2011.02.006

Oliete B, Yassine SA, Cases E, Saurel R (2019) Drying method determines the structure and the solubility of microfluidized pea globulin aggregates. Food Res Int 119:444–454. https://doi.org/10.1016/j.foodres.2019.02.015

Olson DW, White CH, Watson CE (2003) Properties of frozen dairy desserts processed by microfluidization of their mixes. J Dairy Sci 86(4):1157–1162. https://doi.org/10.3168/jds.S0022-0302(03)73698-4

Olson DW, White CH, Richter RL (2004) Effect of pressure and fat content on particle sizes in microfluidized milk. J Dairy Sci 87(10):3217–3223. https://doi.org/10.3168/jds.S0022-0302(04)73457-8

Ozturk OK, Mert B (2018) The use of microfluidization for the production of xanthan and citrus fiber-based gluten-free corn breads. LWT Food Sci Technol. https://doi.org/10.1016/j.lwt.2018.05.025

Palazolo G, Sobral P, Wagner J (2011) Freeze-thaw stability of oil-in-water emulsions prepared with native and thermally-denatured soybean isolates. Food Hydrocoll 25:398–409. https://doi.org/10.1016/j.foodhyd.2010.07.008

Peng S, Zou L, Liu W, Liu C, McClements DJ (2018) Fabrication and characterization of Curcumin-loaded liposomes formed from sunflower lecithin: impact of composition and environmental stress. J Agric Food Chem 66:12421–12430. https://doi.org/10.1021/acs.jafc.8b04136

Penna ALB, Subbarao-Gurrama, Barbosa-Cánovasa GV (2014) High hydrostatic pressure processing on microstructure of probiotic low-fat yogurt of probiotic low-fat yogurt. Food Res Int 40(2007):510–519. https://doi.org/10.1016/j.foodres.2007.01.001

Perrier-Cornet JM, Marie P, Gervais P (2005) Comparison of emulsification efficiency of protein-stabilized oil-in-water emulsions using jet, high pressure and colloid mill homogenization. J Food Eng 66(2):211–217. https://doi.org/10.1016/j.jfoodeng.2004.03.008

Pinheiro-Torres A, Oliveirah F (1998) Residence time distribution studies in continuous thermal processing of liquid foods: a review. J Food Eng 36:1–30. https://doi.org/10.1016/S0260-8774(98)00037-5

Prasad KN, Yang BAO, Zhao M, Ruenroengklin N, Jiang Y (2009) Application of ultrasonication or high-pressure extraction of flavonoids from litchi fruit pericarp. J Food Process Eng 32(2009):828–843. https://doi.org/10.1111/j.1745-4530.2008.00247.x

Qian C, McClements DJ (2011) Formation of nanoemulsions stabilized by model food-grade emulsifiers using high-pressure homogenization: factors affecting particle size. Food Hydrocollids 25(5):1000–1008. https://doi.org/10.1016/j.foodhyd.2010.09.017

Román L, Martínez MM, Gómez M (2015) Assessing of the potential of extruded flour paste as fat replacer in O/W emulsion: a rheological and microstructural study. Food Res Int 74:72–79. https://doi.org/10.1016/j.foodres.2015.04.035

Ryan KN, Zhong Q, Foegeding EA (2013) Use of whey protein soluble aggregates for thermal stability a hypothesis paper. J Food Sci 78(8):R1105–R1115. https://doi.org/10.1111/1750-3841.12207

Sadeghipour Galooyak S, Dabir B, Zolfaghari M (2015) An innovative numerical approach for simulation of emulsion formation in a microfluidizer. Colloids Surf A Physicochem Eng Asp 487:169–179. https://doi.org/10.1016/j.colsurfa.2015.09.059

Salvia-Trujillo L, Rojas-Graü MA, Soliva-Fortuny R, Martín-Belloso O (2013) Effect of processing parameters on physicochemical characteristics of microfluidized lemongrass essential oil-alginate nanoemulsions. Food Hydrocollids 30(1):401–407. https://doi.org/10.1016/j.foodhyd.2012.07.004

Santos J, Calero N, Mun J (2018) Development of food emulsions containing an advanced performance xanthan gum by microfluidization technique. Food Sci Technol Int 24(0):1–9. https://doi.org/10.1177/1082013218756140

Sha XM, Hu ZZ, Tu ZC, Zhang LZ, Duan DL, Huang T, Wang H, Zhang L, Li X, Xiao H (2018) Influence of dynamic high pressure microfluidization on functional properties and structure of gelatin from bighead carp (Hypophthalmichthys nobilis) scale. J Food Process Preserv 2018:e13607. https://doi.org/10.1111/jfpp.13607

Sinaga H, Deeth H, Bhandari B (2018) Effect of microfluidization on casein micelle size of bovine milk. Conf Ser: Earth Environ Sci 122:012084. https://doi.org/10.1088/17551315/122/1/012084

Skurtys O, Aguilera JM (2008) Applications of microfluidic devices in food engineering. Food Biophysics 3(1):1–15. https://doi.org/10.1007/s11483-007-9043-6

Stang BM, Schuchmann H, Schubert H (2001) Emulsification in high-pressure homogenizers. Eng Life Sci 1(4):151–157. https://doi.org/10.1002/1618-2863(200110)1:4<151::AID-ELSC151>3.0.CO;2-D

Sweedman MC, Tizzotti MJ, Schäfer C, Gilbert RG (2013) Structure and physicochemical properties of octenyl succinic anhydride modified starches: a review. Carbohydr Polym 92(1):905,920–905,920. https://doi.org/10.1016/j.carbpol.2012.09.040

Tang CH, Liu F (2013) Cold, gel-like soy protein emulsions by microfluidization: emulsion characteristics, rheological and microstructural properties, and gelling mechanism. Food Hydrocollids 30(1):61–72. https://doi.org/10.1016/j.foodhyd.2012.05.008

Tu Z, Yin Y, Wang H, Liu G, Chen L, Zhang P, Zhang L (2013) Effect of dynamic high-pressure microfluidization on the morphology characteristics and physicochemical properties of maize amylose. Starch-Stärke 65(5–6):390–397. https://doi.org/10.1002/star.201200120

Tunick MH, Van Hekken DL, Cooke PH, Smith PW, Malin EL (2000) Effect of high pressure microfluidization on microstructure of mozzarella cheese. Lebensmittel-Wissenschaft Und-Technologie-Food Science and Technology 33(8):538–544. https://doi.org/10.1006/fstl.2000.0716

Uribe ST, Ibañez E, Decker E, McClements DJ, Zhang R, López-giraldo LJ, Herrero M (2018) Design, fabrication, characterization and in-vitro digestion of alkaloid-catechin-, and cocoa extract-loaded liposomes. J Agric Food Chem 66:12051–12065. https://doi.org/10.1021/acs.jafc.8b04735

Villalobos-Castillejos F, Granillo-Guerrero VG, Leyva-Daniel DE, Alamilla-Beltrán L, Gutiérrez-López GF, Monroy-Villagrana A, Jafari SM (2018) Fabrication of Nanoemulsions by microfluidization. In: Nanoemulsions. Elsevier. https://doi.org/10.1016/B978-0-12-811838-2.00008-4
Villalobos-Espinosa JC, Quintanilla-Carvajal MX, Granillo-Guerrero VG, Alamilla-Beltrán L, Hernández-Sánchez H, Perea-Flores MJ, Azuara-Nieto E, Gutiérrez-López GF (2019) Effect of two-fluid nozzles on the stability characteristics of emulsions prepared by a high-energy method (microfluidization). Rev Mex Ing Quim 18(1):165–180
Wang B, Li D, Wang LJ, Liu YH, Adhikari B (2012) Effect of high-pressure homogenization on microstructure and rheological properties of alkali-treated high-amylose maize starch. J Food Eng 113(1):61–68. https://doi.org/10.1016/j.jfoodeng.2012.05.021
Wang J, Shi A, Agyei D, Wang Q (2017) Formulation of water-in-oil-in-water (W/O/W) emulsions containing trans-resveratrol. RSC Adv 2017(7):35917–35927. https://doi.org/10.1039/c7ra05945k
Wang XM, Zhu XM, Zhang NH, Tu ZC, Wang H, Liu GX, Ye YH (2018) Morphological and structural characteristics of rice amylose by dynamic high – pressure microfluidization modification. J Food Process Preserv 42:e13764. https://doi.org/10.1111/jfpp.13764
Wooster TJ, Golding M, Sanguansri P (2008) Impact of oil type on nanoemulsion formation and Ostwald ripening stability. Langmuir 24(22):12758–12765. https://doi.org/10.1021/la801685v
Xing J, Cheng Y, Chen P, Shan L, Ruan R, Li D, Wang L (2018) Effect of high-pressure homogenization on the extraction of sulforaphane from broccoli (Brassica oleracea) seeds. Powder Technol 358:103–109. https://doi.org/10.1016/j.powtec.2018.12.010
Zhong JZ, Liu W, Liu CM, Wang QH, Li T, Tu ZC, Luo SJ, Cai XF, Xu YJ (2012) Aggregation and conformational changes of bovine beta-lactoglobulin subjected to dynamic high-pressure microfluidization in relation to antigenicity. J Dairy Sci 95:4237–4245. https://doi.org/10.3168/jds.2012-5333
Zhong J, Fu S, Yu H, Zhou L, Liu W, Liu C, Prakash S (2019) Antigenicity of β-lactoglobulin reduced by combining with oleic acid during dynamic high-pressure microfluidization: multi-spectroscopy and molecule dynamics simulation analysis. J Dairy Sci 2013:1–10. https://doi.org/10.3168/jds.2018-14898
Zúñiga R, Kulozik U, Aguilera J (2011) Ultrasonic generation of aerated gelatin gels stabilized by whey protein β-lactoglobulin. Food Hydrocoll 25:958–967. https://doi.org/10.1016/j.foodhyd.2010.09.010

Biopolymer Nanocomposite Based Food Packaging

Sherif M. A. S. Keshk and Ameni Brahmia

1 Introduction

Regular fibers are considered as an environmentally friendly material that exhibits good physical properties comparing with petroleum-based plastics. In 2015, the global natural fiber reinforced polymer composites industry sector was about U$ 5.1 billion. The employment of natural fiber reinforced polymer composites has recently expanded in the shopper goods as developing packing sectors. As showed by evaluations over 2012–2017, the biodegradable fiber reinforced polymer composites industry is expected to grow 15% worldwide (Layth et al. 2015). Packaging is a significant subject in developed civilizations; especially, food packaging has experienced a great growth because most commercialized foodstuffs are marketed in packages (Lopez-Rubio et al. 2004). One of the most important roles of packaging technology is to avoid deterioration of food, thus, to extend shelf-life and maintain quality and safety of the packaged food. Starch is one of the most considered biopolymers owing to its wide availability, biodegradability, non-toxicity, and low cost. Despite advantages, starch films are moisture sensitive and very fragile in comparison to the conventional packaging materials (Flores et al. 2007). In order to recover their usability, incorporation of plasticizers, blending with other materials, chemical modifications, and the combinations of these treatments have been attempted (Colussi et al. 2017). On the other hand, natural fibers are neither artificial nor man-made; they can be extracted from animals or plants (Ticoalu et al. 2010). The use of natural fibers from resources, renewable and non-renewable including oil palm, flax, and sisal, gained significant attention in the last decades. Plants that can

S. M. A. S. Keshk (✉)
Nanomaterials and Systems for Renewable Energy Laboratory, Research and Technology Center of Energy, Technoparc borje cedria, BP, Hammam Lif, Tunisia

A. Brahmia
Laboratoire des Materiaux et de l'Environnement pour le Developpement Durable, University of Tunis El Manar, Tunis, Tunisia

U. Hebbar et al. (eds.), *Nano-food Engineering*, Food Engineering Series,
https://doi.org/10.1007/978-3-030-44552-2_7

produce cellulose fibers are categorized into bast filaments (such as jute, ramie, flax, kenaf, and hemp), seed fibers (such as cotton, kapok, and coir), leaf fibers (such as sisal and abaca), and grass and reed fibers (such as rice, wheat, and corn), and core fibers (hemp, jute, and kenaf) as well as all other kinds as roots and wood (Faruk et al. 2012; Keshk 2014). The fiber-reinforced polymer matrix got extensive consideration in several aspects owing to their good physical properties and superior benefits over artificial fibers in term of its cost, low weight, less damage to handling equipment, good mechanical characterization as tensile and flexural modulus, improved surface finish of composites, being abundant, renewable resources, biodegradability, flexibility during processing, and least health risks (Shalwan and Yousif 2013; Keshk and Yahia 2018; Alghamdi et al. 2019). Natural fiber reinforced composites possess a high specific stiffness and strength that can be realized by adding the tough and lightweight fiber onto the synthetic polymer (thermoplastic and thermoset) (Xie et al. 2010). However, natural fibers are not problems-free, as they have deficits in physical properties. The structure of natural fibers is composed of cellulose, pectin, hemicelluloses, lignin, and waxy substances. They permit moisture absorption from the surroundings that may cause weak bindings between the polymer and fiber. Furthermore, the lacking compatibility between natural polymers and fibers is considered as a challenge since the chemical structures of both polymers and fibers are two different. Modifications of the natural fiber using certain treatments are required in order to compatibilized both polymers. Such modifications are focused on the use of reactive functional groups that have the capacity for reacting with the fiber structures and changing the chemical composition. Thus, fiber modifications may cause reduction of moisture absorption that lead to an excellent improvement of compatibility between the polymer and fiber (Ray and Bousmina 2005). Cellulose has been successfully shaped into nanoforms, namely nanofibrillated cellulose (NFC) and micro-fibrillated cellulose (MFC) (Nogi et al. 2009). NFC can deliver reactive site and possess excellent mechanical and oxygen barrier properties. Whereas, high loading of MFC is commonly required to enhance the mechanical stability of the composite, because of the MFC is larger, and it decreases intrinsic energy and element ratio. MFC films displayed opportunity for food packing owing to its high rigidity together with the essential barrier properties (Zhang et al. 2011; Aulin et al. 2012). Current developments in packaging material have focused on high barrier performance with a minimal amount of material. Therefore, there is a need for functional coatings of cellulosic materials for special or enhanced properties in the field of food packaging (Gouda and Keshk 2010).

2 Starch

In order to recover starch usability, blending with other materials, chemical modifications, and the combinations of these treatments have been evaluated (Colussi et al. 2017; Keshk et al. 2018a, b). Starch films with clay nanoparticles were prepared and their tensile, barrier properties and transition temperature were

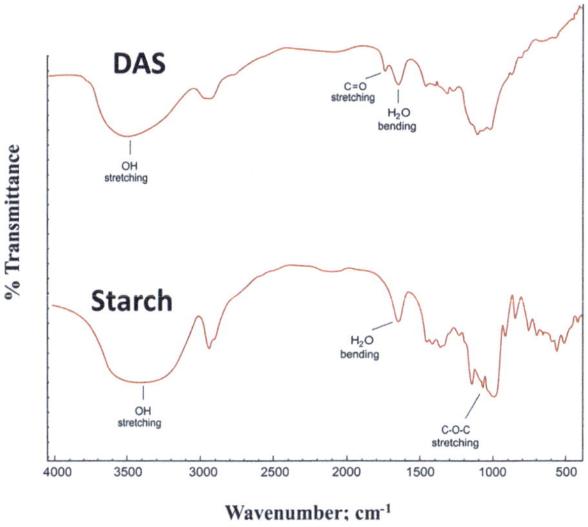

Fig. 1 FT-IR spectra of starch & dialdehyde starch (DAS)

investigated (Souza et al. 2012). The presence of clay affected the tensile and barrier properties; whereas, the glass transition temperature was not affected (Souza et al. 2012). In additional investigation, a biodegradable film obtained from acetylated starch displayed higher moisture content and water solubility in comparison to starch film (Colussi et al. 2017). Among various derivatives of starch, dialdehyde starch (DAS) that produced from the periodate oxidative cleavage of C2–C3 bond of the starch molecule. FT-IR spectroscopy data of starch and DAS are shown in Fig. (1) (Keshk et al. 2016a, b). In the fingerprint region from 800 cm^{-1} to 1500 cm^{-1}, the spectral features of starch and DAS were dissimilar; where the peaks in starch at 1158 cm^{-1} and 930 cm^{-1} that are characteristic to a C–O stretching vibration were missing in DAS (Keshk et al. 2016a, b). Furthermore, a new absorption band at 1735 cm^{-1} appeared in the DAS spectrum, which was assigned to the C=O stretching vibration.

DAS is highly reactive and possesses higher thermal stability (T = 355 °C) than starch (T = 191 °C), hence, it is widely used as a cross-linking agent (Keshk et al. 2016a, b; Wisniewska et al. 2016). The rigidity of cross-linked DAS matrix formed due to hydrogen bonds confers a better thermal stability to DAS (Wisniewska et al. 2016). Nanotechnology can remarkedly recover the quality and safety of food packing. Different food concerns have used nanotechnology to develop safer and more attractive products with longer shelf-life and lower costs. Nanomaterials in food packing technology can alter permeation properties, increase barrier properties, improve mechanical and heat-resistance, confer active antimicrobial and antifungal surfaces, and detect and signal microbiological and biochemical changes (Cichello 2015). Magnetic iron oxide nanoparticles (MNPs) have been used in various arenas owing to their unique properties including large specific surface area and simple

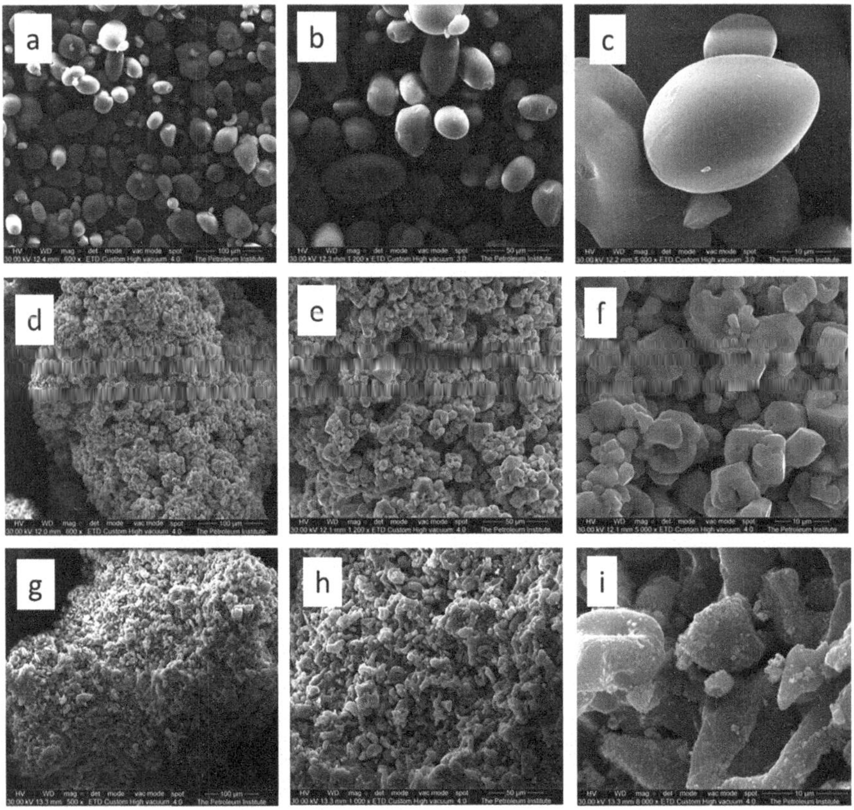

Fig. 2 SEM micrographs at different magnifications for potato starch (**a-c**), DAS, (**d-f**), and MNPs/DAS (**g-i**)

separation with magnetic fields. For food-related applications, they have been used for enzyme immobilization, protein purification, and food analysis (Caoa et al. 2012). The incorporation of MNPs with a polymer packing film will modify barrier properties and prevent the destruction of basic structure of starting materials. Besides this, the addition of MNPs will significantly recover the mechanical (tensile strength, elongation at break) and thermal properties (Rešček et al. 2016). Furthermore, an MNPs/DAS composite showed good thermal stability, small particle size, low biological toxicity, and slow anticancer drug-releasing capability (Saikia et al. 2015). Recently, MNPs/ DAS composite film was fabricated for use in food packaging (Keshk et al. 2018a, b). Scan electron microscope showed that the starch granules were of ellipsoidal morphologies with smooth surfaces and were larger than those of the DAS and MNPs/DAS granules, as the latter were of irregular polygonal shape (Fig. 2).

A comparison of the DAS and MNPs/DAS granules at high magnifications (Fig. 2f and i) indicated that the DAS granules were smoother, while MNPs/DAS

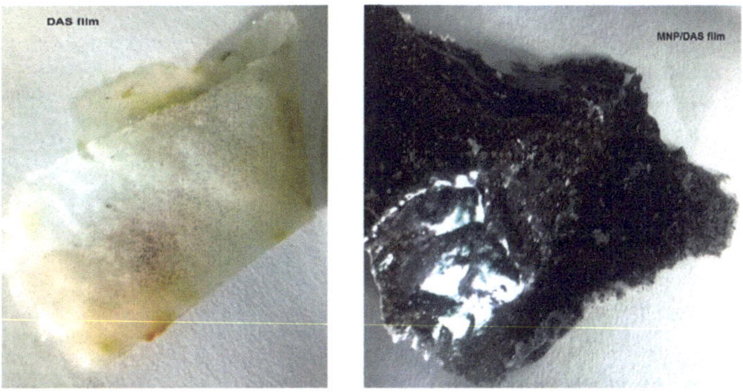

Fig. 3 DAS and MNPs/DAS Films

granules designed agglomerations owing to the magnetic properties of iron. It can also be caused by the crosslinking of DAS and MNPs. The average pore sizes in DAS and MNPs/DAS composite were 4.25 μm and 0.65 μm, respectively. Thus, the MNPs/DAS composite film was more compact with a higher thickness than DAS film (Fig. 3).

Furthermore, thermogravimetric analysis confirmed that the MNPs/DAS film was thermally stable up to 300 °C, and the maximum decomposition happened at 500 °C. Furthermore, SEM micrographs recognized the homogenous distribution of MNPs onto the DAS film. The MNPs/DAS composite film displayed lower water vapor transmission in comparison to DAS film, thus it can be decided that MNPs improved the hydrophobicity and mechanical properties of starch film. Therefore, the prepared MNPs/DAC composite film could be considered for oxygen sensitive foods like Chilled meat, as it contains a highly effective oxygen scavenger (Fe_3O_4) (Keshk et al. 2018a, b).

3 Cellulose

Cellulose fibres offer a large renewable raw material base for various kinds of products, e.g., from paper and board to various composite materials ranging from packaging applications to building materials. The physical modification of cellulose fibres has recently received attention in the field of composite applications (Alghamdi et al. 2019). The physicochemical modification of cellulose is important for making it more compatible with nonpolar polymers by introducing a hydrophobic polymer into its structure (Thielemans et al. 2008). Therefore, there is a need for new derivatives of cellulosic materials for special or enhanced properties in the field of food packaging. Zinc oxide is generally used in several applications, such as in the pharmaceutical, commodity chemical, and glass industries (Arrieta et al. 2010; Jia et al. 2012; Ko et al. 2014; Espitia et al. 2016). ZnO is presently listed as a safe

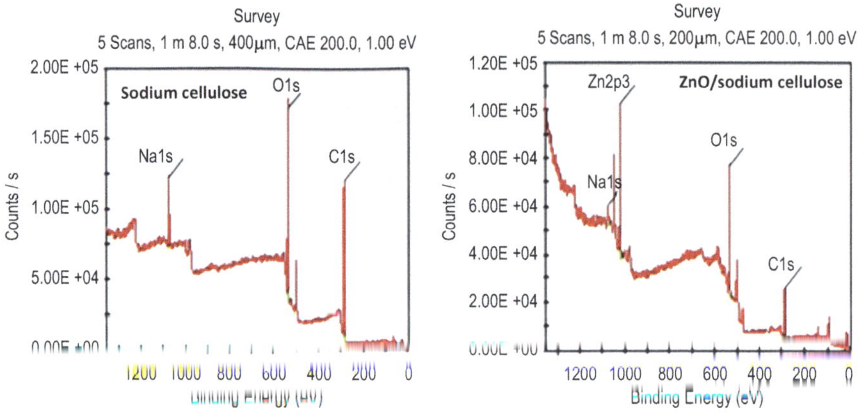

Fig. 4 XPS survey scan spectra recorded with a photon energy of Al Kα (hv = 1486.6 eV)

food additive and preservative by the US Food and Drugs Administration. The reported synthetic methods of metal/cellulose composite preparation have poor metal stability in the composite and heterogeneous distributions of metals (Ma et al. 2013; Al-Sehemi et al. 2014). The preparation of metal/cellulose composites via the reduction of metal salts in aqueous suspensions of cellulose has been investigated (El-Kemary et al. 2013). In this method, a soluble metal salt is used with a suitable reducing agent in the presence of a co-stabilizer to avoid agglomeration and to improve the metal particle distribution on the cellulose (Yue et al. 2013). On the other hand, microwave-assisted synthesis of metal/cellulose composites have been reported, in which ionic liquids were used to obtain cellulose fibres coated with metal particles (Dinand et al. 2002; Keshk et al. 2016a, b). In the fact, the hydrophilicity of cellulose made it difficult to form a good composite with metals owing to hydrogen bonds between cellulose chains that hinder any other metal or metal oxide from forming a good composite. Sodium hydroxide treatment (mercerization) of natural cellulose (cellulose-$(OH)_3$) fibres results in the structural transformation to sodium cellulose (cellulose-$(ONa)_3$) (Keshk 2015; Keshk and Hamdy 2019). During mercerization, cellulose I proceeds through a crystal-to-crystal phase transformation. The intermediate structure between the parallel chains (cellulose I) and anti-parallel chains (cellulose II) is sodium cellulose (Keshk et al. 2018a, b; Keshk and Hamdy 2019). Sodium cellulose increases the interplanar distance between cellulose chains owing to O^- Na^+ group formation. The mercerization affects the twisting and swelling of cellulose because of the presence of Na^+ ions, which play a crucial role in widening the accessible regions between the lattice planes to allow diffusion of the Na^+ ions into those planes (Keshk et al. 2018a, b; Keshk and Hamdy 2019). Zinc oxide/sodium cellulose composite with different ZnO loadings was prepared and characterized by FT-IR, XRD, XPS, SEM, and EDX Keshk and Hamdy 2019). ZnO/sodium cellulose showed a similar XRD pattern to that of cellulose II. The XPS spectrum emphasized the presence of Zn^{2+} and Na^+ ions in the prepared composite (Fig. 4) (Keshk and Hamdy 2019).

Table 1 Barrier properties of cellulose, sodium cellulose and its composite

Sample	Air permeability (cm^3/ (m^2/Pa. S))	Water vapour permeability (g/m s Pa)
Cellulose	1.05	4.68 X 10^{-9}
Sodium cellulose	1.15	4.12 X 10^{-9}
ZnO/sodium cellulose	0.137	2.44 X 10^{-9}

Furthermore, the SEM analysis displayed the growth of ZnO crystals on the sodium cellulose surface. UV-Vis spectrometry of the ZnO/sodium cellulose spectrum exhibited a distinguished absorption band at 360 nm that was attributed to the presence of ZnO particles on the cellulosic fibres (Keshk and Hamdy 2019). Therefore, the ZnO/sodium cellulose fibres swelled, and their diameter increased from 15 μm to 30 μm. Furthermore, ZnO/sodium cellulose exhibited lower air and water vapor permeability than cellulose and sodium cellulose and can be perfectly utilized as food packing film to increase the shelf life of preserved foods (Table 1) (Keshk and Hamdy 2019).

4 Chitosan

Chitosan is considered as a safe natural polymer that possesses biocompatible and biodegradable properties. Chitosan is a polysaccharide produced by chemical or enzymatic deacetylation of chitin. There are at least four prepared methods presented: ionotropic gelation (Calvo et al. 1997), microemulsion (Maitra et al.1999), solvent emulsification diffusion (Niwa et al. 1993) and polyelectrolyte complex (Erbacher et al. 1998). The most widely developed methods are ionotropic gelation and self-assemble polyelectrolytes; as there are simple and mild preparation method without the use of organic solvent or high shear force (Tiyaboonchai 2003; Sudarshan et al. 1992). The nanofibers based on chitosan have large surfaces and high porosity therefore potentially working in the fields of enzyme immobilization (such as chitosan/PVA), filtration (chitosan/PEO, Chitosan/nylon-6...) wound dressing (chitosan/PVA, Chitosan/PET), tissue engineering (Chitosan/collagen) drug delivery (PEG-g-CS/ PLGA...) and catalysis (CS-g-AA) (Sun and Li 2011). Depending on the technique of preparation, chitosan-based nanofibers size can vary from 10 nm to 4 nm (Sun and Li 2011; Van et al. 2013). Chitosan-based nanofibers were prepared via acid hydrolysis of chitin whiskers (Lu et al. 2004; Sriupayo et al. 2005). The average dimensions of the whiskers obtained were 500–417 nm (length) and 50–33 nm (diameter) (Lu et al. 2004; Sriupayo et al. 2005). The suspension consisted of individual chitin fragments that having a spindle shape. As shown in the AFM image of a dilute suspension of chitin whiskers (Fig. 5).

These fragments have a broad distribution in length (L) ranging from 100 to 650 nm and diameter (D) ranging from 10 to 80 nm. SEM images of the fractured

Fig. 5 AFM imaging of a dilute suspension of chitin whiskers

surface of the GSPI sheet and nanocomposites of SPI-5, SPI-15, and SPI-30 are shown in Fig. 6.

The GSPI sheet exhibits a relatively uniform surface. The chitin whiskers, as particles, are easily identified. A relatively uniform distribution of the chitin whisker in the SPI matrix can be observed when the chitin content is lower than 15 wt %. However, as the chitin content increases, the resulting nanocomposites show agglomerates of whiskers. The diameter of whiskers determined by SEM is larger than identified by AFM shown in Fig. 2, which resulted from a charge concentration effect due to the emergence of whiskers from the observed surface.

This suggests that the adhesion between SPI and chitin whisker is strong. Chitosan nanoparticles can be obtained by ionic gelation, where the positively charged amino groups of chitosan form electrostatic interactions with polyanions worked as cross-linkers, such as tripolyphosphate (López et al. 2005). Chitosan–tripolyphosphate (CS–TPP) nanoparticles that incorporated with hydroxy propyl methylcellulose (HPMC) films was significantly improved the mechanical and barrier properties (Caner and Cansiz 2007; De Moura et al. 2009). Chitosan-based nanocomposites have several applications like food industry, medicine and paper manufacturing, etc. These nanocomposites, films and coating based on chitosan, offer capable improvement in food packaging industrial because of chitosan's capability to hold up the life of food products (Shahidi et al. 1999). Different types of nanofillers with different nanoscale have been synthesized and investigated till now (Radhakrishnan et al. 2015). Therefore, physicochemical properties along with the types, size, and surface areas of nanofillers could significantly improve the chitosan. Clay-based nanocomposite enhanced oxygen barricade capacity of chitosan film (Radhakrishnan et al. 2015). Therefore, the integration of clay into the chitosan matrix enhances the transport and barrier characteristics. For the substitute of plastic resources to decrease the ecological contamination, chitosan/methyl cellulose nanocomposite can substitute the plastic in packaging foods (Mura et al. 2011).

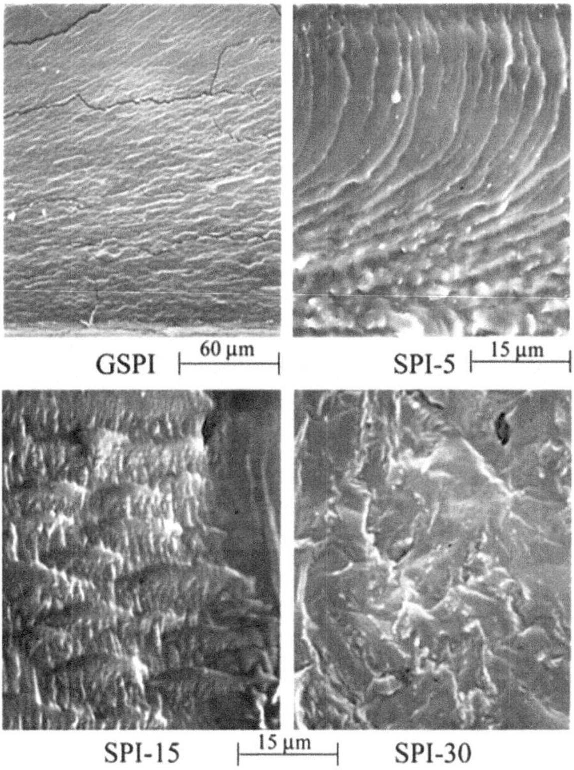

Fig. 6 SEM image of GSPI sheet and SPI/chitin whiskers nanocomposites

5 Conclusion

Nanotechnology has revealed numerous advantages in food packaging industry, including better barriers; mechanical, thermal, and biodegradable properties. However, the use of nanocomposites in food packaging might be challenging owing to the reduced particle size of nanomaterials and the physicochemical characteristics of such nano- materials may be quite different from those of their macro-scale counterparts. Despite several advantages of nanomaterials, their use in food packaging may cause safety problems to human health since they exhibit different physicochemical properties from their macro-scale chemical counterparts. Different research disciplines including identification, characterization, and quantification of the nanoparticles are requested.

References

Alghamdi M, Awwad N, Al Sharaey A, Abd Rabboh H, Keshk SMAS (2019) Physicochemical characterization of the natural hydroxyapatite/ cellulose composite. Indian journal of fiber and textile technology 44:45–50

Al-Sehemi AG, Al-Shihri AS, Kalam A, Du G, Ahmad T (2014) Microwave synthesis, optical properties and surface area studies of NiO nanoparticles. J Mole Struct 1058:56–61

Arrieta MP, López J, López D, Kenny JM, Peponi L (2010) Polym Degrad Stab 95:2126–2146

Aulin C, Salazar-Alvarez G, Lindstr¨ T (2012) High strength, flexible and transparent nanofibrillated cellulose–nanoclay biohybrid films with tunable oxygen and water vapor permeability. Nanoscale 4:6622–6628

Calvo P, Remunan-Lopez C, Vila-Jato JL, Alonso MJ (1997) Novel hydrophilic chitosan-polyethylene oxide nanoparticles as protein carriers J Appl Poly Sci 63:125–132

Caner C, Cansiz O (2007) Effectiveness of chitosan-based coating in improving shelf life of eggs J Sci Food Agric 87:227–232

Caoa M, Lia Z, Wang J, Gea W, Yuea T, Li R, Colvin VL, Yu WW (2012) Food related applications of magnetic iron oxide nanoparticles: Enzyme immobilization, protein purification, and food analysis. Trends Food Sci Technol 27:47–56

Cichello SA (2015) Oxygen absorbers in food preservation: a review. J Food Sci Technol 52:1889–1895

Colussi R, Pinto VZ, El Halal SLM, Biduski B, Prietto L, Castilhos DD, Zavareze ER, Dias ARG (2017) Acetylated rice starches films with different levels of amylose: mechanical, water vapor barrier, thermal, and biodegradability properties. Food Chem 221:1614–1620

De Moura MR, Aouada FA, Avena-Bustillos RJ, McHugh TH, Krochta JM, Mattoso LHC (2009) Improved barrier and mechanical properties of novelhydroxypropyl methylcellulose edible films with chitosan/tripolyphosphatenanoparticlses. J Food Eng 92:448–453

Dinand E, Vignon M, Chanzy H, Heux L (2002) Mercerization of primary wall cellulose and its implication for the conversion of cellulose I→cellulose II. Cellulose 9:7–18

El-Kemary M, Nagy N, El-Mehasseb I (2013) Nickel oxide nanoparticles: synthesis and spectral studies of interactions with glucose. Mater. Sci. Semi.Cond. PRO 16:1747–1752

Erbacher P, Zou S, Steffan AM, Remy JS (1998) Chitosan-based vector/DNA complexes for gene delivery: biophysical characteristics and transfection ability. Pharm Res 15:1332–1339

Espitia PJP, Otoni CG, Soares NFF (2016) In: Barros-Velazquez J (ed) Antimicrobial Food Packaging. Elsevier Academic Press, Cambridge, MA, USA, pp 425–431

Faruk O, Bledzki AK, Fink HP, Sain M (2012) Bio-composites reinforced with natural fibers: 2000-2010. Prog Polym Sci 37:1552–1596

Flores S, Famá L, Rojas JM, Goyanes S, Gerschenson L (2007) Physical properties of tapioca-starch edible films: influence of filmmaking and potassium sorbate. Food Res Int 40:257–265

Gouda M, Keshk SMAS (2010) Evaluation of multifunctional properties of cotton fabric based on metal/chitosan film. Carbohydr Polym 80:504–512

Jia N, Li SM, Ma MG, Sun RC (2012) Rapid microwave-assisted fabrication of cellulose/F--substituted hydroxyapatite nanocomposites using green ionic liquids as additive. Mater Lett 68:44–46

Keshk SMAS (2014) Grafting of cellulose extracted from Kenaf using xanthate method. Journal of Basic & Applied Science 10:339–343

Keshk SMAS (2015) Effect of different alkaline solutions on crystalline structure of cellulose at different temperatures. Carbohydr Polym 115:658–662

Keshk SMAS, Ramadan AM, Sehemi AG, Yousef S, Bondock S (2016a) Peculiar behavior of starch 2, 3-dialdehyde towards sulfanilamide and sulfathiazole. Carbohydr Polym 152:624

Keshk SMAS, Abd-Rabboh HSM, Hamdy MS, Badr IH, (2016b). Physicochemical Characterization of Mercerized Cellulose-Supported Nickel-Oxide. 18th International Conference on Applied Chemistry. Zurich, Switzerland. January 12–13

Keshk SMAS, Syef AFA, El-Zahhar AA, Yousef S, Bondock S (2018a) Physicochemical characterization of calcium lignosulfonate/sodium cellulose composite and its application as food packaging film. Transylvanian Review 16:7609–7619

Keshk SMAS, Yahia IS (2018) Physicochemical characterization of graphene oxide/cellulose I and II composites and their antibacterial activity. Turk J Chem 42(2):561–572

Keshk SMAS, Zahar A, Bondock S (2018b) Synthesis of magnetic nanoparticles /Dialdehyde starch-based composite film for food packaging applications. Starch 1800035:1–7

Keshk SMAS, Hamdy M (2019) Preparation and characterization of zinc oxide/sodium cellulose composite for food packaging. Turk J Chem 43:94–105

Ko H, Mun S, Min S, Kim G, Kim J (2014) Fabrication of cellulose ZnO hybrid nanocomposite and its strain sensing behavior. Materials 7:7000–7009

Layth M, Ansari MNM, Pua G, Jawaid M, M Islam S (2015) A Review on Natural Fiber Reinforced Polymer Composite and Its Applications. Int. J. Polym. Sci.:1–15

López-León T, Carvalho ELS, Seijo B, Ortega-Vinuesa JL, Bastos-González D (2005) Physicochemical characterization of chitosan nanoparticles: electrokinetic and stability behavior. J COLLOID INTERF SCI 283:344–351

Lopez-Rubio A, Almenar E, Hernandez-Mun˜oz P, Lagaron R, Gavara R (2004) Overview of active polymer-based packaging Technologies for Food Applications. Food Rev Intl 20:357

Lu Y, Weng L, Zhang L (2004) Morphology and properties of soy protein isolatethermoplastics reinforced with chitin whiskers. Biomacromolecules 5:1046–1051

Ma MG, Qing SG, Li SM, Zhu JF, Fu LH, Sun RC (2013) Microwave synthesis of cellulose/ CuO nanocomposites in ionic liquid and its thermal transformation to CuO. Carbohydr Polym 91:162–168

Maitra AN, Ghosh PK, De TK, Sahoo SK (1999). US patent 1999, 5,874,111

Mura S, Corrias F, Stara G, Piccinini M, Secchi N, Marongiu D, Innocenzi P, Irudayaraj J, Greppi GF (2011) Innovative composite films of chitosan, methylcellulose, and nanoparticles. J Food Sci 76(7)

Niwa T, Takeuchi H, Hino T, Kunou N, Kawashima Y (1993) Preparations of biodegradable nanospheres of water-soluble and insoluble drugs with D, L-lactide/glycolide copolymer by a novel spontaneous emulsification solvent diffusion method, and the drug release behavior. J Control Release 25:89–98

Nogi M, Iwamoto S, Nakagaito AN, Yano H (2009) Optically transparent nanofiber paper. Adv Mater 20:1–4

Radhakrishnan Y, Gopal G, Lakshmanan CC, Nandakumar KS (2015) Chitosan nanoparticles for generating novel Systems for Better Applications: A review. J Mol Genet Med S4:005. https://doi.org/10.4172/1747-0862.1000S4-005

Ray SS, Bousmina M (2005) Prog Mater Sci 50:962–1079

Reščec A, Ščetar M, Hrnjak-Murgić Z, Dimitrov N, Galić K (2016) Polymer- Plastics Technology and Engineering 55:1450

Saikia C, Hussain A, Ramteke A, Sharma HK, Deb P, Maji TK (2015) Carboxymethyl starch-coated iron oxide magnetic nanoparticles: a potential drug delivery system for isoniazid. Iran Polym J 24:815

Shahidi F, Arachchi JK, Jeon YJ (1999) Food applications of chitin and chitosan. Trends Food Sci Technol 10:37–51

Shalwan A, Yousif BF (2013) In state of art: mechanical and tribological behaviour of polymeric composites based on natural fibre. Materials & Design 48:14–24

Souza AC, Benze R, Ferrão ES, Ditchfield C, Coelho ACV, Tadini CC (2012) Cassava starch biodegradable films: influence of glycerol and clay nanoparticles content on tensile and barrier properties and glass transition temperature. Food Sci Technol 46:110

Sriupayo J, Supaphol P, Blackwell J, Rujiravanit R (2005) Preparation and characterization of chitin whisker-reinforced chitosan nanocomposite films with or without heat treatment. Carbohydr Polym 62:130–136

Sudarshan NR, Hoover DG, Knorr D (1992) Antibacterial action of chitosan. Food Biotechnol 6:257–272

Sun K, Li ZH (2011) Preparations, properties and applications of chitosan based. Express Polymer Letters 5:342–361

Thielemans B, Dufresne A, Chaussy D, Belgacem M (2008) Surface functionalization of cellulose fibres and their incorporation in renewable polymeric matrices. Compos Sci Technol 68:3193–3201

Ticoalu A, Aravinthan T, Cardona F (2010) Proceedings of the Southern Region Engineering Conference (SREC '10), Toowoomba, Australia, pp 113–117

Tiyaboonchai W (2003) Chitosan nanoparticles: A promising system for drug delivery. Naresuan University J 11:51–66

Van SN, Minh HD, Anh DN (2013) Study on chitosan nanoparticles on biophysical characteristics and growth of Robusta coffee in green house. Biocatalysis & Agricultural Biotechnology:289–294

Wisniewska JS, Drzymalska KW, Bajek A, Maj M, Sionkowska A (2016) J Mater Sci Mater Med 27

Xie Y, Hill CAS, Xiao Z, Militz H, Mai C (2010) Silane coupling agents used for natural fiber/polymer composites: A review. Compos A: Appl Sci Manuf 41:806–819

Yue Y, Gungping H, Qinglin W (2013) Bioresources 8:6460–6463

Zhang W, Yang X, Li C, Liang M, Lu C, Deng Y (2011) Mechanochemical activation of cellulose and its thermoplastic polyvinyl alcohol ecocomposites with enhanced physicochemical properties. Carbohydr Polym 83:257–263

Nanotechnology Applications to Improve Solubility of Bioactive Constituents of Foods for Health-Promoting Purposes

Silvana Alfei

1 Introduction

Starting from the consideration that often foods belong to the plant kingdom and that several plants are edible as food and consequently, that many chemicals constituents present in foods are also in plants and vice versa, with "phytochemical" is meant a plant constituent, that can be present also in plant-derived foods, that has beneficial properties, but that is not established as an essential nutrient. Differently, with the term "nutrient" is meant a chemical substance present in food, that must be assumed during the nutrition process, that is essential for the life and the metabolism of humans and that may have animal, plant or fungi source. When a nutrient possesses also health properties is named "nutraceutical". So, for better clarity, nutraceuticals are nutrients, while phytochemicals are non-nutrients but both are edible, can be ingested with the diet and possess properties beneficial for humans health. In this regard, evidences have been extensively reported, asserting that several foods can be associated to a number of health benefits that develop after their ingestion and following, *in vitro* studies have proved that many multifunctional constituents of foods, such as carotenoids, vitamins, phytosterols, polyunsaturated lipids, curcuminoids, flavonoids and polyphenols, in addition to the basic nutritional values, own extra health benefits and are considered "pharmaceutical-grade compounds". In the last decades, consumers perception and preferences toward health-promoting foods, as well as their awareness about the health properties of phytochemicals of plants, have undergone an exponential growth involving the world food market and that one of natural compounds. From this, the desire came to be able to use these beneficial compounds, both as part of a normal food regime and as food supplements or even as less toxic nature-derived therapeutics. As a direct consequence, the increasing demand of the Western societies for availability of

S. Alfei (✉)
Department of Pharmacy, University of Genoa, Genoa, Italy
e-mail: alfei@difar.unige.it

U. Hebbar et al. (eds.), *Nano-food Engineering*, Food Engineering Series,
https://doi.org/10.1007/978-3-030-44552-2_8

nutraceuticals and bioactive natural products, as well as the call for the development of novel functional foods, are today a leading trend in food and plants-derivatives industry. Furthermore, the incorporation in conventional foods and in nutraceutical formulations of natural bioactive compounds, in order to improve food quality, such as its shelf life, taste or aspect, in place of using synthetic additives and the development of novel functionalities, especially related to health promotion, such as anti-oxidant, antiradical and antitumor abilities, are highly desired from manufacturing companies and are object of intensive investigations from food scientists, food engineers and food technologists.

A very dynamic research sector is focused on obtaining and quantifying the various bioactive chemical compounds (BACs) from foods and plants, performing proper and optimized extraction methods (Turrini et al. 2019a, 2020) and suitable and conservative determination techniques (Alfei et al. 2020a). Once extracted, purified and chemically identified, based on their chemical structure and functional groups, BACs are classified into different categories such as poly-hydroxyl acid compounds (polyphenols), indoles, alkaloids, isothiocyanates, saponins, phytoprostanes, furanes, vitamins, carotenoids (Hamri et al. 2011; Jeong et al. 2015; Lim et al. 2017). The most part of food-related BACs have been isolated from vegetables, herbs, fruits, legumes, oils, spices, nuts and whole grains and like phytochemicals, have proved to be endowed with a number of properties that can enhance humans' life. Among the numerous beneficial effects concerning human health aspects, antifungal, antiviral, anti-inflammatory, antibacterial, antiulcer, anti-cholesterol, hypoglycemic, immunomodulatory and antioxidant activities are just a few (Ben Saad et al. 2017; Hsieh et al. 2015; Imm et al. 2014; Kris-Etherton et al. 2002; Bisio et al. 2016, 2017). Vitamin C administered as sodium ascorbate salt proved a strong apoptotic activity on neuroblastoma cells (Carosio et al. 2007). Remarkable healthy activities against a lot of degenerative diseases such as cardiovascular, cancer and central nervous system disabling disorders, including also Parkinson's disease, Alzheimer, multiple sclerosis, amyotrophic lateral sclerosis, have been established by polyphenols such as ellagitannins (ETs) and their metabolic derivatives as gallic and ellagic acids (GA and EA), both in *vitro* and in animal models (Hseu et al. 2012; Ahire and Mishra 2017; Larrosa et al. 2010; Heber 2008; Nejad et al. 2015, 2017; Firdaus et al. 2018). It is now commonly recognized, that the most part of humans diseases are triggered by oxidative stress (OS), in turn caused by the uncontrolled accumulation of reactive chemical species known as Reactive Oxigen Species (ROS) and Reactive Nitrogen Species (RNS) and that the BACs benefits for human health mainly depend on their free radicals scavenging activity, their capability to inhibit the assembly of microtubule and microfilament, to chelate metals and to inhibit protease, among others (Lu et al. 2016; Chang et al. 2016; Son et al. 2016). Concerning whole foods, epidemiological studies have suggested that red wine consumption is related to a reduction in overall mortality caused by cardiovascular events evoked by atherothrombosis (Renaud et al. 1998). Although the exact nature of the protective effect of red wine is unclear, it might be partially attributed to its ability to reduce the progression of atherosclerotic lesions (Hayek et al. 1997). Green tea has proved to have protective effects on cardiovascular diseases (Cabrera et al. 2006). Furthermore, isolated BACs as GA and EA and other more complex natural products such as propolis extracts (which contain phenolic acids, flavonoids, terpenes and essential oils) have demonstrated to be effective in the control of some infectious diseases caused by bacteria also of sporogenic type. In particular, GA proved to be effective against a number of Gram-positive and Gram-negative bacteria as *Paenibacillus larvae*, *Staphilococcus aureus* (Vico et al. 2016; Pu et al. 2019), *Escherichia coli, Pseudomonas fragi, Pseudomonas fluorescens, Pseudomonas putida, Pseudomonas spp. P30–4, Plesiomonas schigelloides* and *Schigella flexneri B* (Díaz-Gómez et al. 2014; Sorrentino et al. 2018; Rattanata et al. 2016). Other polyphenols as 2-(3, 4-dihydroxy-phenyl)-ethanol (DHPE) from olive oil, tumerone and curcumin from *Curcuma longa* L. rhizome, GA again, protocatechuic acid, quercetin and resveratrol from fruits and vegetables in general (Fig. 1) are capable to inhibit platelet aggregation and ROS activity induced by thrombin, collagen or other agonists (Petroni et al. 1995; Lee 2006; Kim et al. 2012; Crescente et al. 2009). Allicin (from garlic), curcumin (from turmeric), catechins (from tea leaves) may help to prevent and treat cancer, cardiovascular illness, neuronal degenerative diseases, diabetes, infectious (Fig. 1) (Pandey and Rizvi 2009; Pham-Huy et al. 2008).

Phytochemicals as terpenoids, that constitute one of the largest group of natural products, mainly extracted from plants, accounting for more than 40,000 individual compounds, with several new compounds being discovered every year, have proved activity against several cancers, including mammary, skin, lung, forestomach, colon, pancreatic and prostate carcinomas. In particular, a large number of triterpenoids have been shown to suppress the growth of a variety of cancer cells without exerting any toxicity in normal cells and to possess chemo-preventive and

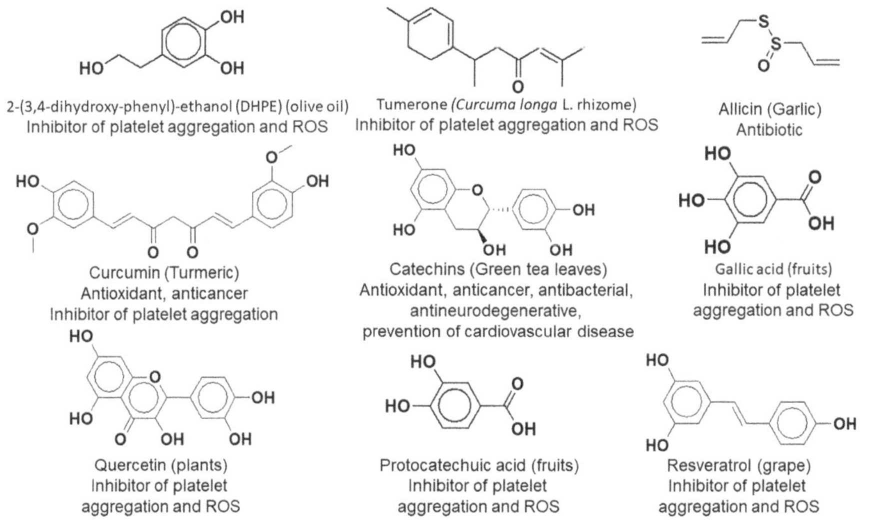

Fig. 1 Some examples of BACs from vegetable foods

therapeutic effects against colon, breast, prostate and skin cancer (Thoppil and Bishayee 2011). Terpenoids are present in a lot of spices and recently di- and triterpenoids from different typology of Salvia, showed a multi-target bioactivity and antibacterial, hypoglycemic and anticancer properties among the others (Bisio et al. 2016, 2017). Food-related bioactive compounds and phytochemicals, not only attract food industry, but are arising an increasing interest also for applications in cosmetics or in medicinal formulations, as skin and hair care products, that could positively interact with the biological function of skin and hair, for enhancing their beauty and health. Unfortunately, the idea of exploiting the strong potentials of foods and related nutraceuticals or phytochemicals for healthy purposes, is not so simply viable for many reasons (Ting et al. 2014). Starting from freshly harvested edible plants and foods, BACs naturally occurring in the food matrix, can be degraded during extraction process or subsequently, from foods processing techniques such as cooking (Bongoni et al. 2013, 2015; Palermo et al. 2014). As a consequence, industrially processed foods often, contain less health-promoting compounds than fresh or frozen foods, and therefore, even if extensively consumed with the daily diet, they would not contribute to reduce the risk of preventable diseases (Liu 2004; Rao and Rao 2007). In this regard, the main cause for the loss of nutraceuticals or phytochemical compounds due to cooking, is thermal decomposition (Palermo et al. 2014). On the other hand, the rational solution of incorporating isolated BACs into commercial food products, in order to increase the bioactive compounds concentration and thus preparing BACs-enriched functional foods, as well as the appealing idea to formulate BACs in pharmaceutical preparations to achieve Food Supplements or food-based therapeutics, are further challenging tasks. BACs poor stability and low rate of solubility strongly limit the feasibility of such resolutions (Lee et al. 2015; Teleki et al. 2013; Yousuf et al. 2016). Furthermore, concerning the actual possibility to exploit BACs as health promoters after ingestion, it has to be remembered that, when orally administered, for an effective therapeutic result, a bioactive compound needs to reach a proper concentration in blood/plasma and then in cells and tissues and ideally, for a protracted time. Solubility in an aqueous environment, permeability through the epithelial cell membrane and molecular interactions in gastro intestinal (GIT) fluids, are therefore additional pivotal factors, that strongly influence the route of BACs towards the bloodstream, and their final distribution to targets and so their actual bio-efficacy. Unfortunately, the majority of these food and plants beneficial constituents are water-insoluble and, in whatever form they are orally introduced, if not properly manufactured, rarely they manage to reach the site of action in therapeutically effective concentration. In addition, nutraceuticals and phytochemicals, after ingestion, are recognized as xenobiotics and promptly degraded by the different digestion chemical, enzymatic and microbial phases in mouth, stomach, small and large intestine (Ting et al. 2014). Then, the survived fraction and the still bioactive metabolites, if they would absorbed, frequently undergo fast metabolism and are excreted via renal, biliary or pulmonary (Xiao et al. 2017). In this sense, BACs bio-efficacy is strongly influenced by their chemical nature and unfortunately, evidences assert that, due to the cited drawbacks, after oral intake, BACs reach only the nano/picomolar concentration range in cells and tissues, i.e. an insignificant dose for getting an efficient response in terms of health promotion. All these considerations explain why when the activity of BACs upon *in vivo* studies is evaluated, the most part of them were not competent to translate the activities previously demonstrated *in vitro*. The first approach to achieve the required bioavailability rates and to take advantage of BACs biological activities, was to resort to higher doses, but even if an improved efficacy was showed it was associated also to systemic toxicity (Simion et al. 2016).

In the last decades, for the progress and prosperity of mankind and inspired by nature's creation, that had shaped the essential building blocks of life in nanoscale, researchers in many fields, had engineered nanomaterials. So, in order to overcome all issues concerning the application of nutraceuticals and phytochemicals as health enhancers, nanotechnology and nanostructures with a magnitude of 10^{-9} m were resort to. By the design of structures, devices and systems with controlled shape, and size arranged at the nanoscale, several food-grade delivery systems have been developed to overcome solubility, permeability and pharmacokinetics drawbacks of bioactive nutrients and non-nutrients (Ting et al. 2014). In the meantime, edible food or plants-related nanoparticles were and are designed and developed for acting as colour, flavor and preservative additives, in order to improve food shelf life, taste and look. In this contest, a number of methods, as particle size reduction, crystal engineering, salt formation, use of surfactants, as well as prodrugs design by physicochemical modifications, solid micro- and nano-dispersions, complexation and complexation inclusion techniques, have been investigated and implemented (Zuccari et al. 2020). Drug Delivery (DD) is a specific engineered technology dealing with the development of drugs carriers known as Drug Delivery Systems (DDSs), that usually resorts to nanotechnology and to the use of solid nanoparticles for its purposes. Nanoparticles (NPs) are heavy involved in the development of devices capable to solve the questions related to BACs formulation and delivery, to the improvement of BACs solubility, bioavailability and cells uptake, to their protection from early degradation and fast metabolism, while minimizing the impact on the sensory and quality properties of the BACs-enriched products (functional foods and beverage). Already in the 2007, the market of nanotechnology products in the food industry concerning food-grade nanosized delivery systems for functional foods and food supplements realization, beverages included, as health promoting products, was worth about \$ 1 billion, but it was foreseen to grow to more than \$20 billion in the next decade (Chau et al. 2007). Today in 2019, the food and beverage sector is a global multi trillion dollar industry and a currently estimate of the global economic impact of nanotechnology is projected to be at least \$3 trillion by 2020, which may employee six million labors in the rising nanotechnology industries worldwide (He et al. 2019).

Nowadays, several diverse types of NPs have been industrialized and tested for various uses and the numerous nanosized carriers developed for delivering bioactive natural products such as nutraceuticals and phytochemicals in a health promotion frame, are mainly natural materials. However, also biocompatible and biodegradable synthetic polymers and copolymers as polyethylene glycol (PEG), poly urethane (PUR), poly capro-lactone (PCL), poly lactic-co-glycolic acid (PLGA), polyvinylalcohol (PVA) or poly 2-vinyl pyridine (P2VP) are extensively under consideration. They are used to form micelles present in many therapeutic devices approved by the Food and Drug Administration (FDA) (Guo and Ma 2014). As an example, modified PVA polymers were used to encapsulate all *trans*-retinoic acid (ATRA) in polymeric micelles, achieving a ATRA formulations characterized by a slow ATRA release, particularly suitable to treat cancer, and endowed wih improved cytotoxic activity on neuroblastoma cells (Zuccari et al. 2005, 2009). Furthermore, also synthetic biocompatible dendrimers as polyester-based ones (Alfei et al. 2017a, Alfei and Castellaro 2017, Alfei and Catena 2018b, 2018c), are extensively under evaluation, as promising nano devices to dissolve, deliver and protect BACs. Recently, non-cytotoxic hydrophilic and amphiphilic dendrimers proved to be successful in physically entrapping (Alfei et al. 2018a, 2019a, 2020c) or covalently linking (Alfei et al. 2020b) bioactive molecules derived by foods or plants as ursolic and oleanolic acids, EA and GA, thus providing them a defensive support or enclosing them in a protective shell. The nanosized core-shell bioactive dendrimer particles so achieved, showed highly improved solubility in water or lipids, enhanced Hydrophilic-Lipophilic Balance (HLB) for a better absorbability, high degree of biocompatibility and biodegradability with consequent very low cell cytotoxicity. In particular, when the BACs were physically entrapped, the BAC-enriched dendrimer was characterized by having a food-related functional *core* and a dendrimer casing (Alfei et al. 2018a, 2019a, 2020c), while when BAC was peripheral covalently bound, the dendrimer formulation was typified by a dendrimer *core* and a bioactive shell (Alfei et al. 2020b). In addition, the drug-loaded dendrimers showed a favorable drug release profile protracted in time (Alfei et al. 2018a, 2020b, 2020c) and improved biological activities (Alfei et al. 2019a, 2019b, 2019c, 2020b, 2020c).

Since the subject of this chapter is very vast, the discussion on synthetic dendrimers, that represent a unique class of polymer NPs and needs a dedicated dissertation, will not be addressed here and only a few other references to the use of dendrimers will be made in the course of this work. This chapter provides an updated overview concerning the contribution and the advances of nanotechnology regarding the "wrap" of nutraceuticals and phytochemicals in more bioavailable food-grade NPs, for allowing their safe and effective oral administration in the frame of medical treatments and for masking eventual unpleasant taste and flavors associated to the isolated BACs. In addition, the use of BACs-enriched NPs as smart foods ingredients for improving look, taste, quality and resistance to spoilage, along storage of nourishment is also disserted. In this regard, nanoemulsions (NEs), nanosuspensions (NSs), nanoencapsulations (NEs), solid NPs-based vehicles as lipid-based, saccharide-based, protein-based, natural inorganic-based or synthetic organic-based ones, their different production methods and some applicative examples are the most topics discussed in this chapter.

2 Solubility of Bioactive Chemical Compounds (BACs): A Pivotal Factor in the Development of Food-Related Healthy Products

Solubility is essential for the GIT absorbability of any bioactive chemical molecule employed for the development of novel therapeutics. It represents a key parameter to achieve the systemic drug concentration necessary for having a significant therapeutic response (Vemula et al. 2010). Unfortunately, low aqueous solubility is the foremost and limiting issue with the majority of bioactive molecules, if considered that more than 40% of the chemicals developed in pharmaceutical industry or naturally occurring in foods and plants are practically water-insoluble (Savjani et al. 2012). The aqueous solubility is the factor which most affects the dissolution of any bioactive compound in the intestinal fluids and which remarkably contributes to their bioavailability (Stegemann et al. 2007). A limited water-solubility translates in a low absorption rate at gut level, in a very poor bioavailability and in a trivial blood and then tissues concentration. In parallel, the solubility issues of a drug, influence also the attempts to formulate them in emulsions for improving their water-dispersibility, asking for an excessive use or for very high concentrations of surfactants with the promotion of unpleasant side effects such as GIT irritation (Sivakumar et al. 2014; Wang et al. 2014). In this context, engineering nanotechnology works extensively to develop powerful solubilizing methods for improving the solubility, absorption and bioavailability of BACs, tries to minimize or even avoid the use of harmful excipients, solvents, co-solvents, emulsifiers or other additives and operates for keeping low the manufacturing costs, at the same time. In the field of Science and Technology of Foods, specifically regarding the development of nutraceuticals, functional foods and food supplements, the basic objective of nanotechnology is to solve the conflict between the exciting and very promising results for humans' health obtained *in vitro*, by the application of food-derived compounds or phytochemicals and the *in vivo* "miserable" ones.

2.1 Factors Affecting Solubility

The amount of solute that can be dissolved in a solvent depends on various factors including temperature, pressure, substances chemical nature and physicochemical properties. A brief description of the main aspects that affect chemical molecules solubility follows.

2.1.1 Particle Size and Shape

The dissolution rate of a solute strongly depends on particle size and molecular symmetry. Unsymmetrical and very small size particles, gifted with large surface area, dissolve better. The reduction of particles size and/or the disruption of molecular symmetry result in an increase in solubility and dissolution rate (Mosharraf and Nyström 1995; Pinal 2004; Ishikawa and Hashimoto 2011).

2.1.2 Temperature

The solute molecules are held together by intramolecular attractions that, during the dissolution process, are to be broken apart by the solvent molecules. Since the solvent kinetic energy increases with temperature, a higher temperature promotes dissolution (Feriyanto et al. 2014).

2.1.3 Molecular Weight

Generally, greater the molecular weight of a substance more poor its solubility. Organic molecules with a large amount of branching in carbon chains show higher solubility and, in a similar way, branched polymers are more soluble of linear ones of the same molecular weight, because the branched-chain molecules exhibit smaller volume/dimension ratio in solution thus dissolving more readily (Jadhav and Pandey 2013; Ravve 2013).

2.1.4 Molecular Polarity

The solubility rate strongly depends on intermolecular attractive forces solute/solvent that must be greater than the intramolecular ones, i.e. solute/solute. Then, generally, polar solutes/substances are dissolved by polar solvents, and nonpolar substances are dissolved by nonpolar solvents.

2.1.5 Physical Form

Molecules arranged in amorphous forms possess greater aqueous solubility than the crystalline ones. Different polymorphs have different solubility and consequently, the study of polymorphic forms is become important due to its strong influence on the physicochemical properties of the bioactive substances, their solubility and subsequently their therapeutic outcomes. (Raza et al. 2014).

2.1.6 pH of the Medium

Both hydrophilic and lipophilic compounds exhibit different solubility at different pH values. Weak acids and weak bases undergo an ionization reaction in solution. The ionized form of substance will be more soluble in water.

2.1.7 Presence of Stabilizers/Emulsifiers

The emulsifiers, are surface-active compounds encompassing a head hydrophilic group and a lipophilic tail and are able to reduce the interfacial tension between the oil and water interface and to increase the both phases solubility (Krog 1977).

3 Techniques and Approaches for Enhancing Solubility and Bioavailability of Poorly Water-Soluble Bioactive Food Constituents and Natural Products (BACs)

BACs solubility can be altered through *Particle Engineering Techniques (PETs)* and *Formulation Approaches (FAs)*. By taking into account of factors influencing molecules solubility (Sect. 2.1), PETs aim at modifying the physicochemical properties of poorly soluble substances by their rearrangement in very smaller sized and so more soluble particles, possibly without employing additives as stabilizers or surfactants.

Smaller particles with improved solubility can be achieved through the use of specific techniques such as mechanical particle-size reduction (wet-milling, dry-milling, high-pressure and ultra-high-pressure homogenization), cryogenic particle engineering (freeze-drying, spray freezing) (Kale et al. 2014; Koshy et al. 2010), nanoprecipitation, nanosuspension and supercritical fluid processing (Kale et al. 2014; Morales et al. 2016).

Formulation approaches (FAs) concern instead the formulation of the BACs in solutions which consist of a mixture of water/oil phases, stabilizers, solvents/co-solvents and other excipients. With such approaches, solid formulations, lipid formulations, as emulsion based drug delivery systems and amorphous formulations, as amorphous solid dispersions (Merisko-Liversidge et al. 2003; Pouton 2006) have been achieved by using spray drying, milling and other techniques.

More in general, emulsion and double-emulsion solvent extraction, spray-drying, freeze drying, liquid anti-solvent and solvent evaporation techniques are the most conventional ones and all of them allow the production of BACs-loaded microparticles (MPs) and nanoparticles (NPs), but, except for spray-drying and freeze drying, they have the disadvantage to be generally associated with high residual solvent content and low encapsulation efficiency.

Furthermore, excluding only freeze drying, all the mentioned techniques could evoke thermal degradations, are multi-stage processes and in most cases, they require a micronisation step in order to generate the desired particle size and size distribution. Micronization techniques, usually by air jet milling, often cause the production of cohesive, electrostatically charged products with occasional crystallographic defects (Majerik et al. 2004).

On the contrary, the untraditional supercritical fluid extraction processes, even if more expensive, offer an alternative single-step advantageous method, that allow to prepare micronized dry powders with controllable physical properties including particle size, morphology, crystallinity and residual solvent content, with reduced operations time (Jung and Perrut 2001).

Highly advantageous and very useful nanomaterials with high degree of solubility can be obtained by the use of Solid Nanoparticles Delivery Systems (SNDSs), either through the BACs physical entrapment or through the formation of covalent bonds. In such a way, BACs-loaded NPs endowed with a number of features, that can promote a valid oral administration, in terms of BACs availability and GIT absorbability, have been prepared.

Thanks to the very small particles size achievable (the most relevant feature), to the surface properties, to the matrix materials and to the compartment structure, by using SNDSs, it is also possible to enhance the bio-efficacy of phytochemicals and of food-related compounds (Kumari et al. 2014). The higher bioactivity is due to an improved dispersion and stability in GIT, to a controlled release rate at the target site, to a better transportation through the endothelial cell layer, to an enhanced systemic spread and still, to the possibility to control the impact of the microbiota metabolism (Xiao et al. 2017).

Furthermore, the reduction of BACs particles size, jointly with the use of polymeric NPs, solid-lipid NPs or nanoemulsions as encapsulating devices, may allow the target release of the BACs and may improve their cellular uptake by decreasing the repulsion factors and/or interfacial tensions (Xiao et al. 2015; Kumar et al. 2015).

Obviously, efficiency and biocompatibility are in general pivotal properties for a well working SNDS, but when the BAC to be loaded is a food constituent, a nutraceutical formulation or a phytochemical for oral administration and food uses, the SNDS has to be also food-grade.

The main features that the delivery systems for food uses should possess are following summarized (Mcclements 2007).

- *Food grade ingredients*. A delivering system usable in food industry should be manufactured with food-grade, possibly all natural ingredients avoiding the use of solvents (Acosta 2009).
- *Food incorporation*. In the design of functional foods preparation, a SNDS must be able to physically incorporate or covalently bind the bioactive compound, compacting it in more soluble and stable NPs, that have to be suitable for being incorporated into the food matrices with high physicochemical stability and minimal impact on the sensory properties of the derivative product (Donsi et al. 2011).
- *Protection against degradation*. For having a positive feedback in terms of health benefits, a BAC should reach the cell in adequate concentration and there, it

should survive for a long time. So, another task of the SNDSs is to protect the loaded BAC, both from the interaction with the food matrix constituents, that might inactivate it and from other environmental degrading agents such as temperature, light, pH, during food manufacturing, storage and processing. In addition, once ingested, the SNDS must go on protecting the loaded BAC from early inactivation by digestion and from fast metabolism (Mcclements 2007).

- *Uptake and bioavailability.* After the intake with food and upon the transport to the sites of action, the BAC-loaded SNDS should promote the cell uptake of the encapsulated compounds (Acosta 2009). In addition, a controlled release in response to specific environmental stimulus may be appreciable (Mcclements 2007).
- *Industrial scalability.* The SNDSs should be easily produced on large scale (Donsi et al. 2010).

3.1 Further Insight Into the Advantages From the Use of SNDSs

Once loaded in food-grade SNDSs, a bioactive phytochemical and/or food constituent initially water-insoluble, can be easily dispersed in aqueous media (Pan et al. 2013) and in parallel, it results protected against oxidation, acidic and alkaline degradative conditions present in stomach and small intestine respectively, thanks to the outer shell of the host material (Hu et al. 2012; Zou et al. 2016).

This can help in preserving also *in vivo*, the bioactivity that BACs have proved in *in vitro* investigations. In fact, because critically affected by poor solubility and stability issues of BACs in GIT conditions, often the results achieved after BACs oral administration don't match the promising ones valuated *in vitro*.

Resorting to SNDSs can help in nullifying this conflict, thus allowing to exploit BACs properties. When loaded in NPs, BACs may more easily reach the specific tissues affected by pathophysiological situations in adequate concentrations and exert their benefits. In designing a SNDSs-based strategy, that aim at the effective oral administration of BACs, the physicochemical features of the materials used to develop NPs in relation to the NPs digestibility in the gastric or intestinal tracts, are key issues that should be addressed to predict and plan the release sites and methods, according to the needs (Xiao et al. 2017).

In addition, it must considered, that the drug loading efficiency, that in turn affects the final release of BACs and their bioavailability and bioactivity, is strongly influenced by the drug-loading method. To better understand the concept, some situations are following illustrated.

Starch-based NPs are digested at oral level by the activity of amylase, while polysaccharide and protein/polysaccharide NPs are degraded in the small intestine, due to variations of pH and salt concentrations, that occur in the GIT lumen during the separate stages of digestion (Zou et al. 2016). Accordingly to these degradative processes the BACs release happen.

Similarly, lipid nanovesicles release phytochemicals in the small intestine simultaneously to the digestion of triglycerides (Xiao et al. 2017). SNDSs can also positively influence the transport of BACs through enterocytes (by transcellular endocytosis for particles between 20 and 1000 nm), by modulating residence time, the efficiency of transportation pathways and by controlling the metabolic reactions undergone by such compounds within endothelial cells.

The surface charge of NPs could be responsible of the formation of hydrogen bonds with the mucosal surfaces, contributing to a momentary retention (Barua and Mitragotri 2014), while the presence of surface cell-penetrating ligands could contribute to enhance transmembrane transport efficiency (Xiao et al. 2017), with a positive influence on the effectiveness and bioactivity of transported material

By using NPs equipped with a lipid phase (o/w or w/o emulsions, liposomes and solid lipid NPs), the bloodstream could be reached via mesenteric lymph and thoracic ducts, avoiding hepatic first-pass metabolism with consequent extended shelf life of BACs-loaded SNDSs.

4 Engineering Nanoparticles (NPs) Suitable for Food Applications: A Nanotechnology Challenge

Today, nanotechnology is widely applied in our everyday life and is changing the entire society. It has begun marching into the agriculture and food industry since 2003, when United States Department of Agriculture published the first roadmap in September 9, 2003. Research on this topic has skyrocketed over the last decades.

The rapid development of nanotechnology has been facilitating the transformations of traditional food and agriculture sectors and the invention of smart and active packaging, nanosensors, nanopesticides and nanofertilizers has been covering now almost every aspect in the food industry.

Numerous novel nanomaterials, have been developed for improving crop growth and monitoring environmental conditions and regarding the topic of this chapter, for enhancing food quality, safety, shelf life and for improving solubility and bioavailability of food bioactive ingredients (Acosta 2009; Magnuson et al. 2011; Wang et al. 2014) and phytochemicals also in the frame of health-promoting uses.

An exemplar case is EA, a polyphenol highly widespread in fruits and vegetables, whose several health-promoting properties, unfortunately associated to a very poor solubility and to a number of pharmacokinetic drawbacks, have inspired several studies focused on the adoption of appropriate SNDSs to ameliorate its solubility, Hydrophilic Lipophilic Balance (HLB) and GIT absorbability and in addition, to protect it from early metabolism (Zuccari et al. 2020).

In this regard, many researchers have resorted to nanotechnology and to the use of NPs. Several EA formulations such as nanosized self-microemulsifying drug delivery systems (SMEDDS) and EA-loaded materials, obtained by microencapsulation, nanoencapsulation, complexation or solid dispersion techniques, have been prepared (Zuccari et al. 2020; Avachat and Patel 2015; Madrigal-Carballoa et al.

2010; Hajipour et al. 2018; Bulani et al. 2016; Mady and Ibrahim 2018; Alfei et al. 2019a).

In particular, very promising results in terms of water-solubility have been achieved using cyclodestrins (Bulani et al. 2016; Mady and Ibrahim 2018) and adopting pectin or polyester-based dendrimers (Alfei et al. 2019a). In order to prepare food-grade edible NPs, nanotechnology resorts to a number of methods, each of them featured by advantages or drawbacks.

4.1 *Nanosuspension-Based Conventional Techniques*

Nanosuspensions (NSs) technology is a formulation approach suitable to improve solubility and bioavailability both of hydrophilic (H) and lipophilic (L) bioactive compounds (HBACs and LBACs). A NS is a colloidal dispersion of NPs in an aqueous media, stabilized by surfactants, sometimes associated with co-surfactants, and polymers (Rabinow 2004).

NSs are gifted with high dispersibility and solubility, thanks to the very small particles dimension and the emulsified BACs gain an increased therapeutic efficacy and a reduced toxicity. NSs-based NPs are suitable for operations of pre- and post-conjugation strategy, which involve the conjugation of proper ligands, such as antibodies or others, before or after the preparation of the BACs-loaded NPs.

In such a way, a sustained, controlled and target delivery of the loaded BAC can be realized, that translates in extra stability in cells and tissues and in better therapeutic effects (Liu et al., 2010). The possible conventional approaches to prepare NSs can be grouped into two methodologies types, i.e. the *top-down* and *bottom-up* methods (Fig. 2) (Gao et al. 2011).

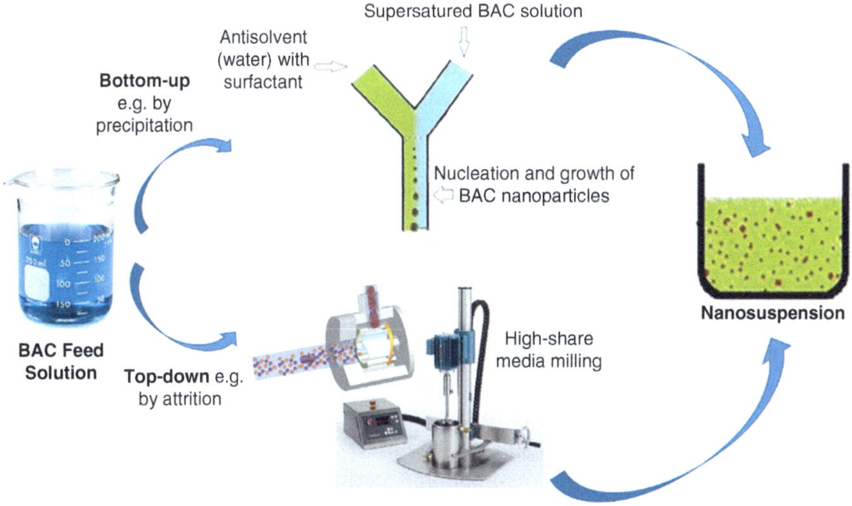

Fig. 2 Example of the *top-down* and *bottom-up* methods to prepare NSs

The *top-down* techniques start from materials made of large dimensions particles and reduce them to nanoscale size by media milling technique, high pressure and ultra-high-pressure homogenization or super critical fluid processes .

The *bottom-up* approaches start instead from the BACs molecules that, subjected to precipitation, melt emulsification, coacervation, inclusion complexation or supercritical fluids extraction, undergo self-association and self-organization forming nanosized materials. Table 1 summarizes the most common conventional *top-down* and *bottom*-up methods.

4.1.1 Some Application of NS-Based Conventional Techniques

NS inclusion complexation technique was successfully applied for encapsulating β-carotene by using *n*-octenyl succinate-modified starch thus improving its dispersibility, coloring strength and bioavailability (De Paz et al. 2012). Quercetin, a very promising antioxidant poorly water-soluble molecule, was instead subjected to high-pressure homogenization (HPH) and the NS achieved showed amorphous particles of nanometric size and enhanced solubility (Karadag et al. 2014). By using HPH, spray-drying and maltodextrin as carrier agent, water re-dispersible powders with radicals scavenging activity, reducing ability, and oxygen radical absorbance capacity, higher than non-treated samples, were easily prepared. These powders proved to be suitable for the development of quercetin-based functional food products. By preparing NS formulations, also α-tocopherol solubility and bioavailability were enhanced and a stable aqueous NPs (150 nm) suspension was obtained by supercritical assisted process (Campardelli and Reverchon 2015).

4.2 Nanosuspension-Based Combined Techniques

In order to ameliorate the performance of the singles *bottom-up* and *top-down* techniques, some combination methods were born by merging the two approaches.

4.2.1 Nanoedge™ Technology

The first combination method was the Nanoedge™ technique, by the company Baxter Healthcare, i.e. an amalgamation of a micro-precipitation step, followed by a phase involving the application of high energy shear forces such as high-pressure homogenization by piston-gap homogenizers or sonication. This technique has the advantage of getting smaller particle size materials, endowed with greater stability in short period of time. Briefly, in this technique, the precipitated suspension (*bottom-up* phase) is further homogenized (*top-down* phase) to get smaller and more stable particles (80–700 nm). Precipitation is performed in water, using water miscible solvents, such as methanol, ethanol or isopropanol to dissolve the BACs. The precipitate particles so obtained, are amorphous or partially amorphous, implying

Table 1 Some common conventional *top-down* and *bottom-up* methods to prepare nanosuspentions (NSs)

Approach type	Technique	Instruments Solvents Additives/polymers	Advantages	Limitations
Bottom-up	Nanoprecipitation (solvent/ anti-solvent)[a]	Mixer (pre-grinding) Organic solvent, water (anti-solvent) Surfactants	Nanosized particles simple low cost equipment	Only for soluble BACs Residual organic solvents
Bottom-up	Supercritical Fluid Extraction (SFE)	Syringe or diaphragm CO_2 pumps Solvents/co-solvents	Nanosized particles selectivity speedy	High costs
Bottom-up	Inclusion complexation[b,c]	Organic solvent, water polymer NPs (β-cyclodextrin, β-lactogloglobulin)	Nanosized particles masks odors/flavors preserves aromas high encapsulation high stability	Use limited to few materials[b]
Bottom-up	Coacervation[b]	Organic/aqueous solvents polymers, chemical/enzymatic cross-linker such (glutaraldehyde, transglutaminase)[b]	Nanosized particles very high payloads (99%)[b] controlled release[b] sustained release[b]	Many influencing variables
Top-down (dissocubes)	High Pressure Homogenization (HPH) at r.t.[d]	Piston-gap homogenizers aqueous media Surfactants	Nanosized particles no material erosion both for organic and aqueous solubility drawbacks	Pre-processing micronization thermic degradation high-cost instruments
Top-down (Nanopure®)	Deep-freeze Homogenization Non aqueous media or in water with water-miscible liquids[d]	Piston-gap homogenizers PEG-400, PEG-1000	Nanosized particles no crystals grow short operation times long term stability	Pre-processing micronization high-cost instruments
Top-down (IDD-P)[e]	Jet stream homogenization[f]	Z-type or Y-type collision chamber aqueous media phospholipids, surfactants, stabilizers	Nanosized particles no crystals grow short operation times long term stability	Pre-processing micronization thermic degradation high-cost instruments
Top-down	Media milling technique collision[g]	High-shear media mills/pearl mills aqueous media excipients	Nanosized particles very short batch-to-batch variation	Long operation times thermic degradation

[a] Patel and Agrawal 2011; [b] Ezhilarasi et al. 2013; [c] hydrogen bonding, van der Waals force or an entropy-driven hydrophobic effect; [d] Keck and Müller 2006; [e] Insoluble Drug Delivery-Particles; [f] Salazar et al. 2014; [g] Ghosh et al. 2011

the risk of an undesired recrystallization during the shelf life of the product, causing the growth of crystals, that negatively affects solubility, oral availability and pharmacokinetic. On the contrary, by a further homogenization in the associated second step, amorphous particles are completely converted to crystalline smaller ones, endowed with greater stability. For the production of final ultrafine sub micronized materials, it was reported a slightly modified procedure, that involves the complete removal of the solvents before high pressure homogenization phase, by including an evaporation step to provide a solvent free modified starting material (Moschwitzer and Muller 2006; Lakshmi and Kumar 2010). By performing a "Nanoedge-like" technique involving microprecipitation followed by sonication, all-trans retinoic acid was formulated in 155 nm sized particles, suitable for oral administration, in 30′ operations time (Zhang et al. 2006).

4.2.2 H 69 Technology

The H 69 process belongs to the SmartCrystal® technology family and is similar to the NanoEdge™ approach. It combines a microprecipitation *bottom-up* step, that involves the dissolution of BACs in a suitable solvent (organic, water-compatible), followed by the addition of an aqueous non-solvent for provoking particles nucleation and precipitation, with a *top-down* phase of HPH for particle size reduction. In H 69 method, differently from Nanoedge, the particle formation takes place in the high-energy zone of a homogenizer, where the just-formed BACs particles are immediately treated with cavitation, particle collision and shear forces. This particular type of precipitation, which happen almost simultaneously to homogenization, is called cavi-precipitation and allows a better control of the particle crystallization and permits to stop nucleation avoiding crystal growth and thus obtaining highly stable drug nanocrystals in the range 20–900 nm (Müller and Möschwitzer 2009). A drawback of this combinative process is that the resulting NSs contain organic solvent residues, that need to be removed before further processing, just as with the conventional Nanoedge™ technology. H 69 process involving a cavi-precipitation step, followed by HPH applied on particles in nucleation, was performed to formulate resveratrol. Resveratrol is a non-flavonoid phenol endowed with remarkable antioxidant power and belonging to the phytoalexins, i.e. molecules naturally produced by several plants. Resveratrol is found in great amount in the grape skin where acts as plant defense against pathogens such as bacteria or fungi. By H 69, it was formulated in particles of 150 nm, suitable for oral administration, after 10 cycles HPH at 1200 bar (Sinha et al. 2012).

4.2.3 H 42 Technology

The H 42 process belongs to the SmartCrystal® technology platform again and combines a *bottom-up* pretreatment step, involving BAC precipitation by spry-drying (SD) with the usual HPH phase belonging to *top-down* approaches, for particle size reduction. Briefly, the poorly soluble BAC is dissolved in an accurately selected organic solvent, possibly endowed with low toxicity, and added with surfactants such as poloxamer or sugars such as mannitol, to improve the results of the SD step and to achieve suitable, more breakable BAC powders for the subsequent comminution process. The obtained spray-dried BAC powder is then dispersed in aqueous media containing surfactants for stabilization purposes and processed to NSs by employing HPH forces using a homogenizer equipment. Among advantages of H 42 combinative technology, there are relatively short processing times, solvent-free dry intermediates and small drug nanocrystals after a reduced number of HPH cycles (170–600 nm). Unfortunately, the employment of high temperatures during SD could make this technology unsuitable to process thermolabile compounds (Möschwitzer 2006; Liu et al. 2012).

A number of experiments were performed by using H 42 technique for achieving a nanosized formulation of the antioxidant resveratrol, eligible for oral administration. The best mean particle size was obtained after only 1 HPH cycle at 1500 bar was 200 nm (Liu et al. 2012).

4.2.4 H 96 Technology

The H 96 combinative technology belongs to the SmartCrystal® technology family (Abbott/Soliqs, Germany). The *bottom-up* pre-treatment step involves the freeze drying (FD) technique that eliminates the organic solvent content as in H 42, while the *top-down* step for particle size reduction is the high-pressure homogenization (HPH) as usually. Poorly soluble BACs are dissolved in a suitable and carefully selected organic solvent or in a mixtures of organic solvents and the solutions are frozen (e.g., with instant freezing or snap-freezing) with liquid nitrogen. Then, the frozen mixtures are subjected to a freeze-drying phase to modify the starting material for improving the particle size reduction effectiveness during the subsequent HPH (Möschwitzer and Lemke 2006).

The H 96 technology is especially suitable to process thermolabile or expensive drugs due to the low temperatures and the high yields of the FD. Its drawback is the extension of the lyophilization step.

4.2.5 Combination Technology (CT)

The CT technology is the only combinative process that does not employ organic solvents and merges two *top-down* approaches by combining a low-energy pearl milling step and the usual HPH phase for particle size further reduction. The *top-down* pre-treatment of BACs involves the milling of their aqueous macrosuspensions achieving, in general, drug particle sizes between 600 nm and 1500 nm. The subsequent homogenization process improves the homogeneity of the NSs, reduces the particles size in the range 250–600 nm and enhances their physical stability during storage by avoiding crystal growth (Keck et al. 2008). Other advantages related to this technology are the possibility to reduce the homogenization pressure and process length, even if particles with sizes relatively bigger if compared to the other combinative technologies, are achieved.

Hesperidin is a glycosylated flavanone found mainly within citrus fruits and especially abundant in the peel and pulp. According to *in vitro* studies, it is an antioxidant and when introduced in the human diet, has proven to be a valid vasoprotector by increasing the efficiency of collagen and connective tissue but, the *in vivo* application of hesperidin for therapeutic purposes is limited by its very poor solubility. By using CT technology, after only five homogenization cycles at 1000 bar, the hesperidin was processed obtaining NPs with a mean particle size of 599 nm, improved solubility and long-term stability (Kobierski and Keck 2008) suitable both for oral administration and topical application. Hesperidin nanocrystals can be found in the Platinum Rare cosmetic product (La Prairie, Switzerland).

Rutin and apigenin are others very poorly soluble BACs belonging to flavonoids class and are endowed with antioxidant power and with potentials for applications in pharmaceutical and cosmetic products. Rutin is found in a wide variety of plants including citrus fruit, while apigenin is found in many fruits and vegetables, but parsley, celery, celeriac, and chamomile tea are the most common sources. To ameliorate their solubility and enhance their skin penetration capability both rutin and apigenin were processed with the CT technology. By a pearl milling step followed by only 1 cycle HPH at 100 bar, more soluble rutin NPs of about 600 nm, suitable both for oral and topical administration, were achieved. Similarly, apigenin 275 nm NPs were obtained after 1 cycle HPH at 300 bar. As an example, rutin nanocrystals were present in a cosmetic product launched by Juvena, Switzerland (Keck et al. 2008).

CT technology could drastically reduce the number of homogenization cycles which is more cost-effective and in addition, is suitable for scaling-up (Al Shaal et al. 2010).

4.3 Emulsion-Based Techniques

Emulsions (EMs) technology is a formulation approach suitable for enhancing solubility and bioavailability of orally administrable HBACs and LBACs, both for pharmaceutical and functional food applications. By using EMs technology, the bioactive molecules are encapsulated in small droplets mixing an aqueous phase (w) and an oil one (o) and obtaining water in oil (w/o), oil in water (o/w) or bi-continuous colloidal dispersions, stabilized by specific additives such as pharmaceutical surfactants, co-surfactants, emulsifiers Generally Regarded As Safe (GRAS). Oils used in EMs preparation include Captex 355, Captex 8000, Witepsol, Myritol 318, Isopropyl myristate, Capryol 90, Sefsol-218, triacetin, isopropyl myristate, castor oil, olive oil, etc. In w/o dispersions, aqueous droplets containing water-soluble HBACs are dispersed in an oil medium; in o/w dispersions, oil droplets containing LBACs are dispersed in an aqueous medium; in bi-continuous colloidal dispersions, micro domains of oil and water are inter-dispersed in the system. While w/o EMs are better choice for HBACs, LBACs are preferably solubilized in o/w EMs. The so obtained emulsion-based delivery systems are able to protect the entrapped compounds and to realize their controlled release at the target site ideally also in a controlled and sustained way (Chakraborty et al. 2009; McClements and Li 2010). Surface charge of NPs obtained by EM-based techniques could affect both the permeability and the integrity of biological barriers. Then, NPs surface ionization, achievable by using properly charged surfactants, is a double-edged sword that has to be considered. Proper surface ionization could be useful, because it can be exploited for enhancing NPs capability in crossing the biological membranes like the Blood Brain Barrier (BBB) in order to achieve higher BACs concentrations at the target site and higher therapeutic efficiency. On the contrary, an excessive number of charged groups on NPs surface, could provoke severe damage to cells by disrupting permanently the protective barriers. In this regard, reported studies have revealed that neutral NPs and low concentration anionic NPs, can be successfully utilized as colloidal drug carriers to brain with good efficiency and trivial damage, while cationic NPs have an immediate toxic effect at the BBB (Lockman et al. 2004).

The emulsion-based delivery systems can be divided into microemulsions (MEs) and nanoemulsions (NEs), based on the emulsion droplets size (Table 2). A third type of EM is known as Self-Emulsifying Drug Delivery System (SEDDS) and a detailed description is reported in Sect. 4.3.3.

4.3.1 Microemulsions (MEs) Preparation Methods

In MEs preparation, the treated BACs arrange in colloidal dispersions by spontaneous self-assembling, that occurs mixing the water phase, the oil phase, BACs dissolved in the suitable phase and stabilizers (> 20%), such as surfactants, often associated to proper co-surfactants, only by applying gently stirring (low energy methods) or by acting on temperature (phase inversion temperature) and without additional application of mechanical devices (Flanagan and Singh 2006). However, in order to speed up the process, or to overcome energy barriers, an additional energy can be applied by using colloidal mills or homogenizers. As already reported, in o/w MEs, oil droplets are dispersed in aqueous media, in w/o MEs water droplets are dispersed in oil media while in the bi-continuous MEs micro domains of oil and water are inter-dispersed in the system. Droplets size is not exceptionally small (< 0.1 mm) but MEs possess good stability thanks to the very poor tendency of droplets to re-aggregate along storage time. ME features depend on pH, temperature, packing ratio, type of additives or physicochemical nature of BACs. The three methods mainly used for MEs formation are reported in Table 3 (Flanagan and Singh 2006).

Table 2 Significant details about Microemulsions (MEs) and Nanoemultions (NEs)

Emulsion type	Average droplet size	Not suitable BACs	Suitable BACs	Advantages	Additives[a]	Solution features
MEs[b]	<0.1 mm	Very high MW[c] BACs	HBACs/LBACs, H/L food additives, H/L bioactive cosmetics	↑solubility[b] ↑bioavailability[b] ↑shelf-life	Surfactants[d] co-surfactans Stabilizers (> 20%)	thermodynamically stable transparent isotropic
NEs[e]	100–500 nm	High melting BACs[e]	HBACs/LBACs, H/L food additives, H/L bioactive cosmetics	↑solubility[e,f] ↑stability[e,f] ↑shelf-life[e] ↑bioavailability[e] ↑drug loading[e]	Surfactants[d] co-surfactans stabilizers (5–10%)	Isotropic, transparent, kinetically stable

[a] Food-grade additives are required; [b] Flanagan and Singh 2006; [c] MW = Molecular Weight; [d] Lecithins (phosphatidylcholine, phospatidylethanolamine, phospatidylinositol), saponins, fatty acids sorbitan esters (sorbatin monostearate/monolaurate); [e] Chime et al. 2014; [f] Odriozola-Serrano et al. 2014

4.3.2 Nanoemulsions (NEs) Preparation Methods

NEs can be obtained both by low energy and high energy techniques. Low-energy techniques don't involve the use of additional mechanical devices and make use of two approaches known as "phase inversion temperature" and "phase transition" methods associated with gently stirring. In this case, the emulsion droplets formation occurs by self-assembling as in MEs, but nano-dimensioned in the place of micro-dimensioned droplets can be achieved. High-energy methods need the use of mechanical devices (Table 3) and strong agitation. However, in both cases, NEs stability results lower than in MEs, because the very small droplets initially obtained, tend to re-aggregate along time with formation and growth of undesired great crystals. Droplets size depends on the type of stirring device used, the operating conditions and the BACs physicochemical properties (Qian and McClements 2011).

4.3.3 Self-Emulsifying Drug Delivery Systems (SEDDS)

SEDDSs are suitable for delivering, *via* oral administration, bioactive food-derived compounds (BACs), food-grade chemicals additives and more in general drugs. These systems are anhydrous pre-concentrated nanodispersion systems, obtained by drying up a mixture made of an oil phase, surfactants, co-surfactants/co-solvents and lipophilic BACs through proper procedures, i.e. spray drying or freeze drying

Table 3 Some common methods for MEs and NEs preparation

MEs preparation methods	Components		High energy mechanical devices	Surfactans Co-surfactans Co-solvents	Stirring	Assembling by
Low energy emulsification	Oil phase Water phase Additives		No	Yes Optional Optional	Mild	Self
Phase Inversion Temperature (PIT)	Oil phase Water phase Additives		No	Ethoxylated non-ionic esters	Mild	Δ temperature
High Pressure Homogenization (HPH)	Oil phase Water phase Additives		Homogenizer equipment	Yes Optional Optional	High	Homogenization High stirring High pressure[a]
NEs preparation methods	Components	Crucial factors	High energy mechanical devices	Surfactans Co-surfactans Co-solvents	Stirring	Assembling by
Low energy	Oil phase Water phase Additives	BAC type, emulsion properties	No	Yes Yes Optional	Mild	Self
High energy	Oil phase Water phase Additives	BAC type, emulsion properties	Microfluidisers Ultrasonicators High-Pressure Homogenizers	Yes Yes Optional	High	Energy devices
Ultra-High-Pressure Homogenization (UHPH)[b]	Oil phase Water phase Additives	BAC type, emulsion properties	Ultra-High-Pressure Homogenizers	Yes Yes Optional	High	Energy devices

[a] Not suitable for highly viscous liquids. [b] Magnuson et al. 2011; Zamora and Guamis 2015

(Ezhilarasi et al. 2013). This powder materials spontaneously will arrange in colloidal EMs when combined with water or with GIT fluids (Lee et al. 2017). The isotropic mixtures of SEDDSs represent a nanotechnology solution adopted for the design of formulations aiming at improving the oral absorption of highly lipophilic drug compounds. SEDDSs can be in turn classified into Self-Nanoemulsifying Drug Delivery Systems (SNEDDSs), whose droplets size is <50 nm and Self-Microemulsifying Drug Delivery Systems (SMEDDSs), whose droplets size ranges from 100 to 200 nm (Singh et al. 2009). These delivery systems can be taken orally through capsules filled with gelatin or together with water (Fig. 3).

When the SEDDS will meet water, by small agitation or GIT fluids, by the digestive motility of the stomach and intestine, the nanosized oil in water droplets (o/w) will form. The efficiency of oral absorption of a BAC from the SEDDS depends on many formulation related parameters, such as surfactant concentration, oil/surfactant ratio, polarity of the emulsion, droplets size and charge, all of which in essence determine the self-emulsification ability. Actually, there are only few products on the pharmaceutical market, formulated as SEDDSs, confirming the difficulty of formulating hydrophobic drug compounds into such formulations. Various physico-chemical properties of the BACs, such as pKa, log P, molecular structure and weight, presence and quantity of ionisable groups have considerable effects on the performances of SEDDSs. BACs gifted with low therapeutic dose are ideal drug candidates for SMEDDSs. The Self-Double-Emulsifying Drug Delivery Systems (SDEDDSs) are novel self-emulsifying formulations, which represent a further evolution of conventional SEDDSs (Singh et al. 2014). Like SEDDSs, SDEDDSs can self-emulsify with GIT fluids, but they form water-in-oil-in-water (w/o/w) or oil-water-in-oil double emulsions. The main advantages associated to the use of

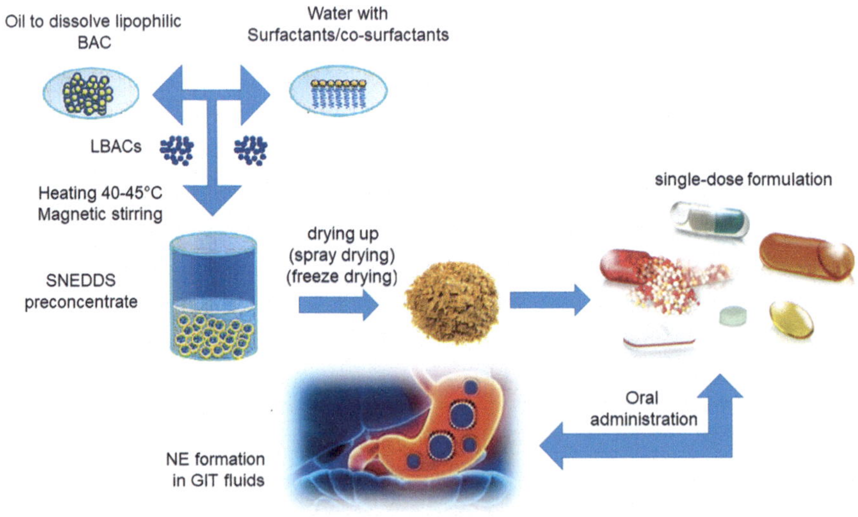

Fig. 3 Schematic process for preparation of SEEDSs and of NE formation

SEEDSs are: oral bioavailability improvement, easiness of manufacture and possibility of easy scale-up, ability to deliver peptides, delivery not effected by lipid digestion process, high drug loading capacity. As an example, a SDEDDS loaded with epigallocatechin-3-gallate (EGCG), i.e. a hydrophilic flavonol, and α-lipoic acid, endowed with improved photo-stability in respect of free EGCG was successfully prepared by Hu et al. (2015).

4.3.4 Some Application of NE-Based Techniques

NEs-based delivery systems have been extensively used for encapsulating herbal drugs, plant extracts or their bioactive chemical entities, as well as vegetable food bioactive constituents that, when administered without proper carriers, are instable in highly acidic pH and liver metabolism. Due to this instability, they are unable to reach therapeutic concentration in the blood, resulting in less or no therapeutic effect. On the contrary, by manufacturing in NEs, degradation or pre-systemic metabolism and severe side effects due to their accumulation to the non-targeted areas, have been minimized, while therapeutic efficacy and safeness have been improved. In addition, the administration also in the pediatric and geriatric patients was authorized (Chime et al. 2014). Lipid NEs containing oils from medicinal plants, have evidenced improved drug solubility, reduced side effects, increased bioavailability and prolonged pharmacological effects in comparison to other traditional formulations. NE techniques were used to encapsulate in nanosized droplets, BACs known as turmeric or curcumin (diferuloylmethane) and dibenzoylmethane (a structural analog of curcumin). Curcumin (E100, in the European code of food additives), is an extract from *Curcuma longa* and, in the food and cosmetic sector, is used primarily as a food coloring. It has a yellow color similar to that of saffron. As cosmetic, it is identified by Color Index: C.I. 75,300 or from Natural Yellow 3. In addition, curcumin can also be considered as a phytochemical to be used as GRAS food supplement according to FDA, because it is endowed with several health properties. Turmeric is an integral part of the folk medicine of many countries in Southeast Asia, where it is commonly used to treat biliary disorders, jaundice, anorexia, cough, diabetic ulcers, liver disorders, rheumatism, inflammation, sinusitis, menstrual disorders, hematuria, and hemorrhage. Curcumin is also used as an antiseptic, analgesic, antimalarial and insect repellent. Triacylglycerol was selected as oil phase and Tween-20 as emulsifier to formulate curcumin in NE. Curcumin NE proved to exert reduced toxicity, improved bioavailability and bioactivity and especially strong anti-inflammatory properties (Wang et al. 2008; Cheng et al. 2001; Simion et al. 2016).

Other polyphenols of pharmacological interest such as tannins, stilbenes and flavonoids (Scalbert and Williamson 2000), that *in vitro* have manifested antioxidant activity exploitable to counteract OS, have been encapsulated in NEs. The poor antioxidant activity showed *in vivo* by epigallocatechin gallate, when formulated in a suitable emulsion composition, has resulted increased without any toxicity issue (Koutelidakis et al. 2012). Bioactive lipids and carotenoids were formulated as NEs

Table 4 Emulsion-based delivery systems for bioactive food-derived constituents (BACs)

BACs	Method	Surfactant(s)	Results	Reference
Carotenoids (Paprika Oleoresin)	Solid self-microemulsifying carotenoid systems (S-SMECS)	Tween 80	↑solubility	Chow et al. 2015
Lutein	MEs-based	Tween 80	↑solubility ↑bioavailability	Amar et al., 2003
	SNEDDS	Labrasol, TranscutolHP/ Lutro-E400[a]		Yoo et al. 2010, Shanmugam et al., 2011
Polymethoxyflavones (PMFs)	High Pressure Homogenization (NEs-based)	Tween 20 Tween 85	↑dissolution rate	Li et al. 2012
β-Carotene	High Pressure Homogenization (o/w NEs-based)	Tween 20	↑emulsion stability ↑solubility ↑bioaccessibility	Yuan et al. 2008 Qian et al. 2012
Lycopene	MEs-based	Ethoxylated sorbitan esters, 3GIO, SML	↑solubility	Spernath et al. 2002, Ha et al. 2015
Quercetin	SNEDDS	Tween 80, PEG 400	↑solubility	Tran et al. 2014

Abbreviations: 3GIO – triglycerol monooleate, SML – sucrose monooleate, PEG – polyethylene glycol; [a] co-surfactants

and when they were delivered, more protection against autoxidation and an increased bioaccessibility were respectively observed. Furthermore, NEs were used to protect lactic acid bacteria and to restore the proper microbiota in diverse intestinal diseases conditions (Martínez-Ballesta et al. 2018).

Some examples of emulsion-based devices up today prepared for delivering some bioactive foods-derived constituents (BACs) are reported in Table 4.

A number of studies were performed aiming at enhancing bioavailability and effectiveness of polyphenols food constituents such as EA form fruits and vegetables, for dermatologic applications with the aim at treating skin diseases or at protecting skin from toxic agents. In this regard, crucial is the BACs capacity to permeate the skin layers deep down. Ethyl acetate fractions of Pomegranate peel polyphenols were emulsified with pomegranate seed oil (as oil phase) achieving polyphenols-loaded NEs suitable for topical applications. Once applied onto the skin, NEs allowed the mixture of polyphenols, including EA, to mange to permeate the skin barrier, to reach the viable epidermis and dermis layers and the deeper skin layers and there, to act as anti-photo-aging cosmetic helping to avoid or delay UV radiation damage (Baccarin and Lemos-Senna 2014).

Lemongrass essential oil (LEO) is a volatile and hydrophobic concentrate extracted from the leaves and stalks of the lemongrass plant, i.e. a tropical and grassy plant commonly used in cooking, but also considered as herbal medicine.

LEO has a powerful citrus scent and it's often found in soaps and other personal care products as flavour. Although lemongrass has traditionally been used to treat digestive problems and high blood pressure, nowadays it's becoming a popular tool in aromatherapy to help relieve stress, anxiety, and depression. LEO, like many other essential oils (EOs), has also proven to be an effective natural antimicrobial agent and gained particular interest and wide acceptance by consumers due to its relative safety, as well as its eco-friendly and biodegradable properties. Unfortunately, LEO and EOs in general, due to the high susceptibility of their labile phytochemical constituents to autooxidation, tend to degrade easily, losing activity and providing smelly or even harmful compounds, responsible for allergic reactions and skin irritation. This fact limits the potential of EOs as therapeutics in their free form and in addition, EOs high volatility, their particular odor/flavor and price restrict EOs introduction in food preparations or in food packaging, with antimicrobial and preservative purposes. By resorting to nanotechnology and by incorporating LEO into NE formulations, its undesired sensory impact has been reduced, while enhancing its antimicrobial activity. In the specific case, edible NEs containing LEO were developed, in order to achieve a coat packaging for protecting plums, that are climacteric perishable fruits, with a short shelf life. More in general, the carnauba wax and LEO NEs prepared, that proved to inhibit the growth of foodborne pathogens as *Salmonella spp.* and *E. coli,* could represent smart nanosized food ingredients for preserving nourishments from degradative spoilage, provoked by bacteria action and for extending their shelf life (Kim et al. 2013).

Similarly to LEO, other EOs have been subjected to modifications by nanotechnology and converted in nanoparticles formulations for improving their antimicrobial activity and for extending food shelf life, minimizing the growth of foodborne pathogens and some examples are reported in Table 8 at the end of the following Sect. 4.4.

4.4 Biocompatible and Biodegradable Solid Nanoparticles Vehicles

Besides poor solubility and low bioavailability, the most bioactive nutraceuticals and phytochemicals, after oral administration, suffer from a number of pharmacokinetic drawbacks, that translate in poor absorption at gastric or intestinal levels, in early degradation, as a result of the GIT digestion and in a fast metabolism, that in the complex, furtherly limit their effectiveness and clinical applications as health promoters.

During last decades, the research was focused on the development of food-grade carriers for overcoming all these barriers. According to the results retrieved in recent years, BACs nanoencapsulation by using solid NPs with diameters ranging from 1 to 100 nm, as delivering carriers, manages to enable higher concentrations of BACs in cells and tissues, by increasing their solubility, by reducing early inactivation, fast metabolism and rapid excretion and by allowing a slowdown delivery (Momin et al. 2013). Given these advantages, in order to increase the benefits for health deriving from a normal intake of BACs with a conventional diet, with an additional ingestions of effective dosages of BACs, the research on the engineering nanotechnology known as Drug Delivery, is unceasingly boosted, to optimize their antioxidant, antiradical and antitumor abilities and their pharmacokinetic profile.

Solid NPs developed, including dendrimer NPs, embody a high number of surface functional groups, available for chemical reactions and host/guest interaction both with HBACs and LBACs (Khan et al. 2017; Alfei et al. 2017a, 2018b, 2018c) and in addition, provide inner cavities able to accommodate hydrophobic bioactive molecules (Fig. 4).

Thanks to the chemical or physical interactions that can occur between the nanocarriers (host molecules) and BACs (guest molecules), high surface area formulations with enhanced solubility may be prepared. In such complexes (dendriplexes if the host molecules are dendrimer solid NPs) or conjugates, NPs act as nanocontainers and protective envelopes that can opposite GIT digestion and as permeability enhancers able to promote GIT absorption and bioavailability of the guest molecules (Bengoechea et al. 2011; Berton-Carabin et al. 2013; Cerqueira et al. 2014). Furthermore, nanosized high molecular weight particles carriers, typically increase the systemic drug/carrier complex retention time, allowing the BACs to reach unaltered the specific sick areas, where their liberation will provide health benefits with higher bio-efficiency (Kaminskas et al. 2011). Moreover, using solid NPs, additional constraints compromising stability, shelf life and effective bioactivity of BACs, such as the effect of light, temperature or oxygen can be minimized (Weiss et al. 2006; Neethirajan and Jayas 2011).

Currently, there are different types of solid nanosized delivery systems, whose classification depends on their physicochemical nature, production method, properties, free energy, interactions type and forces typology etc. (Khan et al. 2017). To date, the most adopted NPs for food-derived bioactive compounds encapsulation are non-synthetic lipid-based delivery systems (solid–lipid NPs, liposomes, micelles and niosomes), protein-based polymeric NPs, oligosaccharides-based (cyclodestrins), polysaccharides-based, metal oxide-based and clay-based NPs. Due to their specific structural properties, low cost and non-toxic nature, these nanosystems are considered more suitable carriers for the delivery of food constituents than synthetic polymeric ones (Harde et al. 2011). Nevertheless, also others biodegradable synthetic organic polymers and biodegradable and biocompatible synthetic dendrimers, as polyester-based ones, known for their nonpareil properties, could be considered eligible as food-grade delivery systems, for realizing a therapeutically effective administration of BACs. These natural or synthetic solid NPs, if necessary, may be used also for masking unpleasant sensory properties often associated to isolated BACs, for authorizing their utilization as tasteless ingredients to prepare food supplements, functional foods or as flavourless additive for preserving food from early degradation, foodborne bacteria attack and for improving food qualities. Polycationic, anionic or uncharged dendrimer NPs are extensively under investigation for formulating nutraceuticals in nanosized materials, both by physical

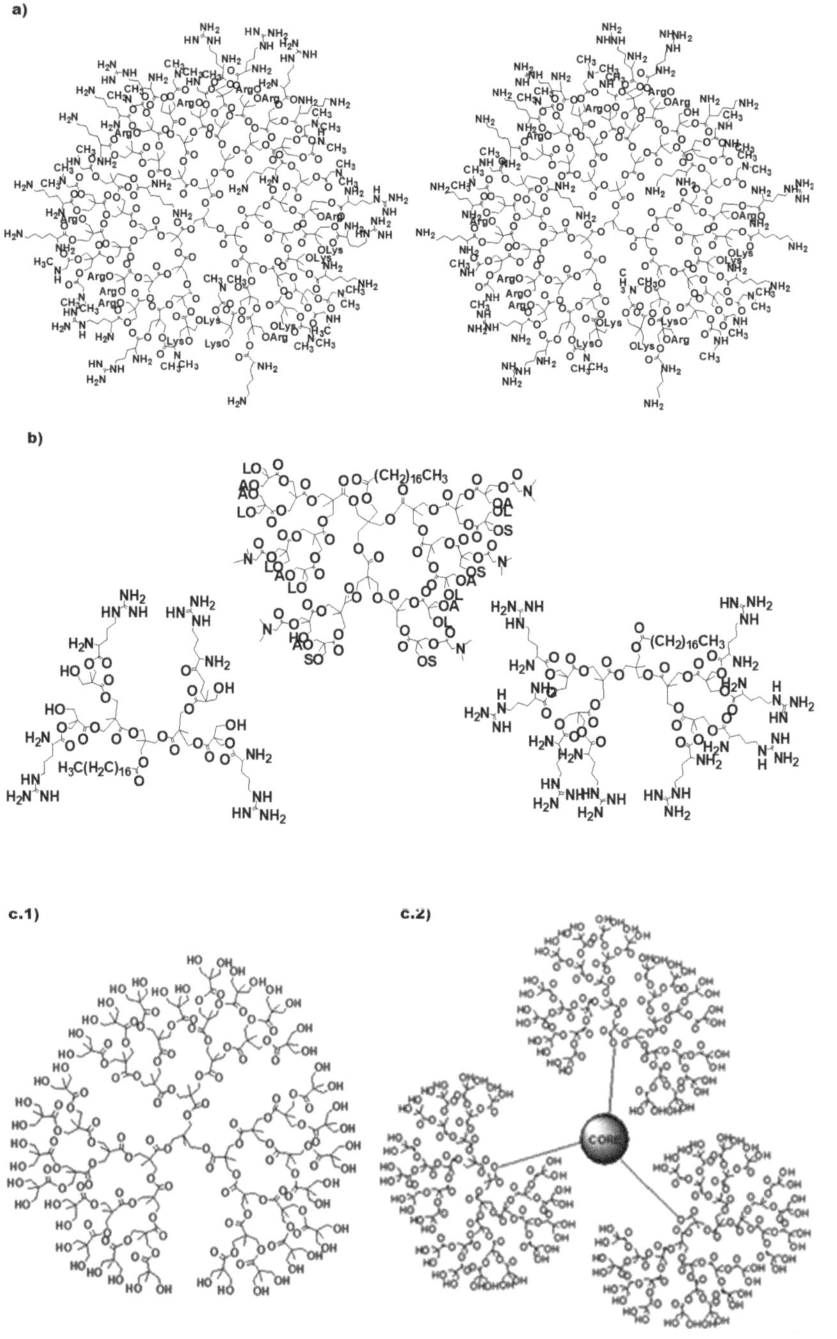

Fig. 4 Examples of polycationic hydrophilic (**a**), polycationic amphiphilic (**b**) and uncharged four generation (**c.1**) and fifth generation (**c.2**) biocompatible dendrimer NPs (Alfei et al. 2017a, 2017b, 2018b, 2018c). In panel (b), A = arginine, S = sarcosine and L 0 lysine

interactions and covalent links, in order to improve their solubility and bioavailability, with healthy purposes (Shadrack et al. 2018; Abderrezak et al. 2012). Very recently, non-soluble triterpenoids extracted by spices and poorly water-soluble polyphenols from fruits as EA and GA, have been successfully loaded on synthetic polyester-based dendrimer nanoparticles, achieving a number of different hydrophilic and amphiphilic water-soluble complexes (Alfei et al. 2018a, 2019a, 2020c) and a hydrophobic nanosized conjugate with GA (Fig. 5). Among the complexes, the EA and GA-enriched dendrimers proved enhanced radical scavenging activity and antioxidant power, while the GA-conjugate, in addition to show an improved antioxidant effect, proved to be more fats-compatible than GA and to be endowed with biological activities higher than those of not conjugated GA (Alfei et al. 2019b, 2019c, 2020b).

In particular, it proved significant antibacterial effects and remarkable activity in inhibiting platelet aggregation and intraplatelet ROS production (Alfei at al. 2019c), thus representing a very promising alternative, as therapeutic devices against bacterial growth, thrombosis, OS and associated diseases. In additon, thanks to its compatibility with fats and oils, it showed to be a well performant preservative additive against auto-oxidative spoilage of some phytochemicals such as essential oils (Alfei et al. 2019b). An overview on the natural solid NPs commonly adopted, either as food-grade delivery systems of BACs with health promotion purposes, or for preparing nanosized edible ingredients, helpful to enhance shelf life, taste, flavour and quality of foods, has been reported below. An additional discussion on the most conventional synthetic organic polymers, employed for similar purposes, is also provided.

4.4.1 Lipid-Based Nanoparticles (LNPs) (Naseri et al. 2015)

Lipid-based NPs include Solid-Lipid Nanoparticles (SLNPs), Liposomes (LPs), Micelles (MICs), that in turn encompass Normal Micelles (n-MICs) and Inverse Micelles (i-MICs) and Niosomes (NIOs) (Fig. 6).

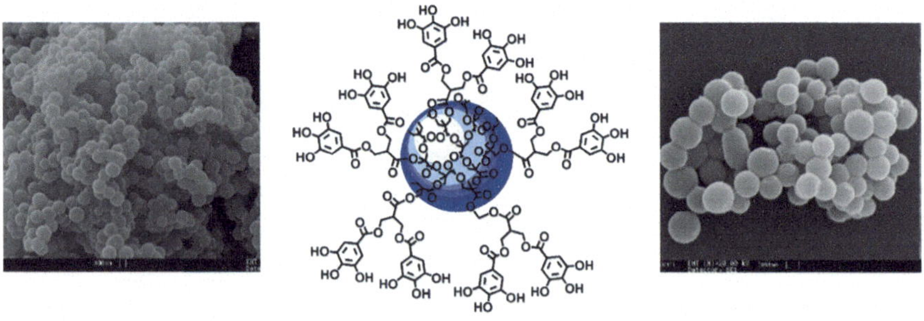

Fig. 5 Schematic representation of the hydrophobic polyester-based dendrimer NPs conjugated with gallic acid (GA) and the relative SEM images, showing a rather perfect spherical morphology

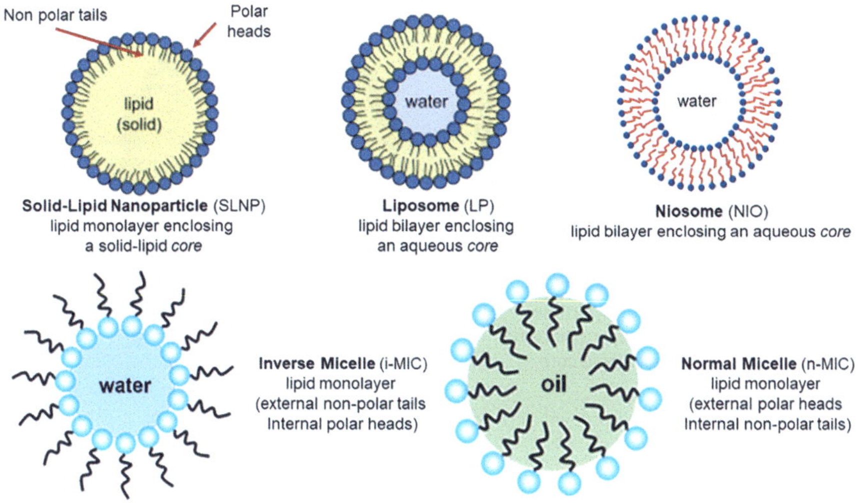

Fig. 6 Examples of Lipid-based Nanoparticles

4.4.1.1 Solid-Lipid Nanoparticles (SLNPs)

The development of solid-lipid NPs (SLNPs) is one of the emerging fields of lipid nanotechnology (Pardeshi et al. 2012; Mashaghi et al. 2013). In the last years, SLNPs have become the most popular and commercially available NPs for delivering lipophilic food-derived functional compounds (LBACs), thanks to their natural composition and biocompatibility. A SLNP has a typical spherical morphology and an average diameter between 10 and 1000 nanometers, easily detectable by Transmission Electron Microscopy (SEM) (Gregoriadis 1990). SLNPs, differently from NEs, that possess an external lipid monolayer shell enclosing a liquid lipid *core*, are made of an analogous external lipid monolayer but with a solid-lipid *core* matrix, that can solubilize lipophilic molecules (Fig. 7) and the lipid *core* is stabilized by surfactants and emulsifiers.

Lipid composition and the surfactant properties could influence SLNPs stability and size. The most common lipids used to get SLNPs include triglycerides (e.g. tristearin), diglycerides (e.g. glycerol bahenate), monoglycerides (e.g. glycerol monostearate), fatty acids (e.g. stearic acid), steroid molecules (e.g. cholesterol) and waxes (e.g. cetyl palmitate) (Blasi et al. 2007). Concerning stabilizers, practically all classes of emulsifiers (with respect to charge and molecular weight) have been adopted to stabilize the lipid dispersion. It has been found, that the combination of more emulsifiers might prevent more efficiently particles agglomeration, which is one of the crucial drawbacks related to SLNPs. A typical percentage fat/aqueous medium ratio of 0.1:30.0 (w/w) is considered the best choice to develop SLNPs (Thatipamula et al. 2011).

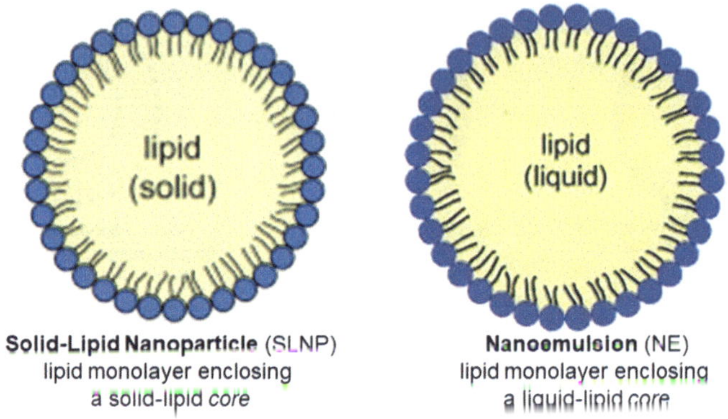

Fig. 7 Comparison between a SLNP and a NEs

4.4.1.2 Liposomes (LPs)

Liposomes (LPs) are nanosized artificial vesicles obtained by mixing phospholipids and cholesterol that, from years, are considered of remarkable interest as immunological adjuvants and drug carriers (Gregoriadis 1990; Alving 1991). LPs are made of a lipid bi-layer enclosing an aqueous *core* (Fig. 6). They proved structural versatility and possess high capacity in encapsulating diverse BACs also with wide range of polarities by including them either into the aqueous *core* of the phospholipid vesicle or at the bilayer interface (Sharma and Sharma 1997). When orally administrated, they can also preserve the included materials from the deleterious activity of external factors, such as enzymes and degrading agents of the GIT (Chaize et al. 2004). Furthermore, LPs obtained from natural lipids, are biodegradable, biologically inactive, non-antigenic, non-pyrogenic and don't present intrinsic toxicity (Vanrooijen and Vannieuwmegen 1980; Campbell, 1983). Until very recently, the use of LPs were primarily directed at the target drug delivery, but today, the abilities of liposomes are being discovered also in other settings. LPs are extensively used by the pharmaceutical industry to control the release of several bioactive compounds as drugs, vaccines and enzymes for the prevention of several pathological situations (Zhou and Neutra 2002; Matteucci and Thrall 2000; Mady et al. 2004). Recently, LPs have attracted rising consideration as delivery systems for food-derived bioactive compounds and nowadays a number of food-related manufacturing companies are pioneering the potentials of this unique science towards this novel application. Presently, LPs are being implemented for the specific oral delivery of certain dietary and nutritional supplements. Unfortunately, also LPs suffer from some drawbacks as instability in plasma (Koo et al. 2005) but, in this regard, new insights into the mechanisms of transendothelial pathways of LPs, improvements in LP-mediated drug delivery evasion systems and sterically stabilized LPs have been developed (Yuan et al. 1994).

4.4.1.3 Micelles (MICs)

Micelles (MICs) are very slim, spherical lipid particles in the range of 10–100 nm, whose formation occurs usually in aqueous solutions (Kataoka et al. 2001), but also in an oil medium MICs can form. MICs, that are formed in aqueous solution, are Normal Micelles (n-MICs), e.g. PEG-PLGA micelles in water medium. They are able to solubilize LBACs in the non-polar inner *core*. Differently, Inverse Micelles (i-MICs) assemble in oil solution, e.g. PLC-P2VP micelles in oleic acid medium, and are capable to solubilize HBACs in the polar inner *core* (Fig. 6). As examples of pharmaceutical applications, micellar delivery systems are extensively employed as successfully carriers of antitumor drugs for allowing the intravenous administration of hydrophobic drugs, avoiding to resort to solubilizing adjuvants, which has been associated with the appearance of toxic symptoms (Kloover et al. 2004). Unimolecular MICs exhibit a maintained distribution thanks to a pH-sensitive drug release and have been described for the administration of rapamycin (Chen et al. 2013). Paclitaxel micellar formulations, consisting of PEG and modified poly-aspartate as hydrophobic block, showed cytotoxicity in a range of human tumors (Matsumura and Kataoka 2009).

MICs can also be exploited to improve bioavailability and systemic residence time of bioactive phytochemicals and nutraceuticals. MICs are able to protect them from early inactivation reactions by surrounding factors, are gifted with high loading capacity and good stability (Kataoka et al. 2001; Gong et al. 2012). Many aspects could influence the release of BACs from MICs, i.e. the MIC intrinsic stability, the BACs diffusion rate, the partition coefficient, the copolymers biodegradation rate, BACs concentration within the MICs, BACs molecular weight and physicochemical features and BACs location within the MICs (Kwon and Okano 1996; Teng et al. 1998). Nevertheless, BACs release can be also voluntarily provoked in the target area by local stimuli, as variation of pH, temperature or by application of ultrasounds or light (Rapoport 2007). In such a way, a more efficient BACs internalization and localization within the separate cell compartments, where the biological activity of the compounds carried is actually required, could be obtained. Entrapping BACs in MICs, by gaining a high BACs concentration specifically at the organic site affected by pathological situations and a low level of side effects relative to systemic administration, very higher therapeutics advantages can be achieved (Zia et al. 2010; Mikhail and Allen 2009).

4.4.1.4 Niosomes (NIOs)

Niosomes (NIOs) are lipid-based lamellar nanostructures (Fig. 6), similar to LPs but not charged, that are formed by merging non-ionic surfactants, such as alkyl or dialkyl polyglycerol ethers and cholesterol (Malhotra and Jain 1994). NIOs represent an alternative to LPs for enhancing oral bioavailability of compounds with limited absorption. NIOs, thanks to their uncharged structure, result in being less toxic for cells providing overall a more beneficial therapeutic effect (Uchegbu and Vyas 1998). NIOs vesicles are osmotically active and stable, act as a reservoir system and could provide the delivery of bioactive compounds in a controlled and sustained way along time. NIOs possess potentials also for different therapeutic applications such as diagnostic imaging agents (Jain et al. 2005; Balasubramaniam et al. 2002; Gude et al. 2002; Shahiwala and Misra 2002; Uchegbu and Vyas 1998).

4.4.2 Some Application of Lipid-Based Nanoparticles (Martínez-Ballesta et al. 2018)

Essential oils (EOs) are important volatile and easily degradable plant derived bioactive compounds with interesting benefits. They are endowed with analgesic, sedative, anti-inflammatory, spasmolytic and anesthetic properties and have showed promising potentials for the prevention of chronic pathologies or infectious diseases. Microemulsion (ME) methods are successful in providing protection to EOs and in inhibiting and/or delaying EOs degradation, but seem to be no competent to enhance their therapeutic activities such as the antimicrobial one. On the contrary, by the nanoscale technology, a slower delivery and a higher cell permeability, especially in the skin layers, was achieved. In particular, by using lipid-based NPs, EOs were entrapped in the *core* of the nanostructure, achieving EOs-loaded lipid NPs that showed the capability to reach different types of cells and proved an increased activity. Similar results (He et al. 2019) were achieved by the nanoencapsulation of ferulic acid, that is a hydroxycinnamic acid phytochemical abundant in plant cell walls, and tocopherol, that instead is an essential nutrient belonging to the lipophilic vitamins, present in many foods and also known as Vitamin E. Both these compounds are endowed with remarkable antioxidant power and if efficiently administered can exert activities to counteract OS and related diseases.

Recently, a new type of compounds has emerged, i.e. plant oxylipins, phytoprostanes and phytofurans, which derived by the non-enzymatic oxidation of linolenic acid. It has been suggested that bioavailability and biological activity of such compounds is based on their structural analogy with human eicosanoids (Pinciroli et al. 2017; Domínguez-Perles et al. 2018). These lipid phytochemicals are capable to migrate into fatty acids-based micelles which in turn contribute to their better solubility and subsequent cellular transportation (Xiao et al. 2017).

A dermal delivery system involving EA-loaded niosomes, was prepared by using a mixture of Span 60 and Tween 60 (2,1), with 15% PEG 400 as solvent. It exhibited very high percentage of entrapment and high efficacy in delivering EA to human epidermis and dermis (Junyaprasert et al., 2012).

4.5 Oligosaccharide-Based Nanoparticles: Use of Cyclodestrins

Cyclodestrins (CDs) represent a form of delivery system suitable for interesting applications in the development of novel formulations for preparing functional foods, food supplements, innovative food-related therapeutics and smart food packaging.

CDs are commonly used as host molecules for implementing the monomolecular inclusion complex technique. CDs are cyclic oligosaccharides consisting of six (α-cyclodextrin), seven (β-cyclodextrin), eight (γ-cyclodextrin) or more glucopyranose units which are linked by α-(1, 4) bonds (Del Valle 2004). They are obtained through enzymatic degradation of amylose by the enzyme cyclodextrin glucosyl transferase, which causes one intramolecular reaction with formation of cyclic α-(1, 4) compounds. CDs has a truncated cone structure and can accommodate hydrophobic molecules inside their hydrophobic interior cavity characterized by the presence of several CH_2 groups. On the contrary, their outer side, rich in OH groups, represent a hydrophilic layer thank to which, CDs are endowed with high water-solubility (Fig. 8).

CDs are able to improve solubility, chemical stability and resistance at the external environment degradative action of the guest molecules transported. Furthermore, CDs allow unpleasant taste and flavor modification and a controlled release of drugs (Linde et al. 2009; Marques 2010). From a toxicological point of view, low doses of CDs are well tolerated by humans, but high doses may cause some adverse effects such as diarrhea and soft stools. Among CDs, β-CDs are currently the most exploited for drug delivery purposes due to their diameter that is suitable for a number of non-polar bioactive guest molecules (Astray et al. 2009; Del Valle 2004; Marques 2010).

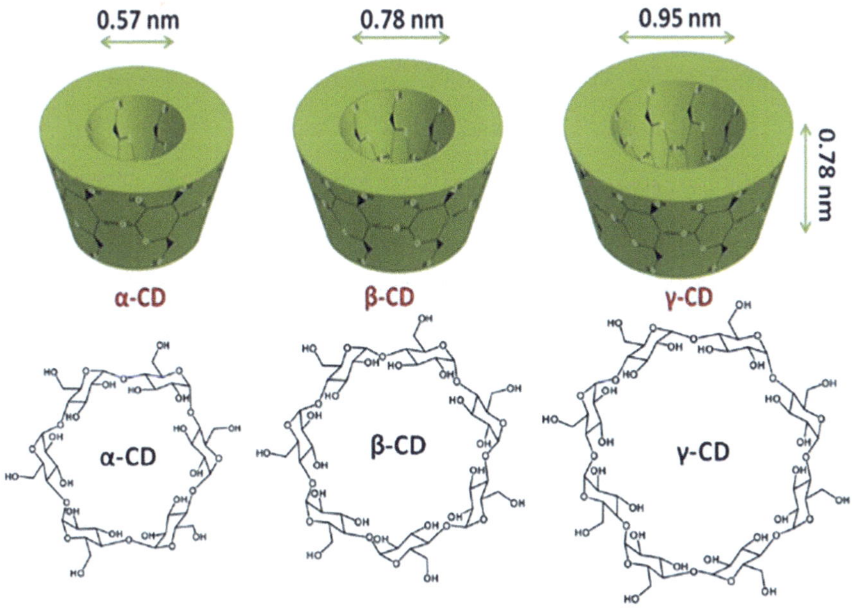

Fig. 8 Chemical structure, spatial arrangement and size of naturally occurring cyclodestrins

4.5.1 Techniques to Prepare CDs Inclusion Complexes (CD-ICPXs)

Different methods are available to prepare the inclusion complexes (ICPXs) of poorly water soluble BACs using CDs as host molecules (Patil et al. 2010) and the basic details are reported in Table 5 (Recharla et al. 2017).

4.5.2 Some Applications of CDs in Food Industry

CDs, and especially β-CDs, are extensively used as carriers suitable for applications in food, pharmaceutical and cosmetics industry. Researchers have been reported several β-CDs applications for forming inclusion complexes with food derived bioactive compounds. The most of these studies report the improvement of stability, shelf life and water solubility of flavonoids or of other plant bioactive compounds by their encapsulation in the CDs inner hydrophobic *core* cavity (Pinho et al. 2014).

Table 5 Most common methods for preparing CD-ICPXs

Method type	Mixing Physical state	Solvent	Drying	Instruments	Preparation steps
Physical blending	mechanical powder	No	No	Mixer	Mixing
Kneading method	mechanical paste	Water Water/alcohol	Yes	Kneading machine	CDs pasting BAC addition mixing drying
Co-precipitation method	mechanical solutions	Solvent (BAC) Water (CD)	Yes	Magnetic mechanical stirrer	dissolutions precipitation drying
Ball Milling technique	mechanical solid state	No	No	Mechanical oscillatory mill	Mixing
Spray drying method[a]	--- solutions	Solvent (BAC) Water (CD)	No	Spray dryer	dissolutions Spray drying
Freeze-drying technique[b]	---- solution	Solvent (BAC) Water (CD)	No	Freeze dryer	dissolutions Freeze drying
Supercritical anti-solvent technique	mechanical solutions/gas CO_2	Solvent (BAC) Water (CD) CO_2 (anti-solvent)	No	Magnetic mechanical stirrer	dissolutions precipitation solvent extraction

[a] non-suitable for thermosensitive compounds; [b] suitable for thermosensitive compounds; [c] by employing CO_2

Linoleic acid (LA) is a polyunsaturated ω-6 fatty acid abundant in many nuts, fatty seeds (flax seeds, hemp seeds, poppy seeds, sesame seeds) and in their derived vegetable oils comprising over half of poppy seed, safflower, sunflower, corn, and soybean oils. The consumption of LA is vital to proper health, as it is an essential fatty acid. Due to its abundant double bonds, LA may be easily oxidized providing molecules such as 9-hydroxyoctadecanoic acid and 13-hydroxyoctadecanoic acid, that have shown to activate the capsaicin receptor, thus playing a major role in hyperalgesia and allodynia, an inflammation form with negative health effects. By using CDs, the thermal stability of the linoleic acid and its resistance against the environmental degradation factors were increased (Hadaruga et al. 2006). The stability and solubility of resveratrol were improved by resveratrol– β-CDs inclusion complexes formation (Lucas-Abellán et al. 2008). Carotenoids, from red bell pepper extracts, were instead encapsulated into 2-hydroxypropyl-β-CD, succeeding in enhancing their water solubility (De Lima et al. 2016). Lycopene (Lyc) (from solanum lycopersicum or tomato) is an acyclic isomer hydrocarbon of β-carotene, containing 11 conjugated double bonds and 2 non-conjugated. Lyc belongs to the carotenoid group and is commonly used as a colorant food additive identified by the abbreviation E160d. It is synthesized by plants and microorganisms, but not by animals and possesses also antioxidant properties. Given the presence of double bonds, Lyc easily undergoes spoilage by *trans-cis* isomerization induced by light, thermal energy and chemical reactions and due to its lipophilicity it is not water-dispersible in GIT fluids. Nano-sized lycopene-β-CD complexes were prepared by solution-enhanced dispersion using supercritical fluids (SEDS) process (Nerome et al. 2013) that promoted dispersion of Lyc in water (Nerome et al. 2013). In addition, several more available polyphenols-enriched nanomaterials were obtained by using CDs and are listed below (Fang and Bhandari 2010).

Hesperetin and hesperidin-loaded HP-β-CD, resveratrol-enriched maltosyl-β-CDs were prepared; olive leaf extract (rich in oleuropein) were loaded into β-CD, quercetin and myricetin were entrapped into HP- β-CD, maltosyl-β-CDs and β-CDs; kaempferol, quercetin and myricetin were encapsulated in HP-β-CD while α- and β-CDs were loaded with 3-hydroxyflavone, morin and quercetin. Moreover, rutin-enriched β-CD, curcumin-loaded HP-β-CD, quercetin and myricetin-loaded HP-β-CD and ferulic acid-loaded α-CD were successfully achieved.

Adopting CDs, EA water solubility was increased by four (Bulani et al. 2016) and later by five (Mady and Ibrahim 2018) times. Concerning the first case, a freeze-dried inclusion complex of EA/β-cyclodextrin (EACD) was prepared by hydrogen bonding interactions and EA accommodation in the β-CD cavity. The solubility and in *vitro* anti-inflammatory activity of EA was significantly enhanced by complexation with β-CD. In the second case, β-CD-based nanosponges (CDNSs) have been utilized, both to improve the solubility of EA and to control its release. Using β-CD and dimethyl carbonate (DMC) as cross-linker, EA nanosponges (EA-NSs) of 423 nm with low polydispersity index (0.409) and high zeta potential (−34 mV), which manifested the construction of a stabilized colloidal nanoformulation were prepared. EA-NSs showed a loading % of about 69, a dissolution efficiency of

50 μg/mL, and a controlled *in vitro* release of about 24 hours and studies on animals displayed an improvement in the oral bioavailability of EA.

Amino acids as well as hydrolyzed soy proteins are endowed with unpleasant bitter taste that affect their use in preparing functional foods or food supplements as beverages, that result acidic. According to a study by Linde et al. (2009) and the judge from a panel of trained tasters, by including the amino acids in α-CDs and by the formation of amino acid inclusion complexes, the bitter taste perception of the amino acids can be altered and the bitter taste from hydrolyzed soy protein can be reduced, thus indicating a general high potential use of CDs for debittering protein hydrolysates in acidic beverages.

4.6 Polysaccharide-Based Nanoparticles

Polysaccharide-based NPs (PNPs) can be considered as another valuable alternative for improving solubility of BACs and for realizing their transport, controlled and protracted release ideally where necessary (Singh and Lillard 2009). PNPs are appropriate also for improving stability of food constituents, in order to use them for developing fortified functional foods (Joye et al. 2015) and for preparing bioactive smart foods ingredients to enhance food shelf-life, quality, look and sensory properties (Yu et al. 2018).

PNPs can be prepared starting from natural hydrophilic polysaccharides such as alginate, chitosan, hyaluronic acid, pectin and cellulose derivatives (hydroxyethyl cellulose and carboxymethylcellulose), that may form macroscopic hydrogels at sufficiently high concentrations in the presence of cross linkers or other substances inducing polymer-polymer interactions (Fig. 9).

PNPs can be achieved by four categories of preparation methods, which include covalent crosslinking, ionic crosslinking, polyelectrolyte complexation, and self-assembly (Liu et al. 2008; Nitta and Numata 2013; Fojan et al. 2006). As cross-linkers, tripolyphosphate sodium salt (PPT) or $ZnCl_2$ anhydrous are commonly used. BACs can be physically entrapped during NPs formation, covalently attached to the precursor materials, or absorbed at the NPs after preparation. In order to achieve long shelf-life bioactive PNPs, NPs are often freeze-dried, in the presence of a suitable cryoprotectant or spray-dried into a microparticulate powder. With the first process, non-applicable to thermolabile BACs, nanomaterials suitable for oral administration as suspensions are achieved, while with the latter one, a fine powder suitable also for direct administration by inhalation or for compression into tablets can be obtained (Vauthier and Bouchemal 2009). A combination of spray-freezing and freeze-drying techniques has been also described, by directly processing the frozen NPs achieved from the first step in a freeze-drying second step.

PNPs possess high affinity to mucosal layers of cell present in the respiratory tract and GIT (Lemarchand et al. 2004) that promotes higher permanence in such compartments, with a high bio-efficiency. PNPs are classified into polyelectrolytes,

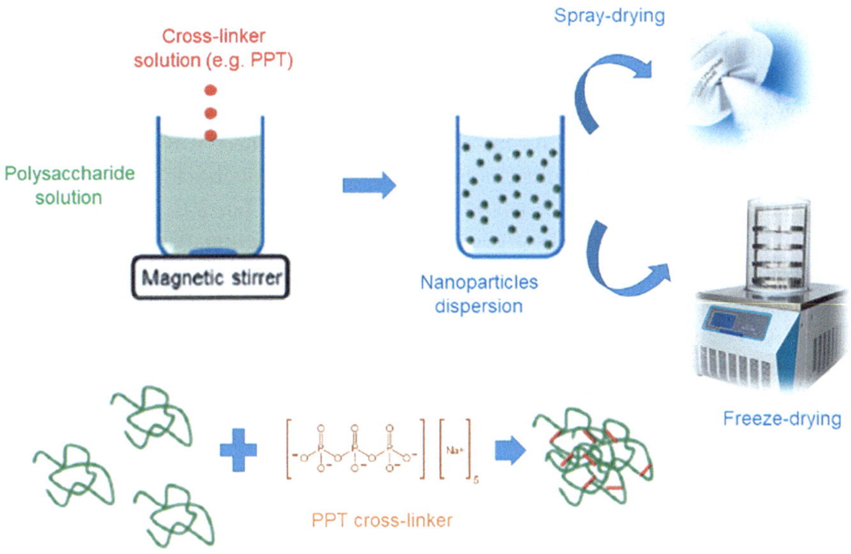

Fig. 9 Schematic representation of the process to prepare polysaccharide-based NPs

in turn divided into cationic, anionic and neutral subtypes, and non-polyelectrolytes (Liu et al. 2008) on the bases of their intrinsic charge.

By encapsulating a bioactive compound using these materials, the transported BAC is covered by a polysaccharide coat that, in addition to exert protection against early degradation, recently has demonstrated to be able to interact with specific receptors in cells and tissues promoting cellular up-take and site-specific controlled targeting (Lemarchand et al. 2004).

4.6.1 Cationic Polyelectrolytes: Chitosan

Chitosan is made of repeated units of D-glucosamine and is non-toxic and biodegradable. It is a bio adhesive material and therefore, thanks to its mucho-adhesive nature, allows a controlled release of encapsulated agents and their prolonged residence time at the site of absorption (Salatin and Jelvehgarim 2017).

Chitosan, thanks to its positive charge, is able to complex negative charged macromolecules, avoiding toxic solvents during preparation, and allows the administration of several types of bioactive compounds, also food derived, for treating diseases in several body compartments such as nasal, oral, ocular and dermal (Amidi et al. 2010). In addition, chitosan NPs can be easily modified by introducing ligands that contribute to a more rapid and efficient interaction with cells membranes for an improved delivery efficiency (Duceppe and Tabrizian 2010).

4.6.2 Anionic Polyelectrolytes: Alginate, Heparin, Pectin, and Hyaluronic Acid

PNPs based on alginate, heparin, pectin and hyaluronic acid sodium salt belong to anionic polyelectrolytes NPs. As far as alginate is concerned, it is endowed with biocompatibility, biodegradability, non-antigenicity, and mucho-adhesive features (You and Peng 2005; Ojea-Jiménez et al. 2012; Sun and Tan 2013). Since BACs encapsulation occurs thank to host/guest electrostatic interaction, these NPs are particularly suitable for incorporating positively charged compounds. A combination of alginate and chitosan is an alternative system for enhancing the circulation time of the BACs delivered. Pectin, is a hydrophilic food compatible polymer, naturally occurring in several fruits (fruits fiber) and often used in food industry as thickener excipient in jams preparations. It is a safe ingredient with no limit on the acceptable daily intake and has reported potential beneficial and prebiotic properties (Chan et al. 2017; Holscher 2017).

Concerning hyaluronic acid NPs, this acid as sodium salt, is highly soluble and stable in aqueous medium, is non-toxic and non-immunogenic. Hyaluronic acid is particularly suitable to target delivery anticancer drugs, because it has affinity to hyaluronan receptors, which are highly expressed in tumor cells.

4.6.3 Neutral Compounds

Neutral PNPs are made of uncharged polysaccharides such as dextran, maltodextrin, pullulan etc. (Salatin and Jelvehgarim 2017).

In these uncharged materials, the presence of hydroxyl groups is the most responsible in modulating the incorporation of BACs in the base skeleton. Dextran was used to develop delivery systems able to escape the reticuloendothelial system and therefore, endowed with enhanced systemic residence time and circulation permanence, which may promote the efficiency of the delivered BACs (Salatin and Jelvehgarim 2017).

Maltodextrin is a water-soluble complex carbohydrate obtained through chemical hydrolysis processes from the breakdown of cereal starches (corn, oats, wheat, rice) or tubers (potatoes, tapioca). If oral administered, it has a neutral and not very sweet taste, it is easily digestible and is usually well tolerated and consequently, it is eligible to encapsulate food related bioactive constituents.

Maltodextrins consist of glucose molecules ordered in more or less long polymer chains. The chains length provides the parameter that allows to identify and classify maltodextrins, basing on their dextrose equivalence (DE), starting from 2–4 up to 19. The higher the DE, the shorter the polysaccharide chain. Maltodextrins have a digestive behavior similar to that of glucose and are excellent substitutes for dextrose and for this reason, in the late eighties, the bodybuilding industry used it massively to increase the intake of carbohydrates in the diet without resorting to sugar.

4.6.4 Some Applications of PNPs in Food Industry

Chitosan has been used as a wall material to encapsulate an olive leaf extract (OLE) and, after spray drying, OLE-loaded microspheres, with a loading percent in polyphenolic compounds of 27% and a smooth surface morphology, were achieved (Kosaraju et al. 2006). OLE resulted physically entrapped into the chitosan matrix. Chitosan was used to encapsulate other polyphenols, endowed with antioxidant properties, such as GA (Cho et al. 2011) and caffeic acid (CA), obtaining bioactive food-grade particles with maintained antioxidant power, while by entrapping EA, chitosan-based NPs eligible as anti-hemorrhagic agents were achieved (Gopalakrishnan et al. 2014).

Yerba mate (*Ilex paraguariensis*) from Aquifoleaceae family, is a South American native species, whose leaves and stems are commonly used by the local population to prepare an infusion beverage with perception of benefits. In effects, yerba mate is gifted with a remarkable bioactive fraction that encompasses the presence of alkaloids as xanthines and methylxanthines, including caffeine and theobromine, saponins and polyphenols, known for their stimulating, anti-obesity, anti-hyperlipidemic, anti-inflammatory, anti-oxidative actions. Studies have shown that yerba mate possesses the beneficial properties of improving lipid metabolism and of counteracting OS, provides protective effect against oxidative damage to lipids and DNA, mitigates lipid peroxidation in blood vessels and ameliorates the metabolic unbalance events associated with diabetes (Rocha et al. 2018). Yerba mate extracts (YME), containing 62 mg of GA/g yerba mate, have been encapsulated with two different matrix systems, i.e. calcium alginate and calcium alginate-chitosan (Deladino et al. 2008). The extracts loading (%) was higher in the alginate beads (85%), than in chitosan coated beads (50%), while the YME release was faster from chitosan-covered formulation, than from the alginate one. The results highlight that polyphenols are better retained into alginate matrix, but after re-dispersion in water, the maximum YME release is allowed in a shorter time by the chitosan coated beads and therefore that, the wall materials play a key role in the release profile of the natural antioxidants of yerba mate and more in general of loaded BACs.

In the recent past, there has been an explosion of probiotic health-based products. Probiotics are microorganisms that improve intestinal microflora when released in specific regions of the GIT. Unfortunately, many reports indicate that, in these products, there is poor survival of probiotic bacteria and in addition that, the survival of these bacteria in the human GIT system is questionable. Providing probiotic living cells with a physical barrier against adverse environmental conditions is therefore appeared a valuable approach of considerable interest. So, the technology of microencapsulation of probiotic bacterial cells was evolved and by alginate-starch encapsulation, probiotics delivery systems, that manage to improve probiotics storage stability, shelf-life and resistance to acidic pH, were prepared. The success of the encapsulation process was further improved by the use of supercritical carbon dioxide method, that allows to avoid the exposure of "active" microbiotics to water, oxygen, solvents or heat that are deleterious for their survival (Vidhyalakshmi et al. 2009).

Maltodextrins are widely used for encapsulating flavours (Bhandari 2007) and polyphenols. The ethanol extracts of black carrots, which contain high levels of anthocyanins, have been spray-dried using maltodextrins of 20–21 DE as carriers and coating agents, achieving at the end of drying process a powder with very high anthocyanins content (Ersus and Yurdagel 2007). Maltodextrins can also be mixed with gum arabic as additional wall material. A mixture of maltodextrins (60%) and gum arabic (40%) has been used for encapsulating procyanidins from grape seeds (Zhang et al. 2007). Procyanidins are members of the proanthocyanidin (or condensed tannins) class of flavonoids that, among others, are gifted with antioxidant powers. The encapsulation efficiency was up to 89%, and the procyanidin properties were not changed during conventional drying, but through spray-dry technique the stability of the products was furtherly improved.

In a recent work, low methoxyl pectin was exploited to improve water solubility of EA, the well-known polyphenol present in several fruits and vegetables and endowed with several recognized heathy properties, among which a strong antioxidant power. In order to ameliorate its insignificant water-solubility that limits its biomedical applications, EA was physically entrapped in a pectin matrix by the spray-drying technique, achieving a solid microdispersion with a drug loading of 22% (w/w) and a water solubility increased by thirty times (Alfei et al. 2019a). This EA formulation is suitable both to prepare fine suspensions for oral administration of EA at a high dosages (i.e. as food supplement) and to be used as an ingredient in functional food preparations as beverages. In *in vivo* evaluations on mice, the EA microdispersion administered thorough EA-enriched pomegranate juice, proved to produce beneficial properties and to be tasty, palatable and well-tolerated (Turrini et al. 2019b).

Pectin, the natural fiber of fruits, proved to be an excellent material carrier for performing the spray drying also of sticky materials. Pectin from citrus fruit, has been used as an encapsulating agent for entrapping bioactive compounds extracted from *Hibiscus sabdariffa* L. by using spray-drying technique. The main bioactive compounds in *H. sabdariffa* L. extract are polyphenols, or more specifically, the anthocyanin complexes. By the encapsulation process were combined into one multipurpose functional food, two bioactive products, i.e. the fruit fiber endowed with beneficial activities on cholesterol metabolism and colonic functions and the antioxidant polyphenols. With this strategy, a novel nutraceutical product suitable for a variety of applications in functional food manufacturing, has been created (Chiou and Langrish 2007).

Gluten protein is a by-product from wheat starch isolation, that has been arising great interest as bioactive ingredient, to develop fortified functional foods. Gliadin, that represent half of the gluten protein, was formulated in NPs by using liquid antisolvent precipitation, that proved to be still unstable in conditions relevant to food products. Then, the gliadin-based NPs prepared, were further coated with different polysaccharides such as low methoxyl pectin (LMP) and high methoxyl pectin (HMP), achieving PNPs with mean particle size of about double that of uncoated ones, and the charge reversed from positive (uncoated) to negative (coated) that, as reported, tends to cause less damage to cells. Thanks to the polysaccharide pectin coating, the stability of the gliadin NPs to environmental stresses, such as pH, ionic strength and thermal treatment, was remarkably improved providing stable functional ingredients for uses in the food industry, as texture modifiers, lightening agents or delivery systems (Joye et al. 2015).

4.7 Protein-Based Nanoparticles

Protein-based NPs (ProNPs) may be prepared through several methods including desolvation, coacervation, emulsification, nanoprecipitation, spray-drying, NP albumin-bound technology, self-assembly, electro-spraying, salting out and crosslinking.

The most used proteins to obtain ProNPs could belong both to animal proteins such as gelatin, collagen, albumin, casein and silk protein and vegetable ones as zein, gliadin and soy protein.

Even if very numerous, all encapsulation methods are based on the precipitation of the protein macromolecules dissolved in a suitable solvent and this occurs when the protein solubility in the adopted solvent decreases by adding a non-solvent or by changing the physicochemical parameters of the protein solution, such as pH, salinity or temperature (Miladi et al. 2014).

To the proteins water solutions, a desolvating agent, like alcohol or acetone, is often added under stirring, in order to promote the dehydration of the protein, that changes its conformation from stretched to coil.

Crosslinking methods allow to achieve a sustained drug delivery and to increase the stability of the protein. Different types of chemical, ionic, thermal and enzymatic cross-linkers are usually used in ProNPs preparation and 8% glutaraldehyde aqueous solution or Calcium phosphate are extensively adopted (Fig. 10).

Unfortunately, often cross-linkers are not removed properly, representing a concern, because residual cross-linkers can induce toxicity to biological systems.

An additional worry is the washing step, which requires dialysis process, thus being time consuming. To address this issue another method of crosslinking, based on irradiation was investigated and introduced. ProNPs were exposed to γ-irradiation in phosphate buffer (pH 7.2), in the absence and/or presence of ethanol and methanol at 30% and 40% (v/v), and the results showed that irradiation has a significant effect on particle size.

By controlling the irradiation dose, crosslinking density and subsequently the particle size could be controlled. This innovative method allows to produce ProNPs in a one-step procedure, along which, simultaneous crosslinking and sterilization may be achieved (Tarhini et al. 2017).

The utility of using of ProNPs as delivery systems for food constituents and phytochemicals is supported by several argumentations and the main ones are listed below.

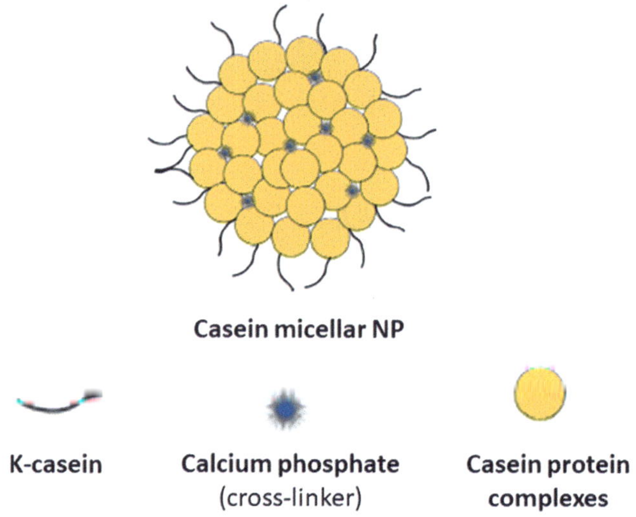

Fig. 10 Schematic representation of a casein-based NP

(1) simple manufacturing
(2) absence of requirement of emulsification performance
(3) compatibility with the high pressure emulsification processes
(4) high freeze–thaw stability (Zhu et al. 2017)
(5) loading capacity trackable by ultraviolet (UV)-spectrophotometry, fluorescence spectrophotometry or high performance liquid chromatography (HPLC)

By resorting to diverse classes of proteins, different types of NPs have been developed, making use of a plethora of laboratory approaches.

Each of them possesses advantages and constraints, but in respect of many other NP carriers suggested for drug delivery, ProNPs present various advantages that were listed below again.

1) abundance of proteins in nature
2) suitability to be transformed
3) absence of strong deleterious effects concerning the biological systems in which they are applied
4) susceptibility to modifications due to the occurrence of functional groups
5) possibility to achieve the desired bio distribution
6) biocompatibility
7) capacity of carrying the selected molecules
8) stability (Tarhini et al. 2017)
9) possibility to add hydrophilic polymers (PEG) for improving circulation residence time (Kaul and Amiji 2004)
10) capacity to stabilize food-grade Pickering emulsions (Zhu et al. 2017)
11) suitability for applications in frozen food (Zhu et al. 2017)
12) suitability for functional food formulations (Zhu et al. 2017)

ProNPs are classified on the base of their origin (animal or plant proteins) and on the base of the diverse types of benefits and constraints associated to proteins provenance. Toxicity and/or infections associated to their application are related only to animal proteins, while plant proteins, due to their hydrophobic characteristics, lack of toxicity and have a lower economic cost (Tarhini et al. 2017)

Table 6 reports schematically the main protein-based nanoparticles features.

4.7.1 Some Applications of ProNPs in Food Industry

Gelatin is an animal protein containing many glycine, proline and 4-hydroxyproline residues. By using gelatin (type A), protein-based microcapsule loaded with epigallocatechin gallate (EGCG), have been produced using the layer-by-layer (LbL) assembly method (Shutava et al. 2009). The EGCG content of the protein/polyphenol film material obtained was as high as 30% w/w and EGCG retained its antioxidant activity also in the LbL assemblies (Shutava et al. 2009). Gelatin was also used to bind GA, thus increasing its stability with a maintained bioactivity (Cirillo et al. 2010). Zein, is a vegetable protein belonging to the prolamine group, contained in the seeds of cereals and in particular in corn. Zein was exploited to prepare EA-loaded biodegradable hollow zein NPs (70 nm) in order to mediate efficient oral

Table 6 The most remarkable ProNPs features

	Protein-based nanoparticles (ProNPs)	
Why as delivery systems	simple manufacturing by just heating protein solution	
	absence of requirement of emulsification performance	
	compatibility with the high pressure emulsification process	
	high freeze–thaw stability	
	good drug loading capacity (DL)	
	DL isoelectric point dependent	
	DL instrumentally predictable and trackable	
Constraints	Toxicity and/or infections (Animal ProNPs)	
Advantages	abundance of proteins in nature	
	suitability to be transformed	
	absence of strong deleterious effects	
	susceptibility to modifications	
	possibility to achieve desired	biodistribution
		biocompatibility
		capacity of carrying selected molecules
		stability
	possibility to add hydrophilic polymers (PEG)	
	capacity to stabilize food Pickering emulsions	
	applicability in frozen food	
	suitability for food formulations	

delivery of EA (Ruan et al. 2018). Firstly EA-loaded NPs made of a nano dispersion of EA/sodium carbonate were prepared by co-precipitation method. Subsequently, these first EA-NPs, were further encapsulated in other hollow-zein-NPs, by using triethyl citrate as a natural plasticizer. The EA/Na_2CO_3 nanodispersion initially prepared became the *core* in the final NPs. The optimized EA-hollow plasticized zein NPs (EA-HTZN) exhibited a dimension of 72 nm, good stability, high drug loading capacity and significantly improved permeation ability in *vitro*. In addition, oral administration of EA-HTZN showed to be effective against inflammation in carrageenan-induced mouse paw edema model and to be endowed with better pharmacokinetic parameters.

Rennet-gelled protein and whey protein gel particles were adopted to encapsulate probiotics microorganisms in order to protect them from degradation and digestion in stomach. Delivery systems were achieved, that manage to improve probiotics storage stability, shelf-life and resistance to acidic gastric pH. Supercritical carbon dioxide extraction method was performed to avoid the exposure of the "active" microbiotics to detrimental water, oxygen, solvents or heat, thus improving the viability of the precious microorganisms (Vidhyalakshmi et al. 2009).

4.8 Inorganic-Based Nanoparticles

An alternative and efficient approach for the encapsulation of BACs is represented by inorganic materials as silica nanoparticles (SiNPs) (Deligiannakis et al. 2012) and gold-nanoparticles (AuNPs) (Rattanata et al., 2016). Inorganic NPs based on metals oxides, such as SiO_2, MgO, and TiO_2, are extensively applied for obtaining bioactive materials to be used as anti-caking agents, food flavor carriers and food color additives, and are approved also by the U.S. FDA (He et al. 2019). Titanium dioxide (TiO_2) or E171, in the European codification of food additives, is a crystalline, colourless compound tending to white, commonly found as additive in many processed foods, but also in personal care and consumer products. It is primarily used as a pigment because of its brightness, high refractive index, and resistance to discoloration. Concerning foods, the highest content of TiO_2 is found in candies, sweets as cake icing and puddings, sauces and chewing gums, while among personal care products, toothpastes, sunscreens and some other crèmes contain from 1% to >10% titanium by weight. Shampoos, deodorants, and shaving creams contain instead the lowest levels of titanium (< 0.01 μg/mg). As asserted by Electron microscopy stability testing of food-grade TiO_2 NPs, approximately 36% of the NPs are less than 100 nm and they readily disperse in water as fairly stable colloids. An anxiety about extensive use of TiO_2, concerns its entering the sewage system after products uses. TiO_2 that enters the environment, may accumulate in the treated effluent discharged to surface waters, the biosolids applied to agricultural land, incinerated wastes or landfill solids and its quantification is quite difficult (Weir et al. 2012). In addition to food products directly serving human beings, animal feeds play a significant role in the global food industry throughout the world. In this sector, copper oxide, di-copper oxide [Cu_2O or copper (I) oxide], iron oxide, and zinc oxide NPs, have been categorized as GRAS by the U.S. FDA and are extensively used to formulate nutritional dietary supplements in animal feeds.

SiO_2-based NPs (SiNPs) are commercially available in the form of fumed silica, also known as pyrogenic silica and are nowadays among the largest industrial nanotechnology products, that find applications in paints, microelectronics, pharmaceuticals, cosmetics and drug delivery. Fumed silica is produced in a flame (Fig. 11) and consists of microscopic droplets of amorphous silica fused into branched, chainlike, three-dimensional secondary particles, which agglomerate into tertiary particles. It appears as powder with an extremely low bulk density and high surface area.

Silica is considered biochemically inert, so it is vastly used as flowing-aid for nutraceutical and pharmaceutical products. SiO_2 NPs with specific functionalities, like antioxidant ones, represent added-value hybrid nanomaterials, that combine the advantages of each component. Since besides to be chemically inert, SiNPs are far more thermally stable, they offer distinct advantages vs. organic polymer matrices. SiNPs provide a nonpareil tool to exploit the health potentials of natural antioxidants present in foods and plants, otherwise unstable and susceptible to easy autoxidation, degradation and inactivation in the presence of water, oxygen, light or heat.

Concerning AuNPs, recent studies have revealed that gold nanoparticles have gained much attention in drug delivery due to their unique dimensions, tunable functionalities on the surface, and controllable drug release. AuNPs can be conjugated with bioactive compounds to prepare bioactive nanomaterials stable, and without toxicity to cells.

Among many novel nanomaterials, clay-based NPs are one of the most widely studied, mainly for their mechanical, thermal, barrier properties and low cost. Nanoclays are developed into several subclasses including montmorillonite, bentonite, kaolinite, hectorite, and halloysite, depending on the physiochemical properties of the nanomaterials. Montmorillonite and bentonite are now listed as GRAS and in Effective Food Contact Substance (FCS) notifications by the U.S. FDA. Polymer/Nanoclay Composites (PNcCs) have been expanded to several

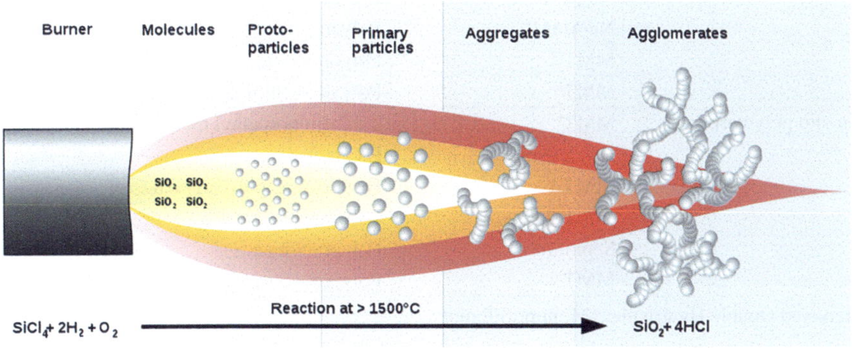

Fig. 11 Preparation of fumed silica in the flame

hybrid, innovative materials including biocomposites and organoclay hybrid films, with properties superior to conventional composite materials and have found application in several industrial sector such as pharmaceutical (as carriers of drugs and penetrants), food (as carrier of nutraceutical and phytochemicals), food packaging and textiles. PNcCs can be prepared thorough three methods i.e. solution-blending, melt-blending and *in-situ* polymerization (Guo et al. 2018).

Table 7 reports the PNcCs achievable by the three methods.

4.8.1 Some Applications of Inorganic-Based Nanoparticles (INPs) in Food Industry

SiNPs with Radical Scavenging Capacity (RSC) have been prepared by using commercially available SiNPs of various sizes (8–30 nm). They were covalently functionalized with a well-known natural antioxidant, i.e. GA, which was grafted on SiNPs surface. In such a way, robust hybrid GA-enriched materials with a remarkable stability were developed, with results comparable to that achievable by the chemical grafting of GA on chitosan.

Table 7 Polymer/nanoclay prepared by the solution-blending, melt-blending method and in-situ polymerization (Guo et al. 2018)

Method	Nanoclay	Polymer
solution-blending	Co-Al-LDH[a]	Polystyrene
	Cloisite® 30B	Chitosan/polylactide
	Mg-Al-LDH[a]	Epoxy
	Mg-Al-LDH[a]	Acrylic resin
	Cloisite and sepiolite	Hydrogenated nitrile butadiene rubber
melt-blending	Cloisite® 30B	Nylon 6
	Na-MMT[b]	Nylon 66
	Na-MMT[b]	Epoxy
	Na-MMT[b]	Polyamide
	MMT[b]	Polypropylene
	MMT[b]	Polyvinyl chloride
in-situ polymerization	MMT[b]	2,2,2-trifluoroethyl methacrylate
	Fe-kanemite	Polypyrrole
	MMT[b]	Bisphenol A polycarbonate
	Vinyl clay	Polystyrene
	Sepiolite	Polypropylene
	MMT[b]	Polypropylene

[a] Layered Double Hydroxide; [b] Montmorillonite

Table 8 Main associations of essential oils (EOs) with solid polymeric NPs

Formulation	BAC	Targeted microorganism	Activity
PLC-based NPs	Tea tree oil	*Tricophyton rubrum*	↑ EO effectiveness against fungal cells infecting nails
Zein-Sodium Caseinate NPs	Thymol	*E. coli* and *Salmonella*	Good antimicrobial activity ttwo-phase release profile
PLGA-based NPs	Cinnamaldehyde and Eugenol	*Salmonella spp.* and *Listeria spp.*	Good antimicrobial activity controlled release profile
Chitosan-based NPs	Carvacrol	*E. coli, S aureus* and *B. cereus*	↑ antimicrobial activity
Chitosan-based NPs	Oregano EO	–	Controlled release profile
Zein-based NPs	Thymol and Carvacrol	*E. coli*	Good antimicrobial activity ↑ solubility
Chitosan-based NPs	Eugenol and Carvacrol	*S. aureus* and *E. coli*	↓ Cytotoxicity toward mice cells
PLGA-based NPs	Carvacrol	*S. epidermidis* biofilms	↓ elasticity and stability of performed biofilm
Methyl and ethylcellulose NPs	Thymol	*S. aureus, E. coli* and *P. aeruginosa*	Good antimicrobial and preventive activity for cosmetic lotions, creams and gels

In order to counteract foodborne pathogens, that are the cause of endemics diseases worldwide, and solve the problem of the increasing antibiotic drug resistance, food-related bioactive compounds, that in *in vitro* tests, proved antibacterial activity as polyphenols, could be a promising solution. It is the case of GA, but its intrinsic low stability and poor bioavailability limit its effectiveness *in vivo*. With the aim at ameliorating GA performances *in vivo*, it was attached to 7 nm diameter fumed silica particles, by performing different functionalization methods that involved the condensation of hydroxyl or carboxyl groups of the acid (Vico et al. 2016). The aim was to increase GA stability and intrinsic antimicrobial activity, for achieving a novel promising bioactive tool against *Paenibacillus larvae*, that affects honeybee. Depending on the type of functional group interested in the linkage, GA-enriched SiNPs named GAO-NPs and GACON-NPs were obtained. The attachment of GA to the inert support led to an improvement of GA stability against auto-oxidation and of its antibacterial activity by 14–27 times, while GA capacity to produce free radicals in alkaline conditions was maintained. GAO-NPs and GACON-NPs represent an alternative to existing antibiotics which, due to an excessive and prolonged use, are increasingly undergoing a progressive loss of activity, caused by the development of resistant bacteria strains.

In order to counteract foodborne pathogens as *Plesiomonas shigelloides* and *Shigella flexneri B*, which are the major cause of diarrheal endemics worldwide, gold nanoparticles (AuNPs) have been chosen as drug delivery agent for GA, which was successfully conjugated with AuNPs, providing core–shell structures.

Conjugation of GA with AuNPs led to the improvement of the antibacterial activity of GA, which proved to act by causing alterations in bacterial membrane compositions, in terms of altered lipids or fatty acids, proteins, and nucleic acids.

Tyrosine-based and GA-based phenolic compounds, highly susceptible of oxidative degradation, were stabilized by using Montmorillonite clay (Laponite) through their incorporation via intercalation into the interlayer space of the clay (Deligiannakis et al. 2012).

4.9 Synthetic Organic Polymer-Based Solid Nanoparticles (SOPNPs)

Synthetic Organic Polymers (SOPs) and copolymers are widely used for preparing drug delivery systems in the form of SOPNPs, because of their biocompatibility and biodegradability, simple design, easy preparation and good efficiency in delivering BACs to the diseased tissues.

The polymers and copolymers properties depend on the physicochemical characteristics of their building blocks.

SOPNPs offer the possibility to load BACs by physical interactions or by covalent conjunction through versatile modifications of their several chemical groups.

By tuning the hydrophilic/hydrophobic balance (HLB), it is possible to achieve NPs endowed with a variety of shapes and morphologies in water solution.

Figure 12 shows only some of the most utilized biocompatible and biodegradable organic polymers approved by the U.S. FDA and European Medicines Agency (EMA). They proved to act as effective carriers for drug delivery in humans, suitable also for oral administration of nutraceuticals and phytochemicals and for producing food-grade smart materials devoted to preserve food quality, looks and taste along storage.

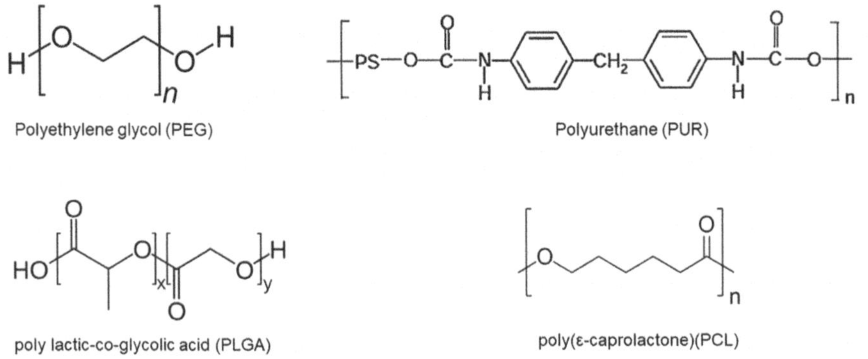

Fig. 12 Structure of the most used organic polymers to produce GRAS NPs-based delivery systems

4.9.1 Polyethylene Glycol (PEG)

PEG is a polyether compound with many applications, from industrial manufacturing to medicine. PEG is also known as polyethylene oxide (PEO) or polyoxyethylene (POE) and its structure is commonly expressed as $H-(O - CH_2 - CH_2)_n - OH$. PEG is generally considered biologically inert and therefor it is used as innocuous component for drugs formulation and in particular is extensively used as excipient in many pharmaceutical products because is unlikely to have specific interactions with biological chemical constituent of drugs.

When used as constituent of any formulation, besides providing a matrix scaffold, PEG may perform different functions, acting as a stabilizer or as a preservative material. PEG derivatives, such as narrow range ethoxylates, are instead often used as surfactants. PEG has been used as hydrophilic block in the synthesis of amphiphilic block copolymers, that were employed either to create polymersomes or to induce complete fusion (mixing of both inner and outer leaflets) in liposomes reconstituted *in vitro*. Polymerosomes are in turn "reservoir" polymeric NPs, in which the outer membrane consists of amphiphilic polymers. They are biomimetic systems with a behavior similar to phospholipids.

PEG is the basis of many skin creams (as cetomacrogol), personal lubricants (frequently combined with glycerin) and is also used in a number of toothpastes as a dispersant, since it manages to bind water and helps to keep xanthan gum uniformly distributed throughout the toothpaste. Concerning food sector, PEG is widely used as an anti-foaming agent both in solids food and in drinks.

PEG is also used as a host polymer for solid polymeric electrolytes and it has proved to be a reliable and well performant polymer to enhance systemic circle residence time, by acting as a steric barrier and by protecting the PEG-coated NPs from opsonization (Aqil et al. 2013). According to these features, PEG is eligible as solvent, co-solvent, container, surfactant, stabilizer, preservative additive, dispersant, lubricant, protecting agent etc. for preparing food-grade nanoformulations of nutraceuticals and phytochemicals. As a polyol, PEG easily reacts with isocyanate derivatives for providing polyurethanes.

4.9.2 Polyurethane (PUR)

PUR is a synthetic alternating copolymer composed of organic units joined by carbamate links, traditionally formed by reacting a di- or tri poly-isocyanate with a polyol such as PEG. Both the isocyanates and polyols used to make PURs contain, on average, two or more functional groups per molecule. Concerning polyols for PURs preparation, the growing interest in sustainable "green" products outstretched attention and research in polyols derived from vegetable oils (Niemeyer et al. 2006), including oils from soybean, cotton seed, neem seed and castor oil or resulting from renewable sources, as dimerized fatty acids from dimerization of unsaturated fatty acids or fatty acids (Rajput et al. 2014). Some bio-based PURs, rather than from poly-isocyanates and polyols, are prepared exploiting the reaction between poly-amines and cyclic carbonates (Bassam et al. 2013).

PURs are chemically inert and no exposure limits have been established in the U.S. by Occupational Safety and Health Administration (OSHA) or by American Conference of Governmental Industrial Hygienists (ACGIH).

PURs are used in the manufacture of high-resilience foam seating, rigid foam insulation panels, microcellular foam seals and gaskets, durable elastomeric wheels and tires, such as roller coaster, escalator, shopping cart, elevator, and skateboard wheels. In addition, PUR can be found in automotive suspension bushings, electrical potting compounds, high performance adhesives, surface coatings and surface sealants, synthetic fibers as Spandex, carpet underlay, hard-plastic parts for electronic instruments, condoms and hoses and, since the most PURs are not melting thermosetting polymers, thermoplastic materials made of PURs are also available.

PURs, especially if made up exploiting aromatic isocyanates, contain chromophores able to interact with light, and therefore they are rising particular interest in the area of coatings, where light stability is a critical factor, as in foods storage. Usage of PUR packaging is calculated in 251millions of pounds, i.e. 4.6% of the total. PURs are used in a variety of food packaging applications, including adhesives materials for rigid and flexible food coating. Furthermore, PURs are also utilized for manufacturing materials eligible for food contact applications as conveyor systems, hoses and tubing, chutes and chute liners, hoppers, gaskets and seals.

On August 10, 2016 the industry group American Chemistry Council (ACC) published a guidance document on the use of PURs in food contact applications [ACC (August 10, 2016). "Guidance for the use of PURs in food contact applications"].

4.9.3 Poly-(lactic-co-glycolic acid) (PLGA)

PLGA is one of the most widely exploited biodegradable polymer for the development of food-grade NPs and several studies report its applications as biomaterial for the controlled delivering and release of BACs (Kumari et al. 2010). PLGA rapid metabolism by physiological hydrolysis, provides lactic acid and glycolic acid, which are in turn easily processed via the Krebs cycle, thus allowing low systemic toxicity. Surface modifications of PLGA using non-toxic and blood compatible material are required and are crucial, in order to allow the increase of the length of blood circulation, for achieving a sustained and protracted delivery of the transported BACs (Boulle et al. 2012; Huang and Reichardt 2001). With this purpose, PEG in combination with hydrophobic biodegradable aliphatic polyesters, is widely used as hydrophilic non-toxic segment. This practice equips PLGA with the capacity to resist against opsonizing and the ulterior phagocytosis, contributing to enlarge its shelf life in the bloodstream and tissues (Martínez-Ballesta et al. 2018).

4.9.4 Polycaprolactone (PCL)

PCL is a biodegradable polyester used in the production of special PURs for imparting good resistance to water, oil, solvent and chlorine. PCLs are often used as additives for resins preparation in order to improve their processing characteristics and their end-use properties (e.g., impact resistance). Being compatible with a range of other materials, PCL can be without problems mixed. As an example, PLC is often mixed with starch to lower its cost and to increase biodegradability or it can be added to polyvinyl chloride (PVC) as a polymeric plasticizer.

The PLCs applications include controlled drug releases, tissue engineering, bone scaffolds, packaging, compost bags etc. A variety of drugs, including bioactive food constituents or phytochemicals, have been encapsulated within PCL beads for achieving a controlled release and target drug delivery (Mady and Shaker 2017).

4.9.5 Some Applications of Organic Polymer-Based Solid Nanoparticles (OPNPs) in Food Industry

A topical ointment was prepared with PEG and 5% standardized pomegranate rind extract, containing high concentration of multi-target bioactive polyphenols including EA (13%). This formulation was essayed in releasing and skin permeation studies and was found to exhibit good physicochemical properties (Mo et al. 2015). When the wound healing activities of this ointment were compared with the equivalent amount of EA (0.65%), the latter was less effective in inhibiting neutrophil infiltration and collagen augmentation in rat skin (Mo et al. 2014). However, both products applied topically, exhibited anti-inflammatory effects in a mouse model of contact dermatitis (Mo et al., 2014). SOPNPs (150–300 nm) made of PLGA, decorated with chitosan and PEG, were loaded with EA (up to 100 μM) achieving EA-loaded PLGA-chitosan-PEG NPs, that were able to potentiate apoptosis-mediated cell death in HepG2 human hepatoma cells (Abd-Rabou and Ahmed 2017).

The hydroxyl groups of food phenols constituents (for instance, anthocyanins) are easily oxidized into quinones, causing the reduction of the biological power of these molecules. In order to minimize this issue, PLGA NPs, which were stabilized resorting to the use of PEG, were used to encapsulate anthocyanins, achieving biodegradable nanosized formulations with 60% efficiency and endowed with improved stability. These anthocyanins-enriched NPs evidenced a biphasic release profile, *in vitro*, characterized by an initial abrupt release followed by a continued supply. In *vivo,* they proved anti-inflammatory and anti-neurodegenerative capacities (Amin et al. 2017; Jiménez-Aguilar et al. 2011), thus preventing memory losses in estrogen-deficient rats (Varadinova et al. 2009) and showed a neuroprotective power against Alzheimer's dementia. The strong efficiency of the anthocyanins, when administered by using NPs, has been further demonstrated by monitoring their capacity to attenuate the expression of clinical inflammatory and apoptotic indicators (Jiménez-Aguilar et al. 2011). Anthocyanins-loaded NPs reduced significantly the level of apoptotic proteins and the expression of various inflammatory markers, cytotoxic compounds and proinflammatory cytokines (Amin et al. 2017). Anthocyanins associated to NPs significantly upregulated endogenous antioxidant genes, thus helping in the prevention of OS with consequent attenuation of the clinical symptoms of the Alzheimer's dementia (Li et al. 2011).

An additional functionality demonstrated for anthocyanins-loaded NPs is related to their capacity to reduce DNA damage in a higher extent than native non-conjugated anthocyanins. The enhanced activities of anthocyanins-loaded NPs derive from an increased stability and prolonged shelf life (Jiménez-Aguilar et al. 2011). By using PCL biodegradable polymer, EA-loaded PLC-NPs (EA-NPs) were prepared through the emulsion–diffusion–evaporation technique. NPs with distinct shape were obtained with high entrapment and loading efficiency.

An *in vivo* study to measure the oral bioavailability of EA-NPs compared to free EA was performed, using New Zealand white rabbits. The results revealed that the oral administration of EA-NPs produced an EA plasma concentration increased by 3.6 times if compared to the administration of free EA. From these results, it can be concluded that incorporation of EA into PCL, as NPs delivery system, enhances significantly its oral bioavailability and activity (Mady and Shaker 2017).

By making use of the "green" approach concerning the adoption of amino polyols from soybean-oil, several soybean-oil-based cationic PURs were prepared in respect of polyols different structures. In the specific case, these materials were employed to developed edible food coatings endowed with antimicrobial properties, for preventing degradation by foodborne bacteria but in general, the edible semi-natural NPs achieved could represent an alternative food ingredient, promising for preserving food from bacteria degradative development and for extending its shelf life. The structure and hydroxyl functionality of adopted amino polyols affected the particle morphology, their mechanical properties, thermal stability, and antibacterial properties. With some strain-specific exceptions, these PURs-based NPs showed good antibacterial properties toward a panel of bacterial pathogens including *Listeria monocytogenes* NADC 2045, *Salmonella typhimurium* ATCC13311 and *Salmonella Minnesota (S. minnesota)* R613). Concerning the activity against the same strains but of wild-type, the PURs-based NPs exhibited better antibacterial properties toward the Gram-positive *Listeria monocytogenes* than the Gram-negative *S. minnesota* and excellent activity against *S. Minnesota* R613 (Xia et al. 2012).

As previously anticipated (Sect. 4.3), Table 8 summarizes the main associations of EOs with NPs developed in the last years, to prepare bioactive materials with considerable antibacterial properties, increased stability, improved pharmacokinetics and bioavailability.

Concluding this overview on nanomaterials for foods industry, the authors report to readers particularly interested, that a detailed table summarizing the most common nanotechnologies performed for encapsulating a number of bioactive compounds with the description of the purposes, of the used materials, solvents, surfactants and of the particles size, associated with the direct reference for any case, is available in an appealing review by Ezhilarasi et al. (2013).

5 Further on the NPs-Mediated Controlled Release

In the last decades, growing attention and resources have been devoted to the development of nanosized carrier systems for an effective delivering of therapeutic agents, including food-related compounds and phytochemicals (BACs). The controlled BACs release, possibly only where necessary, has been identified as a key stage of BACs administration. Studies *in vitro* have proved that by realizing a controlled delivery, a higher local concentration could be reached, while reducing the overall administered dose and consequently the systemic toxicity associated to.

Several internal and external factors can control the specific release of BACs, including pH, temperature, ultrasound or magnetic fields applications, light incidence, type of NPs adopted and their physicochemical features, BACs chemical structure and their physicochemical properties etc. (Fomina et al. 2012).

Nanoencapsulation of BACs, besides to increase their bioavailability, solubility and to prolong their shelf life, could also allow a controlled release. By allowing target delivery and controlled, long-term release, stimuli-sensitive nanocapsules possessing an oil *core*, showed to improve the effects following the oral administration of BACs, thus contributing to decrease the dosage and administration frequency and giving rise to an increased patient compliance (Fomina et al. 2012).

In this contest, the "layer-by-layer self-assembly" of pH-sensitive building blocks has been explored as promising approach to obtain biomaterials with customized properties (Bisby et al. 2000). Such materials showed to be excellent candidates for interesting applications as stimuli-responsive nanocarriers, able to transport BACs through diverse biological barriers and to promote BACs controlled and target release (Morgan et al. 1995). Moreover, by the carful use of biocompatible pH-dependent polyelectrolyte for the building up of the nanocapsules shell, non-toxic nanocarriers with shell permeability were designed (Fomina et al. 2012).

In addition to favour a controlled and target-specific delivery of transported BACs, NPs encapsulation could allow to achieve an increased cellular up-take and a slower BACs release, that translate into an improved bioactivity, thus contributing to exert a sustained therapy (Panyam and Labhasetwar 2004).

Rhodamine-loaded PolyAlkylene Glycol (PAG)-NPs were applied to SH-SY5Y neuroblastoma cells or prostate cancer DU145 cells and were visualized by fluorescence. The results showed PAG-based NPs located in the cytoplasm, suggesting that they have been internalized *via* endocytosis, overcoming, without damage, the phospholipidic barrier of the cell membrane, that represents an impediment for hydrophilic compounds to enter the cells (Bareford and Swaan 2007; Sha et al. 2012). In addition, reported outcomes have highlighted that, by carrying BACs in NPs, target distribution in specific brain areas can be favoured, providing more valuable benefits concerning neuroregeneration and minimizing distribution to the systemic circulation with consequently, less toxic side-effects (Angelova et al. 2013).

6 Biodegradable NPs: The Other Part of the Coin

By using biodegradable and biocompatible NPs to load BACs for preparing bioactive delivery systems or food ingredients, a lot of nonpareil advantages can be achieved, but the risks to human health that could be associated to the long term use of nanosized devices so prepared, are to be more deeply explored (Kumari et al. 2010).

The mechanisms responsible for the deleterious effects of NPs, in the frame of complex biological systems, are surely mainly associated to the formation of pro-oxidants species with increased productions of Reactive Oxygen Species (ROS) and free radicals, that could give rise to OS, inflammation, and consequently to several disabling pathophysiological situations including immune-toxicity. Residual solvents from preparation techniques, as well as polymers intrinsic toxicity, also play a pivotal role in possible immune reactions (Maurer-Jones et al. 2009). From the information available in the literature regarding this, it can be noticed that toxicological outcomes associated to NPs are closely linked to their size, shape and surface charge, as well as to their capacity and modalities to cross biological barriers (Vega-Villa et al. 2008).

In relation with the use of BACs-loaded NPs addressed to be ingested in foods, the constraints associated to their toxicity, following oral ingestion, have been associated to the occurrence of deleterious effects on liver, kidney and spleen and of possible allergic reactions (Lanone and Boczkowski 2006). PEG-grafted liposome infusion has been described to trigger non-IgE-mediated signs of hypersensitivity (Moghimi et al. 2005). On the other hand, the immunogenic characteristics of NPs could be exploited as advantageous tools and NPs that cause hypersensitivity can be used as valuable adjuvants for the development of vaccines.

Fortunately, there is no lack of studies on nanoencapsulation and controlled release from nanostructured carriers, which have demonstrated the safety of these systems. In this frame, different concentrations of free-anthocyanins and anthocyanins loaded in PLGA-based NPs were evaluated concerning the cytotoxic profiles in a model of human neuroblastoma. The results allowed to discard significant cytotoxic effects but the joint application of anthocyanins loaded NPs increased the viability of treated cells, by protecting them from neurotoxic events (Amin et al. 2017). Thus, PLGA-based NPs might constitute promising carrier for phytochemicals with negligible cytotoxicity events.

7 Consideration and Future Perspectives

It seems that, in the coming years, the research trends in the application of edible NPs for the administration of BACs, will be focused on identifying and developing new manufacturing strategies aiming at combining multiple structural designs in order to prepare materials endowed with more specific features and suitable for the diverse types of bioactive compounds.

This strategy would contribute to maximize effectivity and to strive the maximum profit by an extended release and by an *in situ* delivery of BACs-loaded NPs. One of the major constraints concerning the administration of BACs, even when nanotechnology products are adopted, lies in the degradation of these compounds in the GIT lumen before being absorbed. Although reports about *in vitro* results, have revealed promising potentials of the use of NPs for the administration of BACs, in order to forecast in reliable way their actual oral bioavailability, the confirmation by *in vivo* results on simulation models, where the impact of GIT digestion could be evaluated, is required (Minekus et al. 2014). With this aim, in addition to implementing models mimicking the salivary, gastric, and the salts and enzymes concentrations in the intestinal fluids, it is necessary to include also proper simulators of GIT dynamics, structure, and mechanical issues.

The current definition of "bioaccessibility" and "bioavailability", that takes into account only the absorption in the upper GIT emphasizing its value, must be reviewed in the consideration of gut metabolism, that strongly affects the actual biological activity of compounds ingested by oral administration.

A complete information on the features of this physiological process, according to the chemical properties of the BACs of interest and of their metabolites, that come from GIT digestion, is pivotal for the rational design of the specific and well-functioning NPs-based delivering systems. Often, to develop NPs aiming at improving the BACs gastric or small intestine absorption only, may not be a good design. Concerning some phytochemicals, the highest bioefficacy is associated with metabolites that form only in the large intestine, as a result of the metabolism of the local microbiota. It is the case of EA, whose bioactivity is nowadays mainly associated to urolithins, which form only in the last tract of GIT by microbiota metabolism (Espín et al., 2013). In this case GA absorption in the upper GIT turns into being undesirable.

When the effects of NPs and/or nanovesicles (NVs) on the bioefficacy of phytochemicals are evaluated, special attention should be paid to eventual changes in the NPs structure, as a result of the digestive conditions, such as dilution occurring in the GIT fluids, pH and ions environments, as well as the enzymatic activity in these compartments. These factors can change the NPs structure and may provoke modifications in their bioactive functionality. Stability of micelles, liposomes, nanoemulsions or solid NPs can be undermined, thus compromising their success as carriers of bioactive compounds (Xiao et al. 2017).

Among other reflections, it has to be considered, that by loading BACs in NPs, somehow, modification in cellular signaling routes can be evoked and cells responses after BACs-loaded NPs oral administration could differ from the administration of the free forms of BACs. In this regard, it is evident that the interaction between phytochemicals and matrix materials or between them and molecules present in complex biological systems (for instance in mammals) needs further in deep investigations (Xiao et al., 2017).

A considerable amount of topics about NPs-mediated BACs administration remain underexplored, and the application of the gathered knowledge on the use of

NPs, as carriers of BACs, needs to more deeply decode the reciprocal interactions between phytochemicals and GIT and the events, as the gut microbiota activity, for exploiting them with advantages. Despite the rational utility of the nanotechnology to enhance the bioavailability and bioactivity of BACs, NPs are synthesized by physical and chemical methods and by using expensive and hazardous chemicals such as metallic biomaterials, which could limit their actual application *in vivo* and which could have a considerable environmental impact in terms of the residues that derive from the NPs synthesis. Green technologies using no toxic reagents to prepare metal NPs are required and the synthesis of NPs using eco-friendly and biocompatible reagents could be a promising solution and an appealing contribute to minimize the side effect of these processes.

References

Abderrezak A, Bourassa P, Mandeville JS, Sedaghat-Herati R, Tajmir-Riahi HA (2012) Dendrimers bind antioxidant polyphenols and cisplatin drug. PLoS One 7(3):e33102. https://doi.org/10.1371/journal.pone.0033102

Abd-Rabou AA, Ahmed HH (2017) CS PEG decorated PLGA nano-prototype for delivery of bioactive compounds: af novel approach for induction of apoptosis in HepG2 cell line. Adv Med Sci 62:357–367. https://doi.org/10.1016/j.advms.2017.01.003

Acosta E (2009) Bioavailability of nanoparticles in nutrient and nutraceutical delivery. Curr Opin Colloid Interface Sci 14(1):3–15

Ahire V, Mishra KP (2017) Ellagic acid as a potential anti-cancer drug. Int J Radiol Radiat Ther 3:00063. https://doi.org/10.15406/ijrrt.2017.03.00063

Al Shaal L, Müller RH, Shegokar R (2010) Smart crystal combination technology - scale up from lab to pilot scale and long term stability. Pharmazie 65:877–884

Alfei S, Castellaro S, Taptue GB (2017a) Synthesis and NMR characterization of dendrimers based on 2, 2-bis-(hydroxymethyl)-propanoic acid (bis-HMPA) containing peripheral amino acid residues for gene transfection. Org Commun 10:144–177

Alfei S, Castellaro S (2017b) Synthesis and characterization of polyester-based dendrimers containing peripheral arginine or mixed amino acids as potential vectors for gene and drug delivery. Macromol Res 25(12):1172–1186. https://doi.org/10.1007/s13233-017-5160

Alfei S, Taptue GB, Catena S, Bisio A (2018a) Synthesis of water-soluble, polyester-based dendrimer prodrugs for exploiting therapeutic properties of two triterpenoid acids. Chinese J Polym Sci 36(9):999–1010. https://doi.org/10.1007/s10118-018-2124-9

Alfei S, Catena S (2018b) Synthesis and characterization of versatile amphiphilic dendrimers peripherally decorated with positive charged amino acids. Polym Int 67:1572–1584. https://doi.org/10.1002/pi.5680

Alfei S, Catena S (2018c) Synthesis and characterization of fourth generation polyester-based dendrimers with cationic amino acids-modified crown as promising water soluble biomedical devices. Polym Adv Technol 29:2735–2749. https://doi.org/10.1002/pat.4396

Alfei S, Turrini F, Catena S, Zunin P, Parodi B, Zuccari G, Pittaluga A, Boggia R (2019a) Pectin microdispersion vs non-polyamidoamine dendrimer nanodispersions: Two biocompatible approaches to increase ellagic acid water solubility and allow its more ways therapeutic administration. New J Chem 43:2438–2448

Alfei S, Oliveri P, Malegori C (2019b) Assessment of the efficiency of a nanospherical gallic acid dendrimer for long-term preservation of essential oils: An integrated chemometric-assisted FTIR study. ChemistrySelect 4:8891–8901

Alfei S, Signorello MG, Schito A, Catena S, Turrini F (2019c) Reshaped as polyester-based nanoparticles, gallic acid inhibits platelet aggregation, reactive oxygen species production and multi-resistant Gram-positive bacteria with an efficiency never obtained. Nanoscale Adv 1(10):4148–4157

Alfei S, Marengo B, Domenicotti C (2020a) Development of a fast, low cost, conservative and ecological method for quantifying gallic acid in polymeric formulations by FTIR spectroscopy in solution. Chemistryselect 5:4381–4388

Alfei S, Catena S, Turrini F (2020b) Biodegradable and biocompatible spherical dendrimer nanoparticles with a gallic acid shell and a double-acting strong antioxidant activity as potential device to fight diseases from "oxidative stress". Drug Deliv Transl Res 10:259–270

Alfei S, Marengo B, Zuccari G, Turrini F, Domenicotti C (2020c) A biodegradable dendrimer used as improving "excipient" for gallic acid administration and tested on human neuroblastoma cells, showed intriguing results. Pharmaveutics: under review.

Alving CR (1991) Liposomes as carriers of antigens and adjuvants. J Immunol Methods 140:1–13

Amar I, Aserin A, Garti N (2003) Solubilization patterns of lutein and lutein esters in food grade nonionic microemulsions. J Agricult Food Chem 51:4775–4781

Amidi M, Mastrobattista E, Jiskoot W, Hennink WE (2010) Chitosan-based delivery systems for protein therapeutics and antigens. Adv Drug Deliv Rev 62:59–82

Amin FU, Shah SA, Badshah H, Khan M, Kim MO (2017) Anthocyanins encapsulated by PLGA@PEG nanoparticles potentially improved its free radical scavenging capabilities via p38/JNK pathway against a 1–42-induced oxidative stress. J Nanobiotechnol 15:12–27

Angelova A, Angelov B, Drechsler M, Lesieur S (2013) Neurotrophin delivery using nanotechnology. Drug Discov Today 18:1263–1271

Aqil F, Munagala R, Jeyabalan J, Vadhanam MV (2013) Bioavailability of phytochemicals and its enhancement by drug delivery systems. Cancer Lett 334:133–141

Astray G, Gonzalez-Barreiro C, Mejuto JC, Rial-Otero R, Simal-Gándara J (2009) A review on the use of cyclodextrins in foods. Food Hydrocoll 23(7):1631–1640. https://doi.org/10.1016/j.foodhyd.2009.01.001

Avachat AM, Patel VG (2015) Self nanoemulsifying drug delivery system of stabilized ellagic acid-phospholipid complex with improved dissolution and permeability. Saudi Pharm J 23:276–289. https://doi.org/10.1016/j.jsps.2014.11.001

Baccarin T, Lemos-Senna E (2014) Pomegranate seed oil nanoemulsions encapsulating pomegranate peel polyphenol-rich ethyl acetate fraction: development and antioxidant assessment. J Nanopharmaceutics Drug Delivery 2(4):333–343. https://doi.org/10.1166/jnd.2014.1068

Balasubramaniam A, Kumar VA, Pillai KS (2002) Formulation and in vivo evaluation of niosome-encapsulated daunorubicin hydrochloride. Drug Dev Ind Pharm 28:1181–1193

Bareford LM, Swaan PW (2007) Endocytic mechanisms for targeted drug delivery. Adv Drug Deliv Rev 59:748–758

Barua S, Mitragotri S (2014) Challenges associated with penetration of nanoparticles across cell and tissue barriers: a review of current status and future prospects. Nano Today 9:223–243

Bassam N, Laure C, Jean-François B, Celine G, Yann R, Zephirin M (2013) From petrochemical polyurethanes to biobased polyhydroxyurethanes. Macromolecules 46(10):3771–3792. https://doi.org/10.1021/ma400197c

Ben Saad LA, Kim KH, Quah CC, Kim WR, Shahimi M (2017) Anti-inflammatory potential of ellagic acid, gallic acid and punicalagin A&B isolated from Punica granatum. BMC Complement Altern Med 17:47–56. https://doi.org/10.1186/s12906-017-1555-0

Bengoechea C, Jones OG, Guerrero A, McClements DJ (2011) Formation and characterization of lactoferrin/pectin electrostatic complexes: impact of composition, pH and thermal treatment. Food Hydrocoll 25:1227–1232

Berton-Carabin CC, Coupland JN, Elias RJ (2013) Effect of the lipophilicity of model ingredients on their location and reactivity in emulsions and solid lipid nanoparticles. Colloids Surf A Physicochem Eng Asp 431:9–17

Bhandari BR (2007) Spray drying and powder properties. In: Hui YH, Clary C, Farid MM, Fasina OO, Noomhorm A, Welti-Chanes J (eds) Food drying science and technology. DEStech Publications Inc, USA, pp 215–248

Bisby RH, Mead C, Morgan CG (2000) Wavelength-programmed solute release from photosensitive liposomes. Biochem Biophys Res Commun 276:169–173

Bisio A, De Mieri M, Milella L, Schito AM, Parricchi A, Alfei S, Vignola L, Hamburger M, De Tommasi N (2016) Biological activity of constituents of Salvia chamaedryoides. Planta Med 81(S 01):S1–S381. https://doi.org/10.1055/s-0036-1596322

Bisio A, De Mieri M, Milella L, Schito AM, Parricchi A, Russo D, Alfei S, Lapillo M, Tuccinardi T, Hamburger M, De Tommasi N (2017) Antibacterial and hypoglycemic diterpenoids from Salvia chamaedryoides. J Nat Prod 80(2):503–514

Blasi P, Giovagnoli S, Schoubben A, Ricci M, Rossi C (2007) Solid lipid nanoparticles for targeted brain drug delivery. Adv Drug Deliv Rev 59:454–477

Bongoni R, Steenbekkers B, Verkerk R, van Boekel M, Dekker M (2013) Studying consumer behaviour related to the quality of food: a case on vegetable preparation affecting sensory and health attributes. Trends Food Sci Technol 33(2):139–145. https://doi.org/10.1016/j.tifs.2013.08.004

Bongoni R, Verkerk R, Steenbekkers B, Dekker M, Stieger M (2015) Evaluation of different cooking conditions on broccoli (Brassica oleracea var. italica) to improve the nutritional value and consumer acceptance. Plant Foods for Hum Nutr 69(3):228–234. https://doi.org/10.1007/s11130-014-0420-2

Boulle F, Kenis G, Cazorla M, Hamon M, Steinbusch HW, Lanfumey L, van del Hove DL (2012) TrkB inhibition as a therapeutic target for CNS-related disorders. Prog Neurobiol 98:197–206

Bulani VD, Kothavade PS, Kundaikar HS, Gawali NB, Chowdhury AA, Degani MS, Juvekar AR (2016) Inclusion complex of ellagic acid with β-cyclodextrin: characterization and in vitro anti-inflammatory evaluation. J Mol Struct 1105:308–315. https://doi.org/10.1016/j.molstruc.2015.08.054

Cabrera C, Artacho R, Gimenez RJ (2006) Beneficial effects of green tea – a review. Am Coll Nutr 25:79–99

Campardelli R, Reverchon E (2015) A-Tocopherol nanosuspensions produced using a supercritical assisted process. J Food Eng 149:131–136

Campbell PI (1983) Toxicity of some charged lipids used in liposome preparations. Cytobios 37:21–26

Carosio R, Zuccari G, Orienti I, Mangraviti S, Montaldo PG (2007) Sodium Ascorbate induces apoptosis in neuroblastoma cell lines by interfering with iron uptake. Molecular Cancer 6(1):55

Cerqueira MA, Pinheiro AC, Silva HD, Ramos PE, Azevedo MA, Flores-López ML, Rivera MC, Bourbon AI, Ramos OL, Vicente AA (2014) Design of bio-nanosystems for oral delivery of functional compounds. Food Eng Rev 6:1–19

Chaize B, Colletier JP, Winterhalter M, Fournier D (2004) Encapsulation of enzymes in liposomes: high encapsulation efficiency and control of substrate permeability. Artif Cell Blood 32:67–75

Chakraborty S, Shukla D, Mishra B, Singh S (2009) Lipid- an emerging platform for oral delivery of drugs with poor bioavailability. Eur J Pharm Biopharm 73:1–15

Chan SY, Choo WS, Young DJ, Loh XJ (2017) Pectin as a rheology modifier: Origin, structure, commercial production and rheology. Carbohydr Polym 161:118–139

Chang SK, Alasalvar C, Shahidi F (2016) Review of dried fruits: phytochemicals, antioxidant efficacies, and health benefits. J Funct Foods 21:113–132

Chau CF, Wuand SH, Yen GC (2007) The development of regulations for food nanotechnology. Trends Food Sci Technol 18:269–280

Chen YC, Lo CL, Lin YF, Hsiue GH (2013) Rapamycin encapsulated in dual responsive micelles for cancer therapy. Biomaterials 34:1115–1127

Cheng AL, Hsu CH, Lin JK, Hsu MM, Ho YF, Shen TS, Ko JY, Lin JT, Lin BR, Ming-Shiang W, Yu HS, Jee SH, Chen GS, Chen TM, Chen CA, Lai MK, Pu YS, Pan MH, Wang YJ, Tsai CC,

Hsieh CY (2001) Phase I clinical trial of curcumin, a chemopreventive agent, in patients with high-risk or pre-malignant lesions. Anticancer Res 21(4B):2895–2900
Chime SA, Kenechukwu FC, Attama AA (2014) Nanoemulsions—advances in formulation, characterization and applications in drug delivery. In: Sezer AD (ed) Application of nanotechnology in drug delivery. IntechOpen, London, pp 77–111
Chiou D, Langrish TAG (2007) Development and characterization of novel nutraceuticals with spray drying technology. J Food Eng 82:84–91
Cho YS, Kim SK, Ahn CB, Je JY (2011) Preparation, characterization, and antioxidant properties of gallic acid-graftedchitosans. Carbohydr Polym 83:1617–1622
Chow PY, Gue SZ, Leow SK, Goh LB (2015) Solid self-microemulsifying system (S-SMECS) for enhanced bioavailability and pigmentation of highly lipophilic bioactive carotenoid. Powder Technol 274:199–204
Cirillo G, Kraemer K, Fuessel S, Puoci F, Curcio M, Spizzirri UG, Altimari I, Iemma F (2010) Biological activity of a gallic acid-gelatin conjugate. Biomacromolecules 11:3309–3315
Crescente M, Jessen G, Momi S, Höltje H-D, Gresele P, Cerletti C, de Gaetano G (2009) Interactions of gallic acid, resveratrol, quercetin and aspirin at the platelet cyclooxygenase-1 level: functional and modelling studies. Thromb Haemos 102:336–346
De Lima PN, da Silva DD, Costa VG, Falcão DQ, de Lima Araujo KG (2016) Increasing solubility of red bell pepper carotenoids by complexation with 2-hydroxypropyl-β-cyclodextrin. Food Chem 208:124–131. https://doi.org/10.1016/j.foodchem.2016.03.122
De Paz E, Martín A, Estrella A, Rodríguez-Rojo S, Matias AA, Duarte CM, Cocero MJ (2012) Formulation of b-carotene by precipitation from pressurized ethyl acetate-on-water emulsions for application as natural colorant. Food Hydrocoll 26:17–27
Del Valle EMM (2004) Cyclodextrins and their uses: a review. Process Biochem 39:1033–1046
Deladino L, Anbinder PS, Navarro AS, Martino MN (2008) Encapsulation of natural antioxidants extracted from Ilex paraguariensis. Carbohyd Polym 71:126–134
Deligiannakis Y, Sotiriou GA, Pratsinis SE (2012) Antioxidant and antiradical SiO2 nanoparticles covalently functionalized with Gallic acid. ACS Appl Mater Interfaces 4:6609–6617
Díaz-Gómez R, Toledo-Araya H, López-Solís R, Obreque-Slier E (2014) Combined effect of gallic acid and catechin against Escherichia coli. LWT Food Sci Technol 59:896–900
Domínguez-Perles R, Abellán A, León D, Ferreres F, Guy A, Oger C, Galano JM, Durand T, Gil-Izquierdo A (2018) Sorting out the phytoprostane and phytofuran profile in vegetable oils. Food Res Int 107:619–628
Donsi F, Senatore B, Huang QR, Ferrari G (2010) Development of novel pea protein-based Nanoemulsions for delivery of Nutraceuticals. J Agric Food Chem 58(19):10653–10660. https://doi.org/10.1021/jf101804g
Donsi F, Annunziata M, Sessa M, Ferrari G (2011) Nanoencapsulation of essential oils to enhance their antimicrobial activity in foods. Lwt-Food Sci Technol 44:1908–1914
Duceppe N, Tabrizian M (2010) Advances in using chitosan-based nanoparticles for in vitro and in vivo drug and gene delivery. Expert Opin Drug Deliv 7:1191–1207
Ersus S, Yurdagel U (2007) Microencapsulation of anthocyanin pigments of black carrot (Daucus carota L.) by spray drier. J Food Eng 80:805–812
Espín JC, Larrosa M, García-Conesa MT, Tomás-Barberán F (2013) Biological significance of Urolithins, the gut microbial Ellagic acid-derived metabolites: the evidence so far. Evid Based Complement Alternat Med 2013:270418. https://doi.org/10.1155/2013/270418
Ezhilarasi PN, Karthik P, Chhanwal N, Anandharamakrishnan C (2013) Nanoencapsulation techniques for food bioactive components: a review. Food Bioprocess Technol 6(3):628–647. https://doi.org/10.1007/s11947-012-0944-0
Fang Z, Bhandari B (2010) Encapsulation of polyphenols - a review. Trends Food Sci Tech 21:510–523
Feriyanto D, Idris MI, Sebayang D (2014) Effect of Cr to Fe on the solid solubility, lattice parameter and strain of Fe80Cr20 alloy powder. Appl Mech Mater 660:280–284

Firdaus F, Zafeer MF, Ehraz MW, Hossain AMM, Afzal M (2018) Ellagic acid mitigates arsenic-trioxide-induced mitochondrial dysfunction and cytotoxicity in SH-SY5Y cells. J Biochem Mol Toxicol 32:e22024

Flanagan J, Singh H (2006) Microemulsions: a potential delivery system for bioactives in food. Critical Rev Food Sci Nutr 46:221–237

Fojan P, Schwach-Abdellaoui K, Tommeraas K, Gurevich L, Petersen SB (2006) Polysaccharide based nanoparticles and nanoporous matrices. Nano Sci Technol Inst 2:79–82

Fomina N, Sankaranarayanan J, Almutairi A (2012) Photochemical mechanisms of light-triggered release from nanocarriers. Adv Drug Deliv Rev 64:1005–1020

Gao L, Liu G, Wang X, Liu F, Xu Y, Ma J (2011) Preparation of a chemically stable quercetin formulation using nanosuspension technology. Int J Pharm 404:231–237

Ghosh I, Bose S, Vippagunta R, Harmon F (2011) Nanosuspension for improving the bioavailability of a poorly soluble drug and screening of stabilizing agents to inhibit crystal growth. Int J Pharm 409(1):260–268

Gong J, Chen M, Zheng Y, Wang S, Wang Y (2012) Polymeric micelles drug delivery system in oncology. J Control Release 159:312–323

Gopalakrishnan L, Ramana LN, Sethuraman S, Krishnan UM (2014) Ellagic acid encapsulated chitosan nanoparticles as anti-hemorrhagic agent. Carbohyd Polym 111:215–221. https://doi.org/10.1016/j.carbpol.2014.03.093

Gregoriadis G (1990) Immunological adjuvants: a role for liposomes. Immunol Today 11:89–97

Gude RP, Jadhav MG, Rao SG, Jagtap AG (2002) Effects of niosomal cisplatin and combination of the same with theophylline and with activated macrophages in murine B16F10 melanoma model. Cancer Biother Radiopharm 17:183–192

Guo BL, Ma PX (2014) Synthetic biodegradable functional polymers for tissue engineering: a brief review. Sci China Chem 57(4):490–500

Guo F, Aryana S, Han Y, Jiao Y (2018) A review of the synthesis and applications of polymer–nanoclay composites. App Sci 8(9):1696. https://doi.org/10.3390/app8091696

Ha TV, Kim S, Choi Y, Kwak HS, Lee SJ, Wen J, Oey I, Ko S (2015) Antioxidant activity and bioaccessibility of size-different nanoemulsions for lycopeneenriched tomato extract. Food Chem 178:115–121

Hadaruga NG, Hadaruga DI, Paunescu V, Tatu C, Ordodi VL, Bandur G, Lupea AX (2006) Thermal stability of the linoleic acid/α- and β-cyclodextrin complexes. Food Chem 99(3):500–508

Hajipour H, Hamishehkar H, Rahmati-yamchi M, Shanehbandi D, Ahmad SNS, Hasani A (2018) Enhanced anti-cancer capability of ellagic acid using Solid Lipid Nanoparticles (SLNs). Int J Cancer Manag 11:e9402

Hamri SZ, Zeghichi M, Chibane M, Kallithraka S, Benhalima A (2011) What is so special about the mediterranean diet in the maghreb? the role of economics in eating choices and chronic diseases outcomes. Annals Nutr Metab 58:191–216

Harde H, Das M, Jain S (2011) Solid lipid nanoparticles: an oral bioavailability enhancer vehicle. Expert Opin Drug Deliv 8:1407–1424. https://doi.org/10.1517/17425247.2011.604311

Hayek T, Fuhrman B, Vaya J, Rosenblat M, Belinky P, Coleman R, Elis A, Aviramet M (1997) Reduced progression of atherosclerosis in Apolipoprotein E–deficient mice following consumption of red wine, or its polyphenols Quercetin or Catechin, is associated with reduced susceptibility of LDL to oxidation and aggregation. Arterioscler Thromb Vasc Biol 17:2744–2752

He X, Deng H, Hwang H (2019) The current application of nanotechnology in foodand agriculture. J Food Drug Anal 27:1–21

Heber D (2008) Multitargeted therapy of cancer by ellagitannins. Cancer Lett 269:262–268

Holscher HD (2017) Dietary fiber and prebiotics and the gastrointestinal microbiota. Gut Microbes 8:172–184

Hseu YC, Chou CW, Kumar SKJ, Fu KT, Wang HM, Hsu LS, Kuo YH, Wu CR, Chen SC, Yang HL (2012) Ellagic acid protects human keratinocyte (HaCaT) cells against UVA-induced oxidative stress and apoptosis through the upregulation of the HO-1 and Nrf-2 antioxidant genes. Food Chem Toxicol 50:1245–1255. https://doi.org/10.1016/j.fct.2012.02.020

Hsieh YS, Yang SF, Sethi G, Hu DN (2015) Natural bioactives in cancer treatment and prevention. Biomed Res Int 2015:182835. https://doi.org/10.1155/2015/182835

Hu B, Ting Y, Yang X, Tang W, Zeng X, Huang Q (2012) Nanochemoprevention by encapsulation of (−)-epigallocatechin-3-gallate with bioactive peptides/chitosan nanoparticles for enhancement of its bioavailability. Chem Commun 48:2421–2423

Hu C, Zhao G, Xia Q, Sun R (2015) Development and characterization of a selfdouble- emulsifying drug delivery system containing both epigallocatechin-3- gallate and a-lipoic acid. J Mat Sci 50(20):6567–6577

Huang EJ, Reichardt LF (2001) Neurotrophins: roles in neuronal development and function. Annu Rev Neurosci 24:677–736

Imm BY, Kim CH, Imm JY (2014) Effects of partial substitution of lean meat with pork backfat or canola oil on sensory properties of Korean traditional meat patties (Tteokgalbi). Korean J Food Sci An 34(4):496–499

Ishikawa M, Hashimoto Y (2011) Improvement in aqueous solubility in small molecule drug discovery programs by disruption of molecular planarity and symmetry. J Med Chem 54(6):1539–1554

Jadhav PB, Pandey PS (2013) Phase solubility analysis: a technique of purity determination. World Res J Pharm Res 1:5–11

Jain S, Singh P, Mishra V, Vyas SP (2005) Mannosylated niosomes as adjuvantcarrier system for oral genetic immunization against hepatitis B. Immunol Lett 101:41–49

Jeong JY, Park MN, Cho ES, Jang HJ, Park SK, Lee HJ (2015) Epigallocatechin-3-gallate-induced free-radical production upon adipogenic differentiation in bovine bone-marrow mesenchymal stem cells. Cell Tissue Res 362:87–96

Jiménez-Aguilar DM Ortega-Regules AE Lozada-Ramírez JD Pérez-Pérez MCI Vernon-Cartere EJ Welti-Chanesa (2011) Color and chemical stability of spray-dried blueberry extract using mesquite gum as wall material. J Food Compos Anal 24:889–894

Joye IJ, Nelis VA, McClements DJ (2015) Gliadin-based nanoparticles: stabilization by post-productionpolysaccharide coating. Food Hydrocoll 43:236–242

Jung J, Perrut MJ (2001) Particle design using supercritical fluids: literature and patent survey. Supercrit Fluids 20:179–219

Junyaprasert VB, Singhsa P, Suksiriworapong J, Chantasart D (2012) Physicochemical properties and skin permeation of span 60/tween 60 niosomes of ellagic acid. Int J Pharm 423:303–311

Kale BB, Aloorkar NH, Deshmukh SM, Sulake SP, Humbe PV, Mane PP (2014) Recent advancements in particle engineering techniques for pharmaceutical applications. Indo Am J Pharm Res 4:2027–2049

Kaminskas LM, Boyd BJ, Porter CJH (2011) Dendrimer pharmacokinetics: the effect of size, structure and surface characteristics on ADME properties. Nanomed 6:1063–1084

Karadag A, Ozcelik B, Huang Q (2014) Quercetin nanosuspensions produced by high-pressure homogenization. J Agric Food Chem 62:1852–1859

Kataoka K, Harada A, Nagasaki Y (2001) Block copolymer micelles for drug delivery: design, characterization and biological significance. Adv Drug Deliv Rev 47:113–131

Kaul G, Amiji M (2004) Biodistribution and targeting potential of poly(ethylene glycol)-modified gelatin nanoparticles in subcutaneous murine tumor model. J Drug Target 12:585–591

Keck CM, Müller RH (2006) Drug nanocrystals of poorly soluble drugs produced by high pressure homogenisation. Eur J Pharm Biopharm 62:3–16

Keck CM, Kobierski S, Mauludin R, Muller RH (2008) Second generation of drug nanocrystals for delivery of poorly soluble drugs: SmartCrystals technology. Dosis 24(2):124–128

Khan I, Saeed K, Khan I (2017) Nanoparticles: properties, applications and toxicities. Arab J Chem. https://doi.org/10.1016/j.arabjc.2017.05.011

Kim K, Bae ON, Lim KM, Noh JY, Kang S, Chung KY, Chung JH (2012) Novel antiplatelet activity of Protocatechuic acid through the inhibition of high shear stress-induced platelet aggregation. J Pharm Exp Ther 343:704–711

Kim IH, Lee H, Kim JE, Song KB, Lee YS, Chung DS, Min SC (2013) Plum coatings of lemongrass oil-incorporating carnauba wax-based Nanoemulsion. J Food Sci 78(10):E1551–E1559
Kloover JS, den Bakker MA, Gelderblom H, van Meerbeeck JP (2004) Fatal outcome of a hypersensitivity reaction to paclitaxel: a critical review of premedication regimens. Br J Cancer 90:304–305
Kobierski S, Keck CM (2008) Nanocrystal production by BM-HPH combination technology. Controlled Release Society, Abstract, Germany Chapter, Annual Meeting, 45 March:40
Koo OM, Rubinstein I, Onyuksel H (2005) Role of nanotechnology in targeted drug delivery and imaging: a concise review. Nanomed Nanotechnol Biol Med 1:193–212
Kosaraju SL, D'ath L, Lawrence A (2006) Preparation and characterisation of chitosan microspheres for antioxidant delivery. Carbohyd Polym 64:163–167
Koshy P, Pacharane S, Chaudhry A, Jadhav K, Kadam V (2010) Drug particle engineering of poorly water soluble drugs. Der Pharm Lett 2:65–76
Koutelidakis AE, Argyri K, Sevastou Z, Lamprinaki D, Panagopoulou E, Paximada E, Sali A, Papalazarou V, Mallouchos A, Evageliou V, Kostourou V, Mantala I, Kapsokefalou M (2017) Bioactivity of epigallocatechin gallate nanoemulsions evaluated in mice model. J Med Food 20:923–931
Kris-Etherton PM, Hecker KD, Bonanome A, Coval SM, Binkoski AE, Hilpert KF, Etherton TD (2002) Bioactive compounds in foods: their role in the prevention of cardiovascular disease and cancer. Am J Med 113:71–88
Krog N (1977) Functions of emulsifiers in food systems. J Am Oil Chem Soc 54(3):124–131
Kumar VD, Verma PRP, Singh SK (2015) Development and evaluation of biodegradable polymeric nanoparticle for the effective delivery of quercetin using quality by design approach. LWT – Food Sci Technol 61:330–338
Kumari A, Yadav SK, Yadav SC (2010) Biodegradable polymeric nanoparticles based drug delivery systems. Colloids Surf B Biointerfaces 75:1–18
Kumari A, Singla R, Guliani A, Yadav SK (2014) Nanoencapsulation for drug delivery. EXCLI J 13:265–286
Kwon GS, Okano T (1996) Polymeric micelles as new drug carriers. Adv Drug Deliv Rev 21:107–116
Lakshmi P, Kumar GA (2010) Nanosuspension technology: a review. Int J Pharm Pharm Sci 2(Suppl 4):35–40
Lanone S, Boczkowski J (2006) Biomedical applications and potential health risks of nanomaterials: molecular mechanisms. Curr Mol Med 6:651–663
Larrosa M, Garcia-Conesa MT, Espín JC, Tomás-Barberán FA (2010) Ellagitannins, ellagic acid and vascular health. Mol Asp Med 31:513–539
Lee HS (2006) Antiplatelet property of Curcuma longa L. rhizome-derived ar-turmerone. Bioresour Technol 97:1372–1376
Lee HJ, Jayasena DD, Kim SH, Kim HJ, Heo KN, Song JE, Jo C (2015) Comparison of bioactive compounds and quality traits of breat meat from Korean native ducks and commercial ducks. Korean J Food Sci An 35(1):114–120
Lee DR, Ho MJ, Choi YW, Kang MJ (2017) A Polyvinylpyrrolidone-based Supersaturable self-emulsifying drug delivery system for enhanced dissolution of cyclosporine a. Polymers (Basel) 9(4):124. https://doi.org/10.3390/polym9040124
Lemarchand C, Gref R, Couvreur P (2004) Polysaccharide-decorated nanoparticles. Eur J Pharm Biopharm 58:327–341
Li L, Li W, Jung SW, Lee YW, Kim YH (2011) Protective effects of decursin and decursinol angelate against amyloid-protein-induced oxidative stress in the PC12 cell line: the role of Nrf2 and antioxidant enzymes. Biosci Biotechnol Biochem 75:434–442
Li Y, Zheng J, Xiao H, McClements DJ (2012) Nanoemulsion-based delivery systems for poorly water-soluble bioactive compounds: influence of formulation parameters on polymethoxyflavone crystallization. Food Hydrocoll 27:517–528
Lim J, Jeong S, Lee J, Park S, Lee J, Lee S (2017) Effect of shortening replacement with oleogels on the rheological and tomographic characteristics of aerated baked goods. J Sci Food Agric 97(11):3727–3732

Linde GA, Laverde A Jr, Vaz de Faria E, Colauto NB, Faria de Moraes F, Zanin GM (2009) Taste modification of amino acids and protein hydrolysate bya-cyclodextrin. Food Res Int 42:814–818

Liu RH (2004) Potential synergy of phytochemicals in cancer prevention: mechanism of action. J Nutr 134(12):3479S–3485S

Liu Z, Jiao Y, Wang Y, Zhou C, Zhang Z (2008) Polysaccharides-based nanoparticles as drug delivery systems. Adv Drug Deliv Rev 60:1650–1662

Liu Y, Li K, Liu B, Feng SS (2010) A strategy for precision engineering of nanoparticles of biodegradable copolymers for quantitative control of targeted drug delivery. Biomaterials 31:9145–9155

Liu T, Müller RH, Möschwitzer JP (2012) Process optimization of a novel particle size reduction technology H 42. Tag der Pharmazie 2012, FU Berlin, Berlin, abstract P33, booklet: 25

Lockman PR, Koziara JM, Mumper RJ, Allen DD (2004) Nanoparticle surface charges alter blood-brain barrier integrity and permeability. J Drug Target 12:635–641

Lu BY, Li MQ, Yin R (2016) Phytochemical content, health benefits, and toxicology of common edible flowers: a review (2000–2015). Crit Rev Food Sci Nutr 56:S130–S148

Lucas-Abellán C, Fortea I, Gabaldón JA, Núñez-Delicado E (2008) Encapsulation of quercetin and myricetin in cyclodextrins at acidic pH. J Agric Food Chem 56(1):255–259

Madrigal-Carballoa S, Limb S, Rodrigueza G, Vilac AO, Kruegerb CG, Gunasekaranb S, Reed JD (2010) Biopolymer coating of soybean lecithin liposomes via layer-by-layer self-assembly as novel delivery system for ellagic acid. J Funct Foods 2(2):99–106. https://doi.org/10.1016/j.jff.2010.01.002

Mady FM, Ibrahim SRM (2018) Cyclodextrin-based nanosponge for improvement of solubility and oral bioavailability of Ellagic acid. Pakistan J Pharm Sci 31:2069–2076

Mady FM, Shaker MA (2017) Enhanced anticancer activity and oral bioavailability of ellagic acid through encapsulation in biodegradable polymeric nanoparticles. Int J Nanomedicine 12:7405–7417

Mady MM, Ghannam MM, Khalil WA, Repp R, Markus M, Rascher W, Muller R, Fahr A (2004) Efficient gene delivery with serum into human cancer cells using targeted anionic liposomes. J Drug Target 12:11–18

Magnuson BA, Jonaitis TS, Card JW (2011) A brief review of the occurrence, use, and safety of food-related nanomaterials. J Food Sci 76(6):R126–R133. https://doi.org/10.1111/j.1750-3841.2011.02170.x

Majerik V, Horvath G, Charbit G, Badens E, Szokonya L, Bosc N, Teillaud E (2004) Novel particle engineering techniques in drug delivery: review of co-formulations using supercritical fluids and liquefied gas. Hungarian J Ind Chem Veszprm 32:41–56

Malhotra M, Jain NK (1994) Niosomes as drug carriers. Indian Drugs 31:81–86

Marques HMA (2010) A review on cyclodextrin encapsulation of essential oils and volatiles. Flav Fragr J 25(5):313–326. https://doi.org/10.1002/ffj.2019

Martínez-Ballesta MC, Gil-Izquierdo A, García-Viguera C, Domínguez-Perles R (2018) Nanoparticles and controlled delivery for bioactive compounds: outlining challenges for new "smart-foods" for health. Foods 7(5):E72. https://doi.org/10.3390/foods7050072

Mashaghi S, Jadidi T, Koenderink G, Mashaghi A (2013) Lipid nanotechnology. Int J Mol Sci 14:4242–4282

Matsumura Y, Kataoka K (2009) Preclinical and clinical studies of anticancer agent-incorporating polymer micelles. Cancer Sci 100:572–579

Matteucci ML, Thrall DE (2000) The role of liposomes in drug delivery and diagnostic imaging: a review. Vet Radiol Ultrasound 41:100–107

Maurer-Jones MA, Bantz KC, Love SA, Marquis BJ, Haynes CL (2009) Toxicity of therapeutic nanoparticles. Nanomed 4:219–241

Mcclements DJ (2007) Critical review of techniques and methodologies for characterization of emulsion stability. Critical Rev Food Sci Nutr 47(7):611–649. https://doi.org/10.1080/10408390701289292

McClements DJ, Li Y (2010) Review of in vitro digestion models for rapid screening of emulsion-based systems. Food Funct 1:32–59

Merisko-Liversidge E, Liversidge GG, Cooper ER (2003) Nanosizing: a formulation approach for poorly-water-soluble compounds. Eur J Pharm Sci 18(2):113–120

Mikhail AS, Allen C (2009) Block copolymer micelles for delivery of cancer therapy: transport at the whole body, tissue and cellular levels. J Control Release 138:214–223

Miladi K, Ibraheem D, Iqbal M, Sfar S, Fessi H, Elaissari A, Avicenne R (2014) Particles from preformed polymers as carriers for drug delivery. EXCLI J 13:28–57

Minekus M, Alminger M, Alvito P, Ballance S, Bohn T, Bourlieu C, Carrière F, Boutrou R, Corredig M, Dupont D, Dufour C, Egger L, Golding M, Karakaya S, Kirkhus B, Le Feunteun S, Lesmes U, Macierzanka A, Mackie A, Marze S, McClements DJ, Ménard O, Recio I, Santos CN, Singh RP, Vegarud GE, Wickham MS, Weitschies W, Brodkorb A (2014) A standardised static in vitro digestion method suitable for food—an international consensus. Food Funct 5:1113–1124

Mo J, Panichayupakaranant P, Kaewnopparat N, Songkro S, Reanmongkol W (2014) Topical anti-inflammatory potential of standardized pomegranate rind extract and ellagic acid in contact dermatitis. Phytother Res 28:629–632

Mo J, Kaewnopparat N, Songkro S, Panichayupakaranant P, Reanmongkol W (2015) Physicochemical properties, in vitro release and skin permeation studies of a topical formulation of standardized pomegranate rind extract. Pakistan J Pharm Sci 28:29–36

Moghimi SM, Hunter AC, Murray JC (2005) Nanomedicine: current status and future prospects. FASEB J 19:311–330

Momin JK, Jayakumar C, Prajapati JB (2013) Potential of nanotechnology in functional foods. Emir J Food Agric 25:10–19

Morales JO, Watts AB, McConville JT (2016) Mechanical particle-size reduction techniques. In: Williams RO III, Watts AB, Miller DA (eds) Formulating poorly water soluble drugs. Springer, New York, pp 133–170. isbn:978-1-4614-1144-4

Morgan CG, Yianni YP, Sandhu SS, Mitchell AC (1995) Liposome fusion and lipid exchange on ultraviolet irradiation of liposomes containing a photochromic phospholipid. Photochem Photobiol 62:24–29

Möschwitzer J (2006) Method for producing ultrafine submicronic suspensions, WO/002006094808A3

Möschwitzer J, Lemke A (2006) Method for carefully producing ultrafine particle suspensions and ultrafine particles and use thereof. WO/2006/108637

Moschwitzer J, Muller RH (2006) New method for the effective production of ultrafine drug nanocrystals. J Nanosci Nanotechnol 6:3145–3153

Mosharraf M, Nyström C (1995) The effect of particle size and shape on the surface specific dissolution rate of microsized practically insoluble drugs. Int J Pharm 122(1–2):35–47

Müller RH, Möschwitzer J (2009) Method and device for producing very fine particles and coating such particles. US Patent 0297565:A1

Naseri N, Valizadeh H, Zakeri-Milani P (2015) Solid lipid nanoparticles and nanostructured lipid carriers: structure preparation and application. Adv Pharm Bull 5:305–313

Neethirajan S, Jayas DS (2011) Nanotechnology for the food and bioprocessing industries. Food Bioprocess Technol 4:39–47

Nejad KH, Dianat M, Sarkaki A, Naseri MK, Badavi M, Farbood Y (2015) Ellagic acid improves electrocardiogram waves and blood pressure against global cerebral ischemia rat experimental models. Electron Physician 7:1153–1162

Nejad KH, Gharib-Naseri MK, Sarkaki A, Dianat M, Badavi M, Farbood Y (2017) Effects of ellagic acid pretreatment on renal functions disturbances induced by global cerebral ischemic-reperfusion in rat. Iranian J Basic Med Sci 20:75–82

Nerome H, Machmudah S, Diono W, Fukuzato R, Higashiura T, Youn I-S, Lee Y-W, Goto M (2013) Nanoparticle formation of lycopene/β-cyclodextrin inclusion complex using supercritical antisolvent precipitation. J Supercrit Fluids 83:97–103

Niemeyer T, Patel M, Geiger E (2006) A further examination of soy-based polyols in polyurethane systems. Alliance for the Polyurethane Industry Technical Conference, Salt Lake City, UT

Nitta S, Numata K (2013) Biopolymer-based nanoparticles for drug/gene delivery and tissue engineering. Int J Mol Sci 14:1629–1654

Odriozola-Serrano I, Oms-Oliu G, Martín-Belloso O (2014) Nanoemulsionbased delivery systems to improve functionality of lipophilic components. Front Nutr 1:24. https://doi.org/10.3389/fnut.2014.00024

Ojea-Jiménez I, Tort O, Lorenzo J, Puntes VF (2012) Engineered nonviral nanocarriers for intracellular gene delivery applications. Biomed Mater 7:1–6

Palermo M, Pellegrini N, Fogliano V (2014) The effect of cooking on the phytochemical content of vegetables. J Sci Food Agric 94(6):1057–1070. https://doi.org/10.1002/jsfa.6478

Pan K, Zhong QX, Baek SJ (2013) Enhanced dispersibility and bioactivity of curcumin by encapsulation in casein nanocapsules. J Agric Food Chem 61:6036–6043

Pandey KB, Rizvi SI (2009) Plant polyphenols as dietary antioxidants in human health and disease. Ox Med Cell Long 2:270–278

Panyam J, Labhasetwar V (2004) Sustained cytoplasmic delivery of drugs with intracellular receptors using biodegradable nanoparticles. Mol Pharm 1:77–84

Pardeshi C, Rajput P, Belgamwar V, Tekade A, Patil G, Chaudhary K, Sonje A (2012) Solid lipid based nanocarriers: an overview. Acta Pharma 62:433–472

Patel VR, Agrawal YK (2011) Nanosuspension: an approach to enhance solubility of drugs. J Adv Pharm Technol Res 2:81–87

Patil JS, Kadam DV, Marapur SC, Kamalapur MV (2010) Inclusion complex system; a novel technique to improve the solubility and bioavailability of poorly soluble drugs: a review. Int J Pharm Sci Rev Res 2(2):29–34

Petroni A, Blasevich M, Salami M, Papini N, Montedorol GF, Galli C (1995) Inhibition of platelet aggregation and eicosanoid production by phenolic components of olive oil. Thrombosis Res 78(2):151–160

Pham-Huy LA, He H, Pham-Huy C (2008) Free radicals, antioxidants in disease and health. Int J Biomed Sci 4:89–96

Pinal R (2004) Effect of molecular symmetry on melting temperature and solubility. Org Biomol Chem 2:2692–2699

Pinciroli M, Domínguez-Perles R, Abellán A, Guy A, Durand T, Oger C, Galano JM, Ferreres F, Gil-Izquierdo A (2017) Comparative study of the Phytoprostane and Phytofuran content of indica and japonica Rice (Oryza sativa L.) flours. J Agric Food Chem 65:8938–8947

Pinho E, Grootveld M, Soares G, Henriques M (2014) Cyclodextrins as encapsulation agents for plant bioactive compounds. Carbohydr Polym 101:121–135. https://doi.org/10.1016/j.carbpol.2013.08.078

Pouton CW (2006) Formulation of poorly water-soluble drugs for oral administration: physicochemical and physiological issues and the lipid formulation classification system. Eur J Pharm Sci 29(3):278–287

Pu S-B, Ma Z-J, Wang Q (2019) Anti-Staphylococcus aureus evaluation of gallic acid by isothermal microcalorimetry and principle component analysis. J Therm Anal Calorim 136(3):1425–1432. https://doi.org/10.1007/s10973-018-7726-5

Qian C, McClements DJ (2011) Formation of nanoemulsions stabilized by model food-grade emulsifiers using high-pressure homogenization: factors affecting particle size. Food Hydrocoll 25:1000–1008

Qian C, Decker EA, Xiao H, McClements DJ (2012) Nanoemulsion delivery systems: influence of carrier oil on b-carotene bioaccessibility. Food Chem 135:1440–1447

Rabinow BE (2004) Nanosuspensions in drug delivery. Nat Rev Drug Discov 3:785–796

Rajput SD, Mahulikar PP, Gite VV (2014) Biobased dimer fatty acid containing two pack polyurethane for wood finished coatings. Prog Org Coat 77(1):38–46

Rao AV, Rao LG (2007) Carotenoids and human health. Pharm Res 55(3):207–216. https://doi.org/10.1016/j.phrs.2007.01.012

Rapoport N (2007) Physical stimuli-responsive polymeric micelles for anti-cancer drug delivery. Prog Polym Sci 32:962–990

Rattanata N, Klaynongsruang S, Leelayuwat C, Limpaiboon T, Lulitanond A, Boonsiri P, Chio-Srichan S, Soontaranon S, Rugmai S, Daduang J (2016) Gallic acid conjugated with gold nanoparticles: antibacterial activity and mechanism of action on foodborne pathogens. Int J Nanomedicine 11:3347–3356

Ravve A (2013) Free-radical chain-growth polymerization. In: Principles of polymer chemistry (2nd ed.). New York, Springer, pp 41102. Print ISBN 978-1-4613-6898-4. Online ISBN 978–1–4615-4227-8

Raza K, Kumar P, Ratan S, Malik R, Arora S (2014) Polymorphism: the phenomenon affecting the performance of drugs. SOJ Pharm Pharm Sci 1:1–10

Recharla N, Riaz M, Ko S (2017) Novel technologies to enhance solubility of food-derived bioactive compounds: a review. J Funct Foods 39:63–73. https://doi.org/10.1016/j.jff.2017.10.001

Renaud SC, Gueguen R, Schenker J, D'Houtaud A (1998) Alcohol and mortality in middle-aged men from eastern France. Epidemiology 9(2):184–188

Rocha DS, Casagrande L, Model JFA, dos Santos JT, Hoefel AL, Kucharski LC (2018) Effect of yerba mate (Ilex paraguariensis) extract on the metabolism of diabetic rats. Biomed Pharmacother 105:370–376. https://doi.org/10.1016/j.biopha.2018.05.132

Ruan J, Yang Y, Yang F, Wan K, Fan D, Wang D (2018) Novel oral administrated ellagic acid nanoparticles for enhancing oral bioavailability and anti-inflammatory efficacy. J Drug Deliv Sci Technol 46:215–222. https://doi.org/10.1016/j.jddst.2018.05.021

Salatin S, Jelvehgarim M (2017) Natural polysaccharide based nanoparticles for drug/gene delivery. Pharm Sci 23:84–94

Salazar J, Müller RH, Möschwitzer JP (2014) Combinative particle size reduction technologies for the production of drug nanocrystals. J Pharm 2014:265754. https://doi.org/10.1155/2014/265754

Savjani KT, Gajjar AK, Savjani JK (2012) Drug solubility: Importance and enhancement techniques. ISRN Pharm 2012:195727. https://doi.org/10.5402/2012/195727

Scalbert A, Williamson G (2000) Dietary intake and bioavailability of polyphenols. J Nutr 130:2073S–2085S

Sha XY, Guo J, Chen YZ, Fang XL (2012) Effect of phospholipid composition on pharmacokinetics and biodistribution of epirubicin liposomes. J Liposome Res 22:80–88

Shadrack DM, Swai HS, Munissi JJE, Mubofu EB, Nyandoro SS (2018) Polyamidoamine Dendrimers for enhanced solubility of small molecules and other desirable properties for site specific delivery: insights from experimental and computational studies. Molecules 23(6):1419. https://doi.org/10.3390/molecules23061419

Shahiwala A, Misra A (2002) Studies in topical application of niosomally entrapped nimesulide. J Pharm Pharm Sci 5:220–225

Shanmugam S, Baskaran R, Balakrishnan P, Thapa P, Yong CS, Yoo BK (2011) Solid self-nanoemulsifying drug delivery system (S-SNEDDS) containing phosphatidylcholine for enhanced bioavailability of highly lipophilic bioactive carotenoid lutein. Eur J Pharm Biopharm 79(2):250–257

Sharma A, Sharma US (1997) Liposomes in drug delivery: Progress and limitations. Int J Pharm 154:123–140

Shutava TG, Balkundi SS, Vangala P, Steffan JJ, Bigelow RL, Cardelli JA, O'Neal DP, Lvov YM (2009) Layer-by-layer-coated gelatin nanoparticles as a vehicle for delivery of natural polyphenols. ACS Nano 3(7):1877–1885. https://doi.org/10.1021/nn900451a

Simion V, Stan D, Constantinescu CA, Deleanu M, Dragan E, Tucureanu MM, Gan AM, Butoi E, Constantin AM, Maduteanu I, Simionescu M, Calin M (2016) Conjugation of curcumin-loaded lipid nanoemulsions with cell-penetrating peptides increases their cellular uptake and enhances the anti-inflammatory effects in endothelial cells. J Pharm Pharmacol 68(2):195–207

Singh R, Lillard JW (2009) Nanoparticle-based targeted drug delivery. Exp Mol Pathol 86:215–223

Singh B, Bandopadhyay S, Kapil R, Singh R, Katare OP (2009) Selfemulsifying drug delivery systems (SEDDS): Fformulation development, characterization, and applications. Crit Rev Ther Drug Carrier Syst 26:427–451

Singh B, Beg S, Khurana RK, Sandhu PS, Kaur R (2014) Recent advances in self-emulsifying drug delivery systems (SEDDS). Crit Rev Ther Drug Carrier Syst 31(2):121–185

Sinha B, Müller RH, and Möschwitzer J (2012) Precipitation followed by high pressure homogenization as a combinative approach to prepare drug nanocrystals. Tag der Pharmazie FU Berlin, Berlin, abstract V2, booklet page 6, 2012

Sivakumar M, Tang SY, Tan KW (2014) Cavitation technology – a greener processing technique for the generation of pharmaceutical nanoemulsions. Ultrasonics Sonochem 21(6):2069–2083

Son YR, Choi EH, Kim GT, Park TS, Shim SM (2016) Bioefficacy of graviola leaf extracts in scavenging free radicals and upregulating antioxidant genes. Food Funct 7:861–871

Sorrentino E, Succi M, Tipaldi L, Pannella G, Maiuro L, Sturchio M, Coppola R, Tremonte P (2018) Antimicrobial activity of gallic acid against food-related Pseudomonas strains and its use as biocontrol tool to improve the shelf life of fresh black truffles. Int J Food Microbiol 266:183–189

Spernath A, Yaghmur A, Aserin A, Hoffman RE, Garti N (2002) Food-grade microemulsions based on nonionic emulsifiers: media to enhance lycopene solubilization. J Agric Food Chem 50:6917–6922

Stegemann S, Leveiller F, Franchi D, De Jong H, Lindén H (2007) When poor solubility becomes an issue: from early stage to proof of concept. Eur J Pharm Sci 31:249–261

Sun J, Tan H (2013) Alginate-based biomaterials for regenerative medicine applications. Materials 6:1285–1309

Tarhini M, Greige-Gerges H, Elaissari A (2017) Protein-based nanoparticles: from preparation to encapsulation of active molecules. Int J Pharm 522:172–197

Teleki A, Hitzfeld A, Eggersdorfer M (2013) 100 years of vitamins: the science of formulation is the key to functionality. KONA Powder Part J 30:144–163

Teng Y, Morrison ME, Munk P, Webber SE, Prochazka K (1998) Release kinetics studies of aromatic molecules into water from block polymer micelles. Macromolecules 31:3578–3587

Thatipamula R, Palem C, Gannu R, Mudragada S, Yamsani M (2011) Formulation and in vitro characterization of domperidone loaded solid lipid nanoparticles and nanostructured lipid carriers. Daru 19:23–32

Thoppil RJ, Bishayee A (2011) Terpenoids as potential chemopreventive and therapeutic agents in liver cancer. World J Hepatol 3(9):228–249. https://doi.org/10.4254/wjh.v3.i9.228

Ting Y, Jiang Y, Ho CT, Huang Q (2014) Common delivery systems for enhancing in vivo bioavailability and biological efficacy of nutraceuticals. J Funct Foods 7:112–128

Tran TH, Guo Y, Song D, Bruno RS, Lu X (2014) Quercetin-containing selfnanoemulsifying drug delivery system for improving oral bioavailability. J Pharm Sci 103:840–852

Turrini F, Zunin P, Catena S, Villa C, Alfei S, Boggia R (2019a) Traditional or hydro-diffusion and gravity microwave coupled with ultrasound as green technologies for the valorization of pomegranate external peels. Food Bioprod Process 117:30–37

Turrini F, Boggia R, Pittaluga AM, Grilli M, Zunin P (2019b) Microdispersions of ellagic acid and pomegranate extracts as new potential nutraceutical ingredients. SIF monothematic conference: The pharmacological bases of nutraceuticals. Naples, 29–30 March 2019 (poster communication). P 47: p 99 in: Programma ed Abstract

Turrini F, Boggia R, Donno D, Parodi B, Beccaro G, Baldassari S, Signorello MG, Catena S, Alfei S, Zunin P (2020) From pomegranate marcs to a potential bioactive ingredient: a recycling proposal for pomegranate-squeezed marcs. Eur Food Res Technol 246: 273–285

Uchegbu IF, Vyas SP (1998) Non-ionic surfactant based vesicles (niosomes) in drug delivery. Int J Pharm 176:139–172

Vanrooijen N, Vannieuwmegen R (1980) Liposomes in immunology—multilamllar phosphatidylcholine liposomes as a simple, biodegradable and harmless adjuvant without any immunogenic activity of its own. Immunol Commun 9:243–256

Varadinova MG, Docheva-Drenska DI, Boyadjieva NI (2009) Effects of anthocyanins on learning and memory of ovariectomized rats. Menopause 16:345–349

Vauthier C, Bouchemal K (2009) Methods for the preparation and manufacture of polymeric nanoparticles. Pharm Res 26(5):1025–1058

Vega-Villa KR, Takemoto JK, Yáñez JA, Remsberg CM, Forrest ML, Davies NM (2008) Clinical toxicities of nanocarrier systems. Adv Drug Deliv Rev 60:929–938

Vemula VR, Lagishetty V, Lingala S (2010) Solubility enhancement techniques. Int J Pharm Sci Rev Res 5:41–51

Vico TA, Arce VB, Fangio MF, Gende LB, Bertran CA, Mártire DO, Churio MS (2016) Two choices for the functionalization of silica nanoparticles with gallic acid: Characterization of the nanomaterials and their antimicrobial activity against Paenibacillus larvae. J Nanopart Res 18:348

Vidhyalakshmi R, Bhakyaraj R, Subhasree R (2009) Encapsulation "the future of probiotics"—a review. Adv Biol Res 3:96–103

Wang X, Jiang Y, Wang YW, Huang MT, Ho CT, Huang Q (2008) Enhancing anti-inflammation activity of curcumin through O/W nanoemulsions. Food Chem 108:419–424

Wang S, Su R, Nie S, Sun M, Zhang J, Wu D, Moustaid-Moussa N (2014) Application of nanotechnology in improving bioavailability and bioactivity of diet-derived phytochemicals. J Nutr Biochem 25(4):363–376

Weir A, Westerhoff P, Fabricius L, Hristovski K, von Goetz N (2012) Titanium dioxide nanoparticles in food and personal care products. Environ Sci Technol 46:42242–42250. https://doi.org/10.1021/es204168d

Weiss J, Takhistov P, McClements DJ (2006) Functional materials in food nanotechnology. J Food Sci 71:R107–R116

Xia Y, Zhang Z, Kessler MR, Brehm-Stecher B, Larock RC (2012) Antibacterial soybean-oil-based cationic polyurethane coatings prepared from different amino polyols. ChemSusChem 5:2221–2227

Xiao J, Nian S, Huang QR (2015) Assembly of kafirin/carboxymethyl chitosan nanoparticles to enhance the cellular uptake of curcumin. Food Hydrocoll 51:166–175

Xiao J, Cao Y, Huang Q (2017) Edible Nanoencapsulation vehicles for Oral delivery of phytochemicals: a perspective paper. J Agric Food Chem 65:6727–6735

Yoo JH, Shanmugam S, Thapa P, Lee ES, Balakrishnan P, Baskaran R, Yoon SK, Choi HG, Yong CS, Yoo BK, Han K (2010) Novel self-nanoemulsifying drug delivery system for enhanced solubility and dissolution of lutein. Arch Pharm Res 33:417–426

You JO, Peng CA (2005) Calcium-Alginate nanoparticles formed by reverse microemulsion as gene carriers. Macromol Symp 219:147–153

Yousuf B, Gul K, Wani AA, Singh P (2016) Health benefits of anthocyanins and their encapsulation for potential use in food systems: a review. Critical Rev Food Sci Nutr 56(13):2223–2230

Yu H, Park JY, Kwon CW, Hong SC, Park KM, Chang PS (2018) An overview of nanotechnology in food science: preparative methods, practical applications, and safety. J Chem 2018:5427978. https://doi.org/10.1155/2018/5427978

Yuan F, Leunig M, Huang SK, Berk DA, Papahadjopoulos D, Jain RK (1994) Microvascular permeability and interstitial penetration of Sterically stabilized (stealth) liposomes in a human tumor Xenograft. Cancer Res 54(13):3352–3356

Yuan Y, Gao Y, Mao L, Zhao J (2008) Optimisation of conditions for the preparation of b-carotene nanoemulsions using response surface methodology. Food Chem 107(3):1300–1306

Zamora A, Guamis B (2015) Opportunities for ultra-high-pressure homogenisation (UHPH) for the food industry. Food Eng Rev 7(2):130–142

Zhang X, Xia Q, Gu N (2006) Preparation of all-trans retinoic acid Nanosuspensions using a modified precipitation method. Drug Dev Ind Pharm 32(7):857–863

Zhang L, Mou D, Du Y (2007) Procyanidins: extraction and micro-encapsulation. J Agric Food Chem 87:2192–2197

Zhou F, Neutra MR (2002) Antigen delivery to mucosa-associated lymphoid tissues using liposomes as a carrier. Biosci Rep 22:355–369

Zhu SF, Zheng J, Liu F, Qiu CY, Lin WF, Tang CH (2017) The influence of ionic strength on the characteristics of heat-induced soy protein aggregate nanoparticles and the freeze–thaw stability of the resultant Pickering emulsions. Food Funct 8:2974–2981

Zia Q, Farzuddin M, Ansari MA, Alam M, Ali A, Ahmad A, Owais M (2010) Novel drug delivery systems for antifungal compounds. In: Ahmad I, Owais M, Shahid M, Aqil F (eds) Combating fungal infections. Springer, Berlin/Heidelberg, pp 485–528

Zou LQ, Zheng BJ, Zhang RJ, Zhang ZP, Liu W, Liu CM, Xiao H, McClements DJ (2016) Food-grade nanoparticles for encapsulation, protection and delivery of curcumin: Comparison of lipid, protein, and phospholipid nanoparticles under simulated gastrointestinal conditions. RSC Adv 6:3126–3136

Zuccari G, Carosio R, Fini A, Montaldo PG, Orienti I (2005) Modified polyvinylalcohol for encapsulation of all-trans-retinoic acid in polymeric micelles. J Control Rel 103(2):369–380

Zuccari G, Bergamante G, Carosio R, Gotti R, Montaldo PG, Orienti I (2009) Micellar complexes of all-trans retinoic acid with polyvinylalcohol-nicotinoyl esters as new parenteral formulations in neuroblastoma. Drug Delivery 16(4):189–195

Zuccari G, Baldassari S, Ailuno G, Turrini F, Alfei S, Caviglioli G (2020) Formulation Strategies to Improve Oral Bioavailability of Ellagic Acid. Appl Sci 10(10):3353

Emerging Sustainable Nanostructured Materials Facilitated by Herbal Bioactive Agents for Edible Food Packaging

Tabli Ghosh, Monika, and Vimal Katiyar

1 Edible Food Packaging

Since decades, the edible food packaging is considered as a potential candidate for increased consumer need for safe, ready-to-eat food products, and related awareness concerning food waste, and disposable issues of conventional packaging waste. The fresh food products including fruits and vegetables, meat and fish products are perishable by nature; where the quality of the products get deteriorated by several factors including environmental agents such as physiological and microbial entity. The common factors for decreased shelf life of food products include temperature, enzymatic activity, humidity, bacterial attack, etc. However, these specific factors increase the food waste throughout the postharvest supply chain of fruits and vegetables. Further, the growth of pathogenic microorganism in individual food causes serious health issues (foodborne illness) to consumers, especially if food product is distributed under inappropriate conditions. The properties of the food products commonly degraded due to some undesirable reactions, which further alter the texture properties such as flavour, odour, colour. Thus, various preservation techniques are used for improved product quality and safety throughout the life cycle of food products. As shown in Table 1, the conventional preservation methods include thermal treatments (blanching, pasteurization, sterilization), cold processing (chilling, freezing, refrigeration), controlling water (drying, dehydration), use of preservatives (salting, sugaring, acidification), and non-thermal processes (irradiation, ultrasound, pulsed electric field, ohmic heating, microwave heating, filtration, high pressure processing), which are used from early days to reduce the growth rate of pathogenic microorganism in food products, however, these specific techniques often cause some undesirable losses in the nutritional values of food products. To

T. Ghosh · Monika · V. Katiyar (✉)
Department of Chemical Engineering, Indian Institute of Technology Guwahati, North Guwahati, Assam, India
e-mail: vkatiyar@iitg.ac.in

U. Hebbar et al. (eds.), *Nano-food Engineering*, Food Engineering Series,
https://doi.org/10.1007/978-3-030-44552-2_9

Table 1 Available food preservation techniques

Approaches of preservation	Techniques	Advantages	Disadvantages
Thermal treatments	Blanching Pasteurization Sterilization Hurdle technology	Simple and easy technique High heat processing Deactivate enzyme Destroy microorganism	Loss of heat sensitive nutrients Change structural properties of food Food may get overcooked
Cold processing	Chilling Freezing Refrigeration	Reduce rate of microbial action Reduce rate of enzyme action Provide good quality of food products	Loss of nutrients Chilling injury may occur May break cell wall of some foods
Controlling water	Drying Dehydration	Increase storage stability Decrease packaging requirement Off-season product availability Decrease in transportation weight Inexpensive	Loss of nutrients Non-uniform drying More time is required
Chemicals	Chemical preservatives pH controlling	Prevent growth of microbes No loss of nutrients No complex equipment is needed	Alter food texture Change in structure property Increase in concentration of preservatives
Non thermal	Irradiation Ultrasound Electric field method Filtration	Easy to use Prolong shelf life of food products Destroy pathogens	May cause greater nutrient loss May alter chemical properties such as oxidized flavor May alter sensory properties

overcome these shortcomings of the existing techniques, edible food packaging is utilized to maintain the nutritional quality of the food products.

Among available packaging techniques, the edible food packaging is a primary packaging material, which is used globally to protect food products without changing the nutritional quality. However, this specified technique of food packaging is more efficient in preserving food with an additional delivery of nutrients through the packaging material. The worldwide issues in regards to the environment, sustainability, packaging cost, food quality, and storability are considered as crucial factors for consumers, where the safety issues are also imposed by several food packaging related regulations. However, after buying the food products, these are consumed within a few minutes' leaving behind the package based waste that lasts for years to get disposed of. Therefore, edible packaging is considered a remarkable candidate to solve these issues. Further, edible food packaging is very promising and currently one of the most investigated approaches for maintaining the food quality and increased shelf life. The foremost advantages of edible packaging over synthetic/biodegradable packaging are that it can be ingested along with the packaged food products. A food packaging is termed as "edible food packaging" only if it possesses potential qualities similar to the food components and also act as an integral part of food that is designed to be consumed along with food. The primary roles of edible food packaging include as a selective barrier to gas exchange, moisture transfer or oxidation processes by delaying food deterioration, and retaining food property in order to enhance the product life (Fig. 1). In other words, the major functions of food packaging involve preserving and protecting food products from the surrounding environment, and further preventing surface contamination (minimizing exposure to spoilage components; i.e. moisture, microorganism, solutes, vapours, oils and off flavours). The use of food packaging also helps to protect desirable components (flavour volatiles) of food products by protecting them from chemical, physical, and biological activities, which is a promising outcome in fresh food preservations. The prime functions of edible coating/films are represented in Fig. 1. They also provide structural protection to limit mechanical damage during storage, transportation, and handling; and also protect the food against oxidation and other chemical reactions. In addition, they are also utilized to transfer bioactive components such as antimicrobial, antioxidants, colours, flavours, and nutraceutical agents (Fig. 1). Based on this discussion, the chapter discusses the use of edible food packaging for prolonging the shelf life of food products in comparison to available food processing techniques. The utilization of available emerging bio-based nanostructured materials is considered as remarkable agents for improved food properties and shelf life, where the nanostructured materials further acts as a

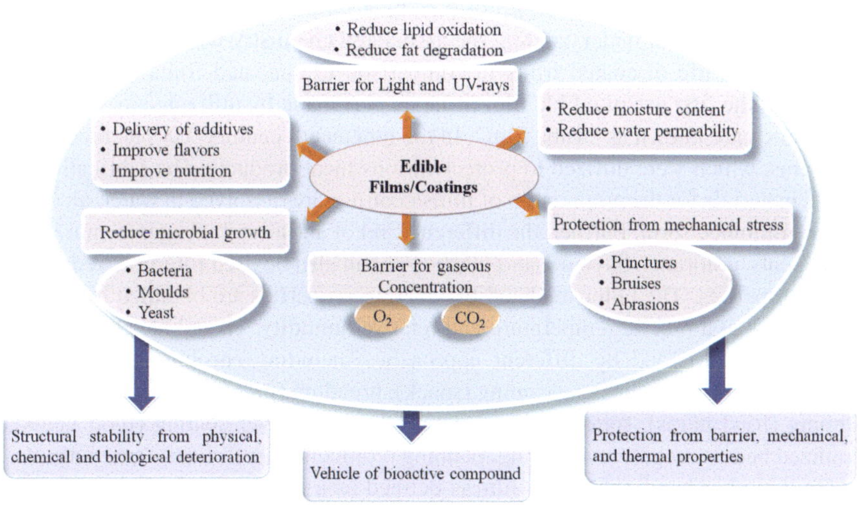

Fig. 1 Functions and trends of edible film and coating in preserving food products

delivery agent for nutraceuticals. Interestingly, the bio-based nanostructured materials can be functionalized and modified according to the targeted food products for the prolonged shelf life.

1.1 Categories and Worldwide Availability of Edible Food Packaging

As mentioned in the earlier section, the various kinds of edible food packages are classified into two groups: (i) edible coatings, and (ii) edible films. Both edible coatings and films are developed based on similar principles, however, they are generally applied in different ways to food products. Moreover, an edible coating can be differentiated from an edible film by the concept that an edible coating is applied on food products as a skinny layer, which remains intact to the food products, whereas an edible film is used to wrap the food product or can be used as sandwich material. The edible coating is defined as a preformed thin layer for primary packaging of food; made up of edible material, which is eatable in nature. Edible coating as a layer must attach easily to the food exterior surface, and also avoid crack or undesirable sensory properties during storage and handling. From very past years, the edible coating is developed on citrus fruit products (dipped coating) using wax to reduce food degradation and extend the shelf life as well as aesthetic appearance. Since the twelfth century, the use of wax coating on citrus fruit products was very familiar in China (Cuq et al. 1995). Further, during the sixteenth century, the fat or lard was used for coating meat in England, which is specially known as "larding" (Donhowe and Fennema 1994). In later days, the various derivatives of sucrose and sugar were also applied on nuts and further waxing/lipid based coating were utilized on vegetables and fruits. In 1930's, hot-melt paraffin waxes were initiated at a commercial level without understanding the associated chemistry which helped to prolong the shelf life of coated fruits in comparison to uncoated fruits (Park 1999). Moreover, the first use of edible film in Japan was made by utilizing soy milk proteins which are known as "Yuba film". In the nineteenth century, US patented gelatine films, which were utilized to protect various meat products. The fabrication of edible materials for the preparation of films/coatings are dissolved in water, alcohol, and in combined form. Further, the different kind of antimicrobial agents, nutraceutical agents, colorants, flavours, and plasticizer can also be used for improved packaging properties. The tailor-made edible coating properties are obtained by drying the solution at a precise temperature and relative humidity. The edible coatings are being applied to food by different approaches including enrobing (chocolates), encapsulation (food items), seasoning (snack), breading (prepared meals), foaming, dipping (food items), spraying (food items), layer by layer coating (food items), fluidized bed coating, screw coating, panning (confectionary), electro-spraying, etc.

On the other hand, the edible film is defined as a thin layer of edible material, which can be placed on or between food components, or even can be sealed into edible pouches. The approach helps to reduce package waste, where the biodegradable food packaging films are non-edible, which may contribute to minimize environmental pollution but create some packaging waste. This problem limits the demands of conventional film to the supply of a barrier to moisture loss and preserves the food from external impurities. Edible films can be used in various way, which is usually utilized to develop single or multilayer films using different techniques such as casting, extrusion, and blow moulding in order to achieve the target performances of edible films. In order to be considered a film, the thickness of film should be less than 250 μm. It is noteworthy to mention that edible coatings/films are considered as promising technologies to preserve the quality and improved the shelf life of foodstuff. Edible film/coating generates a modified atmosphere throughout the food by offering a passive barrier to gases, and water vapour, flavour compounds, and can enhance the structural integrity. Further, the use of edible food packaging provides an attractive alternative to packaging reducing the negative environmental impact and disposals cost.

1.2 Market Analysis of Edible Packaging

The worldwide availability of edible food packaging is segmented on the basis of region (North America, South America, Europe, Asia-pacific, Middle East & Africa and others), material types (protein, polysaccharides, lipids, and others), and applications (Fruits & vegetables, dairy products, nutritional products, bakery and confectionary, meat poultry and fish) (as represented in Fig. 2). According to the

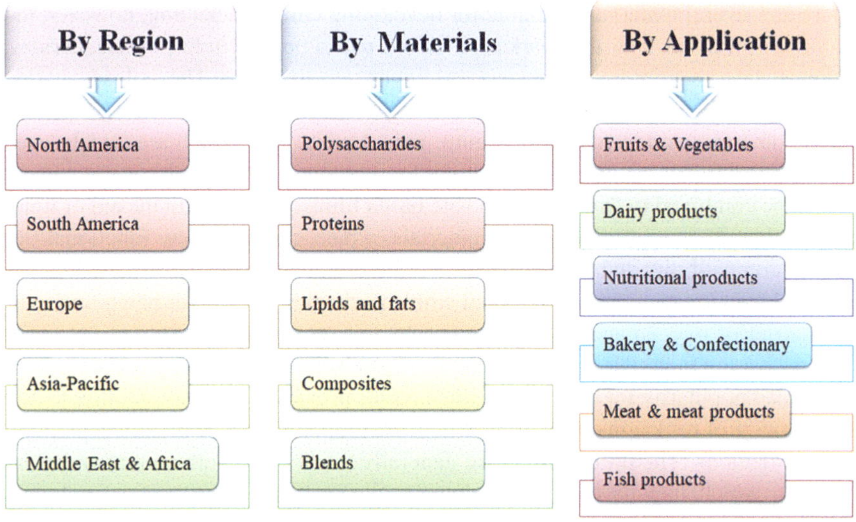

Fig. 2 Market segmentation of edible food packaging

available reports on edible packaging, published by allied market research, the global market of edible packaging was valued at $697 million in 2016 and is projected to reach at $1097 million by 2023. The reduction in packaging wastes increases due to the consumption of processed food & beverages, which encourages the adoption of edible packaging products and thus drive the edible packaging market growth.

On the basis of region, North America region dominates the global edible packaging market, where the market of edible packaging in North America is expected to witness a CAGR of 5.7% during 2017–2023. Further, North America obtain a major shareholder in the global edible packaging market, accounting more than 33.8% share in 2016. The Food & Beverages market dominated the North America edible packaging market by applications in 2016. The rapid growth of pharmaceutical industry and rise in preference of hygienic packaging in this region is the major factor for driving the growth of the edible packaging market. The Pharmaceutical market is expected to witness a CAGR of 6.8% during 2017–2023. Further, the major market segments of North America include US, Canada, and Mexico. The U.S. market has a major shareholder in the North America region, accounting for more than 42% share in 2016. As reported, US is both a prominent consumer and producer of edible packaging in North America market. The Europe edible packaging market size would grow at a growth rate of 5.9% CAGR during 2017–2023. In Europe, edible packaging market is segmented into Germany, UK, France, Russia, Spain, Italy, and Rest of Europe. However, Germany accounted for the largest market share in the edible packaging market in Europe during 2016.

1.2.1 Materials for Edible Packaging

There are several materials available for developing edible packaging, however, the major materials used in the market of edible packaging include polysaccharides, proteins, fats and lipids, plasticizers, additives, etc. (Fig. 3). The polysaccharide based materials for edible packaging include cellulose, chitosan, starch, alginate, pectin, dextrin, carrageenan, gellan gum, gum Arabic, etc., which are widely used for improved functionality and shelf life of food products. Moreover, several materials can be combined to develop composite based films/coatings with the aid of lipid, polysaccharides, and protein-based materials, which can offer new functionalities such as compatibility and biodegradability. The lipid based packaging materials has acquired a major segment in the global edible packaging market for being an excellent barrier of oxygen and water vapour, which is an essential quality required for food packaging materials for preserving food products. Further, the protein segment has gained popularity for easy processing and for providing enormous ability to add a wide range of packaging properties, where the properties can be modified adopting several methods. The protein films are generally made up of both animal and plant proteins, which include corn zein, gluten, collagen, soy, whey protein, gelatine, caseins, and silk fibroin which are considered safe for human consumption. Further, the use of protein based edible packaging is expected to dominate the market in

Fig. 3 Targeted materials for edible packaging

upcoming years for their widespread application in the food and pharmaceutical industry. The value-added benefits of proteins to the body such as high immunity and proper functioning of human body as well as improve the energy levels in humans. The use of plasticizers (monosaccharides, oligosaccharides, polyols, lipids, polyethylene glycols, corn syrup, etc.), additives (antioxidants, antimicrobials, antifungals, antibacterial, flavouring compounds, nutraceuticals, probiotics, etc.) and others materials such as emulsifiers, nanocompounds, nanofillers etc. can improve functionality of edible packaging materials for extending the shelf life of food products. Further, the various polysaccharides materials such as cellulose, chitosan, gum Arabic, starch and other materials are widely used in edible packaging industries. The various polysaccharide materials are widely used in the pharmaceutical sector at large scale. The use of polysaccharide materials provides several advantages including tunable inherent properties, cost effective raw materials, easy handling and durability, which increase the use of polysaccharide and related nanostructured materials over others. The application of polysaccharides and protein for the development of edible packaging films can propose profitable growth opportunities for sustainable solutions due to the reduced dependency on fossil fuel. Further, a detailed discussion has been made on the related materials, their derivatives and related nanostructured materials which are extensively utilized in the edible packaging sector.

1.2.2 Application Analysis of Edible Packaging

The edible packaging materials as edible films/coatings are extensively utilized for various food industries such as fruits and vegetables, meat and meat products, fish products, etc. Besides, edible packaging is also extensively used for packaging pharmaceuticals across the world. In this regards, the Food & beverage industry dominates the edible packaging application segment for increased ready to eat food products. Interestingly, the specific packaging system is used to pack bakery & confectionary, dairy products, fruits & vegetables, meat, poultry & fish, frozen meals, snack foods, etc. (Fig. 2). The materials and packaging techniques may vary depending on the type of food product and storage conditions. In this regards, the bakery and confectionary items are moisture sensitive, where the crispiness of bakery products is an essential property for customer acceptance of the products. The bakery products require packaging materials having better water barrier properties in order to prolong shelf life and quality. The fruits and vegetables, meat, poultry & fish segments cover a large share of the global edible films and coatings market due to their perishable nature. The use of innovative packing design is a kind of strategy to raise product acceptance by the customer.

2 Current Prospects of Emerging Sustainable Biopolymers and Their Nanostructured Materials

Sustainable materials in edible food packaging are commonly obtained from plants, animals, microbes, and other sources. The sustainable materials in edible food packaging include various classes of biopolymers such as carbohydrates, proteins, lipids, and fats, where carbohydrates and proteins are used as matrix materials. Additionally, sustainable nanomaterials/ bionanomaterials are used for edible food packaging applications due to their various noteworthy characteristics such as biocompatibility, non-toxicity, renewable sources, tunable properties, etc. as represented in Fig. 4a. The sustainable nanostructured materials deliver multifunctional activities including antibacterial, antimicrobial, antidiabetic, weight reducing agent, which make them considerable materials to be used in edible food packaging (edible coating and edible films). This class of biomaterials can be used both individually and in combined form for targeted applications. The most focused biopolymers for edible packaging application include the various classes of carbohydrates such as cellulose, chitosan, starch, pectin, dextran, xanthan, agarose, etc. Furthermore, the biopolymers derived from protein based sources such as collagen, keratin, silk fibroin, soy protein, gelatin, have also received much attention. The nanotechnology in food using biopolymers are delivered through nanomaterials, liposomes, nanoemulsion, nanoencapsulation, and nanocomposites. The nanotechnology is widely used in the sectors of packaging, intelligent packaging, active packaging, and in improving food texture, etc. as represented in Fig. 4. Based on this, a detailed discussion on the essential properties of several sources of biopolymers for deriving bionanostructured materials and their properties has been made.

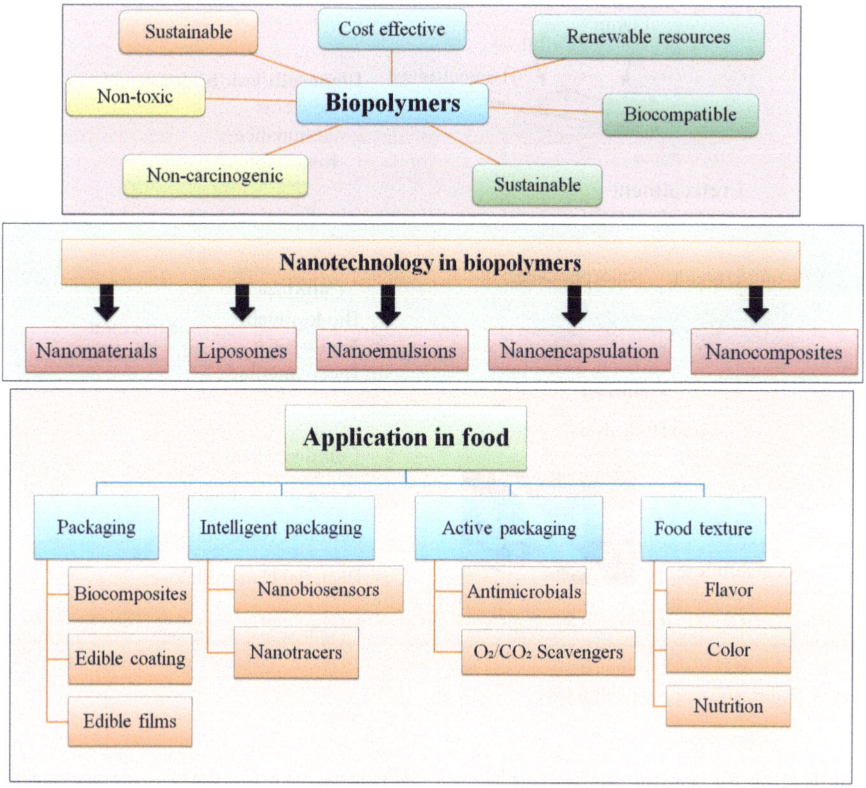

Fig. 4 (**a**) Traits of biopolymers; (**b**) Nanotechnology in biopolymers; and (**c**) Application of biopolymer based nanostructured materials in food

2.1 Cellulose and Its Nanostructured Materials

Cellulose is a class of polysaccharides, consisting of a linear chain of β (1 → 4) linked D-glucose unit, which is one of the most abundant biopolymers available in nature. The vital sources of cellulose are plant cell wall, algae, and oomycetes. However, the rich sources of cellulose mostly include cotton fiber (95–97%) (Chen 2014), hemp fibers (70–78%) (Brinchi et al. 2013), flax fibres (60–81%) (Brinchi et al. 2013), pineapple leaf fibers (79–83%) (Brinchi et al. 2013), rice straw (62%) (Brinchi et al. 2013), wood (40–60%) (Brinchi et al. 2013), etc. Cellulose is tasteless, odorless, biodegradable, biocompatible, and non-toxic in nature. Further, cellulose is crystalline, thermally stable, and provide superior mechanical properties rendering it to be used extensively in various sectors. The promising nature of cellulose allows it to be used in edible food packaging applications including films, coating, and hydrogels (Park et al. 1993; Ghosh et al. 2019; Ghosh and Katiyar 2019). Various reports suggest the use of cellulose and its various derivatives for

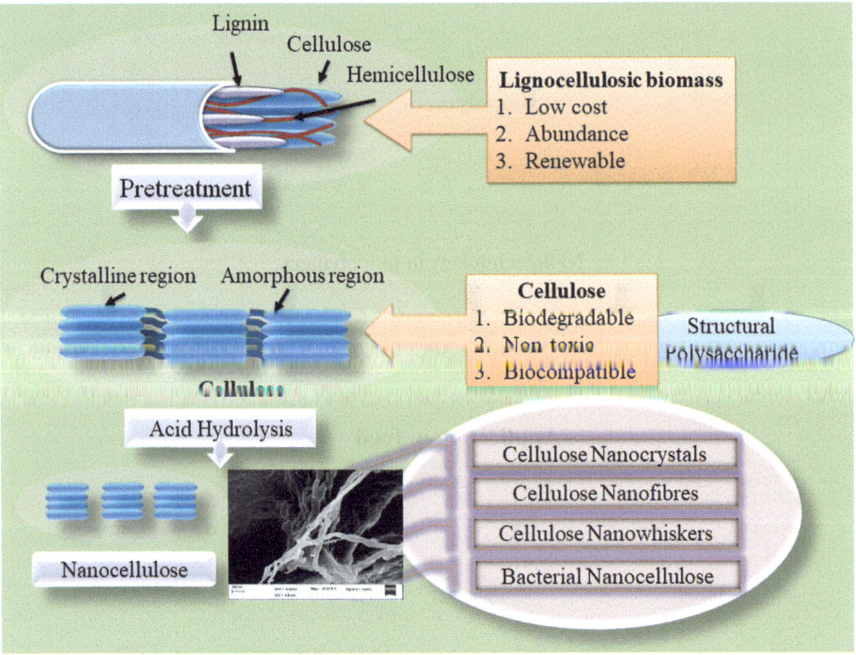

Fig. 5 Fabrication of nanocellulose

edible films and coatings, where cellulose offers positive effectiveness for improved food properties. Cellulose nanostructured materials can be extracted from lignocellulose biomass following (1) Pretreatment method (Chemical pretreatment/enzymatic pretreatment), and (2) Acid hydrolysis method (Fig. 5). The pretreatment method is used to extract cellulose from a particular substrate, whereas the acid hydrolysis treatment yields nanostructured cellulose materials. The nanostructured cellulose can be obtained in the form of nanocrystals, nanofibers, whiskers, which are utilized as edible nano-coating materials for perishable food products (Ghosh et al. 2019). The nanostructured cellulose materials deliver a range of properties depending on the dimensions, crystallinity, surface properties, barrier properties, and mechanical properties, etc. Nanostructured cellulose is also used for food functionalization as thickeners, stabilizers, carrier materials for active agents, etc. Additionally, the nanocellulose material is used for preparing nanocomposite based biodegradable films (active packaging/smart packaging), edible nano-coatings and films, etc.

However, bacterial/microbial nanocellulose has also gained more importance in edible coating and films as compared to plant derived cellulose. Bacterial/microbial cellulose as the name suggests is produced by certain kinds of bacteria (*Acetobacter, Rhizobium, Salmonella, Agrobacterium, Sarcina ventriculi*). The bacterial/microbial cellulose is different from plant based cellulose in terms of its purity, mechanical property, and water holding capacity. Bacterial cellulose has higher tensile strength (in wet state), Young's modulus (in wet state), higher crystallinity and water holding capacity as compared to plant based cellulose. The elastic and Young's modulus of bacterial cellulose is found to be 140 GPa and 172.9 GPa, respectively (Kucińska-Lipka et al. 2015). Among various available bacteria, *Acetobacter xylinum* is widely used to produce bacterial cellulose. Bacterial cellulose is biocompatible, and have better water holding property, remarkable mechanical properties, which makes it suitable for edible food packaging application. However, cellulose and bacterial cellulose do not possess antimicrobial properties and thus need to be used in combination with antibacterial agents so as to serve as ideal materials for packaging application in the sector of active packaging. The incorporation of silver nanoparticles into bacterial cellulose makes it antimicrobial in nature, which, on the other hand, is not seen in neat cellulose material. Bacterial cellulose incorporated with silver nanoparticles has a strong antimicrobial effect for *Escherichia coli* and *Staphylococcus aureus* (Maneerung et al. 2008). The combination of bacterial cellulose and chitosan membrane has an additional benefit in comparison to neat bacterial cellulose membrane for inhibiting *Escherichia coli* and *Staphylococcus aureus* (Lin et al. 2013). Furthermore, it has been reported that the loading of various active agents to cellulose facilitates the food properties due to its biodegradable, biocompatible, and antibacterial property (Laçin 2014). Thus cellulose is found to have immense potential to be used in packaging application with the aid of various value addition strategies.

2.2 *Chitosan and Its Nanostructured Materials*

Chitosan is the second most widely used polysaccharide after cellulose. The biopolymer consists of a linear chain of β-(1 → 4)-D-glucosamine, which is a cationic natural polymer obtained by deacetylation of chitin molecule. The fabrication of chitosan from available sources (crustaceans, insects, microorganism, fungi) generally follows demineralization, deproteinization, and deacetylation process. However, chitosan can be extracted following biological and chemical methods, where biological methods include organic acid producing bacteria, protease producing bacteria, and chitin deacetylase, and chemical methods involve the use of acid and sodium hydroxide solution. Chitosan based source is used in edible packaging application due to its antimicrobial, antifungal, adhesive nature, biocompatible nature, oxygen permeability, etc. As mentioned earlier, chitosan can be used with other biopolymers in order to make it bacteriostatic and fungistatic, which are the essential properties for appropriate food packaging (active packaging). Further, several forms of chitosan such as chitin, chitosan, their oligomers, and monomers can improve the packaging property, whereas the degree of deacetylation, molecular weight of chitosan, and nanoparticles provide various tunable properties for protecting the food products. Besides, chitosan nanoparticles also act as a delivery agent to deliver nutraceuticals, and other bioactive agents for improved food properties. As shown in Fig. 6, the chitosan nanoparticles can be fabricated *via* ionotropic gelation,

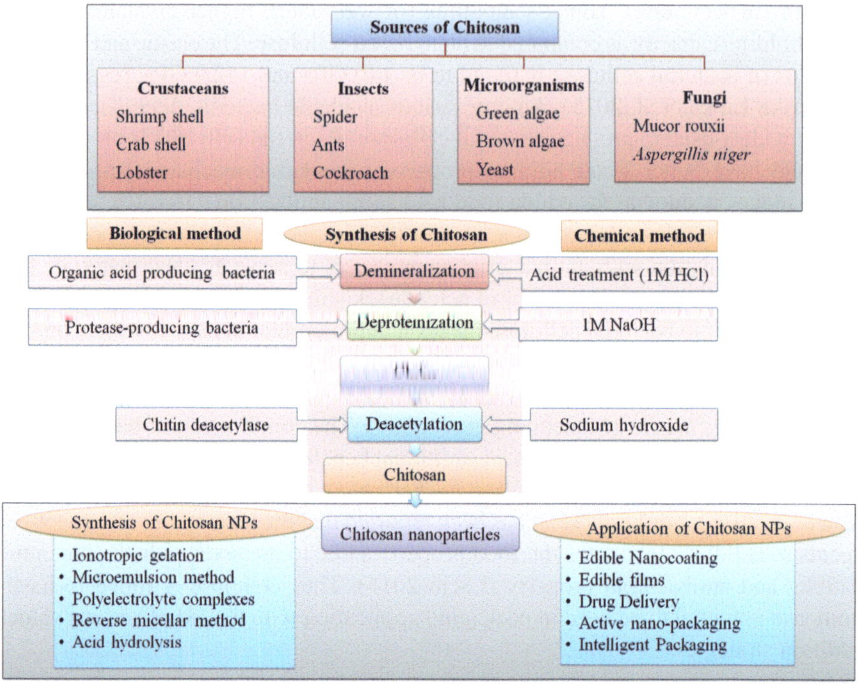

Fig. 6 Routes of synthesis of chitosan nanoparticles for packaging application

microemulsion method, polyelectrolyte complexes, reverse micellar method, acid hydrolysis, etc. The chitosan has a remarkable use in the preparation of active food packaging for the antioxidant, antimicrobial, antifungal properties, etc. Further, chitosan acts as weight reducing agent, which offer an additional benefit for being used as an edible coating material. The hydrogels made from chitosan are considered as potential materials for several applications, however, the chitosan based hydrogel is restricted in food packaging application. The property of chitosan for edible packaging application can be improved by the addition of other biopolymers for developing blends, composites with the aid of collagen, cellulose, starch, alginate, and other biopolymers.

The use of water soluble chitosan based complex and the related nanoparticles also provide a remarkable use in food packaging application, where the water soluble chitosan can easily improve the packaging properties. The complex of water soluble chitosan and other polysaccharides provide an increase in shelf life of food products, when used as an active packaging material, and can be directly applied as edible coating materials. In addition, titanium dioxide provides various medicinal property, which increases the effective use in edible packaging fields. The addition of titanium dioxide nanoparticle to chitosan and pectin materials help in improving antibacterial property, water vapor transmission rate, swelling property, biocompatible nature, etc. which are considered as the desired properties for edible packaging applications. Chitosan is also used with poly (ethylene glycol) and other plasticizers for providing improved packaging properties in prolonging the shelf life. Therefore. chitosan with tailored properties can be used extensively for edible food packaging application.

2.3 Starch and Its Nanostructured Materials

Starch is another kind of polysaccharide consisting of amylose and amylopectin, where the properties of starch are mainly dependent on the composition of both the units. The pure form of starch is insoluble in cold water and it is tasteless and odorless in nature. Starch is cheap, easily available, biocompatible, non-toxic, biodegradable in nature, which makes it a focused polymer for edible packaging application. Besides, starch is widely used biopolymer in various sectors including drug delivery systems, edible coatings, edible films, biodegradable films, food functionalization, etc. Starch is usually modified for improving its inherent properties to be used in edible packaging application wherein, breadfruit derived starch hydrolysates possesses better properties than commercial starch hydrolysate with same ranges of dextrose equivalent from other sources. Similarly, the composites of sago starch, chitosan and silver nanoparticles can be used for edible packaging materials. Interestingly, starches are available in the nanoforms as nanocrystals, nanoparticles, and nano-colloids, which are extensively used in the edible food packaging application (Fig. 7).

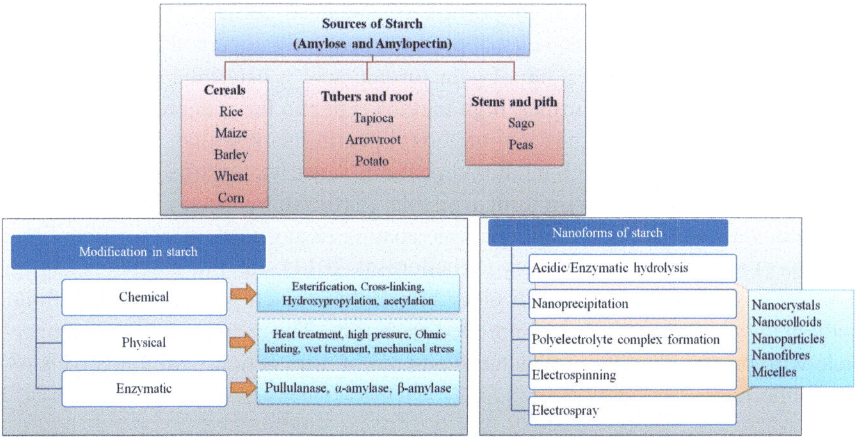

Fig. 7 (**a**) Sources of starch; (**b**) Modification of starch; (**c**) Nanoforms of starch

2.4 *Agarose and Its Nanostructured Materials*

Agarose is a polysaccharide, consisting of agarobiose, where agarobiose is generally consisting of D-galactose and 3,6-anhydro-L-galactopyranose, which are linked by α-(1 → 3) and β-(1 → 4) glycosidic linkages. However, agarose and agaropectin are the two principal components of agar. The specific polysaccharide exists as white powder, gets dissolved in near boiling water and forms a gel like three dimensional structure upon cooling. The properties of agarose may be varied depending on the sources and concentration. Additionally, agarose offers a wide range of thermal, chemical stability, hydrophilic nature, gelling efficiency, favorable biocompatibility, porous structure, which makes it a potential candidate to be used in edible food packaging application. Agar has low biodegradation rate and moderate intension strength (wet condition), which needs to be tuned by adding various kinds of biopolymers. In this regard, various kinds of agarose based edible coatings/films are developed for packaging application with the addition of biopolymers such as chitosan, starch, carboxylated agarose composites. The incorporation of the various derivatives of agarose can be used with other biopolymers for obtaining the tailor-made properties of edible films/coatings.

2.5 *Pectin and Its Nanostructured Materials*

Pectin is a heteropolysaccharide (complex polysaccharide) and an indigestible soluble fiber, which consists of galacturonic acid and is found in some plants such as pears, apples, guavas, plums, oranges, carrots, etc. Pectin is commercially available as high methoxyl and low methoxyl pectin, where high methoxyl pectin needs a high portion of sugar to form gel like structure, however low methoxyl pectin does not require a high portion of sugar to form gel type structure. Additionally, the gelation capacity of pectin is generally governed by its degree of esterification, where the residue of galacturonic acid is esterified to form methyl esters to produce pectin. Further, high methoxyl pectin has a degree of esterification more than 50%, whereas low methoxyl pectin has less than 50% degree of esterification. However, high methoxyl pectin is sensitive towards low pH values, and is the most existing pectin form available. Furthermore, pectin has wide application in various sectors including edible coating (Kang et al. 2007), edible films (Espitia et al. 2014), drug delivery (Sriamornsak 2011), etc. For edible packaging applications, pectin is used with other biopolymers such as chitosan, soy flour, sodium alginate, and others for improved properties. The coating of silver nanoparticle with pectin provides antibacterial and antibiofilm action with tuned packaging properties.

2.6 *Protein Based Biopolymers and Its Nanostructured Materials*

As mentioned earlier, the proteins for edible food packaging application generally include collagen, keratin, soy protein, silk fibroin, etc. Proteins are generally widely available from various animal and plant based sources. Proteins provide better film forming ability, good barrier properties, low cost sources, nutritional benefits, gelation properties, emulsification properties, etc. The collagen is a principal structural component of connective tissue and is commonly found in fibrous tissue including tendons, ligaments, and skin. Collagen consists of triple helix I, which is formed by amino acids such as glycine, arginine, proline, hydroxyproline, hydroxylysine. Collagen is an extensively used biopolymer for edible food packaging applications. Keratin is another kind of structural protein which is a main structural material of hair, horns, outer skin layer, etc. It is used as an essential component in edible packaging application. Protein based biopolymers are thus referred to as promising candidates for edible packaging applications which have been extensively explored for their use and efficacy.

3 Traits of Sustainable Nanomaterials in Edible Food Packaging

The sustainable nanomaterials are extensively used in edible food packaging for nano-food safety, nano-functionality, nano-preservation, etc. as shown in Fig. 8. As mentioned earlier, the sustainable nanomaterials are used to improve the

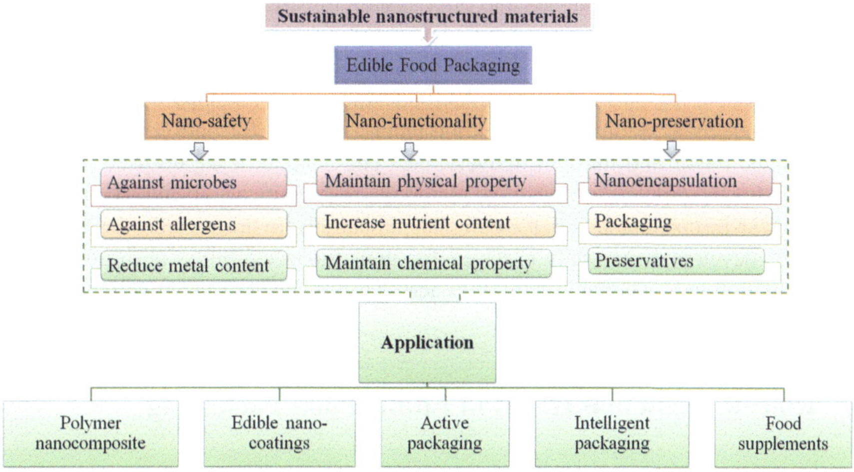

Fig. 8 Traits of sustainable nanomaterials in food sector

functionality of food products. The sustainable nanomaterials are considered as a remarkable source to be used as nano-safety materials, where sustainable nanomaterials are used against microbes, allergens, and to reduce metal content. The chitosan nanoparticles provide an effective antibacterial activity against gram-positive (*Staphylococcus aureus, Bacillus subtilis, Enterococcus faecalis*) and gram-negative bacteria (*Escherichia coli, Salmonella typhi*). Thus, the antimicrobial activity of chitosan NPs provides considerable attention from the past few years, where chitosan and its various derivatives are used to treat many postharvest diseases as antifungal coating materials. However, the antibacterial activity of sustainable nanomaterials is observed due to the repeating units of primary amine groups and the property is further influenced by various factors such as molecular weight of chitosan (low molecu- [illegible] temperature, pH, sources of chitosan, etc. In this regards, the use of antimicrobial/antibacterial for the fabrication of polymeric, and solid lipid NPs, liposomes help in improving product life, particle size, etc. Interestingly, the chitosan based products are commercially more acceptable than cellulose due to the high nitrogen content in comparison to cellulose materials, which is also used as chelating agent. Further, the addressed nanoparticles are used as an additive in improving the qualities of food products in terms of texture, crispiness, flavor, color, etc. Chitosan is used as supplements with other biopolymers for improved antibacterial activity. The cellulose, starch based nanoparticles are extensively used to provide nano-functionality to food products. The cellulose based nanomaterials have attained considerable attention for their tunable rheology property, physical and chemical property, structure-property relationship, etc. As mentioned earlier, cellulose nanostructured materials have a remarkable mechanical property, and biocompatibility, availability, low density, which increase the use of the specific nanomaterials in edible packaging application. The nanoparticles are performed as nanoencapsulation, and preservative materials, where nanoencapsulation helps to increase the bioavailability of active compounds, and protect bioactive compounds and offer sustain release of the active compounds. The wide application area of the specified nanoparticles includes in the area of edible nanocoatings, active packaging, intelligent packaging, and as food supplements to improve product life and acceptance. As food supplements, nanoparticles are added to food products as additives, anticaking agents, systems for delivery of bioactive compounds, flavor and nutrient delivery, etc. The nanoparticles are used as nanoemulsion for the delivery of bioactive materials, nutraceuticals or nutraceutical rich food and can be consumed with nanoemulsion materials. In this regards, the bioavailability of developed materials is tested for their bioaccessibility, absorption, and transformation in biochemical properties across mucus layers and epithelium cells.

4 Role of Medicinal Herbs in Edible Food Packaging

From ancient times, medicinal herbs are being used for the traditional therapeutic cures as homebased remedies. The long-established practices of using herbal medicines for numerous therapies are increasing day by day due to the various reasons including population growth, insufficient drug supply, side effects of synthetic drugs, cost-effectiveness, etc. According to WHO, 80% of people around the world depend on the traditional remedies as a principle healthcare measures (Muthu et al. 2006). Further, plants and its derivatives constitute 25% of medicinal drugs in developed countries (Muthu et al. 2006). India has a very rich diversity of medicinal plants, where 3000 plants are formally documented as medicinal plants, however, more than 6000 plants are used for therapeutics (Seth and Sharma 2004). Many medicinal plants are hindered due to lacking scientific data for efficient use of herbal products, which do not provide drug regulatory activities. The widely used medicinal plants in India include mainly aloe Vera, tulsi, coriander, mint, lemongrass, carom, etc., which are used for treating many diseases including cancer, wound healing, weight reducing agent, diabetes, cholesterol lowering, etc. In this chapter, the main focus has been given to discuss the role of medicinal herbs in edible packaging application. Among available, the generally used medicinal herbs for edible packaging include aloe vera, turmeric, ginger, *Terminalia chebula*, bitter gourd, green tea, onion, and others. Further, honey as a medicinal remedy is another kind of potent agent to be used in edible coating materials, where honey acts as a medium for delivering the medicinal herbs to remove unflavoured conditions.

4.1 Available Sources of Bioactive Agents

Bioactive agents are generally available from many resources such as plant sources, animal sources, marine sources, flavonoids, enzymes, essential oils, and others. As well known, bioactive agents are used for treating many diseases including prevention of cancer, wound healing activity, diabetic, cholesterol lowering, and others. Some of the bioactive compounds may impart some limitations for using as medicinal plants due to the side effects. However, the adequate precaution may help to recover the side effects. Bioactive compounds available from medicinal plants need to be extracted by various means, as the medicinal properties are obtained mainly due to the presence of specific active ingredients. In this regards, the extraction of bioactive agents from medicinal plants are generally done by following conventional and non-conventional techniques. The conventional techniques generally include soxhlet apparatus, maceration, and hydrodistillation, whereas the non-conventional techniques include ultrasonic assisted extraction, microwave assisted extraction, pulsed electric field extraction, enzyme assisted extraction, pressurized liquid extraction supercritical fluid extraction, etc. (Azmir et al. 2013). However, various processes may impart different advantages and disadvantages for the low stability and sensitive property of active compounds against temperature, pH, and other environmental conditions. In this regards, a detailed discussion for some of the available sources of bioactive compounds targeting edible food packaging application has been done in the below section.

4.1.1 Aloe Vera

Aloe vera is a kind of fleshy plant which is easy growing and required low maintenance. Aloe vera is used for multipurp1ose application such as wound healing (Vera 1989), sunburn and suntan (Puvabanditsin and Vongtongsri 2005), relieve from heartburn (Panahi et al. 2015), antitumor property (El-Shemy et al. 2010), reduce blood sugar (Tanaka et al. 2006), dentistry (Mangaiyarkarasi et al. 2015), etc. The medicinal property of aloe vera is generally due to the available active compounds in the leaves of aloe vera, where according to available reports, there are 75 compounds have been identified (Hamman 2008; Raksha et al. 2014). It is supposed that the medicinal effect of aloe vera gel is due to the synergistic effect of active compounds than individual components. The major classes of phytochemical components of aloe vera include tannin, saponin, flavonoids, etc. However, the other phytochemical components of aloe vera generally include 1-tetradecyne, phytol, oleic acid, tridecanoic acid, hexadecanoic acid, squalene, tocopherol, vitamin E, eicosane, etc. (Arunkumar and Muthuselvam 2009). The major sugar of aloe vera is mannose-6-phosphate, which has a positive influence on anti-inflammatory application (Davis et al. 1994).

4.1.2 Ginger

Ginger (*Zingiber officinale*) is an extensively used root spice for various beneficial properties in medicinal sectors. The rhizome or root of the plant provides various health beneficial activities including antioxidant (Kikuzaki and Nakatani 1993), wound healing (Bhagavathula et al. 2009); antidiabetic (Račková et al. 2013), cholesterol lowering (Bhandari et al. 1998); against nausea and vomiting (Ernst and Pittler 2000), antinflammatory property (Podlogar and Verspohl 2012), and others. Besides, ginger is considered a constant component for cooking purpose, health drinks for the hot and fragrant spice. The major nutritional component of ginger includes carbohydrate (sugar and dietary fiber), fat, protein, vitamins (thiamine, riboflavin, niacin, folate, ascorbic acid, tocopherols), minerals (sodium, iron, potassium, phosphorous, calcium), etc. However, the medicinal properties of ginger are mainly obtained due to the availabilities of bioactive compounds. Ginger is mainly used in the form of oleoresin and essential oils for various application, where the properties of oleoresin and essential oils of ginger varies depending on the source, extraction process, solvent used etc. The major bioactive compounds in ginger are 6-gingerol, which has 8-gingerol and 10-gingerol in smaller amounts (Manasa et al. 2013). The other available bioactive compounds in ginger are gingerenone A, zingerone, shogaol, paradol, 1-dehydro-10-gingerdione, zerumbone, etc. The specific bioactive compounds are targeted to affect particular health benefits. In this regards, the bioactive compounds 6-shogoal, 10-gingerol, 8-gingerol, 6-gingerol provide many beneficial effects for human health. As mentioned earlier, antibacterial and antimicrobial properties are desirable for edible packaging application, thus zingerone and zerumbone provide antibacterial and antimicrobial activity, respectively

(Rahmani et al. 2014). Additionally, ginger extract can also help to provide ginger flavored products including edible coating/films/encapsulation. The combined use of curcumin and ginger extract provide additional benefits to promote packaging properties.

4.1.3 Turmeric

Turmeric (*Curcuma longa*) belongs to the ginger family, is extensively used for various medicinal properties. The nutrient compositions of turmeric generally include carbohydrate, protein, fat, minerals, dietary fiber, etc. However, the incorporated medicinal property is mainly due to the presence of various bioactive compounds in turmeric such as three types of curcuminoids such as curcumin, desmethoxycurcumin, bisdesmethoxycurcumin (Wichitnithad et al. 2009) and various volatile oils such as tumerone, atlantone, borneol, zingiberene, sesquiterpenes, etc. Similar to ginger, turmeric extracts as oleoresin and essential oils are used against various diseases. The bioactive in turmeric is used as anticancer agent (Kuttan et al. 1987), antioxidant agent (Jayaprakasha et al. 2002), wound healing (Das and Cohly 1995), antimutagenicity (Jayaprakasha et al. 2002), anti-inflammatory (Aggarwal et al. 2013), antidiabetic (Lekshmi et al. 2012), improve brain function (Ahmed and Gilani 2014), lower blood lipids (Qin et al. 2017), treating depression (Lopresti et al. 2014) etc. However, the use of turmeric is remarkable for edible packaging application mainly due to the presence of curcumin. As home remedies, the paste of turmeric is applied to the skin to treat wounds and itching, which further helps to kill microbes.

4.1.4 Haritaki

Haritaki (*Terminalia chebula)* has the common names myrobalan belongs to the tropical almond class. *T. chebula* is considered as one of the principal ingredients for Ayurvedic medicinal development *viz.* Triphala (A herbal medicine consisting of *T. chebula*, amalaki, bibhitaki), which is used to treat dysfunction in kidney and liver. *T. chebula* fruit is known by various names depending on the origin, and properties, where the preferred variety (*vijaya variety*) is obtained from the Vindhya region of India. Additionally, the fruit is known as the king of medicines in Tibet for its existing biological activities (Lee et al. 2007). The divine fruit is used in various way to treat many health benefits such as antioxidant effect (Cheng et al. 2003), antibacterial effect (Kannan et al. 2009), antidiabetic effect (Kumar et al. 2006), anticancer activity (Saleem et al. 2002), anti-ulcer activity (Raju et al. 2009), wound healing (Li et al. 2011), anti-HIV-1 (Valsaraj et al. 1997), antifungal (Valsaraj et al. 1997), antimalarial activity (Valsaraj et al. 1997), etc. The nutritive composition of *T. chebula* generally consists of starch, reducing sugar, sucrose, protein, ascorbic acid, minerals (Barthakur and Arnold 1991). The minerals of *T. chebula* generally includes phosphorous, potassium, calcium, magnesium, Sulphur, iron, boron, sodium, chloride, silica, selenium, manganese, etc. (Barthakur and Arnold, 1991). The amino acids present in *T. chebula* includes alanine, arginine, aspartic acid, glutamic acid, glycine, histidine, isoleucine, leucine, lysine, methionine, etc. (Barthakur and Arnold 1991). Additionally, *T. chebula* is a good source of phenolic content, tannin content, and triterpenoid content, which provides the medicinal property to food product (Chang and Lin 2012). The active compounds in *T. chebula* are triterpene glycosides, chebuloside I and II, arjunglucoside I, arjungenin, chebulin, ellagic acid, chebulinic acid, gallic acid, luteolin, terflavin A and B, etc. *T. chebula* is effective against dermal wound healing activity in rats, where the treatment of *T. chebula* to wounds provide faster healing and epithelialization (Suguna et al. 2002). Further, there is an increase in total protein, collagen content, hexosamine, uronic acids in tissue due to the effect of *T. chebula* (Suguna et al. 2002).

4.1.5 Other Herbal Sources for Edible Food Packaging Application

As discussed earlier, there are a variety of sources which are used for edible food packaging application for availing with various active ingredients. The extract of bitter gourd provides a positive effect for using as edible food packaging application. Green tea has antioxidant and anti-inflammatory properties and is effective in developing edible coating/film materials. The major polyphenols in green tea polyphenol are epigallocatechin-3-gallate, which is used in treating many diseases. Moreover, chitosan and green tea polyphenols are used to control the release of compounds and help in improved food properties when used as edible food packaging materials. Onion has quercetin, gallic acid, and ferulic acid as main components and is a rich source of polyphenols, flavonoid, anthocyanin.

4.2 Limitations of Medicinal Herbs for Edible Food Packaging Application

As discussed, herbal medicines are widely used for treating many diseases for their cost-effectiveness and local availability, however, the herbal medicines have many side effects limiting the use of herbs. Considerably, herbal medicines effects slowly, so it is considered inappropriate for severe conditions and there is a lack of proper instruction for doses and allergic reaction. However, the unavailability in the proper doses requirement may cause the overdose of active compounds which may cause severe problems in the body. The common disadvantages of medicinal herbs include allergic reactions, irritation, headache, vomiting, gas formation, etc. In this regards, though aloe vera has many health benefits, however excess of everything is not good. The latex content of aloe vera may be a problem for some consumers, as it may impart irritation, skin allergies, burning sensation, discomfort, itching, dehydration, constipation, etc. Further, aloe vera may cause a fall in blood sugar level.

The latex of aloe vera may cause an imbalance in the electrolyte of a diabetic. Additionally, the serious effects of aloe vera include kidney problem, thyroid dysfunction, toxic effects, etc. Similarly, the high intake of ginger may lead to an upset stomach, irritation (skin, eye, mouth), gas, heartburn, diarrhea, lower blood sugar, and it is riskier for bleeding disorder patients. Considerably, ginger intake is not good for pregnant and breastfeeding women. The intake of turmeric may cause kidney stones, upset stomach, diarrhea, allergic reaction, may reduce iron absorption in the body and others. The haritaki has a bitter taste and cause dehydration, diarrhea, less sleep, system fatigue, and a decrease in sexual energy, etc. The green tea intake may lead to stomach problems, iron deficiency, headache, less sleep, vomiting, irregular heart rhythm, muscle contraction, high blood pressure, liver dysfunction, etc. The risk of bitter gourd includes vaginal bleeding, abortion, lowers blood sugar, liver damage, irregular heartbeat, etc. In this regards, the content of medicinal herbs should be checked properly and should consult with an herbalist before the consumption. Further, the herbal medicines act slowly on body thus do not work for severe diseases. Some reports also suggest that the herbal medicines generally consist of a number of phytochemicals, where a single dose contain a less number of active agents which is a problem for proper treatment.

5 Sustainable Nanostructured Materials Facilated by Bioactive Compounds for Edible Food Packaging

The sustainable nanostructured materials are generally extracted from available renewable sources and are considered as commercialized materials for the characteristics of inherent properties. Interestingly, sustainable nanostructured materials are used for the delivery of nutraceuticals and bioactive compounds. As shown in Fig. 9, the bioactive compounds are generally obtained from available herbs proving various health beneficial properties. Edible coatings and films are a suitable way for the delivery of the active compounds to consumer, where sustainable nanostructured materials act as a delivery agent. Additionally, the bioactive compounds help to enrich the food products with natural health beneficial agents. The edible packaging facilitated by bioactive agents is widely used to improve the shelf life of perishable food products. The perishable food products such as fruits and vegetables are essential parts of a healthy diet, where a single fruit product cannot provide complete nutrients. As discussed in the previous section, fruits and vegetables, meat and meat products, fish and fish products are perishable food products, having a tendency to degrade faster if stored in unfavorable condition. Further, fruits and vegetables are a rich source of natural antioxidant compounds, which are required for good health of consumer. In this regards, there is a problem relating to the seasonal availability of perishable fruits and vegetables. Thus various postharvest techniques are used to prolong shelf life of fruit products such as low temperature storage (freezing, chilling), high temperature storage (sterilization, pasteurization, drying,

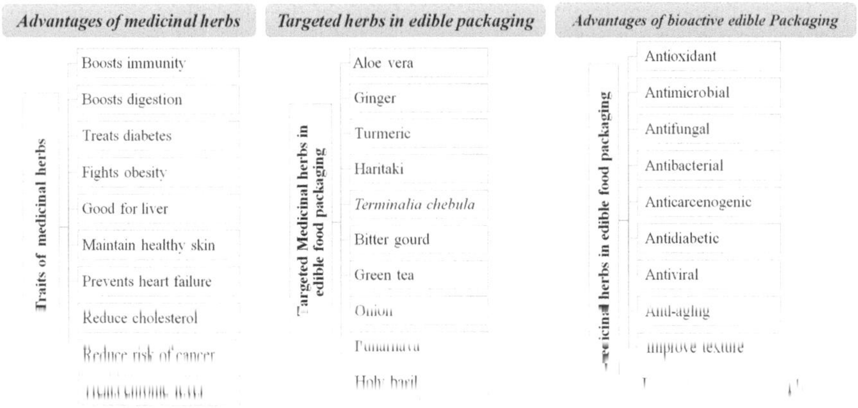

Fig. 9 Significance of bioactive compounds in edible food packaging

etc.), applied packaging technology (modified atmospheric packaging, controlled atmospheric packaging, active packaging, intelligent packaging), salting, etc. However, the use of high temperature and low temperature processing for the improved shelf life of food products may change the food properties due to undesirable chemical and biochemical changes. So the use of edible coatings and films with the addition of bioactive compounds can improve the product life of food products.

The aloe vera can be used to develop edible coating materials in addition to applied packaging technology, where aloe vera coated kiwifruit slices provide improved fruit quality when stored under modified passive atmosphere (Benítez et al. 2013). The edible coating based on aloe vera gel maintains the property of sweet cherry in terms of weight loss, reduce respiration rate, color, firmness, total soluble solid content, etc. during cold storage (Martínez-Romero et al. 2006). The aloe vera is generally used as an antimicrobial coating for improved shelf life and quality of papaya fruits (Marpudi et al. 2011). The properties of aloe vera based coating for papaya fruits are further improved when papaya leaf extracts are added, which help to maintain physiochemical properties. However, the available biopolymers can be used in addition to aloe vera for improved food quality. In this regards, the use of chitosan and aloe vera are very effective in maintaining the quality of blueberries in terms of delayed fungal decay and reduce weight loss, etc. at cold storage (Vieira et al. 2016). The limonene liposomes have a tendency to improve the shelf life of strawberries during transportation (Dhital et al. 2017). In this regards, the transportation of fruits and vegetables are a crucial reason for fruit degradation, where vibration during transportation may cause abrasion, bruising, scuffing, punctures, cuts, etc. The application of edible nano-coating can improve the property of food products during mechanical operation, storage, transportation, etc. The limonene based liposomes help to reduce the fungal growth on strawberries, which further improve the shelf life of fruit products. The nanotechnology based packaging help to improve the quality of fruits and vegetables by maintaining the preservation properties of fruits and vegetables for a longer period of time. The edible packaging incorporating essential oils such as ginger or cinnamon with sodium caseinate and soy protein isolate provides tunable film property for edible packaging (Atarés et al. 2010a; Atarés et al. 2010b). Further, the nanoemulsion based edible coating from sodium caseinate and ginger essential oils can maintain the properties of chicken breast in comparison to emulsion based edible coating (Noori et al. 2018). The nanoemulsion coating materials with ginger essential oils (6%) have a tendency to decrease the total aerobic psychrophilic bacteria of specified products. Further, the addition of essential oils to edible coating materials have a tendency to increase opacity, antimicrobial activity, antifungal activity of coating materials. However, the sources of essential oils include flowers, herbs, resins, citruses, spices, etc. In this regards, the various essential oils from available sources are widely used to improve the food quality during storage due to the available activities such as antioxidant, antimicrobials, antibacterial, antifungal activity, etc. The properties of the essential oils have attained various application in the area of food additives, edible packaging and pharmaceutical industries. The curcumin is a bioactive compound available in turmeric, which is used to make bioactive coatings incorporating curcumin in protein for developing nanofiber mats (Alehosseini et al. 2019).

6 Conclusion

The development of edible food packaging with the incorporation of bio-based nanostructured materials is very effective in improving the food properties. However, the postharvest management system is very crucial in extending the shelf life of perishable food products. The inclusion of various unit operations can improve the food properties, however, the techniques may alter the food texture. The use of edible packaging in terms of edible films and coatings can improve the product life without altering the food properties. Moreover, the addition of various bioactive compounds for developing edible packaging materials help to provide additional benefits such as antimicrobial property, antioxidant property, antifungal activity, antibacterial activity, etc. In this regards, the combined use of biopolymers and nutraceuticals are a potential candidate to improve the product life of food products.

References

Aggarwal BB, Yuan W, Li S, Gupta SC (2013) Curcumin-free turmeric exhibits anti-inflammatory and anticancer activities: identification of novel components of turmeric. Mol Nutr Food Res 57(9):1529–1542. https://doi.org/10.1002/mnfr.201200838

Ahmed T, Gilani AH (2014) Therapeutic potential of turmeric in Alzheimer's disease: curcumin or curcuminoids? Phytother Res 28(4):517–525. https://doi.org/10.1002/ptr.5030

Alehosseini A, Gómez-Mascaraque LG, Martínez-Sanz M, López-Rubio A (2019) Electrospun curcumin-loaded protein nanofiber mats as active/bioactive coatings for food packaging applications. Food Hydrocoll 87:758–771. https://doi.org/10.1016/j.foodhyd.2018.08.056

Arunkumar S, Muthuselvam M (2009) Analysis of phytochemical constituents and antimicrobial activities of Aloe vera L. against clinical pathogens. World J Agric Sec 5(5):572–576

Atarés L, Bonilla J, Chiralt A (2010a) Characterization of sodium caseinate-based edible films incorporated with cinnamon or ginger essential oils. J Food Eng 100(4):678–687. https://doi.org/10.1016/j.jfoodeng.2010.05.018

Atarés L, De Jesús C, Talens P, Chiralt A (2010b) Characterization of SPI-based edible films incorporated with cinnamon or ginger essential oils. J Food Eng 99(3):384–391. https://doi.org/10.1016/j.jfoodeng.2010.03.004

Azmir J, Zaidul ISM, Rahman MM, Sharif KM, Mohamed A, Sahena F et al (2013) Techniques for extraction of bioactive compounds from plant materials: a review. J Food Eng 117:426–436. https://doi.org/10.1016/j.jfoodeng.2013.01.014

Barthakur NN, Arnold NP (1991) Nutritive value of the chebulic myrobalan (Terminalia chebula Retz.) and its potential as a food source. Food Chem 40(2):213–219. https://doi.org/10.1016/0308-8146(91)90105-W

Benítez S, Achaerandio I, Sepulcre F, Pujolà M (2013) Aloe vera based edible coatings improve the quality of minimally processed 'Hayward'kiwifruit. Postharvest Biol Tec 81:29–36. https://doi.org/10.1016/j.postharvbio.2013.02.009

Bhagavathula N, Warner RL, DaSilva M, McClintock SD, Barron A, Aslam MN et al (2009) A combination of curcumin and ginger extract improves abrasion wound healing in corticosteroid-impaired hairless rat skin. Wound Repair Regen 17(3):360–366. https://doi.org/10.1111/j.1524-475X.2009.00483.x

Bhandari U, Sharma JN, Zafar R (1998) The protective action of ethanolic ginger (Zingiber officinale) extract in cholesterol fed rabbits. J Ethnopharmacol 61(2):167–171. https://doi.org/10.1016/S0378-8741(98)00026-9

Brinchi L, Cotana F, Fortunati E, Kenny JM (2013) Production of nanocrystalline cellulose from lignocellulosic biomass: technology and applications. Carbohydr Polym 94(1):154–169. https://doi.org/10.1016/j.carbpol.2013.01.033

Chang CL, Lin CS (2012) Phytochemical composition, antioxidant activity, and neuroprotective effect of Terminalia chebula Retzius extracts. Evid Based Complement Alternat Med 2012:1–7. https://doi.org/10.1155/2012/125247

Chen H (2014) Chemical composition and structure of natural lignocellulose. In: Biotechnology of Lignocellulose. Springer, Dordrecht, pp 25–71. https://doi.org/10.1007/978-94-007-6898-7_2

Cheng HY, Lin TC, Yu KH, Yang CM, Lin CC (2003) Antioxidant and free radical scavenging activities of Terminalia chebula. Biol Pharm Bulletin 26:1331–1335. https://doi.org/10.1248/bpb.26.1331

Cuq B, Gontard N, Guilbert S (1995) Edible films and coatings as active layers. In: Active food packaging. Springer, Boston, MA, pp 111–142

Das SK, Cohly HHP (1995) U.S. Patent No. 5,401,504. Washington, DC: U.S. Patent and Trademark Office

Davis RH, Donato JJ, Hartman GM, Haas RC (1994) Anti-inflammatory and wound healing activity of a growth substance in Aloe vera. J Am Podiatric Med Assoc 84(2):77–81. https://doi.org/10.7547/87507315-84-2-77

Dhital R, Joshi P, Becerra-Mora N, Umagiliyage A, Chai T, Kohli P, Choudhary R (2017) Integrity of edible nano-coatings and its effects on quality of strawberries subjected to simulated in-transit vibrations. LWT 80:257–264. https://doi.org/10.1016/j.lwt.2017.02.033

Donhowe IG, Fennema O (1994) Edible films and coatings: characteristics, formation, definitions, and testing methods. In: Edible coatings and films to improve food quality. CRC Press, Boca Raton, Florida1-24

El-Shemy HA, Aboul-Soud MAM, Nassr-Allah AA, Aboul-Enein KM, Kabash A, Yagi A (2010) Antitumor properties and modulation of antioxidant enzymes' activity by Aloe vera leaf active principles isolated via supercritical carbon dioxide extraction. Curr Med Chem 17(2):129–138. https://doi.org/10.2174/092986710790112620

Ernst E, Pittler MH (2000) Efficacy of ginger for nausea and vomiting: a systematic review of randomized clinical trials. Br J Anaesth 84(3):367–371. https://doi.org/10.1093/oxfordjournals.bja.a013442

Espitia PJP, Du WX, de Jesús A-BR, Soares NDFF, McHugh TH (2014) Edible films from pectin: physical-mechanical and antimicrobial properties-a review. Food Hydrocoll 35:287–296. https://doi.org/10.1016/j.foodhyd.2013.06.005

Ghosh T, Katiyar V (2019) Cellulose-based hydrogel films for food packaging. In: Mondal M (ed) Cellulose-based superabsorbent hydrogels. Polymers and polymeric composites, A Reference Series. Springer, Cham, pp 1061–1084. https://doi.org/10.1007/978-3-319-77830-3_35

Ghosh T, Teramoto Y, Katiyar V (2019) Influence of nontoxic magnetic cellulose nanofibers on chitosan based edible nanocoating: a candidate for improved mechanical, thermal, optical, and texture properties. J Agric Food Chem 67(15):4289–4299. https://doi.org/10.1021/acs.jafc.8b05905

Hamman J (2008) Composition and applications of Aloe vera leaf gel. Molecules 13(8):1599–1616. https://doi.org/10.3390/molecules13081599

Jayaprakasha GK, Jena BS, Negi PS, Sakariah KK (2002) Evaluation of antioxidant activities and antimutagenicity of turmeric oil: a byproduct from curcumin production. Zeitschrift für Naturforschung C 57(9–10):828–835. https://doi.org/10.1515/znc-2002-9-1013

Kang HJ, Jo C, Kwon JH, Kim JH, Chung HJ, Byun MW (2007) Effect of a pectin-based edible coating containing green tea powder on the quality of irradiated pork patty. Food Control 18(5):430–435. https://doi.org/10.1016/j.foodcont.2005.11.010

Kannan P, Ramadevi SR, Hopper W (2009) Antibacterial activity of Terminalia chebula fruit extract. Afr J Microbiol Res 3(4):180–184

Kikuzaki H, Nakatani N (1993) Antioxidant effects of some ginger constituents. J Food Sci 58(6):1407–1410. https://doi.org/10.1111/j.1365-2621.1993.tb06194.x

Kucińska-Lipka J, Gubanska I, Janik H (2015) Bacterial cellulose in the field of wound healing and regenerative medicine of skin: recent trends and future prospectives. Polym Bull 72(9):2399–2419. https://doi.org/10.1007/s00289-015-1407-3

Kumar GPS, Arulselvan P, Kumar DS, Subramanian SP (2006) Anti-diabetic activity of fruits of Terminalia chebula on streptozotocin induced diabetic rats. J Health Sci 52(3):283–291. https://doi.org/10.1248/jhs.52.283

Kuttan R, Sudheeran PC, Josph CD (1987) Turmeric and curcumin as topical agents in cancer therapy. Tumori J 73(1):29–31. https://doi.org/10.1177/030089168707300105

Laçin NT (2014) Development of biodegradable antibacterial cellulose based hydrogel membranes for wound healing. Int J Biol Macromol 67:22–27. https://doi.org/10.1016/j.ijbiomac.2014.03.003

Lee HS, Jung SH, Yun BS, Lee KW (2007) Isolation of chebulic acid from Terminalia chebula Retz. and its antioxidant effect in isolated rat hepatocytes. Arch Toxicol 81(3):211–218. https://doi.org/10.1007/s00204-006-0139-4

Lekshmi PC, Arimboor R, Indulekha PS, Nirmala Menon A (2012) Turmeric (Curcuma longa L.) volatile oil inhibits key enzymes linked to type 2 diabetes. Int J Food Sci Nutr 63(7):832–834. https://doi.org/10.3109/09637486.2011.607156

Li K, Diao Y, Zhang H, Wang S, Zhang Z, Yu B et al (2011) Tannin extracts from immature fruits of Terminalia chebula Fructus Retz. promote cutaneous wound healing in rats. BMC Complement Altern Med 11(1):86. https://doi.org/10.1186/1472-6882-11-86

Lin WC, Lien CC, Yeh HJ, Yu CM, Hsu SH (2013) Bacterial cellulose and bacterial cellulose–chitosan membranes for wound dressing applications. Carbohydr Polym 94(1):603–611. https://doi.org/10.1016/j.carbpol.2013.01.076

Lopresti AL, Maes M, Maker GL, Hood SD, Drummond PD (2014) Curcumin for the treatment of major depression: a randomised, double-blind, placebo controlled study. J Affect Disord 167:368–375. https://doi.org/10.1016/j.jad.2014.06.001

Manasa D, Srinivas P, Sowbhagya HB (2013) Enzyme-assisted extraction of bioactive compounds from ginger (Zingiber officinale Roscoe). Food Chem 139(1–4):509–514. https://doi.org/10.1016/j.foodchem.2013.01.099

Maneerung T, Tokura S, Rujiravanit R (2008) Impregnation of silver nanoparticles into bacterial cellulose for antimicrobial wound dressing. Carbohydr Polym 72(1):43–51. https://doi.org/10.1016/j.carbpol.2007.07.025

Mangaiyarkarasi SP, Manigandan T, Elumalai M, Cholan PK, Kaur RP (2015) Benefits of Aloe vera in dentistry. J Pharm Bioallied Sci 7(Suppl 1):S255. https://doi.org/10.4103/0975-7406.155943

Marpudi SL, Abirami LSS, Srividya N (2011) Enhancement of storage life and quality maintenance of papaya fruits using Aloe vera based antimicrobial coating

Martínez-Romero D, Alburquerque N, Valverde JM, Guillén F, Castillo S, Valero D, Serrano M (2006) Postharvest sweet cherry quality and safety maintenance by Aloe vera treatment: A new edible coating. Postharvest Biol Tec 39(1):93–100. https://doi.org/10.1016/j.postharvbio.2005.09.006

Muthu C, Ayyanar M, Raja N, Ignacimuthu S (2006) Medicinal plants used by traditional healers in Kancheepuram District of Tamil Nadu, India. J Ethnobiol Ethnomed 2:43–53. https://doi.org/10.1186/1746-4269-2-43

Noori S, Zeynali F, Almasi H (2018) Antimicrobial and antioxidant efficiency of nanoemulsion-based edible coating containing ginger (Zingiber officinale) essential oil and its effect on safety and quality attributes of chicken breast fillets. Food Control 84:312–320. https://doi.org/10.1016/j.foodcont.2017.08.015

Panahi Y, Khedmat H, Valizadegan G, Mohtashami R, Sahebkar A (2015) Efficacy and safety of Aloe vera syrup for the treatment of gastroesophageal reflux disease: a pilot randomized positive-controlled trial. J Tradit Chin Med 35(6):632–636. https://doi.org/10.1016/S0254-6272(15)30151-5

Park HJ (1999) Development of advanced edible coatings for fruits. Trends Food Sci Technol 10(8):254–260. https://doi.org/10.1016/S0924-2244(00)00003-0

Park HJ, Weller CL, Vergano PJ, Testin RF (1993) Permeability and mechanical properties of cellulose-based edible films. J Food Sci 58(6):1361–1364. https://doi.org/10.1111/j.1365-2621.1993.tb06183.x

Podlogar JA, Verspohl EJ (2012) Antiinflammatory effects of ginger and some of its components in human bronchial epithelial (BEAS-2B) cells. Phytother Res 26(3):333–336. https://doi.org/10.1002/ptr.3558

Puvabanditsin P, Vongtongsri R (2005) Efficacy of aloe vera cream in prevention and treatment of sunburn and suntan. J Med Assoc Thail 88:S173–S176

Qin S, Huang L, Gong J, Shen S, Huang J, Ren H, Hu H (2017) Efficacy and safety of turmeric and curcumin in lowering blood lipid levels in patients with cardiovascular risk factors: a meta-analysis of randomized controlled trials. Nutr J 16(1):68. https://doi.org/10.1186/s12937-017-0293-y

Račková L, Cupáková M, Ťažký A, Mičová J, Kolek E, Košťálová D (2013) Redox properties of ginger extracts: Perspectives of use of Zingiber officinale Rosc. as antidiabetic agent. Interdiscip Toxicol 6(1):26–33. https://doi.org/10.2478/intox-2013-0005

Rahmani H, Al Shabrmi FM, Aly SM (2014) Active ingredients of ginger as potential candidates in the prevention and treatment of diseases via modulation of biological activities. Int J Physiol Pathophysiol Pharmacol 6(2):125–136

Raju D, Ilango K, Chitra V, Ashish K (2009) Evaluation of Anti-ulcer activity of methanolic extract of Terminalia chebula fruits in experimental rats. J Pharm Sci Res 1(3):101

Raksha B, Pooja S, Babu S (2014) Bioactive compounds and medicinal properties of Aloe vera L.: an update. J Plant Sci 2(3):102–107. https://doi.org/10.11648/j.jps.20140203.11

Saleem A, Husheem M, Härkönen P, Pihlaja K (2002) Inhibition of cancer cell growth by crude extract and the phenolics of Terminalia chebula retz. Fruit. J Ethnopharmacol 81(3):327–336. https://doi.org/10.1016/S0378-8741(02)00099-5

Seth SD, Sharma B (2004) Medicinal plants in India. Indian J Med Res 120:9–11

Sriamornsak P (2011) Application of pectin in oral drug delivery. Expert Opin Drug Deliv 8(8):1009–1023. https://doi.org/10.1517/17425247.2011.584867

Suguna L, Singh S, Sivakumar P, Sampath P, Chandrakasan G (2002) Influence of Terminalia chebula on dermal wound healing in rats. Phytother Res 16(3):227–231. https://doi.org/10.1002/ptr.827

Tanaka M, Misawa E, Ito Y, Habara N, Nomaguchi K, Yamada M et al (2006) Identification of five phytosterols from Aloe vera gel as anti-diabetic compounds. Biol Pharm Bull 29(7):1418–1422. https://doi.org/10.1248/bpb.29.1418

Valsaraj R, Pushpangadan P, Smitt UW, Adsersen A, Christensen SB, Sittie A et al (1997) New anti-HIV-1, antimalarial, and antifungal compounds from terminalia bellerica. J Nat Prod 60(7):739–742. https://doi.org/10.1021/np970010m

Vera A (1989) Wound healing, oral & topical activity of Aloe vera. J Am Podiatric Med Assoc 79:559–562

Vieira JM, Flores-López ML, de Rodríguez DJ, Sousa MC, Vicente AA, Martins JT (2016) Effect of chitosan–Aloe vera coating on postharvest quality of blueberry (Vaccinium corymbosum) fruit. Postharvest Biol Tec 116:88–97. https://doi.org/10.1016/j.postharvbio.2016.01.011

Wichitnithad W, Jongaroonngamsang N, Pummangura S, Rojsitthisak P (2009) A simple isocratic HPLC method for the simultaneous determination of curcuminoids in commercial turmeric extracts. Phytochem Anal 20(4):314–319. https://doi.org/10.1002/pca.1129

Nano-Films for Food Packaging

M. L. Zambrano-Zaragoza, R. M. González-Reza, D. Quintanar-Guerrero, and N. Mendoza-Muñoz

1 Definitions of Nano-Films

At present, efforts have been made to improve the quality and safety of food products. One of the best-known strategies is the formation of films in which different materials and/or substances with a great impact on the shelf- life of the food can be transported. In addition, these materials can be nanostructured in form, which confers different or potentiated properties in contrast to their initial size. In this respect nano-films have been attractive in food packaging, since their improved functional properties, such as mechanical strength and barrier properties, have even antimicrobial and antioxidants activities to maintain the quality and extend shelf-life in various ways in different food applications (González-Reza et al. 2018; Sothornvit 2019).

Nano-films can be defined as a barrier formed by polymeric materials that contain at least one of their components in nanostructured form, and which fulfills a specific function of interaction with the environment and the food, as well as being able to transport substances of high biological impact. The use of nanocomposites for food packaging began in the 1990s, and the use of montmorillonite clay as a

M. L. Zambrano-Zaragoza · R. M. González-Reza
Laboratorio de Procesos de Transformación y Tecnologías Emergentes en Alimentos, Universidad Nacional Autónoma de México, FES-Cuautitlán, Cuautitlán Izcalli, Edo. de México, Mexico
e-mail: luz.zambrano@unam.mx

D. Quintanar-Guerrero
Laboratorio de Posgrado en Tecnología Farmacéutica, Universidad Nacional Autónoma de México, FES-Cuautitlán, Cuautitlán Izcalli, Edo. de Mexico, Mexico
e-mail: quintana@unam.mx

N. Mendoza-Muñoz (✉)
Laboratorio de Farmacia, Facultad de Ciencias Químicas, Universidad de Colima, Colima, Mexico
e-mail: nmendoza0@ucol.mx

U. Hebbar et al. (eds.), *Nano-food Engineering*, Food Engineering Series,
https://doi.org/10.1007/978-3-030-44552-2_10

nanocomponent has been utilized in a wide range of polymers, such as polyethylene, nylon, polyvinyl chloride, and starch. An example is the transparent nanocomposite coating and plastic film known as Durethan®, produced by Bayer® and that contains clay nanoparticles dispersed throughout the plastic (Majid et al. 2018).

The packaging of food with nano-films is a new-generation technology based on nanomaterials, which has become one of the most developed areas of nanotechnology and represents a radical alternative to conventional food packaging (Bradley et al. 2011). Nanomaterials are entities that are incorporated into traditional packaging materials, mainly in low fractions, to improve the properties of the original material (Huang et al. 2015b). Polymer nanomaterials can potentially be employed as food packaging materials and can be classified into four types according to purpose of use: (a) improved packaging; (b) active packaging; (c) intelligent packaging, and (d) degradable or compostable biopolymers.

In this regard depending on the application one wishes to have, or a combination of those mentioned previously it is necessary to carry out compatibility studies with the product to be conserved, considering aspects of its useful life, in addition to the existing regulations on nanomaterials in food packaging.

2 Materials in Nano-Films Preparations

Until recently, the majority of food packaging materials had been derived from petroleum due to its machinability, barrier properties, transparency, and availability. However, at present, the development of biodegradable and eco-friendly materials is necessary; these materials also possess a good barrier to O_2 and CO_2, and water vapor and increase the shelf-life of food. In this respect, nanotechnology has represented an excellent alternative to modified properties of packaging prepared with biodegradable polymers and inorganic fillers, exploiting the properties of nanometric size, such as chemical reactivity, superficial area, and uniform distribution on nano-film. The principal materials utilized in nano-film preparation are as follows: (1) biopolymers of vegetal and animal origin, such as polysaccharides, including starch, chitin, pectin, cellulose, and lignin, or proteins, including gelatin, carrageenan, agar, gluten, etc.; (2) biopolymers obtained from fermentation and biomass, such as xanthan gum, polyhydroxy alko-nates, poly-lactic acid, etc., these two classifications of materials obtained from sustainable and renewable sources; (3) synthetic biodegradable polymers, such as polyvinyl alcohol, polycaprolactone, poly-hydroxybutyrate, among others; (4) waxy lipid, solids, and liquid, such as beeswax, candelilla wax, candeuba wax, etc., used in the preparation of solid lipid nanoparticles, and (5) inorganic materials utilized as fillers and modified in their thermal, barrier, optical, and mechanical properties, such as clays (nano-montmorillonite and nano-bentonite) and inorganic compounds with antimicrobial properties, such as nano-TiO_2, nano-Ag, nano-Au, among others. Natural biopolymers are preferred to prepare nano-films due to that these are sensitive to pH since there are anionic polymers (pectin, carrageenan, and hyaluronic acid). cationic polymers (chitosan and polylysine), or amphiphile polymers (collagen and gelatin) or neutral polymers (dextran and pullulan). Hence, depending on the function to be developed in the nano-film, these can have a manner of control release depending on pH, temperature, ionic strength, food-lipid content, and humidity (Arora and Padua 2010; Hoseinnejad et al. 2018; Ghaderi et al. 2019; Huang et al. 2019).

Currently, and in order to obtain nano-films for food packaging with improved properties in relation to thermal resistance, and barrier and mechanical properties, as well as having better control of the release of antioxidant, antimicrobial, and flavor substances in active food packaging, composite materials are employed, including mixtures of polymeric nanostructures filled with inorganic clays or oxides. Also, intelligent food packaging requires nano-films capable of changing color by means of the effect of pH, temperature, enzymatic activity, or protein modifications (microbial growth and protein degradation). In the fabrication of nano-films, the principal structures utilized are nanofibers, nanowires, nanocrystals, nanocapsules, solid lipid nanoparticles, nanoparticles, or combinations of these.

Moreover, the materials employed to prepare nano-film matrices and substances with antimicrobial, antioxidant and flavoring properties are used in the new packaging products, preferably of natural origin. In this latter regard, we note the following: (1) essentials oils, such as clove, cinnamon, rosemary, lime, orange, and others, are used because these have antioxidant properties due to the effect of the polyphenols that these contain and their antimicrobial capacity due to a volatile component; (2) plant extracts that contain anthocyanins, quercetin, epicatechin, gallic acid, tannin, and other substances with a high antioxidant capacity; moreover, these components also possess antimicrobial properties; (3) microorganisms and enzymes, such as lactase and cellulase, and (4) conductive materials, such as polymers, magnetic nanoparticles, employed to detect pathogenic microorganisms and sporulation used at low concentrations (Helmy 2018; Vasile 2018). Thus, nanotechnology in nano-films renders it possible to obtain structures with functions at the molecular level.

3 Compatibility of Nano-Films and Food Products

Food composition is very varied and includes a multicomponent system in which moisture and lipid content exert an important effect on the functionality of food packaging, such as the release of active substances trapped in nano-films that are composed of different polymers, and probably other materials, such as lipid, oxides, or clays. The particle size of nanostructured materials comprises an important parameter in terms of chemical reactivity, ionic strength, therefore the potential interactions between the organic and inorganic molecules responsible for potential interaction with the food. In this respect, the surface charge determines the stability of the colloidal systems, their structure, and the possibility of remaining in the matrix of the nano-film. Measurement of the surface charge is represented by the zeta potential (ζ) and determines a large part of the interactions between the food surface and the nano-film. These interactions are affected by pH, temperature, ionic strength, humidity, fat content; the ζ, based on electrophoretic mobility, is expressed in mV of the superficial charge (Sharma et al. 2019b).

Foods are considered solid matrices due to their composition, structure, and nature. It should be considered that cellular structures in foods are composed of colloidal systems in which the organelles are embedded. These colloidal systems have superficial charges, viscosity, and size that interact to produce food deterioration. However, depending on the composition of the nano-films, food can be slowed in its deterioration. Also, the different interactions among the components of food directly affect the sensory changes associated with the food's structure, exerting an influence in the first place on color and texture, and finally, on flavor and aroma. The ζ of food complexes will be important in the superficial phenomena between the food and the nano-film materials (Hernández-Montelongo et al. 2016, Cano-Sarmiento et al. 2018, Souza et al. 2019). The characteristics of the materials used are central, since different polysaccharides, proteins, and lipids are utilized in the manufacture of nano-films. Moreover, because mechanical and thermal reinforcers or modifiers are also employed as nano-clays, oxides and other inorganic materials, some of these also have antioxidant and antimicrobial properties (Hong and Chen 2014). It is also necessary to consider the positive, negative, or non-polar superficial charges involved; the materials that contribute these properties must be selected according to the function that is required in the nano-film for food preservation. If any active substance (antioxidant, antimicrobial, enzymatic) is added it is necessary to consider how long the surface concentration should be sustained to increase shelf-life and to maintain the physicochemical, microbiological, and sensory characteristics of the packaged food (Deutsch et al. 2019; Ghaderi et al. 2019).

4 Preparation Methods Nano-Films Dispersions

One of the first steps after the selection of the materials that will form the nano-films is the dispersion of their components. The dispersion process of the nanomaterials in the solutions or dispersions of the support materials (generally biopolymers) should be of high quality, that is, the nanomaterials should be perfectly distributed, in order to generate a homogeneous film that maximizes its functional properties (mechanical, barrier, optics, etc.).

In the case of polymeric nanoparticles, liposomes, solid lipid nanoparticles, and similar systems that can be obtained in the form of aqueous nanosuspensions, their incorporation into polymeric support matrices is a relatively simple task, and generally low-shear mechanical agitation is sufficient to incorporate and homogenize the nano-systems efficiently.

However, in the case of inorganic nanomaterials obtained as nano-powders, their large surface area, high surface energy, and the electrostatic charge in the breaking up of agglomerates is the major challenge. The clustering of particles in polymer nano-films could reduce their functional characteristics; therefore, efficient integration should be the goal. Surface modification of inorganic particles could comprise a solution because it produces excellent integration and improved interface interaction between the inorganic filler and the polymer matrix (Hong and Chen 2014). Direct introduction of functional groups by chemical treatment or grafting of functional polymeric molecules onto the hydroxyl groups are two of the approaches to modify the surface of the particles.

On the other hand, high-shear homogenization by rotor-stator systems could aid in disrupting the aggregates and dispersing the nano fillers based on clays or metal oxides in the polymeric solution. There is a special case considered for nanoclays and for hydrophilic clays; mixing them in water with water-soluble polymers results in good dispersion and homogenous films. In this case, the degree of exfoliation can be improved by the use of conventional shear devices, such as extruders, mixers, ultrasonicators, ball milling, fluidizations, etc., (Vartiainen et al. 2010). Another special case is that of the dispersion of nano-organoclays; in this situation, it is necessary to add a polar activator prior to the dispersion. For complete dispersion, the first step is the wet-out; in this step, organoclay is mixed with the polymeric solution under vigorous stirring, and big agglomerates are broken up; however, at this point, dispersion is not complete. Complete dispersion is achieved by the addition of a polar activator; the function of polar activators is to disrupt the weak Van der Waal forces, which tend to hold the clay platelets together (Patel et al. 2006). Once these platelets are separated, the dispersion is complete, and the nano-film can be formed by means of the appropriate method.

5 Functionality of Nano-Films as Control Release Systems

An important function of the packaging lies in the protective effect and trapping substances that generate benefits in the foods, achieving this to increase the shelf-life of the food and in addition, for it to be an attractive commercial strategy that allows to conserve, for more prolonged time, the color, flavor, aroma, and texture of food, limiting the oxidation reactions that give rise to rancidity, discoloration, and off-flavors. The employment of the nanostructures that generally encapsulate the different components has for increasing the functionality of the substances added intentionally during the manufacture of nano-films, with the advantage of the benefits of the nanometric size, in terms of greater surface area, reactivity, and better distribution on the packaging, hence better control of the release of active substances.

In addition, with regard to the global trend, it has been necessary to take into consideration the manufacture of new materials with biodegradable and eco-friendly substances that meet the requirements for use in contact with food in accordance to FDA regulations. There are many functions of active packaging in food preservation, such as the use of components that comply with antioxidant, antimicrobial, and other protective functions. Moreover, nano-films can be utilized as electronic devices and sensors with changes in color during storage time and the production of reaction products that can be used as intelligent nano-packaging.

5.1 Antimicrobial Packaging

The food industry faces important challenges in order to manufacture products that meet the requirements that consumers demand today. One of its main challenges is to provide safe products free of pathogenic microorganisms and their toxins, as well as other microorganisms that cause food spoilage and deterioration of food, decreasing their shelf-life- nano-food packaging represent a way to help protect food from microbial recontamination, taking into consideration that, given that the concentration of oxygen and optimal conditions for the growth of pathogens and other microorganisms is found on the surface of the food, where they develop more easily, it is possible to take advantage of nano films loaded with antimicrobial substances, although there are also some polymers that have antimicrobial activity, such as chitosan. Also, there are nano-clays that are employed to improve the mechanical and thermal properties in the polymeric matrix, but that also have antimicrobial properties (Baysal and Çelik 2019). When nano-films are incorporated into natural or synthetic antimicrobial substances in a nanostructure, with the purpose of these remaining during storage time, with the antimicrobial being released into the headspace and fulfilling its function on the surface of the food, then the nano-film is denominated an active nano-film. An important effect that is utilized when nanocapsules, nanowires, nanotubes, nanoparticles, etc., loaded with active substances are incorporated into the polymeric support matrix, is that the antimicrobial substance remains for a longer time in the polymeric matrix and lower concentrations of these are necessary and with best effectiveness, such as biocides or antimicrobials, due the protective effect of the wall of the nanostructure.

Nano-film preparations use both organic and inorganic antimicrobial substances: the organic substances include natural compounds of vegetal or animal origin, and the synthetic substances, polymers such as chitosan and inorganic compounds mainly include nano-titanium, nano-selenium, nano-silver, nano-fold, nano-copper, and nano-magnesium. The mechanisms of action of the compound with the antimicrobial effect is different, and among the main ones, we find the following: (1) damage to the cell wall and inhibition of nucleic-acid generation; (2) modification and damage of the cell membrane, and (3) coagulation of the cellular cytoplasm.

Essentials oils are natural substances employed for their antimicrobial effect that is attributable to a compound such as monoterpenes, diterpenes, triterpenes and phenolic compounds (Huang et al. 2019). Many essential oils have been tested as antimicrobial agents, including clove essential oil nanostructured in an inclusion complex, and nanocellulose in a chitosan matrix was tested with respect to Gram-positive bacteria (*Bacillus subtilis* and *Staphylococcus aureus*) and Gram-negative bacteria (*Escherichia coli* and *Pseudomonas aeruginosa*), finding that the high eugenol in clove is responsible for the antimicrobial effect. Due to that it can denature proteins and react with cell-membrane phospholipids. Changing their permeability and inhibiting the growth of numerous microorganisms (Adel et al. 2019).

5.2 *Antioxidant Packaging*

Active and intelligent packaging with antioxidant capacities is employed in food packaging. Nano-films, then, comprise a viable alternative to incorporating, as fillers, nanoencapsulated active substances in a polymer matrix. Moreover some polymers have also been found with certain antioxidant properties, such as gelatin. The antioxidant capacity of nano-films is utilized as indicative of oxidation in meat and meat products, and substances with antioxidant capacities are able to adsorb free radicals by changing the color and indicating that the product has expired or has significantly decreased its shelf life. As an example of this, the *Lycium ruthenicum* Murr. extract, with its high content of anthocyanins that are sensitive to changes in pH, was used to determine the freshness of pork meat; the extract is sensitive to microbial growth and other oxidative reactions (Qin et al. 2019). Gallic acid has been used to maintain the oxidative stability of walnuts. The phenol was encapsulated in soft flour/polyethylene oxide nanofibers (Aydogdu et al. 2019). In general, polyphenolic natural extracts are the natural substances that possess the greatest antioxidant capacity, and these are incorporated into biodegradable and non-biodegradable polymers. Among the materials that have been considered due to their loading behavior are proteins such as gelatin, zein, and casein that, when prepared by different methods, such as nanoparticles and nanowires, and incorporated into a polymeric matrix, modified the mechanical properties and the release kinetic (Luecha et al. 2010; Barbosa-Pereira et al. 2014; Souza et al. 2019).

5.3 *Retention and Release of Flavors*

Aroma and flavor are the most important sensory characteristics of food and a criterion in the purchase decision. Recently, the benefits of nano-sized systems have been used to encapsulate and protect compounds that are volatile and sensitive to environmental conditions (light, temperature, O_2, pH ionic strength) such as aromas and flavors. The incorporation of nanostructures into the polymeric matrix this represents an alternative that contributes to the retention of aromas, mainly those prepared with hydrophilic polymers, such as proteins and polysaccharides, it being important to emphasize that these must the selection of the polymers for encapsulation depends to a great extent on the characteristics of the food. The biopolymers used form a network with the nanoparticles that fill and modify the diffusion properties of the film, adsorbing and retaining aromas or else releasing the flavor entrapped in the nanostructure and that acts as a barrier that allows a sustained release of flavor, thus preserving the characteristics of the food during storage (Saifullah et al. 2019).

5.4 *Nano-Films as Biosensors*

Another application area of nano-films is as nanosensors in intelligent packaging. The use of materials that react in function of food condition during storage, transport and commercialization, contributed to establish the shelf life of food, considering color changes by reaction with vapors, gases and products of microbial growth (Huang et al. 2015a). The nano-films containing substances that serve to inform to the consumer on the internal changes in the product respect to environment conditions. In general, nanosensors targeting improved the pathogen detection and many of these are development considering the electrically conducting polymers (Wesley et al. 2015).

6 Applications Methods of Nano-Films

6.1 *Electrospinning*

Electrospinning is a single method enabling the production of nanofibers from a polymer solution, dispersion, or melt. In this technique, a strong electrical field creates the stretching of the droplet of fluid in a single direction, thus originating a continuous line of polymer solution or melting polymer into a fine fiber followed by its deposition in a grounded collector (Wen et al. 2017; Vlachou et al. 2019). Figure 1 schematize the basic setup of the electrospinning, Layer-by-Layer, co-extrusion, and plasma techniques for the preapration of nano-films.

Electrospinning is a technique with enormous potential because it can be used to produce fibers with oriented nanomaterials inside the polymeric matrix, any solid particle (e.g., nanoclays, nanotubes, inorganic nanoparticles) or dispersion in the fluid can will aligned and stretched in the same flow direction. Electrospinning also permits the easy incorporation of bioactive compounds into nanofibers with high encapsulation efficiency.

The fabrication of nanofibers via electrospinning is affected by many different parameters. Among these, the solution properties are the most important. The latter these include the following: the polymer concentration; molecular weight and viscosity; the solution conductivity and relative volatility; volatility of the solvent; surface tension, and the dielectric constant (Vlachou et al. 2019).

Nanofibers obtained by electrospinning can be used directly into the food or included as a mat into the supporting materials of the nano-films, food-grade polymers such as proteins and polysaccharides are examples of supporting materials.

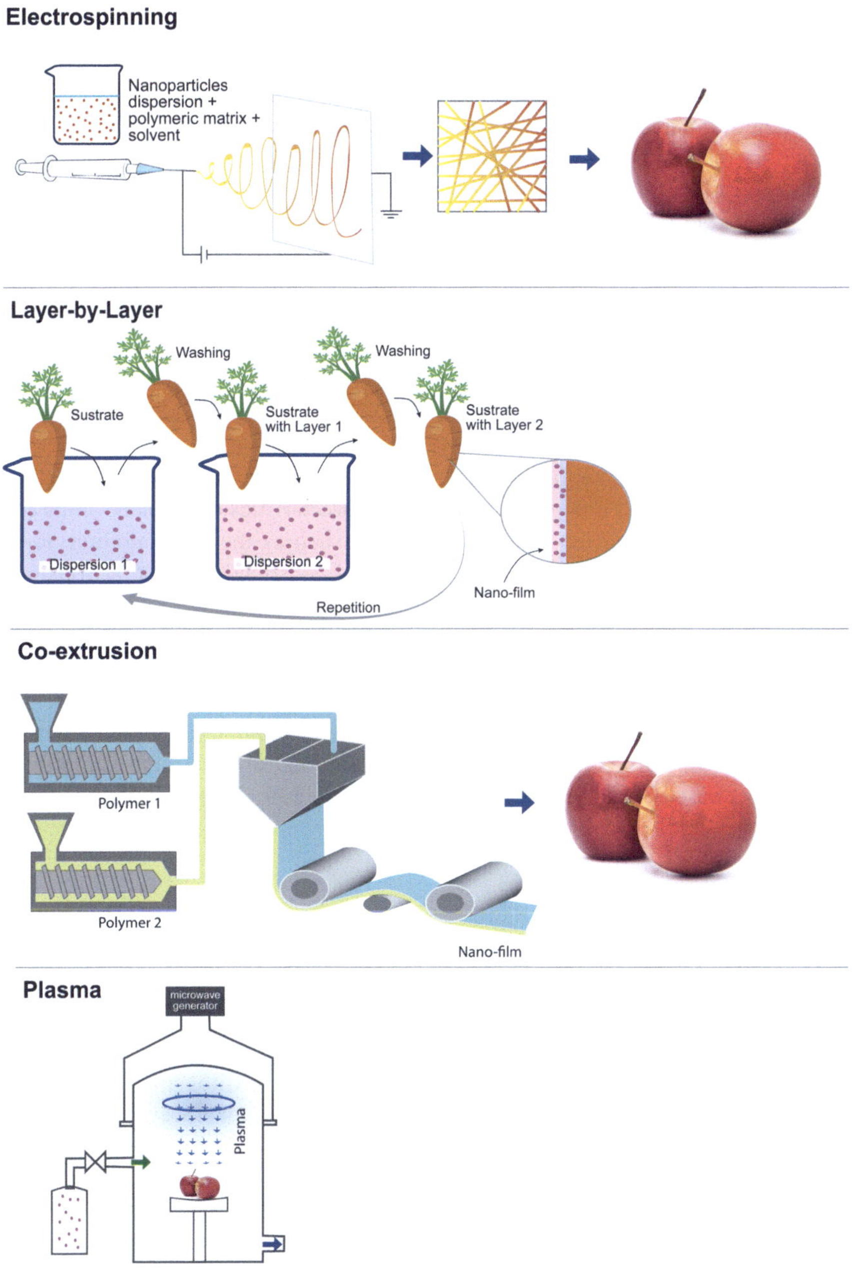

Fig. 1 Schematization of the main techniques for the preparation of nano-films for food applications

6.2 *Layer-by-Layer*

Layer-by-Layer (LbL) assembly is a cyclical process that includes the sequential adsorption of complementary molecules on a substrate surface, driven by multiple interactions involving electrostatic and/or nonelectrostatic interactions (Zhang et al. 2018). Typically, a charged material is adsorbed onto a substrate, and after washing, an oppositely charged material is adsorbed on top of the first layer. This constitutes a single bilayer with a thickness generally on the order of nanometers, and the deposition process can then be repeated until a multilayer film of desired thickness has been assembled (Richardson et al. 2015). Figure 1 schematize the process of films obtention by LbL approach.

The LbL assembly method can produce films on various forms of substrates, and includes different techniques such as immersion, spinning, spraying, the electromagnetic approach, and fluidic assembly. LbL has numerous advantages over other preparation methods of nano-films; it includes fine control of composition, morphology, thickness, and topography. The final characteristics of the nano-films can be adapted by adjusting the solution properties such as the concentration, ionic strength, and the pH, and process parameters such as washing time, interaction time, temperature, and drying conditions.

The LbL method possesses the advantage that the nano-film is fabricated directly on the surface-of-interest, for example, fruits, vegetables or other food substrates (Medeiros et al. 2012; Arnon-Rips and Poverenov 2018), eliminating the need to produce the film outside of the food and its later application. Charged biopolymers such as chitosan, carboxymethyl cellulose, pectin, and alginate are examples of film-formers based on electrostatic forces used in the LbL formation of nano-films intended for food preservation.

6.3 *Co-Extrusion*

Co-extrusion is the production procedure in which a plastic film containing two or more distinct plastic layers is formed without requiring any intermediate steps (Selke and Hernandez 2001). The melted polymers are fed together carefully to produce a layered melt, which is then processed in conventional ways to produce a plastic film or sheet (Lagarón 2011) (Fig. 1). Two modalities are possible: *feed block co-extrusion*, where the streams are merged into a single laminar melt flow in a feedblock that is positioned immediately upstream of the extrusion die, and *die co-extrusion*, which employs a complex die construction in which separate melt-path manifolds are arranged to merge at a point close to the die exit (Ebnesajjad 2016).

The co-extrusion method is one of the easiest ways to obtain economical and functional films. Reinforced or functional nano-films by co-extrusion can be produced incorporating nanoparticles (fillers) in one or more of the melted polymers; once they are cold, the fillers are dispersed throughout the polymer forming the composite film. Six nano-fillers constitute the panel of nanoparticles under the scope of the NanoSafePack project, including clay nanoparticles, metal oxide nanoparticles, and promising applications such as zeolites and calcium carbonate on the nanometer scale (Fito 2013). The inclusion of fillers enhances the rigidity, barrier properties, and clarity, or provides functionality to the film. Recently, the use of nanoclays, a type of layered silicate with nanometric thickness and a diameter of 50–200 nm, as fillers in food-grade nano-films, has gained popularity because they can be obtained from nature and entertain a low cost.

6.4 Plasm

Recently, plasm deposition technology has been applied to enrich, with functional materials, the top of films. These processes usually include two steps: i) an initial phase, carried out either with low- or atmospheric-pressure plasma, aimed to enrich the topmost surface of the substrates with functional groups (e.g., grafting oxygen- or nitrogen-containing reactive groups or depositing functionalized films), and ii) the process continues with the adsorption or the covalent binding of the bioactive molecule to the substrate surface (Palumbo et al. 2015). At present, a strategy is preferred that involves one-step direct deposition of the plasma at atmospheric pressure, because this limits damage to the deposited structures, preserving the functionality of the active molecules (Heyse et al. 2011). Direct plasma deposition at atmospheric pressure allows the efficient spraying of low vapor- pressure pure liquids, salts solutions, or even nanoparticle suspensions (Palumbo et al. 2018). During the application of the plasma, some aspects of the feeding solution should be considered, for example, the bioactive molecule prone to be deposited should be dispersed in water or a suitable buffer solution, and also, additional gas or vapor can be added. Figure 1 schematizes the basic setup of direct plasma deposition.

7 Properties Nano-Films

7.1 Barrier Properties

Barrier properties include permeability of gases (such as O_2, CO_2, and N_2), water vapor, aroma compounds, and light, and these acquire vital importance during the design of advanced food packinging. The introduction of polymer-based films, particularly polysaccharide-based films, as packaging materials has been increasing over the last few decades because of their biodegradability, innocuousness, and low cost. However, due to the hydrophilic nature, they present inherent permeability to low-molecular-weight substances, including permanent gases, water, organic vapors, aromas, and flavors (Siracusa 2012). The incorporation of nanomaterials into films can improve the barrier performance of films. The inclusion of inorganic materials such as nanoclays, in particular, montmorillonite, laponite, hectorite, and cloisite (Luecha et al. 2010; Nagarajan et al. 2014; Qi et al. 2016; Wu et al. 2018), has proved to increase the water and gas-barrier properties of nanocomposite films due to the presence of dispersed silicate layers in an organized manner with large aspect ratios in the polymer matrix (Nagarajan et al. 2014). Also, organic polymers such as PolyLactic Acid (PLA), poly(ε-caprolactone), and chitosan in the form of nanoparticles have been tested and showed to reinforce barrier properties.

7.2 Mechanical Properties

The mechanical properties of nano-films include the elastic modulus, tensile strength, and the percentage of elongation as the most important parameters. The adequate mechanical strength of films is necessary to protect the integrity of packaging throughout manipulation. Despite the use of natural polymers in the formation of food-grade films, there are some strong limitations: for example, starch-based films present poor tensile properties and high water-vapor permeability when compared to synthetic conventional films (Souza et al. 2012). The low mechanical property of composite films, their hydrophilic nature, and their sensitivity to moisture content can be reinforced by cross-linking with nanoclays, carbon nanotubes, or other inorganic nanomaterials. However, in practice, the results of the increase in the mechanical properties of nano-films are not highly predictable. It has been observed that reinforcement with nanoclays works well at a specific range of percentage of use; if the proportion of nanoclays is high, films can lose their integrity (Souza et al. 2012; Nagarajan et al. 2014). High rigidity in films in the presence of nanoclay can be mediated by the presence of strong hydrogen interactions and electrostatic interaction between matrix and clay.

7.3 Optical Properties

The optical properties of the films not only correspond to properties of appearance, such as transparency and color, but also include the transmission of UV light, an optical property that entertains importance in oxidation or catalytic processes within the food (Sanchez-Garcia and Lagaron 2010) showed that nano-films composed of Poly(Lactic Acid) (PLA) PolyHydroxyButyrate-*co*-Valerate (PHBV), and PolyCaproLactone (PCL) in which an organo-modified mica-based clay grade was included, enhanced barrier properties to UV light. At 5 wt% of clay, light transmission decreased by up to 90% at 300 nm in the UV region. Transparency is often affected by the incorporation of clays or other types of nanoparticles; Luminosity, L*, which forms the CIELAB color system, can be utilized as a parameter for measuring clarity. The L* parameter is expected to decrease with an increasing nanoparticle content (Vicente et al. 2011).

7.4 Functional Properties

Antimicrobial and antioxidant activity are two of the most sought-after functions in the design of nano-films. The incorporation of antimicrobial substances into the packaging system delays or prevents microbial growth during postharvest, storage, and the retail display of the food. Nanomaterials can provide active functions for food packaging when they are incorporated into nano-films. Special emphasis has been given to metal-oxide nanoparticles (silver, zinc oxide, titanium oxide, and magnesium oxide), which can be used in active packaging to inhibit the growth of microorganisms (Azeredo et al. 2012). Polymeric nanoparticles (nanospheres and nanocapsules) and other colloidal carriers embedded in natural film-forming polymers can allow the formation of nano-films with functional properties, despite that some nanoparticle-forming polymers present antimicrobial activity themselves, such as chitosan. The encapsulation of plant extracts (essential oils, saponins, vegetal proteins, lipids, etc.) with antimicrobial activity, such as algicides, bactericides, and fungicides, is currently a promising resource to prevent the growth of microorganisms inside the package (Azeredo et al. 2012; Leceta et al. 2013), In the same manner, plant extracts rich in polyphenolic and proanthocyanidin compounds possess potential antioxidant properties that inhibit the lipid oxidation of food ingredients (Perumalla and Hettiarachchy 2011).

7.5 Control Release

Active compounds such as antioxidants, antimicrobials, vitamins, essential oil and plant extracts are incorporated into food formulations to improve performance; however, sometimes these active compounds are consumed in reactions. As a result, the protection stops, and food quality diminishes. Controlled release technology can overcome this limitation, because if the zero-order release kinetic is reached, the concentration of active compounds can be maintained constant in the package to provide sustained food protection.

The incorporation of active compounds directly into a polymeric film structure can be an alternative to achieve protection in food systems; however, rapid release by diffusion into the food can render the failure of the strategies to prolong food shelf-life (Mastromatteo et al. 2010). Nanoencapsulation of the active compounds in polymeric nanoparticles, liposomes, solid lipid nanoparticles, nanostructured lipid carriers, and later inclusion in the polymeric matrix of nano-films allows a protection effect of the active compounds against degradation and also permits the package to release these active compounds in a controlled manner to improve food safety and quality and, in combination, a superior effect can be achieved (Chen et al. 2018).

8 Recent Development in Nano-Films Packaging

Among the new techniques in the packaging of food, we find active packaging, smart packaging, and bioactive packaging, which involve an intentional interaction with food or its environment and which an influence on consumer health have been the main innovations in the field of packaging technology. These new techniques aid to prolong the useful life of food, improving or maintaining its quality, providing indications, and regulating the freshness of the food product.

The new technologies of food packaging, in addition to the basic function of containment, increase the margin of quality and the safety of food. The innovative techniques of food packaging help meet the demands of the entire food-supply chain when preparing for people's lifestyle. In this context, nano-films of natural proteins have been developed. An example of this is biodegradable zein nano-film compounds reinforced with chitosan nanoparticles and cinnamon essential oil (Vahedikia et al. 2019) Here it was found that the addition of nanostructures has an adequate antimicrobial potential to be applied against Gram-positive bacteria (*Staphylococcus aureus*) and Gram-negative bacteria (*Escherichia coli*), in addition to improving the physical, mechanical, and structural attributes of the obtained nano-films.

The functional properties of biopolymer nano-films have been improved by incorporating mesoporous silica nanoparticles loaded with curcurmin for food packaging (Wu et al. 2019b). The improved properties of this nano-film include controlled release, increased mechanical properties, and excellent antimicrobial activity against *S. aureus* and *E. coli*.

Regarding the nanofibers contained in nano-films, one of the innovative developments is a biomechanical process for the production of cellulose nanofibers as a more ecological and energy-saving solution and energy's (Liu et al. 2019). In this respect, enzymatic pretreatment of the fibers saved around 60% of the electrical energy in mechanical refining subsequent to the production of cellulose nanofibers, in addition to the previous treatment with enzymes. Mechanical resistance is not significantly affected, and the thermal stability can remain elevated.

Nanocompound films based on k-carrageenan reinforced with cellulose nanocrystals (Kassab et al. 2019) have been developed, revealing a Young modulus, tensile strength and greatly improved tenacity. In this regard, films of nanocomposites that are ecologically strong and flexible on terms of the environment were obtained. Another study of-interest is that of chitosan/ε-polylysine bionanocomposite self-assembly films that were evaluated *in situ* (Wu et al. 2019a), highlighting that bionanocompound films showed excellent antimicrobial efficacy against *E. coli* and *S. aureus* due to increasing the proportion of ε-polylysine. In addition, ε-polylysine exhibited sustained and slow-release behavior that was related to the concentration of sodium tripolyphosphate, and this could lead to an antimicrobial activity of long-term films.

9 Regulation in Nano-Films for Food Applications

Nanotechnology has shown great potential for applications in different knowledge areas, presenting products on the market that are identified as containing declared nanomaterials. Documents on emerging nanotechnologies state that food and beverages were found (Taylor 2008), however, the regulations of nanotechnology for food-industry applications are limited; thus, the global market faces many obstacles in the regulation of these products. In the United States, several federal agencies oversee products based on nanotechnology. In 2006, the Food and Drug Administration (FDA) (FDA 2011) formed the Working Group on Nanotechnology, which is responsible for the development of regulatory approaches for nanotechnology-based products that will ensure safety and efficacy while facilitating beneficial technological innovation. The group issued a report recommending the evaluation of the orientation of the agency that could indicate which data are those that the manufacturer is required to inform the FDA in terms of the nanoproducts (Huang et al. 2015b).

The Institute of Food Science and Technology (IFST) has recommended that nanomaterials be treated as new and potentially harmful materials until proven safe. The European Food Safety Authority (EFSA) mentions that an in-depth risk assessment is needed for deliberately designed substances for the size of particles that exhibit functional, physical, and chemical properties that differ significantly from those on a larger scale until more information is available in terms of nanoscience and nanotechnologies (EFSA 2009). The EFSA presents guidance information on the assessment of the potential risk of nanomaterials for food-related uses, which provides information on the requirements for the detection, identification, and characterization of nanomaterials (EFSA 2011). Studies on toxicity *in vitro* and *in vivo* are required if it cannot be demonstrated that there is no nanomaterial migration after the formulation, and that the nanomaterial is not transformed before or during digestion. In the specific case of nano-films, plastic packaging nanomaterials require a pre-marketing permit with a safety assessment, according to the chemical and toxicity data presented by the manufacturer based on the data requirements guidelines (Bradley et al. 2011).

European legislation controls the compounds that can be used in the manufacture of containers destined to contain food (Regulation EC 1935/2004), as well as the limitations of plastic additives in terms of limits in their migration into food products; in principle, these are also applicable to nanomaterials (European Commission 2004). Current legislation does not distinguish materials produced by nanotechnology from those usually developed by standard manufacturing methods. Nanomaterials used for the manufacture of nano-packaging materials are not evaluated as new chemical products. Therefore, the products of nanotechnology are treated by means of a combination of the general food legislation of the European Union and more specific controls on specific substances. However, due to possible differences in physicochemical or biological properties, the maximum safe migration levels determined for the macrocomponents will not be applicable in the case of their nano-equivalent material. In addition, not only the number of elements must be considered when using nanomaterials, but also the specific migration of the particles must also be considered. Due to its highly developed surface, specific toxicological problems could arise, and the migration of nanomaterials, which depends on their structure and size, should be quantified instead of merely determined with regard to their constituent elements (Huang et al. 2015a).

10 Trends in Nanofilms on Food Packaging Applications

As has been described previouslynano-films in food packaging can play an important role in increasing a product's shelf-life. There are several aspects to consider for their design, and these mark their trends, which can be classified in the following three ways: (a) nano-film components in packaging; (b) functionality of the film in the packaging, and (c) controlled release of the active substances from the nano-film to the food.

Nano-film components in packaging: A current trend in food technology is to offer natural products with "healthy" labels and with the best food quality and safety. Thus, consumer preferences are linked to aspects, such as environmentally friendly production and the food's contribution to decrease CO_2 emissions, reduction of energy-inputs for production, transport, and storage, degradable packaging, biodegradable, biocompatible, or renewable components, edibility, antioxidant benefits, etc. (Silvestre et al. 2011). The use of natural ingredients (antimicrobials, antifungals, antioxidants, and nutrients, etc.) or polymers (chitosan, starch, cellulose, carrageenan, pectin, etc.) rather than those based on artificial or chemical ingredients, in particular petroleum-based, is a current and emerging issue in the development of nano-films (Ferreira et al. 2018; Ogunsona et al. 2018; Aydogdu et al. 2019; Jiang et al. 2019; Mujtaba et al. 2019; Souza et al. 2019; Theodoratou et al. 2019).

Functionality of the film in the packaging: This aspect is related to the capacity of the nano-film to interact with the packaging and to work adequately. Thus, the nano-film needs to comply with several characteristics, such as good adherence, low density, non-abrasiveness, non-toxicity, biodegradable, low combustibility, economic, low water absorption, remaining inside the packaging, good barrier ability to water vapor, resistance to food- component attack, and having the best mechanical properties to adapt to the use conditions of the packaging, for example, in folding or reusable packaging, etc. (Ferreira et al. 2018; Ogunsona et al. 2018; Wu et al. 2018; Jiang et al. 2019; Qin et al. 2019; Theodoratou et al. 2019). It is important to point out that some polymers that have poor properties as bulk materials can function better in composites at nanometric size (e.g., cellulose) (Sharma et al. 2019a).

Furthermore, the use of novel intelligent/smart nano-films can monitor the condition of packaged food or the environmental conditions and can also indicate fraudulent imitation (Rol et al. 2019).

Controlled releaseof active substances from nano-film into food: In general, nano-films contain active substances that need to be released adequately into the food in a sufficient amount to guarantee the nano-film's function, for example, to control or delay microbial growth (e.g., silver nanoparticles), UV blockers (e.g., titanium dioxide, silicon oxide, or magnesium oxide), and lipid oxidation at the food surface. A promising technology to encapsulate active substances into natural and synthetic polymers is electrospinning. This is based on the grounds of the stretching of the polymer solution and active substances under the effect of an electric field (Munhuweyi et al. 2018; Aydogdu et al. 2019). The system obtained are nanofibers that can be incorporated into macromolecular dispersions to prepare active and intelligent food-packaging nano-films (Qin et al. 2019).

11 Conclusions

Nanotechnology applied in the development of nano-films represent a viable alternative to incorporate natural ingredient to be employed in food preservation. Considering that these give the possibility of developing biodegradable packaging, eco-friendly and with the advantage of these nano-films serve as vectors for releasing of substances that can interact with the food through the headspace with the release of volatile compounds to act as antimicrobial or substances that released from the container to be able to act on the surface of food product avoiding the oxidations reactions. The use of fillers based on nano-clays has also made it possible to develop of materials with greater mechanical and thermal resistance that also have the possibility of having a better control release of active substances used to increase the shelf life of food.

References

Adel AM, Ibrahim AA, El-Shafei AM, Al-Shemy MT (2019) Inclusion complex of clove oil with chitosan/β-cyclodextrin citrate/oxidized nanocellulose biocomposite for active food packaging. Food Packag Shelf Life 20:100307. https://doi.org/10.1016/j.fpsl.2019.100307

Arnon-Rips H, Poverenov E (2018) Improving food products' quality and storability by using Layer by Layer edible coatings. Trends Food Sci Technol 75:81–92. https://doi.org/10.1016/j.tifs.2018.03.003

Arora A, Padua GW (2010) Review: nanocomposites in food packaging. J Food Sci 75:43–49. https://doi.org/10.1111/j.1750-3841.2009.01456.x

Aydogdu A, Yildiz E, Aydogdu Y, Sumnu G, Sahin S, Ayhan Z (2019) Enhancing oxidative stability of walnuts by using gallic acid loaded lentil flour based electrospun nanofibers as active packaging material. Food Hydrocoll 95:245–255. https://doi.org/10.1016/j.foodhyd.2019.04.020

Barbosa-Pereira L, Angulo I, Lagarón JM, Paseiro-Losada P, Cruz JM (2014) Development of new active packaging films containing bioactive nanocomposites. Innov Food Sci Emerg Technol 26:310–318. https://doi.org/10.1016/J.IFSET.2014.06.002

Baysal G, Çelik BY (2019) Synthesis and characterization of antibacterial bio-nano films for food packaging. J Environ Sci Heal – Part B Pestic Food Contam Agric Wastes 54:79–88. https://doi.org/10.1080/03601234.2018.1530546

Bradley EL, Castle L, Chaudhry Q (2011) Applications of nanomaterials in food packaging with a consideration of opportunities for developing countries. Trends Food Sci Technol 22:604–610. https://doi.org/10.1016/j.tifs.2011.01.002

Cano-Sarmiento C, Téllez-Medina DI, Viveros-Contreras R, Cornejo-Mazón M, Figueroa-Hernández CY, García-Armenta E, Alamilla-Beltrán L, García HS, Gutiérrez-López GF (2018) Zeta potential of food matrices. Food Eng Rev 10:113–138. https://doi.org/10.1007/s12393-018-9176-z

Chen X, Chen M, Xu C, Yam KL (2018) Critical review of controlled release packaging to improve food safety and quality. Crit Rev Food Sci Nutr 8398:1–14. https://doi.org/10.1080/10408398.2018.1453778

De Azeredo HMC, Agroindustry ET, Mesquita RDS (2012) Antimicrobial nanostructures in food packaging. Trends Food Sci Technol 30:56–69. https://doi.org/10.1016/j.tifs.2012.11.006

Deutsch M, Schoo C, Knoll M (2019) Time-dependent switching of temperature-sensitive nanofilm processors. Sensors Actuators A Phys 289:19–25. https://doi.org/10.1016/j.sna.2019.02.023

Ebnesajjad S (2016) Extrusion. In: Ebnesajjad S (ed) Fluoroplastics, 2nd edn. Elsevier, pp 282–347

EFSA (2009) The Potential Risks Arising from Nanoscience and Nanotechnologies on Food and Feed Safety- Scientific Opinion of the Scientific Committee

EFSA (2011) Guidance on risk assessment of the application of nanoscience and nanotechnologies in the food and feed chain. EFSA J 9:1–36. https://doi.org/10.2903/j.efsa.2018.5327

European Commission (2004) Regulation (EC) No 1935/2004 of the European Parliament and of the Council of 27 October 2004 on materials and articles intended to come into contact with food and repealing Directives 80/590/EEC and 89/109/EEC. Off J Eur Union 338:4–17

FDA (2011) Food Ingredients & Packaging

Ferreira FV, Dufresne A, Pinheiro IF, Souza DHS, Gouveia RF, Mei LHI, Lona LMF (2018) How do cellulose nanocrystals affect the overall properties of biodegradable polymer nanocomposites: a comprehensive review. Eur Polym J 108:274–285. https://doi.org/10.1016/j.eurpolymj.2018.08.045

Fito C (2013) Development of a best practices guide for the safe handling and use of nanoparticles in packaging industries D1. 1. Report on the nanoclays and metal oxide nanoparticles employed on the packaging industry

Ghaderi J, Hosseini SF, Keyvani N, Gómez-Guillén MC (2019) Polymer blending effects on the physicochemical and structural features of the chitosan/poly(vinyl alcohol)/fish gelatin ternary biodegradable films. Food Hydrocoll 95:122–132. https://doi.org/10.1016/j.foodhyd.2019.04.021

González-Reza R, García-Betanzos C, Sánchez-Valdes L, Quintanar-Guerrero D, Cornejo-Villegas M, Zambrano-Zaragoza M (2018) The functionalization of nanostructures and their potential applications in edible coatings. Coatings 8:160. https://doi.org/10.3390/coatings8050160

Helmy E (2018) Nano-biotechnology breakthrough and food-packing industry- a review. Microb Biosyst 1:50–69. https://doi.org/10.21608/mb.2016.5257

Hernández-Montelongo J, Nascimento VF, Murillo D, Taketa TB, Sahoo P, De Souza AA, Beppu MM, Cotta MA (2016) Nanofilms of hyaluronan/chitosan assembled layer-by-layer: an antibacterial surface for Xylella fastidiosa. Carbohydr Polym 136:1–11. https://doi.org/10.1016/j.carbpol.2015.08.076

Heyse P, Van Hoeck A, Roeffaers MBJ, Raffin JP, Steinbüchel A, Stöveken T, Lammertyn J, Verboven P, Jacobs PA, Hofkens J, Paulussen S, Sels BF (2011) Exploration of atmospheric pressure plasma nanofilm technology for straightforward bio-active coating deposition:

enzymes, plasmas and polymers, an elegant synergy. Plasma Process Polym 8:965–974. https://doi.org/10.1002/ppap.201000170
Hong RY, Chen Q (2014) Dispersion of inorganic nanoparticles in polymer matrices: challenges and solutions. In: Susheell K, Yuvaraj H (eds) Organic-inorganic hybrid nanomaterials. Springer International Publishing, pp 1–38
Hoseinnejad M, Jafari SM, Katouzian I (2018) Inorganic and metal nanoparticles and their antimicrobial activity in food packaging applications. Crit Rev Microbiol 44:161–181. https://doi.org/10.1080/1040841X.2017.1332001
Huang J-Y, Li X, Zhou W (2015a) Safety assessment of nanocomposite for food packaging application. Trends Food Sci Technol 45:187–199. https://doi.org/10.1016/j.tifs.2015.07.002
Huang JY, Li X, Zhou W (2015b) Safety assessment of nanocomposite for food packaging application. Trends Food Sci Technol 45:187–199. https://doi.org/10.1016/j.tifs.2015.07.002
Huang T, Qian Y, Wei J, Zhou C (2019) Polymeric antimicrobial food packaging and its applications. Polymers (Basel) 11:560. https://doi.org/10.3390/polym11030560
juan Jiang S, Zhang T, Song Y, Qian F, Tuo Y, Mu G (2019) Mechanical properties of whey protein concentrate based film improved by the coexistence of nanocrystalline cellulose and transglutaminase. Int J Biol Macromol 126:1266–1272. https://doi.org/10.1016/j.ijbiomac.2018.12.254
Kassab Z, Aziz F, Hannache H, Ben Youcef H, El Achaby M (2019) Improved mechanical properties of k-carrageenan-based nanocomposite films reinforced with cellulose nanocrystals. Int J Biol Macromol 123:1248–1256. https://doi.org/10.1016/j.ijbiomac.2018.12.030
Lagarón JM (2011) Multifunctional and nanoreinforced polymers for food packaging. In: Lagarón J-M (ed) Multifunctional and nanoreinforced polymers for food packaging. Woodhead publishing, pp 1–28
Leceta I, Guerrero P, De Caba K (2013) Functional properties of chitosan-based films. Carbohydr Polym 93:339–346. https://doi.org/10.1016/j.carbpol.2012.04.031
Liu X, Jiang Y, Song X, Qin C, Wang S, Li K (2019) A bio-mechanical process for cellulose nanofiber production – towards a greener and energy conservation solution. Carbohydr Polym 208:191–199. https://doi.org/10.1016/j.carbpol.2018.12.071
Luecha J, Sozer N, Kokini JL (2010) Synthesis and properties of corn zein/montmorillonite nanocomposite films. J Mater Sci 45:3529–3537. https://doi.org/10.1007/s10853-010-4395-6
Majid I, Ahmad Nayik G, Mohammad Dar S, Nanda V (2018) Novel food packaging technologies: innovations and future prospective. J Saudi Soc Agric Sci 17:454–462. https://doi.org/10.1016/j.jssas.2016.11.003
Mastromatteo M, Mastromatteo M, Conte A, Del A (2010) Advances in controlled release devices for food packaging applications. Trends Food Sci Technol 21:591–598. https://doi.org/10.1016/j.tifs.2010.07.010
Medeiros BGDS, Pinheiro AC, Teixeira JA, Vicente AA, Carneiro-da-Cunha MG (2012) Polysaccharide/protein Nanomultilayer coatings: construction, characterization and evaluation of their effect on 'Rocha' pear (Pyrus communis L.) shelf-life. Food Bioprocess Technol 5:2435–2445. https://doi.org/10.1007/s11947-010-0508-0
Mujtaba M, Morsi RE, Kerch G, Elsabee MZ, Kaya M, Labidi J, Khawar KM (2019) Current advancements in chitosan-based film production for food technology; a review. Int J Biol Macromol 121:889–904. https://doi.org/10.1016/j.ijbiomac.2018.10.109
Munhuweyi K, Caleb OJ, van Reenen AJ, Opara UL (2018) Physical and antifungal properties of β-cyclodextrin microcapsules and nanofibre films containing cinnamon and oregano essential oils. LWT – Food Sci Technol 87:413–422. https://doi.org/10.1016/j.lwt.2017.09.012
Nagarajan M, Benjakul S, Prodpran T, Songtipya P (2014) Characteristics of bio-nanocomposite films from tilapia skin gelatin incorporated with hydrophilic and hydrophobic nanoclays. J Food Eng 143:195–204. https://doi.org/10.1016/j.jfoodeng.2014.06.038
Ogunsona E, Ojogbo E, Mekonnen T (2018) Advanced material applications of starch and its derivatives. Eur Polym J 108:570–581. https://doi.org/10.1016/j.eurpolymj.2018.09.039
Palumbo F, Camporeale G, Yang YW, Wu JS, Sardella E, Dilecce G, Calvano CD, Quintieri L, Caputo L, Baruzzi F, Favia P (2015) Direct plasma deposition of lysozyme-embedded

bio-composite thin films. Plasma Process Polym 12:1302–1310. https://doi.org/10.1002/ppap.201500039
Palumbo F, Treglia A, Lo Porto C, Fracassi F, Baruzzi F, Frache G, El Assad D, Pistillo BR, Favia P (2018) Plasma-deposited nanocapsules containing coatings for drug delivery applications. ACS Appl Mater Interfaces 10:35516–35525. https://doi.org/10.1021/acsami.8b11504
Patel H, Somani R, Bajaj H, Jasra R (2006) Nanoclays for polymer nanocomposites, paints, inks, greases and cosmetics formulations, drug delivery vehicle and waste water treatment. Bull Mater Sci 29:133–145
Perumalla AVS, Hettiarachchy NS (2011) Green tea and grape seed extracts — Potential applications in food safety and quality. Food Res Int 44:827–839. https://doi.org/10.1016/j.foodres.2011.01.022
Qi G, Li N, Sun XS, Shi YC, Wang D (2016) Effects of glycerol and nanoclay on physiochemical properties of camelina gum-based films. Carbohydr Polym 152:747–754. https://doi.org/10.1016/j.carbpol.2016.07.068
Qin Y, Liu Y, Yuan L, Yong H, Liu J (2019) Preparation and characterization of antioxidant, antimicrobial and pH-sensitive films based on chitosan, silver nanoparticles and purple corn extract. Food Hydrocoll 96:102–111. https://doi.org/10.1016/j.foodhyd.2019.05.017
Richardson JJ, Bjornmalm M, Caruso F (2015) Technology-driven layer-by-layer assembly of nanofilms. Science 348(80):aaa2491–aaa2491. https://doi.org/10.1126/science.aaa2491
Rol F, Belgacem MN, Gandini A, Bras J (2019) Recent advances in surface-modified cellulose nanofibrils. Prog Polym Sci 88:241–264. https://doi.org/10.1016/j.progpolymsci.2018.09.002
Saifullah M, Shishir MRI, Ferdowsi R, Tanver Rahman MR, Van Vuong Q (2019) Micro and nano encapsulation, retention and controlled release of flavor and aroma compounds: a critical review. Trends Food Sci Technol 86:230–251. https://doi.org/10.1016/j.tifs.2019.02.030
Sanchez-Garcia MD, Lagaron JM (2010) On the use of plant cellulose nanowhiskers to enhance the barrier properties of polylactic acid. Cellulose 17:987–1004. https://doi.org/10.1007/s10570-010-9430-x
Selke SE, Hernandez RJ (2001) Packaging: polymers in flexible packaging. Encycl Mater Sci Technol:6652–6656. https://doi.org/10.1016/B0-08-043152-6/01176-1
Sharma A, Thakur M, Bhattacharya M, Mandal T, Goswami S (2019a) Commercial application of cellulose nano-composites – a review. Biotechnol Reports 21:e00316. https://doi.org/10.1016/j.btre.2019.e00316
Sharma S, Jaiswal S, Duffy B, Jaiswal A (2019b) Nanostructured materials for food applications: spectroscopy, microscopy and physical properties. Bioengineering 6:26. https://doi.org/10.3390/bioengineering6010026
Silvestre C, Duraccio D, Cimmino S (2011) Food packaging based on polymer nanomaterials. Prog Polym Sci 36:1766–1782. https://doi.org/10.1016/j.progpolymsci.2011.02.003
Siracusa V (2012) Food packaging permeability behaviour: a report. Int J Polym Sci 2012:1–11. https://doi.org/10.1155/2012/302029
Sothornvit R (2019) Nanostructured materials for food packaging systems: new functional properties. Curr Opin Food Sci 25:82–87. https://doi.org/10.1016/j.cofs.2019.03.001
Souza AC, Benze R, Ferrão ES, Ditchfield C, Coelho ACV, Tadini CC (2012) Cassava starch biodegradable films: influence of glycerol and clay nanoparticles content on tensile and barrier properties and glass transition temperature. LWT – Food Sci Technol 46:110–117. https://doi.org/10.1016/J.LWT.2011.10.018
Souza VGL, Pires JRA, Vieira ÉT, Coelhoso IM, Duarte MP, Fernando AL (2019) Activity of chitosan-montmorillonite bionanocomposites incorporated with rosemary essential oil: from in vitro assays to application in fresh poultry meat. Food Hydrocoll 89:241–252. https://doi.org/10.1016/j.foodhyd.2018.10.049
Taylor MR (2008) Assuring the safety of nanomaterials in food packaging: the regulatory process and key issues. Woodrow Wilson International Center for Scholars
Theodoratou A, Costa L, Bonnet L, Blanc C, Lapinte V, Etienne P, Milhiet PE, Robin JJ, Oberdisse J, Chopineau J, Aubert-Pouëssel A (2019) Physico-chemical properties and surface char-

acterization of renewable hybrid nanofilms interacting with model proteins. Eur Polym J 111:161–169. https://doi.org/10.1016/j.eurpolymj.2018.12.018
Vahedikia N, Garavand F, Tajeddin B, Cacciotti I, Jafari SM, Omidi T, Zahedi Z (2019) Biodegradable zein film composites reinforced with chitosan nanoparticles and cinnamon essential oil: physical, mechanical, structural and antimicrobial attributes. Colloids Surf B 177:25–32. https://doi.org/10.1016/j.colsurfb.2019.01.045
Vartiainen J, Tuominen M, Nättinen K (2010) Bio-hybrid nanocomposite coatings from sonicated chitosan and nanoclay. J Appl Polym Sci 116:3638–3647. https://doi.org/10.1002/app.31922
Vasile C (2018) Polymeric nanocomposites and nanocoatings for food packaging: a review. Materials (Basel) 11:1834. https://doi.org/10.3390/ma11101834
Vicente AA, Cerqueira MA, Hilliou L, Rocha CMR (2011) Protein-based resins for food packaging. In: Multifunctional and nanoreinforced polymers for food packaging. Elsevier, pp 610–648
Vlachou M, Siamidi A, Kyriakou S (2019) Electrospinning and drug delivery. In: Electrospinning and Electrospraying – Techniques and Applications. IntechOpen
Wen P, Zong M-H, Linhardt RJ, Feng K, Wu H (2017) Electrospinning: a novel nano-encapsulation approach for bioactive compounds. Trends Food Sci Technol 70:56–68. https://doi.org/10.1016/j.tifs.2017.10.009
Wesley SJ, Raja P, Raj AA, Tiroutchelvamae D (2015) Review on – nanotechnology applications in food packaging and safety. Int J Eng Res 3:645–651. https://doi.org/10.17950/ijer/v3s11/1105
Wu Z, Huang X, Li YC, Xiao H, Wang X (2018) Novel chitosan films with laponite immobilized Ag nanoparticles for active food packaging. Carbohydr Polym 199:210–218. https://doi.org/10.1016/j.carbpol.2018.07.030
Wu C, Sun J, Lu Y, Wu T, Pang J, Hu Y (2019a) In situ self-assembly chitosan/ε-polylysine bionanocomposite film with enhanced antimicrobial properties for food packaging. Int J Biol Macromol 132:385–392. https://doi.org/10.1016/j.ijbiomac.2019.03.133
Wu C, Zhu Y, Wu T, Wang L, Yuan Y, Chen J, Hu Y, Pang J (2019b) Enhanced functional properties of biopolymer film incorporated with curcurmin-loaded mesoporous silica nanoparticles for food packaging. Food Chem 288:139–145. https://doi.org/10.1016/j.foodchem.2019.03.010
Zhang S, Xing M, Li B (2018) Biomimetic layer-by-layer self-assembly of nanofilms, nanocoatings, and 3D scaffolds for tissue engineering. Int J Mol Sci 19:1641. https://doi.org/10.3390/ijms19061641

Spray Drying as for Food-Grade Nanomaterial

Afroza Sultana, Asmaliza Ghani, Shuji Adachi, and Hidefumi Yoshii

1 Introduction

In Food Technology (1995), the cross-cut structure of spray-dried powder was depicted as structure model of emulsified fish-oil powder as shown in Fig. 1. In the powder formation of functional food compounds, which are usually fat-soluble, emulsification process is essential to produce feed solution to spray dryer. In this figure, the emulsified oil-droplet diameter in the powder was about 1–2 μm and powder diameter was about 300 μm. The oil-droplet diameters in emulsion and spray-dried powder were about μm size almost 20 years ago. We could see two emulsification methods with small oil-droplet size; low- and high-energy methods (Wang et al. 2015). In high-energy method such as high-pressure homogenization (Microfluidizer®), oil-droplet diameters ranged from 50 to 600 nm. Encina et al.

A. Sultana
Department of Food Processing and Engineering, Faculty of Food Science and Technology, Chattogram Veterinary and Animal Sciences University, Chattogram, Bangladesh
e-mail: afroza-sultana@cvasu.ac.bd

A. Ghani
School of Food Industry, Faculty of Bioresources & Food Industry, Universiti Sultan Zainal Abidin, Kuala Terengganu, Malaysia
e-mail: asmaliza@unisza.edu.my

S. Adachi
Department of Agriculture and Food Technology, Faculty of Bioenvironmental Science, Kyoto University of Advanced Science, Kyoto, Japan
e-mail: adachi.shuji@kuas.ac.jp

H. Yoshii (✉)
Department of Food Science and Nutrition, Faculty of Agriculture, Setsunan University, Hirakata, Osaka, Japan
e-mail: hidefumi.yoshii@setsunan.ac.jp

U. Hebbar et al. (eds.), *Nano-food Engineering*, Food Engineering Series,
https://doi.org/10.1007/978-3-030-44552-2_11

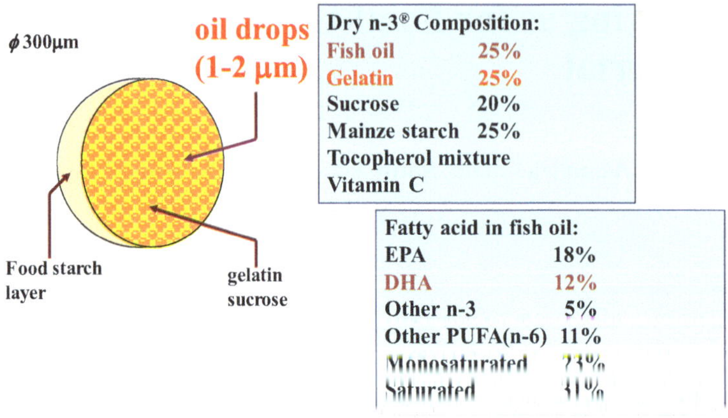

Fig. 1 Schematic model diagram of spray-dried fish oil powder from Andersen (1995)

(2016) summarized the conventional spray-drying and future trends for microencapsulation of fish oil.

Recently, the diameter of emulsified fish-oil droplet became smaller than 300 nm. Kuo et al. (2008) indicated that nano-emulsion of anti-oxidant synergy formulation containing γ-tocopherol has enhanced bioavailability and anti-inflammatory properties than their suspension. Lane et al. (2013) showed the comparison of bioavailability of vegetarian ω-3 fatty acids in nano-emulsion or bulk oil. Dey et al. (2012) concluded that nano-emulsion significantly enhanced the absorption of lipids from the emulsion system in cannulated small intenstine of rats. Oil-droplet size and emulsion structure of the lipid oil affected the rate of lipolysis and bioavailability (Garaiova et al. 2007; Klinkesorn and McClements 2009). Functional active compounds have been extracted from natural products such as plants and fish to examine their physiological functionalities. However, many of those functional compounds are unstable to heat, light, oxygen, and environmental conditions. Encapsulation technologies have increased in importance in food industry to develop and design the stable functional compounds in food powder. Microencapsulation is the most common and widely used method in food and pharmaceutical industries. Desai and Park (2005) defined that "encapsulation" is based on coating or entrapping the core material such as functional food compounds, fish oil. In order to use those compounds in a stable form, the encapsulation of functional compounds has been studied as comprehensive powdering. Among the various encapsulation techniques, spray drying is cost-effective since the product can be obtained continuously in a one-step operation. One of the key steps in the spray drying technique is emulsion preparation, which has a marked influence on the surface-oil content in the powder. Those processes are two-step microencapsulation process (emulsification of the core material and drying of the emulsion) for transforming liquid into powdery solids to improve their handling properties and to protect them from oxidation by means of a wall material that acts as a physical barrier for diffusion of oxygen to the

Table 1 Commercial microencapsulated fish oils to fortify foods [Referred from Encina et al. (2016)]

Commercial product	Process	
Arjuna/Omega-3Powder ZP (India)	Spray drying	
BASF/Dryn-318:12and 5:25 (Germany)	Spray drying	coated with starch granules
BASF/Dryn-318:12and 5:25 (Germany)	Spray drying	coanstarch coated matirix
The Wright Group/Supercoat (USA)	Spray drying	
National Starch-Omega	Spray drying	
Kievit/Vana-Sana (Netherlands)	Spray dryied	by spray drying on moving belt
Nu-Mega/Driphorm HiDHA (Australia)	Spray dryied	powder with Maillard products
Wacker/OmegaDry (Germany)	Complexation	in water followed by vacuum drying

core material. Functional foods were the largest application segment of omega-3 ingredients in 2013, accounting for over 55% of the global demand. Global omega-3 market expected to reach US $3.77 Bn by 2025. The commercial microencapsulated fish oils to fortify foods are almost formed by spray drying as shown in Table 1.

This paper summarizes the effects of emulsifiers, excipient types, emulsification conditions and spray drying conditions on the properties of spray-dried powders and the stability of functional compounds.

2 Effect of Oil-Droplet Size

Figure 2 shows the cross-cut scanning electron microscope photos of emulsified *d*-limonene spray-dried powders using an equal weight mixture of gum arabic and maltodextrin (MD). The average oil-droplet diameters were about 4.1 and 0.84 μm. Diameters of both powders were about 60–65 μm. Photo magnification was 6500 times for both the powders. Which oil-droplet diameter is better to keep the functional compounds stable in spray-dried powder? In the formation of spray-dried powder such as Fig. 2, the emulsification process and atomization of feed solution of wall material and emulsified oil were very important to control the oil-droplet diameter.

In the oil-existence-place in spray-dried powder, total oils are categorized inside oil (encapsulated oil), which is located at the closures place surrounded in wall material and open pore route to vacuole, and surface oil, which is adsorbed on spray-dried powder or contact to the powder surface. (Red pin points indicate surface oil and blue pin points indicate encapsulation oil in Fig. 3.) Vacuole in spray-dried powder also affect the oil-droplet distribution and surface oil content.

The oil-droplet diameter affected surface-oil ratio and encapsulation yield. Figure 4 shows the effect of oil droplet diameter on the encapsulation yield and surface oil ratio based on the data in the reported papers. The surface-oil ratio and the encapsulation yield are defined by Eqs. (1) and (2), respectively.

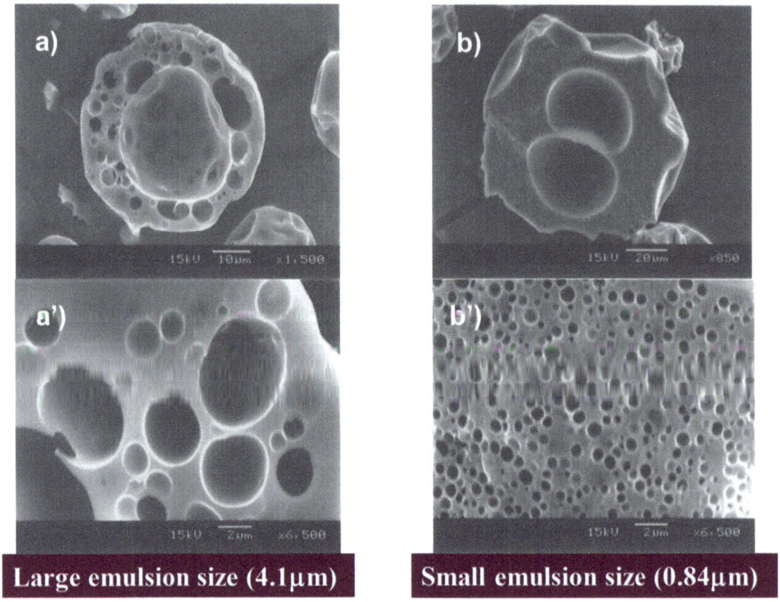

Fig. 2 SEM photographs of cross-cut spray-dried powder with emulsified *d*-limonene (a, a') average oil-droplet diameter = 4.1 μm, (b,b') average oil-droplet diameter = 0.84 μm. Capsule material was the blend of gaum Arabic and maltodextrin. Powder sizes were about 60–65 μm

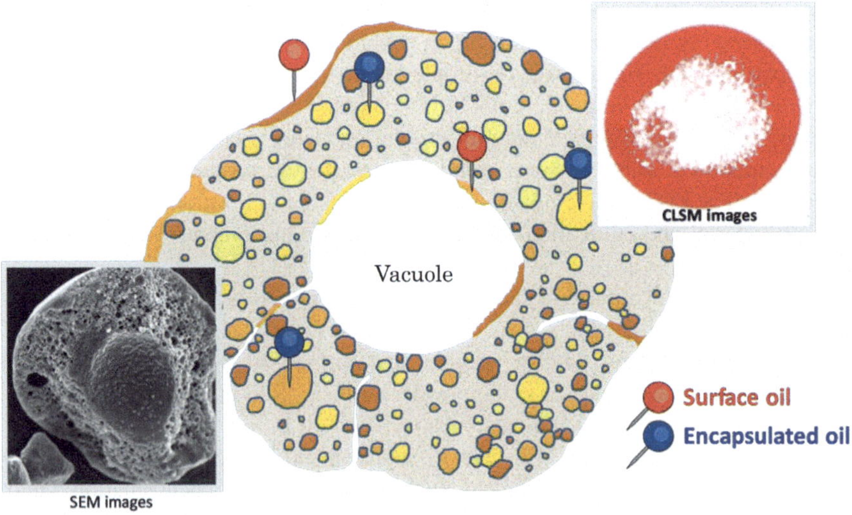

Fig. 3 Cross-cut structure of emulsified fish oil powder by spray drying

$$\text{Surface} - \text{oil ratio} = (\text{weight of surface oil in powder}) / (\text{weight of total oil in powder}) \quad (1)$$

$$\text{Encapsulation yield} = \left(\begin{array}{c} \text{weight of total oil in powder} - \text{weight of surface oil in powder}) / \\ (\text{weight of initially added oil in powder} \end{array} \right) \quad (2)$$

A bigger oil-droplet diameter showed higher surface-oil ratio in spite of the different conditions. Oil-droplet diameter significantly affected the surface-oil ratio and encapsulation yield. Vega and Roos (2006) reviewed the structure of spray-dried dairy and dairy-like emulsions. They discussed the importance of surface-oil content (free fat) in the oxidative stability of oil in the powder and summarized the measurement method of free fat in spray-dried powder. The surface-oil contents (extractable oil) of fish oil encapsulated in spray-dried powder were measured by Drusch et al. (2006), Serfert et al. (2009) with modified starch as wall material, Drusch and Berg (2008) with modified starch or glucose, Jafari et al. (2008) with modified starch and whey protein and Chen et al. (2013) with whey protein and casein. Hogan et al. (2001) reported surface-oil content of soy oil in spray-dried powder using sodium caseinate and MDs. They showed the microencapsulation efficiency increased from 0 to 88.4% when dextrose equivalent (DE) of MD increased from 0 to 28. The reason for this fact was considered that powders with

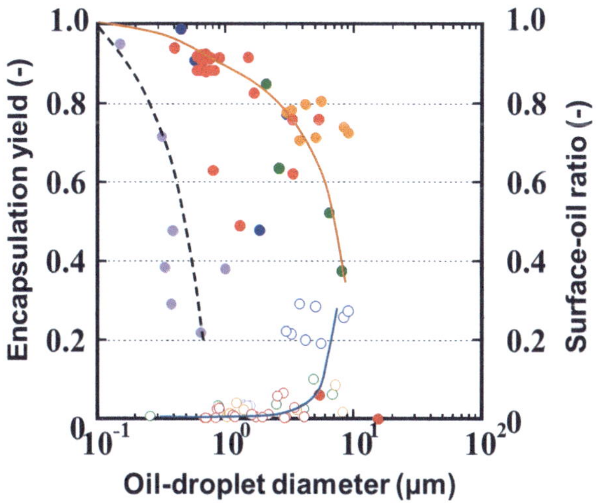

Fig. 4 Correlation between average reconstituted oil-droplet diameter and encapsulation yield or surface oil ratio in spray-dried powder. Encapsulation yield: closed keys; Surface-oil ratio: open keys (This figure is refereed with Abd Ghani et al. 2017)

high DE of MD might form less porous and more uniform matrices during drying. As shown in Fig. 4, the oil-droplet diameter is very important factor in the physicochemical properties of spray-dried powder, especially the surface-oil content on spray-dried powder.

3 Oil-Droplet Diameter before and after Spray Drying

Oil-droplet diameter in feed solution, which is emulsified with homogenizer, might change during drying, mainly at atomizer. The feed solution disperses to small liquid droplets in size and this atomization process is essential in spray drying. Abd Ghani et al. (2017) investigated the effect of oil-droplet diameter and DE of MD on the surface-oil ratio of microencapsulated fish oil by spray drying. They changed the solid content (40, 50 and 60 wt%) in feed emulsion. The solid content affected viscosity of feed emulsion and oil-droplet size in the emulsion. Figure 5 shows the relationship between fish-oil droplet diameters in feed emulsion and those after spray drying. These data were obtained by the spray dryer (L-8 type, Ohkawara Kakohki Co., Ltd., Yokohama, Japan) using rotary atomizer (30, 000 rpm) to form fish oil powder. (Abd Ghani et al., 2017). This figure shows that larger oil-droplets with a diameter larger than 1 μm became smaller after drying. On the other hand, the smaller droplets became bigger after spray drying.

This implies that the larger emulsion might be broken down during atomization owing to the high shear force on the rotary disc and the smaller oil droplets below 500 nm increased in diameter due to oil-droplet coalescence, promoted by the shrinkage of the droplets in the atomized emulsion solution during spray drying. These coalescence might affect the surface-oil ratio, as well as the encapsulation efficiency.

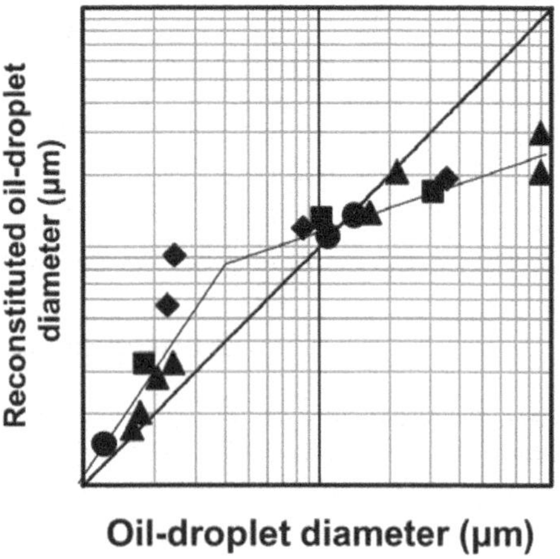

Fig. 5 Correlation between fish oil-droplet diameters in feed emulsion and those after spray drying; Oil-loading percentage to solid [%]: ● 30, ▲40, ■50, ♦60. Diameter values represent the average of triplicate measurements with a standard deviation below 10%

4 Surface-Oil Ratio

The oil encapsulation yield, y is defined by Eq. (3). In the measurement of surface-oil by solvent washing of oil, the thickness, δ, of the removal surface-oil on the powder surface neighborhood film is assumed, and y could be expressed.

$$y=\frac{\frac{\pi}{6}d_p^3C_0-\left\{\frac{\pi}{6}d_p^3C_0-\frac{\pi}{6}\left(d_p-2\delta\right)^3C_0\right\}}{\frac{\pi}{6}d_p^3C_0}=\frac{\left(d_p-2\delta\right)^3}{d_p^3}$$
$$=\left(1-2\frac{\delta}{d_p}\right)^3\cong\left(1-2\frac{d_e}{d_p}\right)^3=\left(1-2E\right)^3 \tag{3}$$

Here, C_0 is the initial weight of oil in the powder volume (g/m^3), d_p is the average powder diameter. If the thickness, δ, was assumed to be a similar order of the average oil-droplet diameter, d_e, then E is the ratio of d_e to d_p. By using Eq. (3), surface oil ratio, s is calculated by the following equation:

$$s=1-\left(1-2E\right)^3 \tag{4}$$

The encapsulation yield and surface-oil ratio were plotted as a function of the ratio of the oil-droplet diameter to the reconstituted emulsion to powder particle diameter in Fig. 6. Equation (4) assumes no effect of oil fraction ratio on surface-oil content. Kikuchi et al. (2013) calculated surface-oil content of microcapsules containing various oil fractions and oil-droplet sizes using percolation theory (Stauffer and Aharony 1985).

Abd Ghani et al. (2017) investigated the effects of oil and powder diameters in the solid content from 30 to 60 wt%, and measured the surface oil ratio, s, on

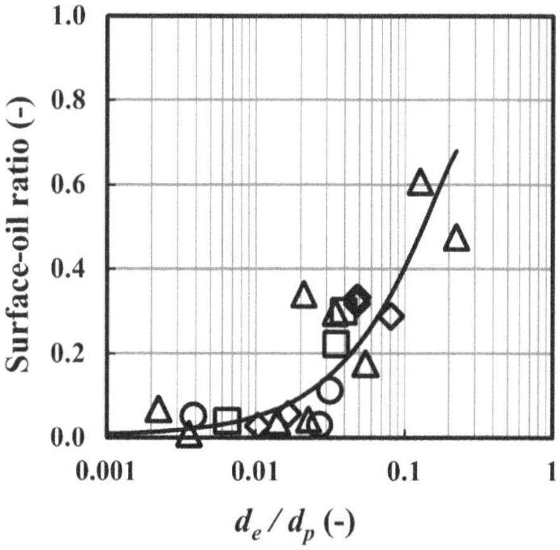

Fig. 6 Relationship between surface-oil ratio of emulsified fish oil spray-dried powder and diameter ratio of oil droplet diameter, d_e to powder diameter, d_p, d_e/d_p. Keys show solid content concentration. ○: 30 wt%, △: 40 wt%, □: 50 wt%, ◇: 60 wt%. Solid line is calculated line with $s = 1-(1-2\,d_e/d_p)^3$

spray-dried powder of emulsified fish oil at various emulsification conditions in the emulsification process. Figure 6 shows the relationship between surface-oil ratio of emulsified fish oil spray-dried powder and the ratio in diameter of oil droplet, d_e to powder diameter, d_p, d_e/d_p. The surface-oil ratios could be well correlated by Eq. (4). In some commercial spray-dried powders of emulsified fish oil, the powder and oil-droplet diameter were about 300 μm and 0.3 μm, respectively.

5 Morphology of Spray-Dried Powder

The surface oil ratio is an important factor to estimate the stability of functional compounds in spray-dried powder since non-encapsulated functional compounds might be easily oxidized compared with the encapsulated ones in spray-dried powder. Morphology of spray-dried powder is also important factor in the stability of encapsulated compounds. Walton and Mumford (1999)investigated the morphology of spray-dried particles using a single droplet drying method in the effect of process variables upon the morphology of spray-dried powder. They indicated the formation model of vacuole during spray drying. Walton (2000) also reviewed the morphology of spray-dried powder in qualitative view. They categorized three distinct categories of particle morphology, namely, crystalline, skin forming and agglomerate. When MD was used as wall material, skin formation rate and solid content were important in the formation of spray-dried powder. Effect of additives on the morphology of spray-dried powder was investigated by using confocal laser scanning microscope (CLSM) and scanning electron microscopy (SEM) (Paramita et al. 2010). Using CLSM, the vacuoles of powder could be observed easily without ripping off the particle. The vacuole size and the percentage of powders with vacuole were important factors to estimate the surface oil content as discussed in the previous section. Abd Ghani et al. (2017) investigated the effects of oil-droplet diameter and DE of MD on the surface-oil ratio of microencapsulated fish oil by spray drying. A mixed aqueous solution of emulsifier (sodium caseinate, SC: 3–9 wt% in solid) and wall material (MD): 31–67 wt% in solid) was prepared. Fish oil (30–60 wt% in solid) containing 30% eicosapentaenoic acid (EPA) and 25% docosahexaenoic acid (DHA) was added to the mixed solution. This mixed solution of fish oil, wall material and emulsifier were emulsified with mechanical homogenizer (Polytron® PT−6100, Kinematica AG, Luzern, Swiss) to prepare a micro-emulsion. This emulsion was furthermore two-times emulsified to form nano-sized emulsion with a high-pressure emulsifier at 20 MPa (LAB2000, SMT Co., Ltd., Tokyo, Japan) for 4 min or 100 MPa (Starburst Mini HJP-25001, Sugino Machine Limited, Uozu, Japan). The emulsified solution was spray-dried by the L-8 type spray dryer to prepare various emulsified fish oil spray-dried powders. The spray-dried powders had a lot of hollow particles. This hollow (vacuole) formation depended on DE of MD since a film formation rate during drying was affected with moisture diffusion rate in MD.

Existence probabilities of vacuole in the particles of spray-dried powder for various DE of MD were measured using CLSM containing a florescent agent (sodium fluorescein) feed emulsion. Furthermore, the vacuole sizes were measured for 30

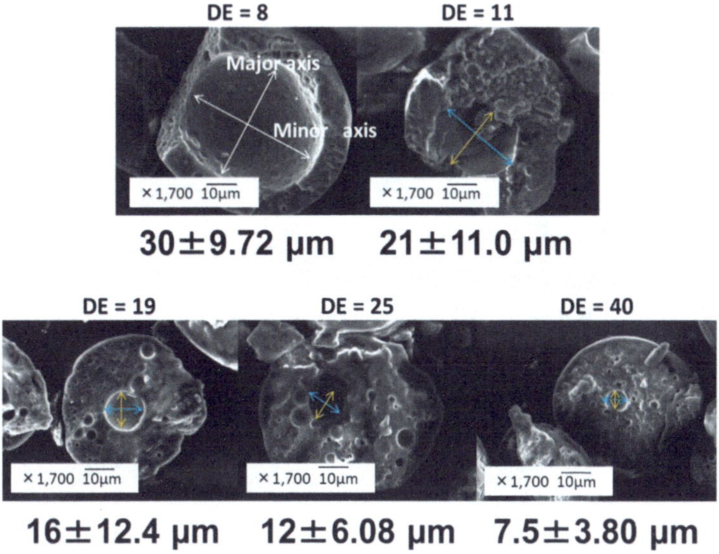

Fig. 7 SEM images of cross-cut section of fish oil spray-dried powders with different DE values of MD; oil load was 40 wt%, a homogenization at 20 MPa, and a spray-dryer atomizer speed of 30,000 rpm. Values under SEM image was vacuole diameter

particles for cross-cut section of the powder. Figure 7 shows SEM images of fish oil spray-dried powders with different DE values of MD. The particles were almost rounded and not spherical, with wrinkles of almost the same size as the particle diameter. Vacuoles in the powder could not be observed for DE = 40.

Keogh et al. (2001) indicated that the vacuole volume was a factor affecting the shelf life of the encapsulated fish oil. Probability of vacuole in spray-dried particles was counted by using CLSM images. Figure 8 shows CLSM images of the same powders. MD (DE = 8) had high vacuole existence percentages of about 64% and MD (DE = 40) had that of about 25%. The vacuole structures of spray-dried powder might affect the moisture diffusivity in MD and film formation rate during spray drying. Figure 9 shows the effect of the ratio, d_v/d_p of vacuole diameter, d_v to powder one, d_p on the surface oil ratio of powder. The values of d_v/d_p and surface oil ratio, s correlated linearly well. It shows that DE of MD affected significantly the structure of the powder.

6 Stability of Functional Compound

6.1 *Fish Oil*

The spray-dried powders (5 g) were placed in test tubes at 60 °C. Surface oil was washed with hexane and concentrated using a rotary evaporator. Encapsulated oil was measured using the washed powder. Using the solubilization of the spray-dried powders through incubation in N,N-dimethylformamide (DMF), the peroxide value,

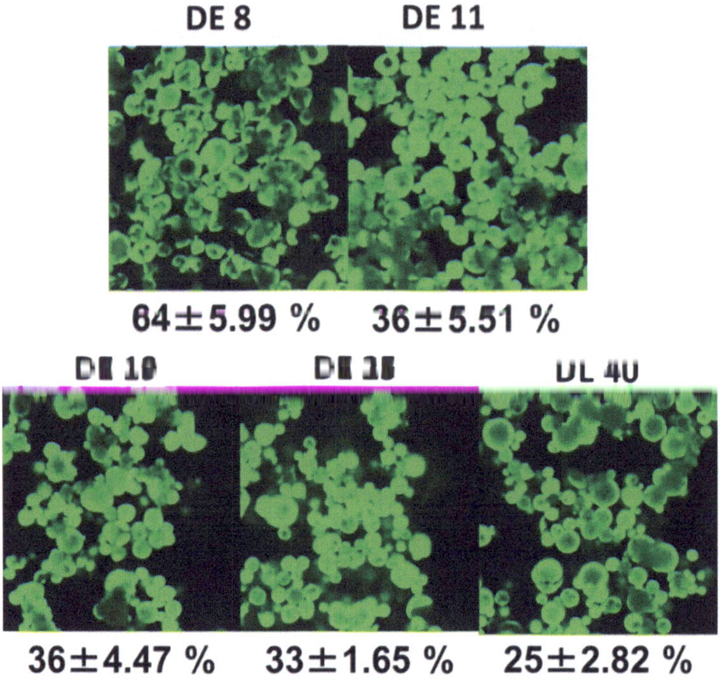

Fig. 8 CLSM images of cross-cut section of fish oil spray-dried powders with different DE values of MD; oil load was 40 wt%, a homogenization at 20 MPa, and a spray-dryer atomizer speed of 30,000 rpm. Values under CLSM image was the percentage of vacuole existence probability which may affect the surface-oil content

Fig. 9 Effect of DE of MD on surface oil ratio of emulsified fish oil dry powder. Each key number was the DE of MD. △:8, ○: 11, □: 19, △: 24, ●: 40

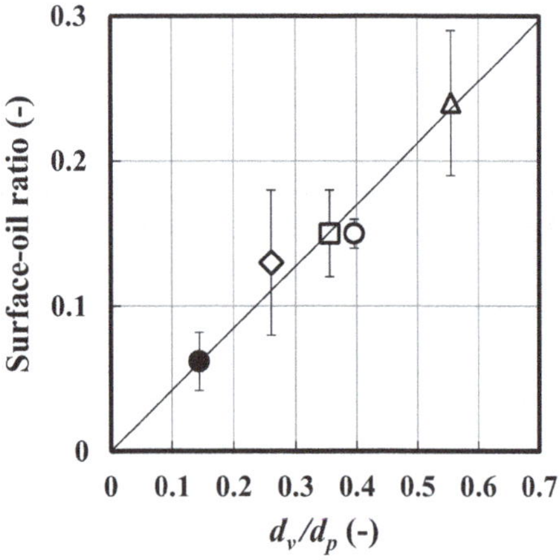

PV of the powders was easily measured without extracting the fish oil from the powder. PV was determined using the DMF method based on the AOCS Cd 8–53 acetic acid-chloroform procedure (AOCS 1997). As shown in Fig. 10, surface oils oxidized significantly faster than encapsulated oils for three MDs (DE = 11, 19, and

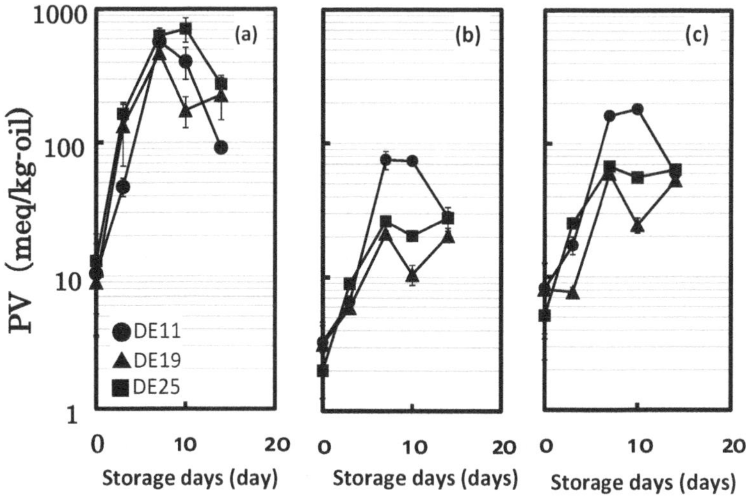

Fig. 10 PV changes of fish oil in emulsified fish oil spray dried powder at 60 °C. (**a**): surface oil, (**b**): encapsulated oil, (**c**): total oil. ●: DE = 11, ▲: DE = 19, ■: DE = 25

25). Ahn et al. (2008) also found higher PVs of surface oil in encapsulated seed oil kept at 60 °C for 30 days. They concluded that optimizing the microencapsulation condition could reduce surface-oil oxidation. Surface-oil oxidation was not affected by the DE of MD and was much the same during the storage time. On the other hand, the DE of MD affected the PVs of the encapsulated oils. Encapsulated oil for MD of DE = 11 had higher oxidation than those for MD of DE = 19 and 25. These data indicated that the surface-oil ratio and selection of wall material are important in forming stable fish-oil powder by spray drying. Especially, surface oil ratio might significantly affect the quality degradation due to odor formation in the initial oxidation stages of fish oil encapsulated spray-dried powder.

6.2 *Flaxseed Oil*

Effect of oil-droplet diameter on functional food compounds in spray-dried powder was also investigated by using flaxseed oil. Flaxseed oil is a good model oil to investigate the stability of polyunsaturated acid in oil. The average diameters of the oil droplets in the emulsion prepared using the rotor-stator homogenizer and that of those prepared using the high-pressure homogenizer ranged from 3.7 to 4.1 μm and 0.10 to 0.28 μm, respectively.

Spray-dried powders contained flaxseed oil (30 wt% to solid), MD (DE = 25) (67 wt%), and emulsifier (3 wt%) of sodium caseinate (SC) or polymerized sodium caseinate (PSC) with transglutaminase (3 wt%). Six types of spray-dried powders with oil droplets of various combinations of sizes were prepared with SC and

Table 2 Physical properties of spray-dried powder containing flaxseed oil

Emulsifier	Emulsion condition (Cycle number of high-pressure homogenization)	Powder diameter (μm)	Reconstituted oil-droplet diameter (μm)	Moisture content (wt%)
SC	Large oil-droplet (0 cycle)	23.1	4.21 ± 0.54	1.39
	Small oil-droplet (2 cycles)	20.0	0.22 ± 0.13	1.73
PSC	Large oil-droplet (0 cycles)	22.0	3.50 ± 0.54	1.41
	Small oil-droplet (2 cycles)	18.1	0.25 ± 0.11	1.27

The values are presented as the mean ± standard deviation of three measurements

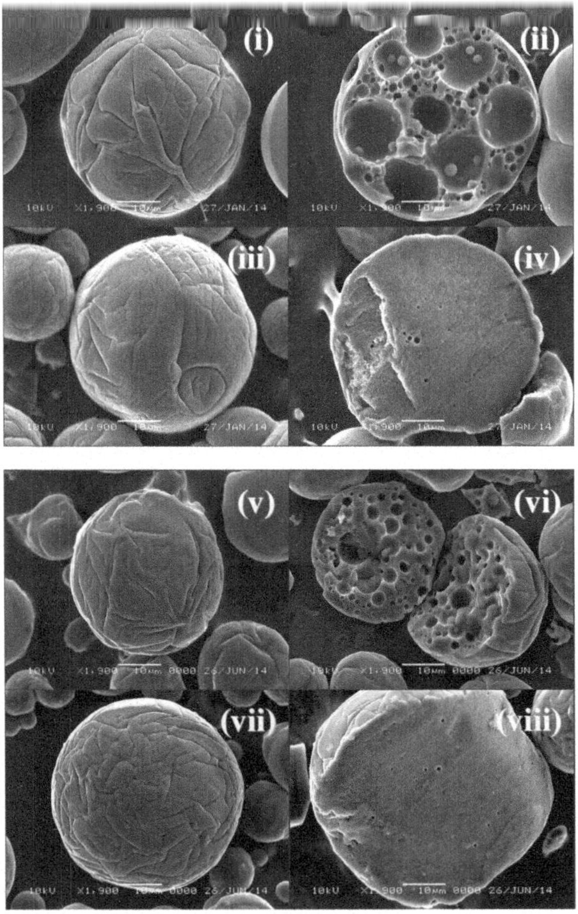

Fig. 11 SEM micrographs of spray-dried powders containing (i, ii, iii, iv) SC and (v, vi, vii, viii) PSC with (i, ii, v, vi) large and (iii, iv, vii, viii) small oil droplets. Cross-sectional structures (ii, iv, vi, viii) of solid particles and external structures (i, iii, v, vii)

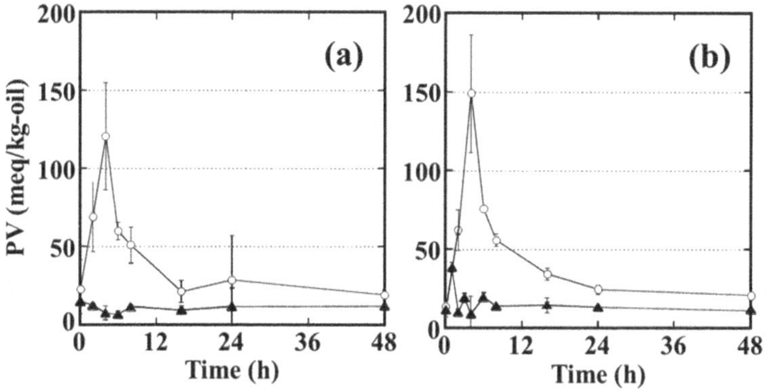

Fig. 12 Changes in PV of flaxseed oil in spray-dried powders with SC (**a**) and PSC (**b**)○represents the emulsion of large oil droplets by rotor-stator homogenization only and ▲ does the emulsion of small oil droplets by high-pressure homogenization

PSC. Table 2 shows properties of spray-dried powders containing SC or PSC prepared by rotor-stator or high-pressure homogenization. Figure 11 shows SEM photographs of spray-dried powders with small and large oil-droplets of flaxseed oil. The surface oil ratios for the powders were 0.046 for SC and 0.049 for PSC with small oil droplets and those values were 0.148 for SC and 0.217 for PSC with large oil droplets. As mentioned above in the surface-oil ratio, the oil droplet size had a remarkable effect on the surface oil ratio of the powders.

Oxidative stability of spray-dried powder containing flaxseed oil was evaluated by PV measurement at 105 °C incubation. After the incubation of spray-dried powder, a spray-dried powder was dissolved in 0.9 mL DMF under sonication. This solution was used to measure PV. Figure 12 shows changes in PV for flaxseed oil in spray-dried powders with and SC (a) and PSC (b).

All flaxseed oils in spray-dried powders at time zero were characterized by a low level of oxidation from 11.5 to 22.6 meq/kg-oil. PVs of the powders with small-oil droplets by high-pressure homogenization remained almost constant at the initial values for SC and PSC throughout incubation for 7 days. PVs of the powders with large oil-droplets represented the highest PV at 4 h and reached 120.6 meq/kg-oil for SC and 150 meq/kg-oil for PSC. These results might be due to the oxidation of surface oil in spray-dried powder. Stability of flaxseed oil in spray-dried powder was also investigated by the conductometric determination method. Two grams of spray-dried powder was placed in a test tube held at 105 °C and the air flowed continuously into this test tube at 100 mL/min. The outlet air from the test tube was bubbled into the concatenated water. The induction times and the conductivities at 1000 min were shown in Fig. 13. Induction times for the powder with large oil droplets were about 2 h. On the other hand, the induction times for the powder with small oil droplets were 0.42 h for SC and 0.57 h for PSC. These data indicated that oxidation rates of small-oil droplets were faster than those of large-oil droplets. The conductivities at 1000 min were about 150 μS/cm for large-oil droplets and 50 μS/cm for small-oil droplets. This fact means that spray-dried powders containing large-oil droplets

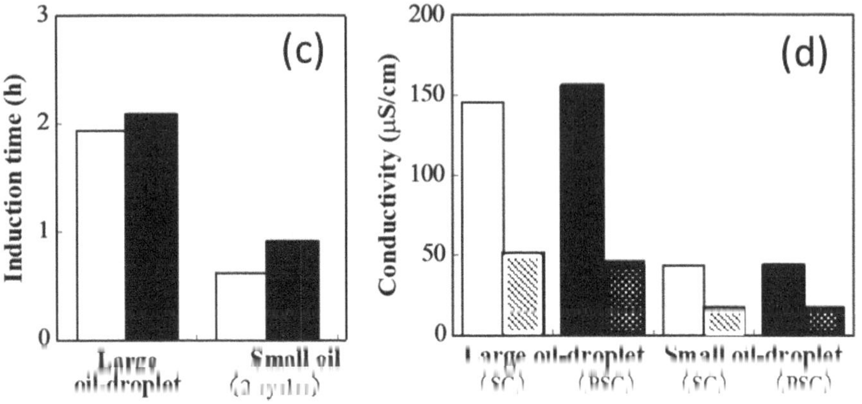

Fig. 13 Induction times (c) and conductivities (d) in the conductometric determination test. Unfilled and filled bars represent the use of SC and PSC. The grid line pattern indicated the data of hexane washed spray-dried powder

formed a larger quantity of volatile oxidation products. By removal of the surface oil of the powder by washing with hexane, the conductivity increase due to oxidation of flaxseed oil in spray-dried powder was significantly suppressed.

The surface-oil ratios of the powders with small-oil droplets were significantly lower than those of the powders with large-oil droplets. The oil-droplet size had a significantly effect on the surface oilratio and affected the initial oxidation stability of flaxseed oil in spray-dried powder.

6.3 Squalene

Table 3 shows physical properties of spray-dried powder containing emulsified squalene oil. Evaluation of oxidation stability of fish oil and flaxseed oil in spray-dried powder was very difficult to observe PV since the PV values at the initial stage are usually a constant value, or those values increase and decrease with the oxidation of polyunsaturated fatty acids with the beginning of polymerization of oxidative compounds, respectively. Therefore, it is very difficult to accurately evaluate the oxidation behavior of unsaturated fatty acids using PV values. On the other hand, squalene was used to evaluate the oxidation rate of functional compound in spray-dried powder since squalene could be analyzed by gas-chromatography. Two kinds of spray-dried powder, which were 0.27 μm and 0.69 μm of average oil-droplet diameter, were used to evaluate the squalene stability. Figure 14 shows the changes in retention of squalene at the powder incubation temperatures of 105 °C and 50 °C. Squalene stability depended remarkably on the incubation temperature. At 105 °C for 28 days' incubation, the retention of squalene decreased about 0.3 for the spray-dried powder with large oil-droplet diameter, 0.69 μm and about 0.6 for that with small oil-droplet diameter, 0.27 μm. At 50 °C for the same incubation days, the retention of squalene was almost the same for both powders.

Table 3 Physical properties of spray-dried powder containing emulsified squalene oil

Homogenization pressure [MPa]	Powder diameter [μm]	Reconstituted oil-droplet diameter [μm]	Moisture content [wt%]	Retention of squalene [%]	Surface-oil ratio [%]
100	27.7	0.27	2.95	90.1	3.4
20	29.0	0.69	2.82	95.5	7.0

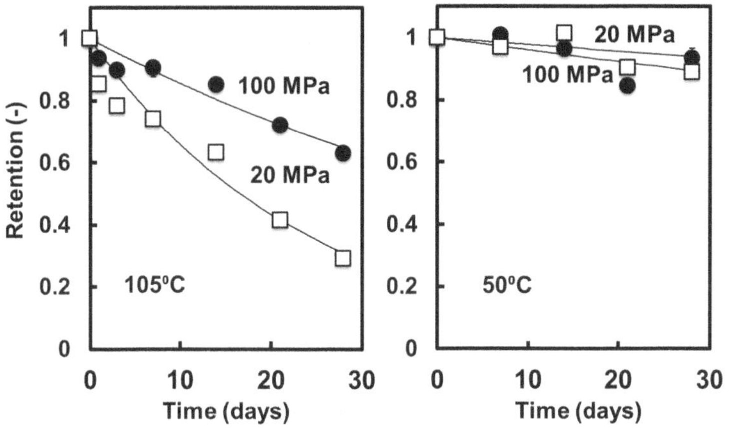

Fig. 14 Squalene retention changes at powder incubation temperature, 105 °C and 50 °C. Oil-droplet diameter ●: 0.27 μm, □: 0.69 μm

Squalene stability in spray-dried powder depended on oil-droplet diameter and this dependency changed with the incubation temperature. When the oil droplet diameter was small at high incubation temperature, the oxidation was triggered by the unsaturated fatty acid in the oil droplet and the radical might quickly transfer to other unsaturated fatty acid in oil droplets. However, this radical transfer limited in a small oil droplet, and might not propagate to other oil-droplets. In a large oil-droplets, the radical trigger might propagate the oxidation of the unsaturated fatty acid in large oil-droplet. The dependency of oil-droplet diameter might be due to this volume of oil-droplet and the radical propagation mechanisms.

7 Conclusion

As a result of recent advances in technology of the emulsification for reducing oil droplet size, the technique of spray-drying of a functional lipid has attracted attention as a method for producing functional food powders. The formation process of functional food powders constitutes two steps, the emulsification step and spray-drying one. The first emulsification operation of the functional compounds is very important to obtain the stable emulsion, and to control oil-droplet diameter and its distribution in emulsion. Therefore, the selection of the emulsifier and wall material is also very important. In this chapter, effect of oil-droplet diameter on the stability of functional lipids such as fish oil 0was discussed based on our previous publications. A nano-sized oil droplet of about 1 μm or less was prepared with high-pressure homogenizer and a micrometer-sized oil droplet was formed by mechanical homogenizer. The smaller oil droplets could obtain lower the surface oil ratio and the higher encapsulated ratio of functional lipid in spray-dried powder. The initial oxidation of functional lipid occurred at the surface oil. The surface oil ratio greatly depended on the ratio of oil-droplet diameter to powder diameter and the powder morphology. At high storage temperature of the powder, the stability of the functional lipid was higher for smaller oil-droplet. This phenomenon might be due to the radical transfer in oil-droplets. This chapter mentioned the importance of diameter ratio of oil-droplet to spray dried powder and effect of oil droplet size to the stability of functional lipid in spray-dried powder.

References

Abd Ghani A, Adachi S, Sato K, Shiga H, Iwamoto S, Neoh TL, Adachi S, Yoshii H (2017) Effects of oil-droplet diameter and dextrose equivalent of maltodextrin on the surface-oil ratio of microencapsulated fish oil by spray drying. J Chem Eng Jpn 50(10):799–806. https://doi.org/10.1252/jcej.17we048

Ahn JH, Kim YP, Lee YM, Seo EM, Lee KW, Kim HS (2008) Optimization of microencapsulation of seed oil by response surface methodology. Food Chem 107(1):98–105. https://doi.org/10.1016/j.foodchem.2007.07.067

Andersen S (1995) Microencapsulated Marine Omega-3 Fatty Acids for Use in the Food Industry, Food Tech. Europe 104–106

AOCS (1997) Official methods and recommended practices of the American Oil Chemists' Society Cd 8–53. Peroxide value acetic acid-chloroform. 4th ed. Champaign (IL)

Chen Q, McGillivray D, Wen J, Zhong F, Quek SY (2013) Co-encapsulation of fish oil with phytosterol esters and limonene by milk proteins. J Food Eng 117(4):505–512. https://doi.org/10.1016/j.jfoodeng.2013.01.011

Desai KGH, Park HJ (2005) Recent developments in microencapsulation of food ingredients. Dry Technol 23(7):1361–1396. https://doi.org/10.1081/DRT-200063478

Dey TK, Ghosh S, Ghosh M, Koley H, Dhar P (2012) Comparative study of gastrointestinal absorption of EPA & DHA rich fish oil from nano and conventional emulsion formulation in rats. Food Res Int 49(1):72–79. https://doi.org/10.1016/j.foodres.2012.07.056

Drusch S, Berg S (2008) Extractable oil in microcapsules prepared by spray-drying: localisation, determination and impact on oxidative stability. Food Chem 109(1):17–24. https://doi.org/10.1016/j.foodchem.2007.12.016

Drusch S, Serfert Y, Heuvel AVD, Schwarz K (2006) Physicochemical characterization and oxidative stability of fish oil encapsulated in an amorphous matrix containing trehalose. Food Res Int 39(7):807–815. https://doi.org/10.1016/j.foodres.2006.03.003

Encina C, Vergara C, Gimenez B, Oyarzun-Ampuero F, Robert P (2016) Conventional spray-drying and future trends for the microencapsulation of fish oil. Trends Food Sci Tech 56:40–60. https://doi.org/10.1016/j.tifs.2016.07.014

Garaiova I, Guschina IA, Plummer SF, Tang J, Wang D, Plummer NT (2007) A randomised cross-over trial in healthy adults indicating improved absorption of omega-3 fatty acids by pre-emulsification. Nutr J 6(4):1–9

Hogan SA, McNamee BF, O'Riordan ED, O'Sullivan M (2001) Emulsification and microencapsulation properties of sodium caseinate/carbohydrate blends. Int Dairy J 11(3):137–144. https://doi.org/10.1016/S0958-6946(01)00091-7

Jafari SM, Assadpoor E, Bhandari B, He Y (2008) Nano particle encapsulation of fish oil by spray drying. Food Res Int 41(2):172–183. https://doi.org/10.1016/j.foodres.2007.11.002

Keogh MK, O'Kennedy BT, Kelly J, Auty MA, Kelly PM, Fureby A, Haahr AM (2001) Stability to oxidation of spray-dried fish oil powder microencapsulated using milk ingredients. J Food Sci 66(2):217–224. https://doi.org/10.1111/j.1365-2621.2001.tb11320.x

Kikuchi K, Yamamoto S, Shiga H, Yoshii H, Adachi S (2013) Surface oil content of microcapsules containing various oil fractions and oil-droplet sizes. Jpn. J Food Eng 14(4):169–173. https://doi.org/10.11301/jsfe.14.169

Klinkesorn U, McClements DJ (2009) Influence of chitosan on stability and lipase digestibility of lecithin-stabilized tuna oil-in-water emulsions. Food Chem 114(4):1308–1315. https://doi.org/10.1016/j.foodchem.2008.11.012

Kuo F, Subramanian B, Kotyla T, Wilson TA, Yoganathan S, Nicolosi RJ (2008) Nanoemulsions of an anti-oxidant synergy formulation containing gamma tocopherol have enhanced bioavailability and anti-inflammatory properties. Int J Pharm 363(1–2):206–213. https://doi.org/10.1016/j.ijpharm.2008.07.022

Lane KE, Li W, Smith C, Derbyshire E (2013) The bioavailability of an omega-3-rich algal oil is improved by nanoemulsion technology using yogurt as a food vehicle. Int J Food Sci Tech 49(5):1264–1271. https://doi.org/10.1111/ijfs.12455

Paramita V, Iida K, Yoshii H, Furuta T (2010) Effect of additives on the morphology of spray-dried powder. Dry Technol 28(3):323–329. https://doi.org/10.1080/07373931003627098

Serfert Y, Drusch S, Schmidt-Hansberg B, Kind M, Schwarz K (2009) Process engineering parameters and type of *n*-octenylsuccinate-derivatised starch affect oxidative stability of microencapsulated long chain polyunsaturated fatty acids. J Food Eng 95(3):386–392. https://doi.org/10.1016/j.jfoodeng.2009.05.021

Stauffer D, Aharony A (1985) Introduction to percolation theory. Taylor and Francis, London. https://doi.org/10.4324/9780203211595

Vega C, Roos YH (2006) Invited review: spray-dried dairy and dairy-like emulsions–compositional considerations. J Dairy Sci 89(2):383–401. https://doi.org/10.3168/jds.S0022-0302(06)72103-8

Walton DE (2000) The morphology of spray-dried particles a qualitative view. Dry Technol 18(9):1943–1986. https://doi.org/10.1080/07373930008917822

Walton DE, Mumford CJ (1999) The morphology of spray-dried particles: the effect of process variables upon the morphology of spray-dried particles. Chem Eng Res Des 77(5):442–460. https://doi.org/10.1205/026387699526296

Wang Z, Neves MA, Isoda H, Nakajima M (2015) Preparation and characterization of micro/nano-emulsions containing functional food components. Jpn. J Food Eng 16(4):263–276. https://doi.org/10.11301/jsfe.16.263

Nanosensors for Foods

Gamze Toydemir, Sema Demirci Cekic, Gulay Ozkan, Seda Uzunboy, Asli Neslihan Avan, Esra Capanoglu, and Resat Apak

1 Introduction

A serious threat to public health emerging as one of the main concerns is the accessibility to safe food. Residual chemicals in foods above safe levels and the possible presence of biological toxins may cause foodborne illnesses. As stated in World Health Organization (WHO) reports, approximately 75% of the diarrheal episodes around the world stem from biologically contaminated foods. Each year, nearly three million people die from diseases caused by unhealthy food and water, making food safety a top priority of WHO (2002). Not only in the underdeveloped countries but even in the United States millions of people get sick due to contaminated food (CDC 2010). The growing population of the world and urbanization of farmlands increase the demand for effective consumption of foodstuffs. In this regard, food industry has a major role in supplying high quality, safe, healthy, nutritious and sustainable food. Additionally, it is important to produce fresh, flavorful, authentic foods with a satisfactory shelf life and to prevent adulteration and fraud (Sarkar et al. 2017).

Modern food industry comprises a number of applications such as processing, encapsulation, packaging, control and conservation of the quality and safety of foodstuffs. Nanotechnology may help food industry in all these areas. Nanotechnology can be defined as the science and technology of matter within a size interval of

G. Toydemir
Department of Food Engineering, Alanya Alaaddin Keykubat University, 07400, Alanya, Antalya, Turkey

S. D. Cekic · S. Uzunboy · A. N. Avan · R. Apak
Department of Chemistry, Faculty of Engineering, Istanbul University-Cerrahpaşa, Avcilar 34320, Istanbul, Turkey

G. Ozkan · E. Capanoglu (✉)
Department of Food Engineering, Faculty of Chemical and Metallurgical Engineering, Istanbul Technical University, 34469, Maslak, Istanbul, Turkey
e-mail: capanogl@itu.edu.tr

U. Hebbar et al. (eds.), *Nano-food Engineering*, Food Engineering Series,
https://doi.org/10.1007/978-3-030-44552-2_12

1–100 nm. Owing to the large specific surface area and other superb physicochemical characteristics of nanoparticles, nanotechnology has recently found numerous applications in different industries. It also has a great potential to generate pioneering processes and products that can be useful to different areas including the food sector. Nanotechnology can modify color, flavor, taste and texture of food products depending on consumers' preferences. It is also possible to insert ingredients having nutritional values so as to prepare functional foods. Additionally, nanotechnology can help to increase shelf life and to bring new advantages for tracking spoilage of foods using smart packaging or labeling techniques (Nagarajan, 2008).

The basic superiorities of nanomaterials are an outcome of the following distinct features: (i) high surface area to volume ratio, (ii) increased and alterable surface reactivity, (iii) rapid reaction kinetics, (iv) comparable Debye length of nanoparticles to the dimensions of analyte ions/molecules enabling closer interactions, (v) precise chemical composition and dislocation-defect free structures of nanocrystals (vi) high stability depending on high crystallinity (vii) compatibility for use in microelectronic devices (viii) high integration density, (ix) dimension-based low power consumption (x) relative low-cost (Ramgir 2013).

Nanotechnology helps to improve characteristics and functionalities of food products (Roco and Bainbridge 2001). The term "nanofood" has entered our lives a while ago, pointing out foods which are planted, cultivated, processed, packaged using nanotechnology or foods containing nanomaterials (Joseph and Morrison 2006). There are number of different application areas of nanotechnology related to the food sector such as intelligent packaging, storage, quality monitoring or production of foods modified considering consumers' demands. Nanosensors are important tools for ensuring safety and hygiene of foodstuffs by monitoring microbes, toxins and contaminants through all steps of the production and consumption chains. (Neethirajan and Jayas 2011).

The contribution of nanotechnology in the food sector can be summarized under a few main topics, one of these is intelligent packaging technologies. Due to their extremely small dimensions compared to classical packaging materials, nanoparticles do not cause a big difference in density and processability. The aims of using advanced packaging materials are mostly maintaining food safety and restraining microbial growth (Wesley et al. 2014). Food packaging is one of the key parameters to ensure a long shelf life by providing protection against spoilage and/or bacteria. Another important role of packaging is to restrain nutrient loss and to help to trace food products. Polymer nanocomposites may lead to obtain intelligent packaging materials having self-healing feature and responding to the changes in atmospheric parameters such as temperature and humidity. Owing to intelligent packaging technology, consumers can be timely warned against spoilage. Immediately after the detection of the first signals of spoilage, specific components can be released by active packaging material. Nanotechnology helps to improve the chemical, mechanical and antimicrobial parameters of these kinds of materials (Brody 2003).

Since oxygen leads to oxidation of fats/oils and propagation of microorganisms, O_2 should be kept out of the package to avoid rancidity and spoilage. Oxygen may additionally cause changes in the color, texture, and flavor inside packaging material. Using nanotechnology, effective oxygen scavengers can be produced. Using a nanotechnological approach, Rivett and Speer fabricated a film for providing active packaging that can selectively control transfer of oxygen and aroma effecting enzymes (Rivett and Speer 2009). It should be noted that packaging materials are never thoroughly impermeable for oxygen or humidity, and actually no material bears all the expected properties. Polymer nanocomposites provide a good alternative for this purpose, also providing a solution for obtaining environmentally friendly packaging materials. Bionanocomposites also display faster biodegradation rates than conventional polymers, which may be important in short life-time applications in which recycling has practical and economic difficulties (Ray et al. 2002). There are also commercial products enriched with nanoparticles which decrease oxygen inlet and lead loss of moisture.

Noble metal nanoparticles may find extensive use in the food sector. Especially, due to their antibacterial properties, silver nanoparticles are very common materials used for storage of foods and beverages (Duncan 2011). Silver has superior properties as antibacterial agent and is relatively a low-cost product for disinfection. These characteristics make it highly suitable to be used as water disinfectant in developing countries (Solsana and Méndez 2003). According to modified regulation of U.S. Food and Drug Administration, silver nitrate may be added to bottled waters at a concentration lower than 17 μg/kg (FDA 2009). It is a long-time known fact that silver nanoparticles and colloids have antimicrobial properties against a number of bacteria (Duncan 2011). Although both silver nanoparticles and silver nitrate seems to be potential sources of Ag^+, nanoparticles can have extra antimicrobial properties. For example, Navarro et al. (2008) stated that when equivalent doses of silver nanoparticles and silver nitrate were compared, nanoparticles were found to be more effective against algae. There are different studies in the literature explaining the mechanism of antimicrobial effect of silver nanoparticles, e.g., silver nanoparticles can be activated as antibacterial agents by using external electrical field or light (Akhavan and Ghaderi 2009; Fuertes et al. 2011). This knowledge can be applied to obtain self-sterilizing food containers with the addition of silver nanostructures (Duncan 2011).

Another commonly used nanomaterial in the food sector is titanium dioxide, which can be used not only as a food additive but also as an antimicrobial agent. TiO_2 particles smaller than 100 nm are used in industrial scale for packaging and storage containers of food (Alfadul and Elneshwy 2010). However, the antibacterial effect of TiO_2 involves a mechanism of photoactivation that leads to free radical generation, which may create potential risks for certain physiological targets in the human body (Long et al. 2006). The variety of the nanoparticle samples can be increased such as zinc (or zinc oxide) (Brunner et al. 2006), silicon dioxide and carbon at nanoscale (Chen and von Mikecz 2005). On the other hand, nano-wires made of gold and platinum are used for producing biosensors. Nano-sized proteins, polysaccharides and carbon can be mentioned among the other common nanoparticles used for the food industry (generally for food packaging). It should be noted that introduction of a new nanomaterial to be used in the food sector requires rigorous toxicity studies (Aigbogun et al. 2017).

Nanotechnology also helps to track foodstuff starting from the production area until reaching the consumers. For example, invisible nanobarcodes can be attached not only onto package but directly on the food products for providing a tool to trace the products by the authorized personnel of a certain brand. This technology allows to inhibit counterfeiting and adulteration (Neethirajan and Jayas 2011).

Nanoencapsulation system is another example of the benefits of nanotechnology in the food sector. Encapsulation helps to extend the stability of a food sample by preventing oxidation and rancidity. Additionally, it can provide the conservation of chemically/physically unstable ingredients of food, or control the release of the active compound (Ubbink and Krüger 2006). Nanocapsules protect the active ingredient through the digestive system and provide them to be released in the target area (Assadpour and Jafari 2019). There are specific bioactive compounds in certain foods such as β-carotene, lycopene, conjugated linoleic acid and isoflavones. Although the relative amount of these compounds is generally small, encapsulation may help to develop functional foods containing these substances. Calcium alginate, casein, α-lactalbumin are among the materials used for obtaining nanocapsules (Aigbogun et al. 2017). In spite of the numerous scientific contributions, there are only a few commercial applications of nanoencapsulation, e.g., an Australian food company applied this technique for incorporating tuna fish oil into bread. The nanoencapsules including fish oil did not open before reaching the stomach, omega 3 fatty acids in forms of capsules were used for masking the unpleasant taste and flavor. Owing to this controlled release of active ingredients, organoleptic perceptions, bioavailability and efficiency of the foods could be enhanced (Nu Mega Personal communication with company representative November 2007).

When it comes to nanosensors, actually this concept is as old as life on earth, possibly through the evolution of natural nanobiosensors such as enzymes, antibodies, and DNA. These nanobiomaterials can magnificently sense and response the metabolic activities and stimuli around them, and can store and compute the information and provide a continuous energy supply (Vogel 2019). It is very well known that nanoparticles have superior properties compared to bulk materials such as large surface volume ratio that provides powerful interaction with target analytes, and remarkable electrical properties enabling the development of electrochemical sensors. Because of these features, nanomaterials are very suitable using in recognition elements. In January 2000, National Nanotechnology Initiative (NNI) was established in the United States, constituting an important factor to the acceleration of nanoscience (Roco 2003). Three years later, United States took another step for which a road map was announced by the Department of Agriculture (US DOA 2003).

The US NNI was followed by different establishments in Europe and Asia shortly after (Roco 2003).

Use of nanotechnology helps to develop promising novel chemosensors or biosensors for various areas of the food industry. Nanosensors are the devices engineered using nano-sized materials to detect spoilage and contaminations in foodstuff. Some of the significant roles of nanosensors in the food production chain can be mentioned as to specify possible pathogens and trace signs of deterioration. Making use of the nanosensor technology, a number of nanoparticles can be inserted on a single device, helping to improve sensitivity (Sarkar et al. 2017). In addition to previous achievements of nanosensors in the health sector, their scope in food and agriculture has been expanded to include the determination of food pollutants such as pesticides and pathogen bacteria besides monitoring temperature, humidity, gaseous aroma constituents of foodstuff (Srivastava et al. 2018). Nanosensors can be designed to meet a special need, e.g., Zhao et al. (2004) produced fluorescent nanoparticles at 60 nm diameter using antibody-conjugated silica to detect pathogens in ground beef, where the authors developed a fluorescence-based immunoassay to determine a single bacterial cell in a relatively short time period (Zhao et al. 2004). In another study, Fu et al. (2008) manufactured a rapidly responding biosensor to detect *Salmonella* in food samples, where they bound fluorescent dye biosensor particles to anti-*Salmonella* antibodies located on silicon/gold nanorod array; hence, they obtained biosensor particles becoming visible in the presence of *Salmonella* in the food sample (Fu et al. 2008).

One of the large subgroups of nanosensors are nanobiosensors in which biologically active compounds are immobilized or bonded to the sensing system. These devices consist of recognition and sensing elements (Srivastava et al. 2018). The task of the recognition system is basically to serve as a specific receptor; transducers made of nanomaterials are for sensing. During the operation of biosensors, the physico-chemical changes occurring through the interaction between target analyte and biologically active material is converted into an output signal by using transducers (Chaubey and Malhotra 2002). Different types of nanoparticles can be used as materials for sensor fabrication. The flexible morphology of nanostructures improves the optical, mechanical, electrical, and thermal features of the products. The other important contribution of the nanomaterials is to increase transduction capability. According to their transduction mechanism, nanosensors can be divided into two main groups as optical and electrochemical nanosensors. In this point of view, optical nanosensors are mostly made of metallic nanoparticles and quantum dots which are known to enhance optical properties, whereas carbon nanotube electrodes are often used to produce electrochemical sensors. Besides these two categories, mass nanosensors are also known but can hardly find an application area in the food sector (Biju et al. 2008; Chao et al. 2012).

Finally, although nanotechnology and nanomaterials provide distinct advantages in the food industry, it should be borne in mind that the potential toxicities of nanomaterials are not fully understood yet, possible awaiting further investigations in the future based on a long time interval. Therefore, caution should be paid when using nanotechnology in the food sector (He and Hwang 2016).

2 Types and Principles of Nanosensors

2.1 *Electrochemical Nanosensors*

Nanomaterials help to form a solid medium for analyte sensing by providing a satisfactory detection limit and superior sensitivity (Willander et al. 2014). Nanomaterials have also attracted attention as a building material for electrochemical sensors. Owing to their superior specifications, nanomaterials can be incorporated to sensor technologies for enhancing current forms (Barry and O'Riordan 2016). Being chemically compatible and inert, noble metals (Guo and Wang 2011) and carbon are especially used in this area (Sharp and Burkitt 2015). In their paper on nanoscale sensor usage for assuring food safety, Wang and Duncan (2017) presented a number of studies on nanosensing in foods between 2005 and 2016. The authors generally observed a logarithmic increase in the overall number of related studies, and a superior number of electrochemical sensors in preference to optical sensors in recent years (Wang and Duncan 2017).

Electrochemical nanosensors work on the main principles of electrochemistry. During the interaction between nanomaterial and target analyte, an electron exchange takes place which generates electrical signals measured with the aid of electrochemical methods (Srivastava et al. 2018) The measured quantities can be voltage, current, or impedance. Current electrochemical transducers are compatible with miniaturization/nanofabrication; they also have superior properties such as simplicity, rapidity, minimal power requirement, economical cost and low detection limit, making them suitable as sensing devices. Electrochemical nanosensors can be classified as amperometric, voltammetric, and potentiometric sensors according to their working principles. Amperometric sensors are the devices that measure current produced by the redox reactions of electroactive species. During the system operation, the potential keeps constant and the Faradaic current is measured for quantitative determination of the analyte (Chaubey and Malhotra 2002). Owing to the contribution of nanotechnology, amperometric biosensors became more reliable, sensitive and yet cheap. To give a few examples, Yan et al. (2013) designed an acetylcholinesterase biosensor by means of multiwall carbon nanotubes to detect organophosphorus pesticides. In another study, Lin et al. (2013) investigated the electrochemical behavior of selected antioxidants (butylated hydroxyanisole (BHA), butylated hydroxytoluene (BHT) and butylated hydroquinone (TBHQ)) using a gold nanoparticle modified glassy carbon electrode. The authors reported that they could separate the oxidation peaks of the analytes; hence they could determine three antioxidants simultaneously and the developed method was successfully applied to determine their presence in edible oil samples (Lin et al. 2013).

Another electrochemical method of choice is voltammetry; e.g. in voltammetric methods the potential is altered in a controlled way and the subsequently produced current is measured. There are various applications in the literature using voltammetry, e.g., Nesakumar et al. (2016) developed an electrochemical acetylcholinesterase biosensor involving modified platinum electrode using ZnO nanocuboids and applied this technique to determine carbosulfan in rice. An electrochemical DNA sensor was produced by direct electro-deposition of gold nano-aggregates to determine *Salmonella typhi* (Singh et al. 2015). In another study, Yavuz et al. (2016) detected heavy metal contamination in food samples. For this purpose, the researchers modified the glassy carbon electrode by attaching graphene oxide covalently; then this modified electrode was used for the determination of Pb^{2+} and Cd^{2+} via differential pulse anodic stripping voltammetry (Yavuz et al. 2016).

Nanoparticles are commonly used for modification of conventional electrodes. For example, Karimi-Maleh et al. (2014) modified carbon-paste electrode using (9,10-dihydro-9,10-ethanoanthracene-11,12-dicarboximido)-4-ethylbenzene-1,2-diol (DEDED) and NiO nanoparticles (NiO/NPs/DEDED/CPE) for the determination of ascorbic acid and a synthetic azo dye, Sudan I, simultaneously using voltammetric methods (Karimi-Maleh et al. 2014). In another electrochemical study, Elyasi et al. (2013) fabricated an electrochemical sensor to detect Sudan I in food samples. For this purpose, the authors modified carbon-paste electrode with Pt/carbon nanotubes and an ionic liquid, resulting in oxidation current enhancement.

Different research groups reported that when the performances of nanoparticle-modified electrodes and bare electrodes were compared, a significant signal enhancement was observed in favor of modified electrodes. For example, Cheraghi et al. (2017) modified carbon-paste electrode using single wall carbon nanotubes (SWCNTs)-decorated cadmium oxide (CdO) nanoparticles using a binder ionic liquid (1,3-dipropylimidazolium bromide, DPIB) (CPE/CdO/SWCNTs/DPIB), and used this modified electrode in the determination of vanillin and folic acid in food samples by square wave voltammetry (SWV) (Cheraghi et al. 2017).

Potentiometry is another commonly used electrochemical technique that is both simple and inexpensive. It has a short response time with high selectivity, low detection limit, and adaptability to miniaturization. Potentiometric sensors measure the potential generated between the working and reference electrodes. Potentiometric sensors can be examined in two groups, i.e., field effect transistors and ion selective electrodes (Düzgün et al. 2011). In this regard, field-effect transistors (FETs) are potentiometric devices that rely on the effective charge of a binding analyte over the channel surface, which induces a change in the effective gate potential as well as the carrier concentration in the transistor channel through capacitive coupling. For example, a quasi-two-dimensional semiconductor oxide, indium(III) oxide-based pH sensor, displayed a low detection limit as much as 0.0005 pH and was useful in detecting glucose at femtomolar levels (Chen et al. 2017).

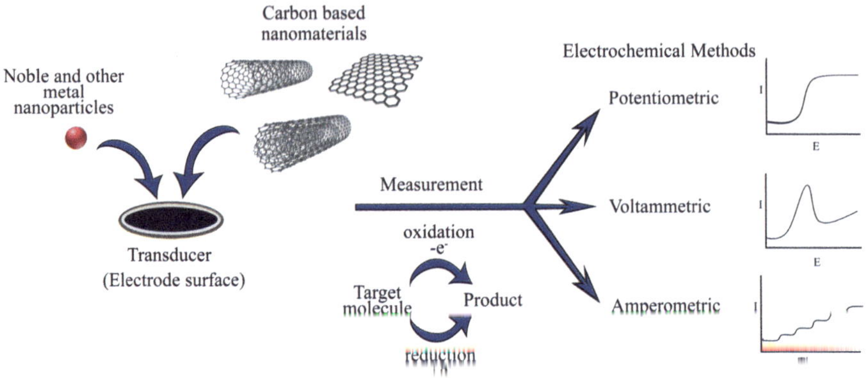

Fig. 1 Different types of nanomaterial(s) used in development of electrochemical sensors and basic measurement methods

Different types of nanomaterial used in development of electrochemical sensors and basic measurement methods are summarized at Fig. 1.

2.2 *Optical Nanosensors*

Use of optical nanosensors involves the transduction of optical signals into electrical signals for analytical purposes. These sensors are highly analyte-specific and generally use a small sample volume (Sarkar et al. 2017). The commonly used spectroscopic methods are highly coherent with biosensor technology. Almost all spectroscopic methods, i.e., absorption, fluorescence, phosphorescence, Raman, surface-enhanced Raman scattering, refraction and luminescence (fluorescence and phosphorescence) can be applied to nanosensor technology. Optical nanosensors consist of an optical receptor (or a sensing unit) and an inert matrix, and find use in the detection of certain properties of the tested sample. These nanosensors utilize optical signals to transduce biological or chemical information obtained from the sample for analytical knowledge. The introduction of optical nanosensors dates back to the late 1990s. Keiji et al. (1996) entrapped fluorescein in polyacrylamide nanoparticles for sensing pH. The basic analytical principle of the nanosensor was based on luminescence. Fluorescence-based nanosensors with their ease of measurement, high sensitivity and fast response ability are highly popular in this area, though their selectivity may sometimes be lower than those of absorption spectroscopy-based sensors. Gold and silver nanoparticles and quantum dots can be associated with fluorescence detection to produce nanosensors. There are numerous studies in the literature using turn-on or turn-off (i.e., based on fluorescence appearance or decay) fluorescence-based nanosensors (Srivastava et al. 2018).

Among optical nanosensors, fiber-optic nanosensors can be mentioned as an important group. In order to fabricate an optical nanosensor, the required chemical or biological recognition element consists of a sensitive layer covalently integrated to an optical transducer. Fiber-optics have a number of superior features in the production of optical nanosensors such as being of light-weight and easy configuration. When the receptor contacts with the surface of the analyte, the generated physicochemical perturbation can be converted to an electric signal. The role of the optical probe in the sensor is to receive this signal which is subsequently transmitted to the database. Measurement with a fiber-optic nanosensor can be realized directly (i.e., optical properties received from the analyte is measured directly) or indirectly (i.e., the magnitude of the measured response comes from optically detectable probes/bioprobes). In food science, the first fiber-optic sensor was used to detect *Escherichia coli* in a complex food matrix. (DeMarco et al. 1999). In other examples, a commercial fiber-optic sensor was developed to detect various pathogens, where a fluorophore-labeled antibody was used by the company (Nanduri et al. 2006; Valadez et al. 2009). Sinha et al. (2005) used a fiber-optic nanosensor to detect benzo[*a*]pyrene, a well-known carcinogen that may reach elevated levels in well-cooked meat and chicken skin, inside a single cell.

A noteworthy group among optical sensors is 'probes encapsulated by biologically localized embedding', known as PEBBLEs. Actually, the term PEBBLE refers to a series of nano-fabrication techniques which can be used for different optical sensing technologies. Cross-linked polyacrylamide and sol-gel silica are the mostly exploited matrices for producing PEBBLEs, and are used for H^+, some alkali metal ions (K^+, Na^+), alkaline earth ions (Ca^{2+}, Mg^{2+}) and Zn^{2+}, Cu^{2+}, Cl^-, and also for some non-ionic species (Buck et al. 2004a). The radius of the smallest nanospheric sensor may be as low as 10 nm. A sensor probe like this can be encapsulated using biologically localized embedding technique. These are suitable to insert into a cell (Clark et al. 1998). PEBBLEs can have a spherical shape, the diameter of the spheres generally laying between 20 and 100 nm. These are especially designed to be applied to a biological medium, consisting of an encapsulated dye (specific for targeted analyte) and a reference dye in an inert matrix. Since PEBBLEs are very small, they may be considered to be physically harmless to the cellular medium. The encapsulation of dyes within an inert matrix ensures minimum chemical interference (Buck et al. 2004b). There is a limited number of studies on optical PEBBLE nanosensors for food analysis. One was realized by Martin-Orozco et al. (2006) for the selective detection of Mg^{2+} involving encapsulation of a hydrophilic dye (coumarin 343, C343) and a commercially used reference dye (Texas Red, TxRed) in polyacrylamide nanoparticles. The main purpose was to determine Mg^{2+} inside human macrophage cells to investigate the attack of *Salmonella* which is a gram-negative bacterium. It is known that *Salmonella* can be present in certain foodstuff of animal origin such as meat, poultry and milk and it can pass from contaminated food to humans (Martin-Orozco et al. 2006).

Another optical method, surface-enhanced Raman scattering (SERS), is a relatively old but unique vibrational spectroscopic technique which can provide very specific vibrational profiles for analytes having small molecules (Fleischmann et al. 1974). It may allow the separation of even structural isomers with certain similarities such as fructose and glucose. Raman spectroscopy can be used for the identification of molecular structure, but with growing interest for production of nanoscale metals and nanomaterial surfaces, the SERS version of this technique has become more popular. With the progress achieved both in Raman technology and production of metal surfaces at nanoscale, SERS has become one of the most promising techniques for the analysis of food and pharmaceuticals and in intracellular or intercellular imaging. The Raman spectrometers used in applications of industrial food processing are generally coupled with fiber optic cables which can transmit light. With the use of optical fibers as long as 10 m, easy access was realized to distant samples (Li and Church 2014). In this regard, Talley et al. (2004) developed a SERS-based pH sensor consisting of silver and gold-silica core nanoparticles. The authors produced a pH sensitive nanosensor by mixing silver nanoparticles (50–80 nm) with *p*-mercaptobenzoic acid. The sensor gave a sensitive response between pH 6–8, and was used for the quick determination of *Bacillus subtilis* spore (Goodrace et al. 2000). He et al. (2011) used traditional Raman spectroscopy combined with sensors technology for the determination of antifungal effect of ZnO nanoparticles (70 ± 15 nm). The authors worked on two postharvest pathogenic fungi *Penicillium expansum* and *Botrytis cinerea*, and besides Raman spectroscopy, conventional microbiological plating and scanning electron microscopy (SEM) were also used to examine antifungal activities of ZnO nanoparticles against these fungi. Prevention of fungal activity is very important to prevent economical loss after fruit harvest. Especially, *B. cinerea* is one of the most important sources of certain diseases of table grapes, while *P. expansum* is known as the major reason of rot of stored apples and pears. As a result, He et al. (2011) reported that ZnO nanoparticles could be used as an efficient fungicide in food safety. Müller et al. (2014) detected thiabendazole, a chemical fungicide, with a SERS nanosensor. On the other hand, Saute and Narayanan (2011) detected a dithiocarbamate fungicide, thiram, in acetonitrile-water solution at ultralow levels by using dogbone-shaped gold nanoparticles. For this purpose, two dogbone-shaped gold nanoparticles at different nanometric sizes were synthesized *via* seed-mediated growth method and fully characterized. After the SERS spectra of thiram-bound nanoparticles were taken, two different LOD values for thiram were reported for different-sized nanoparticles, i.e., the lower LOD was obtained with larger nanoparticles as colloidal substrates (Saute and Narayanan 2011). Perchlorate anion, another food contaminant, could also be detected with SERS combined with gold nanoparticles and nanorods. Gu et al. (2009) determined perchlorate in water at 0.1 $\mu g\ L^{-1}$ levels using gold nanoparticles modified with dimethylaminoethanethiol as SERS substrate, which they suggested to be used *in situ* detection with a portable Raman spectrometer.

Since waterborne or foodborne microorganisms such as *E. coli*, *Legionella pneumophila*, *S. typhimurium* are the main reasons of serious (even fatal) bacterial infections, their rapid, selective and sensitive detection in food or water is critically important.

In food safety applications, determination and discrimination of bacteria play an important role. For identification of the pathogens, generally conventional methods are used, and these biochemical and microbiological tests need an isolated bacterial culture, expensive equipment and chemicals. On the other hand, polymerase chain reaction (PCR) and immunological detection can be alternative but these are generally criticized in terms of accuracy, specificity and sensitivity. It is known that Raman spectroscopy can be useful for biological samples, and Raman signals can be further intensified *via* SERS incorporating metallic nanoparticles through surface plasmon resonance (SPR). A series of studies depending on SERS method for determination, imaging, and identification of bacteria have been published in the literature in the last decades (Galvan and Yu 2018; Xie et al. 2013). Chen et al. (2015) synthesized silver nanoparticles (Ag NPs) within bacterial cell suspensions to detect foodborne bacteria, where they have reported to detect certain widespread bacteria that can cause diseases, such as *E. coli, Pseudomonas aeruginosa*, and *Listeria* in a short time using label-free NIR-SERS method. Additionally, they could identify two *Listeria* species (*L. monocytogenes* and *L. innocua*) by using very small sample volumes (Chen et al. 2015). Zhang et al. (2012) used magnetic–plasmonic Fe_3O_4–Au core–shell nanoparticles in the presence of external point magnetic field to determine bacteria. While *Bacillus cereus* is a bacterium related to food poisoning, *B. anthracis* cause a mortal disease anthrax. This bacterium may exist in the animals eating grass such as sheep, goat and cattle and can pass to the human body *via* infectious meat. Zhang et al. (2005) reported to have developed a quick determination method for anthrax spores using SERS on silver film over nanosphere substrates. The researchers preferred to use *B. subtilis* (the harmless stimulant) instead of *B. anthracis* (Zhang et al. 2005). Liu et al. (2009) developed a SERS-active substrate consisting of Ag-nanoparticles array embedded in anodic aluminum oxide to determine chemical specifications of bacterial cell wall to help identification of drug resistant bacteria. The nanochannels in the substrate having uniform narrow gaps between Ag-nanoparticles provided a reproducible Raman signal enhancement factor (Liu et al. 2009). Additionally, Lin et al. (2014) used antibody conjugated gold nanoparticles with single-walled carbon nanotubes for detection of Salmonella DT104.

'Surface Plasmon Resonance' (SPR) is another main optical method used in combination with nanotechnology. The sign of the real part of dielectric functions change sign across the interface of two different materials and the surface plasmons are the vibrations of the coherent delocalized electrons in this interface. Surface plasmon resonance (SPR) sensors are the optical sensors based on excitation of surface plasmons. SPR (bio)sensors technology can be used in the determination of pathogens, toxins and drug residues related to food quality or safety (Homola 2008). SPR nanosensors have the essential components of an optical sensor, a transducer, that provides a correlation between chemical or biological features of the analyte and the optical output, and an electronic database. The first two elements of the sensor determine the basic properties of the sensor such as stability, sensitivity and

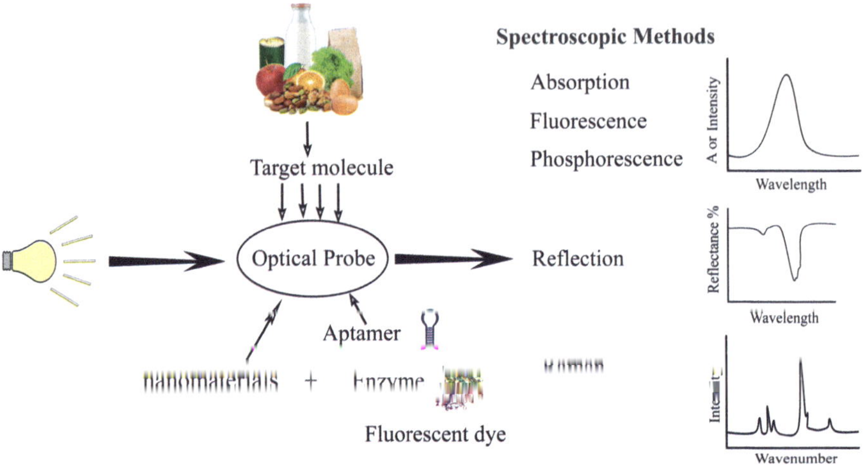

Fig. 2 Basic spectroscopic methods used in nanosensor based determinations of food-related analysis

resolution (Homola et al. 1999). For the selective determination of a target analyte, biorecognition molecules such as antibodies, enzymes or receptors are immobilized onto the surface of the sensor. With specific retention of the target analyte through the recognition element on the sensor, refraction intensity changes (Vollmer and Arnold 2008). Kara et al. (2013) developed a SPR nanosensor for determination of chloramphenicol in honey, and used molecularly imprinted nanoparticles attached to the SPR nanosensor. Yılmaz et al. (2017) produced a SPR nanosensor by imprinting a triazinic pesticide for herbicide determination. SPR biosensors were successfully applied by Mazumdar et al. (2007) for the determination of *Salmonella*, i.e., one of the main causes of food poisoning in milk In their study, Mazumdar et al. (2007) spiked the milk samples with *S. typhimurium* cells and reported to detect *S. typhimurium* at 1.25×10^5 cells mL^{-1}without a matrix interference for milk. The authors used a SPR chip having eight channels and applied a Kretschmann configuration and cuvette-based system.

Basic spectroscopic methods can be used in nanosensor based determinations, these are schematized in Fig. 2.

2.3 Nanoparticle-Based Nanosensors

Two of the most used noble metals in nanotechnology are gold and silver. The properties of noble metal NPs can be altered by changing size, shape, environment, and the method of synthesis. AuNPs can be used for different purposes such as being carriers for the biorecognition elements or for signal transduction and amplification. Absorption-based colorimetric sensing depends on aggregation-induced interparticle surface plasmon coupling of AuNPs. During the aggregation of AuNPs, the color turns from red to blue accompanying a bathochromic shift in the absorption maximum wavelength within the visible spectrum. This concept enables the detection of certain analytes which cause aggregation or redispersion of AuNPs (Saha et al. 2014) AuNPs are very useful tools for enhancing the surface area and conductivity of electrochemical sensors (Bülbül et al. 2015). Phillips et al. (2008) used AuNPs for the determination of bacterial food contamination, where they utilized non-covalent conjugates between AuNPs and poly(*para*-phenyleneethynylene). The mechanism of bacteria detection for this fluorometric sensor involved the interactions between bacteria and nanoparticles, releasing the bound fluorescent polymer from the gold-nanoparticle quencher, thereby recovering the initial fluorescence of the polymer. In other words, the (−) – charged bacteria were themselves attached to the (+) – charged AuNPs, thereby releasing the (−) – charged polymer from the surface and restoring fluorescence (Phillips et al. 2008). Saha et al. (2014) published a comprehensive review on AuNP-based sensing.

The working principles of silver nanoparticles (AgNPs) based sensors are very similar to those of AuNPs, involving a color change between dispersed and aggregated AgNPs (from yellow to brown). This change can be associated with the concentration of analyte. Vilela et al. (2012) discussed the colorimetric assays depending on aggregation of gold and silver nanoparticles. When the performances of AuNPs and AgNPs are compared, AgNPs have both advantages and limitations. Higher extinction coefficients and low cost are the superiorities of AgNPs, but oxidation of AgNPs surfaces is quite easy (Bülbül et al. 2015).

Xiong et al. (2008) modified AgNPs using p-sulfonatocalixarene (pSC_4) and produced a colorimetric histidine (His) probe. The authors took the advantage of color change in the solution due to pSC_4-AgNPs aggregation triggered by His.

Cerium oxide NPs (also known as nanoceria) are also used frequently. Cerium has two different oxidation states (Ce(III)/Ce(IV)), and with the change of the oxidation states, nanoparticle surface properties also change. Through its peroxidase-, superoxide- and oxidase-like activities, nanoceria can be used for replacing these enzymes mentioned above. With certain interactions at the surface, nanoceria exhibits unique color patterns owing to this specialty, making it available for colorimetric determinations. Different researchers used cerium nanoparticles for the determination of food antioxidants and glucose. Sharpe et al. (2013) developed a portable nanoparticle-based assay for antioxidant determination. The authors immobilized ceria nanoparticles, then these nanoparticles interacted with antioxidants tested. The antioxidants acted as reducing agents and caused a color change through the redox and surface chemistry reactions. The authors succeeded in determining a number of food antioxidants such as ascorbic acid, gallic acid, vanilic acid, quercetin, caffeic acid, and epigallocatechin gallate (Ornatska et al. 2011; Sharpe et al. 2013; Saha et al. 2014). In another study, Ozdemir-Olgun et al. (2018) developed a cerium oxide nanoparticles–based colorimetric sensor using tetramethyl benzidine (TMB) for the determination of total antioxidant capacity in food and plant extracts.

The proposed colorimetric sensor consisted of a poly (acrylic acid) sodium salt-coated cerium oxide nanoparticles. The authors utilized the oxidase-like mimetic effect of ceria nanoparticles, and exploited the color change of TMB as redox mediator in the absence and presence of tested antioxidants instead of the intrinsic color change of nanoceria, as the latter yielded a much lower extinction coefficient (Ozdemir Olgun et al. 2018). In a similar study, Kamer et al. (2019) used a nanoceria-based sensor for the determination of reactive species scavenging activity of antioxidants, where they exploited the peroxidase-like activity of nanoceria. By using {Ce(III) + aqueous NH_3} in conjunction with a zahter (*Thymbra spicata* L.) extract, nanoceria was synthesized by 'green chemistry' to generate reactive species (RS) upon reaction with hydrogen peroxide, and the RS on nanoceria surface caused a coloration on the N,N-dimethyl-p-phenylenediamine (DMPD) reagent acting as a redox mediator, finally, the DMPD color intensity was attenuated in the presence of antioxidants, enabling an indirect antioxidant activity assay (Kamer et al. 2019).

Magnetic Nanoparticles (MNPs) can be easily functionalized with a number of surface groups and can be used for different analyses. MNPs are often used for detection and removal of food contaminants. Kim et al. (2010a) developed fluorescence-coded MagPlex microsphere sets coated with antibodies for different bacteria, and then used them to capture these bacteria and toxins from spiked sample matrices including apple juice, green pepper, tomato, ground beef, alfalfa sprouts, milk, lettuce, spinach, and chicken washes. Dominguez et al. (2015) integrated magnetic nanobeads into a flow-based sensor system, and used it for the determination of organophosphate insecticides.

2.4 *Nanotube-Based Nanosensors*

Although different materials can be used to produce nanotubes, the most popular ones are made of carbon. Carbon nanotubes (CNTs), discovered in 1990's, rapidly became the most well-known material worldwide, due to their superior features such as electronic conductivity, flexible physical structure, high mechanical strength and folding abilities (Malik et al. 2013). Nanotubes can be defined as long and narrow tubes having a honeycomb-like structure. They have an extremely large specific surface area and can immobilize numerous biomolecules. CNTs can be single-walled (SW) or multi-walled (MW). While SWCNTs consist of a single layer of graphite bended as cylindrical shape, MWCNTs have a concentric array of single nanotubes. Microchips and other recent electrochemical devices can be used alongside nanotubes for determination of different food ingredients (such as sugars, vitamins and isoflavones) or food additives. CNTs have a superior surface chemistry compared to many other substances, and furthermore, their mechanical properties and electrical conductivities are perfectly suitable to be used as electrode material (Sarkar et al. 2017). Cesarino et al. (2012) developed an acetylcholinesterase (AChE) biosensor for the determination of carbamate pesticides in cabbage, broccoli and apple. For this purpose, a glassy carbon electrode was modified by using polyaniline and multi-walled carbon nanotubes as core-shell and AChE immobilization (Cesarino et al. 2012).

2.5 *Quantum Dots (QDs)*

QDs are nano-sized crystalline clusters having semiconductor properties (Murphy 2002). QDs are usually used for the determination of food contaminants such as pesticide residues and bacterial toxins, and have certain superior features such as long lasting photostability, broad absorption, and specific emission spectra. QDs have been used to develop different biosensors by integration of various biomolecules (Sarkar et al. 2017). Cadmium sulphide, cadmium selenide, cadmium telluride, indium phosphide and gallium arsenide are generally used as semiconductors for producing QDs. They have remarkable features such as high quantum yield and molar extinction coefficients, and additionally are highly resistant to photo- and chemical disruption. It is known that QDs are more stable than enzymes or fluorescence dyes (Perez-Lopez and Merkoçi 2011). Ji et al. (2005) developed a novel biosensor based on (CdSe)ZnS core-shell QDs; the authors coupled an organophosphorus hydrolase (OPH) to the core shell *via* electrostatic interactions. The developed nanobiosensor was used for the determination of an organophosphorus pesticide, paraoxon, which may be important for food (Ji et al. 2005). 2,4-Dichlorophenoxyacetic acid is a herbicide highly harmful for human health even at very low concentrations, and for its detection in food, Vinayaka et al. (2009) used CdTe QDs.

2.6 *Nanofibers*

Nanofibers with their large surface areas and porosity have attracted great attention in recent years. Electrospinning is one of the most preferred technique to produce nanofibers. Electrospinning is to spin fibers with the aid of the electrostatic forces, making it possible to produce fibers at nano lavels (Subbiah et al. 2005). According to specific requirements, it is also possible to change pore sizes and shapes of the nanofibers. The large surface-to-volume ratio of the material provides a number of available immobilization sites for biorecognition elements. These specialties lead to the manufacture of highly sensitive sensors for different applications including food safety analysis. A large number of biocompatible materials can be incorporated to nanofibers with different methods for obtaining desired functions (Matlock-Colangelo and Baeumner 2012). One of the most used material to produce nanofibers is cellulose, because it is a readily available and cheap by-product of food and agriculture industry. On the other hand, the high crystallinity of pure cellulose may obstruct electrospinning which may require the addition of other polymers. Cellulosic nanofibers produced by electrospinning technology can be used for different applications such as immobilization of biologically valuable materials, packaging technology, production of biosensors and filtration (Rezaei et al. 2015). Wang and Hsieh (2004) produced electrospun cellulose acetate nanofiber membrane-based polymer thin-film optical sensors. In another study, Yao et al. (1992) reported that highly photosensitive photochromic nanofibers are very convenient for producing optical devices and/or biosensors. Thanks to their favorable electronic and mechanical properties carbon nanofibers (CNFs) are one of the most striking materials in recent years. Tang et al. (2011) used electrospun CNFs modified electrode for non-enzymatic determination of xanthine (Xa). Since it is an intermediate of the purine metabolism, amount of Xa is evaluated as an indicator of different diseases. Additionally, Xa is the first indicator of ATP decomposition in dead fish, so it is an important parameter of the freshness of fish (Tang et al. 2011). There are different types of nanofibers containing various materials and electrospun nanofibers are widely used in food and agriculture industry. In their comprehensive review Mercante et al. (2017) summarized electrospinning-based (bio)sensors and their applications.

Different nanostructures used in production of food nanosensors with the selected applications are summarized at Table 1.

2.7 Electronic Nose and Electronic Tongue

In sensor studies, evaluation and determination of human-related senses occupy an important place. The first occurrence of commercial electronic noses (also known as e-noses) dates back to 1993. Electronic noses are a kind of gas sensor, usually consisting of metal-oxide semiconductors and conducting polymers, and their analysis techniques depend on a kind of artificial neural network (Gardner and Bartlett 2000). Volatile compounds can be detected in the headspace of the sample through electronic noses, the basic concept of which is mimicking human nose for differentiation of the odors using an array of sensors composed of thin or thick-film semiconductors. The conductance of the sensors can be changed by exposing volatile compounds. It is known that different gas samples including hydrogen sulfide, chlorine and alcohols could be detected at room temperature by using zinc oxide nanoparticles. While working with this kind of devices, samples in microliter volumes is sufficient for analysis, where the sample is evaporated and its vapor injected to the system (Sozer and Kokini 2012). Nanomaterials can be produced at controlled morphology and dimension, and their surface activity can be altered, enabling their usage for the manufacture of different devices including electronic noses. Nanomaterials-incorporated sensitive devices display a fast response, and are suitable for miniaturization and integration. The improvements in the nanoscience of novel nanomaterials can be used for electronic nose applications (Ramgir 2013). In the analysis of aromas or odors, it is possible to work with a group of trained expert panelists but it should be noted that the differences between individuals may affect the result. The other technique used for this purpose is gas chromatography-olfactometry suitable for odor-active substances. In recent years, electronic noses began to replace these conventional techniques to overcome subjectivity problems.

Table 1 Different nanostructures used in production of food nanosensors with the selected applications

Nanomaterial	Example	Structure	Method Used	Basic Principle	Tested Sample	Ref
Nanoparticles	– Gold Nanoparticles (AuNPs)	- AuNP-poly(*p*-phenyleneethynylene	- Fluorimetry	- The fluorescence of polymer is quenched by NPs and restored by bacteria	- Bacteria determination	- Phillips et al. (2008)
	– Silver Nanoparticles (AgNPs)	- *p*-sulfonatocalixarene (pSC_4) modified AgNPs	- Colorimetry	- Inducing of the aggregation between pSC_4 and AgNPs in the presence of histidine	- Histidine determination	- Xiong et al. (2008)
	– Cerium Oxide Nanoparticles (nanoceria)	- Poly (acrylic acid) sodium salt coated cerium oxide NPs (PAANa)-coated CeO-NPs	- Colorimetry	- Oxidase-like mimetic activity of CeO-NPs. Tetramethyl benzidine is oxidized by CeO-NPs and in the presence of antioxidants color intensity decreases	- Antioxidant activity	- Ozdemir-Olgun et al. (2018)
	– Magnetic nanoparticles (MNPs)	- Fluorescence-coded MagPlex microsphere sets coated with antibodies	- Fluorimetry	- Bonding between target antigens and antibodies	- Bacteria and toxin determination	- Kim et al. (2010a)
Nanotube	– Multi-walled carbon nanotubes (MWCNTs)	- Acetylcholinesterase (AChE) biosensor on polyaniline and MWCNTs core–shell modified glassy carbon electrode	- Chronoamperometry	- Inhibition of enzyme by addition of pesticide	- Carbaryl and methomyl pesticides in food sample	- Cesarino et al. (2012)
Quantum Dots (QDs)	– (CdSe)ZnS core-shell QDs	- Organophosphorus hydrolase (OPH) coupled core-shell	- Photoluminescence (PL)	- Decrease of PL intensity of the OPH/QDs bioconjugate in the presence of paraoxon	- Paraoxon determination	- Ji et al. (2005)
Nanofibers	– Carbone nanofibers (CNFs)	- Carbon nanofibers modified carbon paste electrode (CPE)	- Amperometry	- Electrocatalytic activity of CNF-CPE towards the oxidation of Xa	- Xanthine (Xa) determination	- Tang et al. (2011)

Electronic noses are the devices that can detect the odors coming from the tested sample with great precision, sensitivity and selectivity. These devices comprise a group of chemical sensors which can mimic the sense of smell, and are capable of replacing human olfactory system. Electronic noses are so suitable to use in food, cosmetic, pharmaceutical and clinical applications. Use of electronic sensors, which are committed to different odor-sensitive biological or chemical materials, eases to differentiate complex odors. To analyze unknown odors, a database is created by using the characteristic fingerprints of known odors, hence a recognition system is formed by training the system. Electronic noses have three main elements, namely sampling, detection and data processing systems. In food industry, electronic noses can be used for investigating shelf life, freshness, authenticity, quality and tracking process (Peris and Escuder-Gilabert 2009). For example, bioconversion during fermentation step of grape can be controlled by aroma analysis. Pinheiro et al. (2002) used an electronic nose to monitor the aroma profile on-line and real-time. In the study, the authors detected the muscatel aroma to track fermentation and reported to concentrate the aroma components to enhance selectivity (Pinheiro et al. 2002). In another study, García et al. (2005) used the technique to detect spoiled Iberian hams; the sensor used consisted of a mixture of tin-oxide semiconductor thin film with or without metal catalyst. The authors also used statistical methods to differentiate spoiled and unspoiled samples. Electronic noses are perhaps mostly used to investigate ripening process of fruits and vegetables from harvest to consumption as the quality determiner. In a sample study, the maturity of mango fruits was examined by using a commercial electronic nose device having 18 pieces of metal-oxide sensors (Lebrun et al. 2008). Freshness is another important parameter of quality for consumers, because during the shelf life, different volatile organic compounds evolve, presenting a great potential for electronic nose detection. Especially, seafood and meat release certain volatiles during degradation by bacterial processes. Another main issue for this method is to identify geographical origin by analysis of aroma through electronic nose (Peris and Escuder-Gilabert 2009). A great advantage of the electronic nose concept is its nondestructive testing ability (Sarkar et al. 2017).

The electronic tongues (e-tongues) are the devices designed to simulate human taste. While electronic noses serve to detect volatile compounds, electronic tongues are for detection of nonvolatile compounds in the food samples. Volatile compounds are characterized having chemicals with a low boiling point, small molecular weight and higher vapor pressure. The volatile organic compounds such as pyruvic acid, acetic acid and citrinin, highly occurring in food products, can be detected by electronic sensing devices such as e-nose and e-tongue. Both e-tongue and e-nose with their satisfactory accuracy, reliability, selectivity, sensitivity and quick response are suitable devices for sensing applications in the food industry. In recent applications, Torri and Piochi (2016) used electronic nose and electronic tongue to determine the possible aroma transfer from plastic packages. Men et al. (2014) used electronic nose and tongue to analyze mixed edible oil. Electronic tongue systems can be at different types: using impedance spectrum sensors, potentiometric electronic tongue and voltammetric systems. In the presented study, the authors preferred to use a voltammetric system consisting of an array of sensors, electrochemical work station and a computer for data analysis (Men et al. 2014).

2.8 *Nanobarcode Technology*

Barcode technology is used to symbolize encoded data or information about the food product (Srivastava et al. 2018). In its simplest definition, barcodes are readable tags containing tremendous amount of information. The term of "bar-code" comes from the lines (bars) on conventional barcodes. The information contained in the barcode is very useful for monitoring and identifying the products. In present time, with improvements in nanotechnology, micro and nanoscale barcodes have begun to be used alongside with macrobarcodes. The nanobarcodes with their small size can be used for identifying small targets. By the beginning of the nanobarcode technology, the shapes of barcodes also varied, e.g., as nano-beads, −wires, −clusters, −rods, −tubes, −sheets and nano-scale metal organic frameworks (NMOF). Nanobarcodes can be used for different applications, including the detection of biomolecules (nucleic acid and proteins), imaging, security and drug delivery. There are different encoding techniques for production of nanobarcodes and decoding of the signals, in conjunction with fluorescence microscopy or spectra for a fluorescent encoding nanobarcode. This situation brings their own limitations for each technology; for instance, the drawback of fluorescent nanobarcodes is the spectral overlap of fluorescent elements. In their comprehensive review, Shikha et al. (2017) mentioned encoding techniques under a number of main groups, such as optical, graphical, sequence, morphological, magnetic, phase change based encoding techniques and hybrid approaches.

3 Use of Nanosensors in Foods

As food is vital for human life and as the food industry is the largest manufacturing sector in the world (Yaktine and Pray 2009), food safety and quality remains a major concern worldwide in respect to public health. Consumers, with increasing knowledge and health awareness, are now looking for food products that are free from contaminants. Food contaminants, which can broadly be grouped as chemical contaminants (pesticides, veterinary drug residues, heavy metals, dioxins, melamine, etc.) and biological contaminants (foodborne pathogenic bacteria, naturally occurring toxins, etc.), occur through processing, handling, and distribution of most foods and show adverse effects on health (Kuswandi et al. 2017; Sonawane et al. 2014).

Many useful technologies have traditionally been proposed for food detection, including the chromatographic methods, enzyme hydrolysis, immunological and biological assays, and sensor tests (Heo et al. 2012; Peng et al. 2012; Jadzinsky et al. 2007; Sonawane et al. 2014). However, while progress has been achieved, these tests are still costly and complicated, require time-consuming processes and expensive instrumentation, and are not portable which strongly limit their performance and wider application (Bülbül et al. 2015; Li and Sheng 2014; Qu et al. 2011). In this context, nanosensors take the advantage of nanotechnology as a more efficient tool to detect food contaminants in an easy and quick manner, and enable assessing the quality and safety of foods on-site (Warriner et al. 2014). In continuation of this chapter, the integration of nanomaterials in the analysis methods of major toxic chemical and microbial food contaminants, as novel nanosensors and nanobiosensors, will be discussed as promising tools for providing enhanced food safety and quality.

Table 2 provides examples to the applications of various nanoparticle-based sensors for the detection of the most predominant chemical and biological food contaminants.

Table 2 Nanoparticle-based sensors used for the detection of chemical and biological food contaminants

Type of contaminant	Analyte	Nanoparticles used	Method/Sensor	References
Pesticides	Carbaryl methomyl	MWCNT/PANI	Electrochemical AChE based nanobiosensor; glassy carbon electrode	Cesarino et al. (2012)
	Acephate and methyl-paraoxon	Fe_3O_4 MNPs	MNP peroxidase mimetic-based colorimetric assay	Liang et al. (2013)
	Carbofuran	AuNPs and PB-MWCNTs-CTS	Electrochemical immunonanosensor	Sun et al. (2012)
	Dimethoate, dipterex, carbaryl, chlorpyrifos, and carbofuran	Lum-AgNPs	Nanoparticle-based chemiluminescent sensor	He et al. (2015)
Antibiotics	Ampicillin	AuNPs	Apta-nanosensor; AuNP-based dual fluorescence-colorimetric method	Song et al. (2012a)
	Chloramphenicol	CAP imprinted PEDMAH nanoparticles	Gold surface plasmon resonance nanosensor	Kara et al. (2013)
	Neomycin, kanamycin, and streptomycin	Acid-functionalized AuNPs	Surface plasmon resonance spectroscopy	Frasconi et al. (2010)
Heavy metals	Pb(II)	Aminothiophenol-functionalized MWCNT	Potentiometric sensor	Guo et al. (2011a), (2011b)
	Pb(II), Hg(II), Cd(II)	Triphenylphosphine-modified MWCNT	Electrochemical sensor; MWCNT-modified carbon screen-printed electrode	Bagheri et al. (2013)
	Cu(II), Pb(II), Fe(II), Ni(II), and Zn(II)	SPIMP-MWCNT	Flame atomic Absorption spectrometry	Ghaedi et al. (2013)
	Hg(II)	Bipy/AuNPs	UV-vis spectrometry	Chen et al. (2015)
	Cu(II)	DMAP-AuNPs	Electrochemical; cysteine modified gold nanoparticle film/ paper-based sensor modified with cysteine	Liana et al. (2013)

(continued)

Table 2 (continued)

Type of contaminant	Analyte	Nanoparticles used	Method/Sensor	References
Mycotoxins	Aflatoxin B1	C-AuNPs	Immunoelectrode	Sharma et al. (2010)
	Aflatoxin B1	Citrate-protected AuNPs	Colorimetric aptasensor	Hosseini et al. (2015)
	Aflatoxin B1 and ochratoxin A	Antigen-modified MNPs and antibody-functionalized UCNPs	Immunosensing probes; competitive Upconversion fluorescence immunoassay	Wu et al. (2011)
	Ochratoxin A	Nanostructured zinc oxide	BSA/r-IgGs/Nano-ZnO/ITO immunoelectrode	Ansari et al. (2010)
	Ochratoxin A	Aptamer-functionalized AuNPs	Colorimetric aptamer AuNP probes	Soh et al. (2015)
Pathogenic bacteria and bacterial toxins	*Escherichia coli*	AuNPs functionalized with quaternary ammonium ligands	Colorimetric enzymatic nanocomposite biosensor	Miranda et al. (2011)
	Escherichia coli O157	Colloidal AuNPs	Immunochromatographic strip	Jung et al. (2005)
	Escherichia coli	CNTs decorated with AuNPs	Multi-array, electrochemical aptasensor	Maurer et al. (2012)
	Escherichia coli O157:H7	Oligonucleotide-functionalized AuNPs	Piezoelectric biosensor	Chen et al. (2008)
	Salmonella	Immuno-modified MNPs and antibody-immobilized titanium dioxide nanocrystals	Optical nanocrystal probes; UV-Vis absorption spectrometry	Joo et al. (2012)
	Cholera toxin	Lactose-stabilized AuNPs	Colorimetric aggregation bioassay, UV-Vis spectrometry	Schofield et al. (2007)
	Staphylococcal enterotoxin B	AuNPs	Chemiluminescence immunosensor	Yang et al. (2009)

(continued)

Table 2 (continued)

Type of contaminant	Analyte	Nanoparticles used	Method/Sensor	References
Adulterants	Melamine	AuNPs	Colorimetric sensor; UV-Vis spectrometry	Li et al. (2010)
	Melamine	Dopamine-stabilized AgNPs	Colorimetric sensor; UV-Vis spectrometry	Ma et al. (2011)
	Melamine	Aptamer-modified AuNPs	Colorimetric aptasensor; UV-Vis spectrometry	Yun et al. (2014
	Pork meat	Colloidal AuNPs	Colorimetric sensor; UV-Vis spectrometry	Ali et al. (2011)
	Horse meat	Thiolated [illegible] functionalized AuNPs	AuNP probes	Houhoula [illegible]

MWCNT/PANI, multiwalled carbon nanotubes/polyaniline; *AchE*, acetylcholinesterase; *Fe_3O_4 MNPs*, iron oxide magnetic nanoparticles; *AuNPs*, gold nanoparticles; *PB-MWCNTs-CTS*, Prussian blue-multiwalled carbon nanotube-chitosan; *Lum-AgNPs*, luminol-functionalized silver nanoparticles; *CAP*, chloramphenicol; *PEDMAH*, poly(ethylene glycol dimethacrylate-N-methacryloyl-L-histidine methylester); *SPIMP-MWCNT*, 2-((3-silylpropylimino) methyl) phenol functionalized multiwalled carbon nanotube; *Bipy*, 2,2′-bipyridyl; *DMAP*, 4-(dimethylamino) pyridine; *C-AuNPs*, cysteamine-functionalized AuNPs; *UCNPs*, upconversion nanoparticles; *BSA/r-IgGs/Nano-ZnO/ITO*, bovine serum albumin/rabbit-immunoglubin antibodies/nanostructured zinc oxide/indium-tin-oxide; *CNT*, carbon nanotube

3.1 Analysis of Chemical Food Contaminants

3.1.1 Pesticides

Pesticides are the chemicals used for pest control in plant protection (Health and Safety Executive 2009; Kuswandi and Mascini 2005) and their presence in related foodstuffs should be monitored because of their potential health risks (Caetano and Machado 2008; Liu et al. 2008a, b). Organophosphates and carbamates are the most widely employed ones among this class of compounds (Kumar et al. 2010; Singh and Walker 2006).

The mode of action of pesticides involves the inhibition of the enzyme acetylcholinesterase (AChE), by binding to the serine moiety within the active site of the enzyme, leading to the deacetylation of the neurotransmitter acetylcholine (Kuswandi et al. 2008; Kuswandi and Suwandari 2007; Andreescu and Marty 2006; Pope 1999). Hence, the detection of the AChE enzyme inhibition -by the measurement of the enzyme activity before and after exposure to the pesticides- is indicated as the most common biosensor method (Kuswandi et al. 2017; Bülbül et al. 2015).

The working principle of the biosensor includes the formation of thiocholine (TCh) and acetic acid as a result of the catalysis of acetylthiocholine (ATCh) by AChE, which is followed by further oxidation of the electroactive TCh in the reaction. Gold nanoparticles (AuNPs) and multiwalled carbon nanotubes (MWCNTs) are used in AChE-based nanobiosensors to facilitate the electron transfer owing to the electrooxidation of TCh (Kuswandi et al. 2017). The level of carbamate pesticides in cabbage, broccoli, and apple samples were determined using electrochemical AChE-based nanobiosensor, developed on polyaniline (PANI) and MWCNT core-shell-modified glassy carbon electrode, and consistent results with those from HPLC measurements were obtained (Cesarino et al. 2012). Here, PANI acts as a matrix for enzyme immobilization and helps for electrocatalysis of TCh (Somerset et al. 2007, 2009). Carbon nanotubes (CNTs), having strong adsorption ability and large surface area, are also promising candidates in solid phase extraction (SPE) technology (Li and Sheng 2014). Ravelo-Perez et al. (2008) detected 8 different organophosphorus pesticide residues (ethoprophos, diazinon, chlorpyriphos-methyl, fenitrothion, malathion, chlorpyriphos, fenamiphos, and buprofezin) in apple, grape, orange, and pineapple juice samples by using MWCNTs as SPE sorbents.

A colorimetric method using both AChE and choline oxidase (ChO) enzymes, and iron oxide magnetic nanoparticles (Fe3O4 MNPs) that mimick the peroxidase activity, was carried out for the detection of organophosphate group pesticides. In this method, choline was formed as a product of the AChE activity, which was further converted to hydrogen peroxide (H_2O_2) by ChO activity. The occurrence of H_2O_2 was monitored by a color change as a result of the catalytic action of the MNPs on the oxidation of 3,5,3′,5′-tetramethylbenzidine. Pesticides inhibited AChE activity, which led to a concurrent decrease in H_2O_2 production and color intensity. The decrease in the color intensity was used to detect AChE inhibition by pesticides (Liang et al. 2013).

Non-AChE-based nanobiosensors have also been reported for the detection of pesticides. An electrochemical immunonanosensor, comprised of AuNPs, Prussian blue-MWCNT-chitosan (PB-MWCNT-CTS), and staphylococcal protein A (SPA) layer-by-layer assemble technology has been developed for carbofuran detection and quantification. In this method, a number of amino and carboxyl groups were provided by the porous three-dimensional PB-MWCNT-CTS nanocomposite film which also offered a large specific surface area to immobilize SPA. The binding capacity of the antibody increased by the formation of a self-assembled SPA layer on the electrode surface (Sun et al. 2012).

Functionalized NPs have also been developed for assays with chemiluminescent detection. Recently, luminol-functionalized silver NPs (Lum-AgNPs) were used as functional nanomaterials for the detection of organophosphate and carbamate pesticides (i.e., dimethoate, dipterex, carbaryl, chlorpyrifos, carbofuran). A hydrogen peroxide (H_2O_2)-based chemiluminescent detection was used together with the Lum-AgNPs to generate a chemiluminescent fingerprint for each specific pesticide (He et al. 2015).

3.1.2 Antibiotics

Antibiotics are extensively used as veterinary drugs to treat bacterial infections and to support animal growth. The misuse of antibiotic drugs in animal husbandry leads to the presence of undesirable residues in food products and when these food products are ingested too often this could result in the formation of antibiotic resistant bacteria in the gut, reducing the efficacy of drugs used to treat human infections (Huet et al. 2009; Virolainen et al. 2008).

Ampicillin, a well-known antibiotic used against a broad spectrum of bacteria (e.g., *Haemophilus influenzae*, *E. coli*, Salmonella, and Shigella), has been detected by a recently developed apta-nanosensor using a AuNP-based dual fluorescence-colorimetric method (Song et al. 2012a). Among the selected [illegible] (AMP4 (5′-[illegible]-3′), AMP17 (5′-[illegible] TAGCGG-3′), and AMP18 (5′-TTAGTTGGGGTTCAGTTGG-3′)), having high sensitivity and specificity to ampicillin, AMP17 ssDNA (single-stranded DNA) aptamers were used for adsorption onto the surface of AuNPs. As a result of the interaction between ampicillin and the aptamer, AuNPs were released from the aptamer and were aggregated in the presence of salt, leading to a color change from red to purple. The higher the concentration of ampicillin in the sample gives a more intense color in the solution, making this method a promising alternative for the detection of ampicillin in the food chain (Song et al. 2012a).

Chloramphenicol (CAP) is another broad spectrum antibiotic used to cure infections in bees and causes contamination in honey (Fedorova et al. 2010). Kara et al. (2013) developed a surface plasmon resonance (SPR) nanosensor for the detection of CAP by using molecularly imprinted nanoparticles. The CAP imprinted nanoparticles were attached onto the surface of gold SPR nanosensor. The imprinted nanoparticles showed a monolayer attachment onto the SPR nanosensor and selectively recognized the CAP molecules. The CAP imprinted SPR nanosensor showed reproducible results for honey samples indicating its potential use for food safety purposes.

A series of antibiotics, including neomycin (NE), kanamycin (KA), and streptomycin (ST), having vicinal hydroxyl groups in their molecular structure, were detected in milk samples by using AuNPs modified with a composite capping monolayer consisting of the electropolymerizable thioaniline units and phenylboronic acid ligands that act as the functional material for developing sensitive and selective matrixes for this series of antibiotic compounds. The electropolymerization of the functionalized nanoparticles on a Au surface in the presence of the selected antibiotic substrates, NE, KA, and ST, yielded molecularly imprinted AuNP matrixes with high sensitivity and specificity for the detection of these imprinted antibiotics. Surface plasmon resonance (SPR) spectroscopy was used for probing the binding processes of the antibiotic analytes to the imprinted sites (Frasconi et al. 2010). SPR is a widely-used method in optical sensors and biosensors (Homola 2008) that monitors refractive index changes on thin metal films (i.e., Au) arising from recognition events or chemical transformations (Baba et al. 2001; Knoll 1998). The antibiotic substrates in milk samples were analyzed by different sensing matrixes, based on the changes in the dielectric properties of the matrixes as a result of the binding of the antibiotics to the imprinted sites (Frasconi et al. 2010).

3.1.3 Heavy Metals

Heavy metal contamination in food mostly arises from soil contamination due to discharge of industrial effluents or repeated applications of commercial fertilizers and pesticides for agriculture (Kuswandi et al. 2017). Several metal ions, including arsenic, cadmium, lead, and mercury, can accumulate in the plants grown in such contaminated areas and can affect human health adversely by consumption of crops containing heavy metals (Kuswandi et al. 2017; Atafar et al. 2010).

Currently, various nanomaterials, including MWCNTs (Bagheri et al. 2013; Guo et al. 2011a, 2011b), AuNPs (Chen et al. 2015; Wan et al. 2015; Zhou et al. 2014), and AgNPs (Ravindran et al. 2012; Zhou et al. 2011a, 2011b) have been utilized to improve the sensitivity of biosensors used in heavy metal detection. Guo et al. (2011a, 2011b) developed aminothiophenol-functionalized MWCNT to monitor Pb(II) ions in water, using a potentiometric sensor. The strong covalent bonding between the carbonyl group of MWCNTs and the amino group of aminothiophenol provided the solid framework for Pb(II) binding, indicating this functionalized sensor as an excellent tool for heavy metal detection. An electrochemical sensor was also developed for the detection of Pb(II), Hg(II), and Cd(II) ions in soil, fish, tap water, and waste water samples, using MWCNT modified with triphenylphosphine, in which the heavy metal ions were bound to the phosphorus group of triphenylphosphine monolayer on the MWCNT-modified carbon screen-printed electrode. This system was promising for routine analytical applications for trace metal detection, with low detection limits (Bagheri et al. 2013). MWCNTs, which were chemically functionalized with 2-((3-silylpropylimino) methyl) phenol, were applied as the solid phase extraction sorbents of a number of metal ions, including Cu^{2+}, Pb^{2+}, Fe^{2+}, Ni^{2+}, and Zn^{2+}, in a variety of food samples (leek, radish, pine apple, and banana), providing successful results for the preconcentration and determination of these analytes in different food materials (Ghaedi et al. 2013).

Colorimetric detection of Hg(II) ion in water by using AuNPs has been reported as another widely-accepted method. In the principle of this method, when no Hg(II) was present, color change from ruby red to royal purple occurs as a result of 2,2′-bipyridyl (bipy) induced aggregation of AuNPs, which could be reverted when Hg(II) exists in the solution (Chen et al. 2015). Gold nanoparticles stabilized with 4-(dimethylamino)pyridine were used in paper-based electrochemical sensors where sintered AuNP films were utilized as a sensing surface. Cysteine was self-assembled on gold, which resulted in the formation of self-assembled monolayers of thiols on the gold surface, and this cysteine-modified gold surface was used to detect copper ions electrochemically. The cysteine-sintered gold nanoparticle films on paper was able to successfully complex copper ions, forming a basis for a simple and a more economical process for the analysis of relevant species (Liana et al. 2013).

Besides AuNPs, AgNPs have also been utilized for optical nanosensor construction for the analysis of Pb(II), Cr(VI), and Hg(II) ions in water (Kumar and Anthony 2014; Farhadi et al. 2012; Ravindran et al. 2012; Zhou et al. 2011a, 2011b). A colorimetric nanosensor has been developed by Zhou et al. (2011a, 2011b) in which 4-mercaptobenzoic acid-functionalized AgNPs were employed as a receptor for the detection of Cu(II) ions. The principle of this method was based on the aggregation of 4-mercaptobenzoic acid-functionalized AgNPs in the presence of Cu(II) ions which led to a color change from bright yellow to purple.

3.2 *Analysis of Biological Food Contaminants*

3.2.1 Naturally Occurring Fungal Toxins

Toxic fungal metabolites, named as mycotoxins, are known to contaminate a wide range of agricultural commodities (at least 25% of the grain produced worldwide), inducing significant health problems as a result of their carcinogenic, teratogenic, mutagenic, and immunosuppressive properties (Kuswandi et al. 2017; Bülbül et al. 2015; Cheli et al. 2008; Sharma et al. 2010). Aflatoxins are the most studied group of mycotoxins (Hussein and Brasel 2001), including aflatoxin B1 (AF B1) as the most common mycotoxin in food crops that is produced by the strains of *Aspergillus parasiticus* and *A. flavus* during crop production and storage (Kuswandi et al. 2017; Sonawane et al. 2014; Sharma et al. 2010). Sharma et al. (2010) developed an immunosensor for the detection of AFB1 that was based on antibody attached AuNPs. In this study, fabrication of AuNPs-based immunoelectrode was achieved by utilizing amino terminal of cysteamine-functionalized AuNPs (C-AuNPs) for immobilization of AF B1 antibody (aAFB1) in the solution phase (aAFB1-C-AuNP) in order to enhance the antibodies loading onto AuNP surface, aAFB1, that was covalently attached to C-AuNPs, was immobilized onto 4-mercaptobenzoic acid (MBA) based self-assembled monolayer on gold electrode (MBA/Au), for the fabrication of bovine serum albumin/aAFB1-C-AuNP/MBA/Au immunoelectrode. The electrode was further explored for AFB1 detection by monitoring the specific interactions between aAFB1 and AFB1 (Sharma et al. 2010). In another study, AuNP-based aptasensor was developed for AFB1 detection. Here, the aptamer-target interaction led to the desorption of the AFB1 binding aptamer from the surface of AuNPs, inducing the aggregation of AuNPs and a change of the color of AuNPs from red to purple (Hosseini et al. 2015).

Besides AFB1, ochratoxin A (OTA) is another widely-studied mycotoxin, produced by *A. ochraceus*, *A. carbonarius*, and *P. verrucosum*, that can be present in different food products including cereals, dried fruits, coffee, cocoa, spices, beer, wine, and meat (Xiao et al. 2015; Fernandez-Cruz et al. 2010). A competitive fluorescence immunoassay was developed for monitoring AFB1 and OTA in naturally contaminated maize samples by using artificial antigen-modified magnetic nanoparticles (MNPs) as immunosensing probes and antibody-functionalized upconversion nanoparticles (UCNPs) as multicolor signal probes; in which the antibody-functionalized UCNPs were attached to the surface of the MNPs by antibody–antigen affinity (Wu et al. 2011). The principle of this assay was based on a gradual decrease in the fluorescent signals of nanocomposites as the concentrations of AFB1 and OTA increased. The use of MNPs provided the advantage of rapid separation and purification of the immunocomplex, as well as a strong fluorescence signal from the UCNPs was observed. This study suggested the UCNP labeling-based competitive fluorescent immunoassay as a promising method for the sensitive detection of small molecule toxins (Wu et al. 2011). Ansari et al. (2010) worked on an impedimetric sensor for OTA detection using sol-gel derived nanostructured zinc oxide (nano-ZnO) film deposited onto indium–tin–oxide (ITO) glass plate for co-immobilization of rabbit-immunoglubin antibodies (r-IgGs) and bovine serum albumin (BSA). This method was based on the electrostatic interactions between positively charged nano-ZnO and carboxyl groups of r-IgG; where the free amino-terminal sites of r-IgG preferably bind to the carboxylic group of the OTA molecules. Strong r-IgGs-OTA interactions that occurred at the nano-ZnO immunoelectrode surface resulted in an amplified electrochemical signal (Ansari et al. 2010). A colorimetric aptasensor was developed for monitoring OTA in red wine that was based on disassembly of the preformed AuNP dimers (Xiao et al. 2015). In the absence of OTA, most of the AuNPs are formed into dimers, whereas in the presence of OTA, the preformed AuNP dimers diassembled into monomers, leading to a color change from blue to red (Xiao et al. 2015). Another study also demonstrated a colorimetric assay for OTA detection in red wine samples, based on aptamer-target recognition and aptamer controlled growth of AuNPs (Soh et al. 2015). In this assay, the interaction of OTA with aptamer resulted in desorption of the aptamer strands that were attached to the surface of aptamer-functionalized AuNPs; thus modulating the amount of aptamer strands attached to the surface of AuNPs. The resulting aptamer coverage on AuNPs surface gave rise to different colored solutions as a result of the growth of AuNPs into nanostructures with varying morphologies. While low aptamer coverage led AuNPs to grow into spherical NPs with resultant red-colored solutions; high aptamer coverage gave rise to AuNPs that grow into branched NPs with resultant blue-colored solutions (Soh et al. 2015).

3.2.2 Foodborne Pathogenic Bacteria and Bacterial Toxins

The presence of pathogens in food or water can lead to food poisoning with symptoms including diarrhea, fever, abdominal cramps, nausea, and vomiting. The pathogenic bacteria that are responsible for the majority of foodborne illness outbreaks are listed as Campylobacter spp., Salmonella spp., *E. coli* O157:H7, *L. monocytogenes*, *Staphylococcus aureus*, and *B. cereus* (Poltronieri et al. 2014; Jokerst et al. 2012; Gandhi and Chikindas 2007; Murphy et al. 2006; Park et al. 2001). The traditional method used in pathogen detection is colony forming unit assay which is time consuming and laborious (Bülbül et al. 2015; Poltronieri et al. 2014). The alternative methods, including polymerase chain reaction and enzyme-linked immunosorbent assay, are also accepted as slow, expensive, and labour-intensive methods which creates the need for developing rapid and cheap methods that will also enable on-site analysis (Bülbül et al. 2015; Law et al. 2015; Sonawane et al. 2014). At this point, nanobiosensors take place as promising alternatives to the existing traditional methods, which provide rapid, sensitive, and specific results in monitoring foodborne pathogens.

A colorimetric enzyme-nanoparticle conjugate system was developed by Miranda et al. (2011) which they applied for detection of *E. coli*. In the principle of this assay, AuNPs, functionalized with quaternary ammonium ligands, were used for inhibiting β-galactosidase (β-Gal) enzyme activity as a result of electrostatic interactions between their head group and the enzyme. Herein, anionic surface of E. coli, which was selected as a model analyte, bound to the cationic surface of AuNPs, releasing β-Gal with simultaneous restoration of its activity. The active enzyme, that comes off as a result of this binding event, converted the pale-yellow substrate into a red product, providing an enzyme-amplified colorimetric readout (Miranda et al. 2011). In another study, an immunochromatographic (IC) test strip was used for monitoring *E. coli* O157 in enriched raw beef, pork, bovine feces, and swine feces samples (Jung et al. 2005). In this method, murine monoclonal antibody to *E. coli* O157:H7 lipopolysaccharide was conjugated with colloidal AuNPs that yielded a visually observable signal in the presence of the target bacteria (Jung et al. 2005). Maurer et al. (2012) worked on the development of a nano-featured multi-step biosensor which was based on carbon nanotubes grown on a graphite substrate that were decorated with AuNPs which were directly synthesized on the nanotube surface and modified with E. coli-specific thiolated RNA (attached to the bound nanoparticles). By the utilization of the compounded nano-materials for sensor development, the electrical behaviour property of CNTs could be retained with an additional advantage of increased surface area providing additional analyte attachment sites. The coating of AuNPs with a selective RNA sequence was shown to enhance E. coli detection compared to their uncoated counterparts (Maurer et al. 2012). In another study, Chen et al. (2008) designed a piezoelectric quartz crystal microbalance (QCM) DNA sensor which utilized sequence-specifically oligonucleotide-functionalized AuNPs for amplification (as "mass enhancers") for real time detection and verification of *E. coli* O157:H7 in real food samples (apple juice, milk, and ground beef). The target DNA of *E. coli* O157:H7 eaeA gene was captured by the single-stranded DNA (ssDNA) of a synthesized thiolated probe which self-assembled on the QCM sensor surface. Exposure of the immobilized probe to the *E. coli* O157:H7 eaeA gene fragment -that was amplified by PCR-induced hybridization which resulted in a mass change and a consequent frequency shift of the piezoelectric biosensor. A second thiolated probe, conjugated to the AuNPs and complementary to the target sequence, was used to amplify the frequency change of the piezoelectric biosensor, leading to a highly sensitive detection of *E. coli* O157:H7 (Chen et al. 2008).

Immunomagnetic separation by the use immuno-modified magnetic NPs (MNPs) was reported to be one of the most common NP-based methods used for monitoring another widely distributed pathogenic bacterium, Salmonella. Salmonella in milk were selectively captured by antibody-conjugated MNPs which were subsequently concentrated and separated from analyte samples by applying an external magnetic field. The MNP-Salmonella complexes were exposed to antibody-immobilized titanium dioxide nanocrystals (TNs), that absorb UV-light, for detecting the complexes by absorbance measurement. Magnetic separation of MNP-Salmonella-TN complexes from solution enabled monitoring the UV-Vis absorption spectrum of the unbound TN solution. Salmonella concentration was determined by the decrease in light absorption intensity (Joo et al. 2012). Huang et al. (2010) worked on developing amine-functionalized MNPs (AF-MNPs) for rapid capture and efficient removal of bacterial pathogens, including E. coli and Salmonella, from water, grape juice, green tea, and urine samples. The principle of this assay was based on the strong electrostatic interactions between the positive charges on the surface of AF-MNPs and negatively charged sites on the surface of target pathogens leading to an efficient adsorption of bacteria on the particle surface (Huang et al. 2010). An electrochemical impedance immunosensor was developed for Salmonella detection by using glassy carbon electrode (GC) modified by covalently conjugated antibodies to single-walled carbon nanotubes (SWCNTs) (Jain et al. 2012). SWCNTs, providing high surface area, enabled both antibody immobilization and electrochemical measurements. The insulating properties of bacterial cell membrane led to the changes in charge transfer resistance and impedance, before and after the formation of antigen-antibody complex, and these changes were monitored by the sensor for determining the presence of Salmonella pathogenic bacterium (Jain et al. 2012).

Cholera toxin (CT), which is secreted by the bacterium *Vibrio cholera*, is the major virulence factor cholera disease (Minke et al. 1999), caused by consuming contaminated food or water (Sack et al. 2004). A colorimetric sensing system was established by Schofield et al. (2007) for CT which was based on lactose-stabilized AuNPs. In the principle of this colorimetric bioassay, the red color of AuNPs that were conjugated with the specifically synthesized lactose derivative shifted to a deep purple color upon aggregation of the nanoparticles as a result of binding of CT. The thiolated lactose derivative mimicked the function of the receptor to which CT binds in the small intestine (Schofield et al. 2007).

Another group of bacterial toxins are Staphylococcal enterotoxins (SEs) that form a group of 21 heat-stable toxins causing foodborne diseases (Asao et al. 2003). Yang et al. (2009) worked on developing an enhanced chemiluminescence (ECL) immunosensor modified with gold nanoparticles for detecting Staphylococcal enterotoxin B (SEB) in mushroom, tomato, and baby food samples using a point-of-care cooled charge-coupled device (CCD) detector that was established for microbial toxin analysis. In this assay, gold nanoparticles conjugated with anti-SEB antibodies were immobilized onto a polycarbonate surface where SEB was monitored by a "sandwich-type" ELISA assay in which SEB was first bound to the immobilized anti-SEB primary antibody followed by binding to the horseradish peroxidase (HRP)-labeled secondary antibody. An ECL assay was utilized for detecting bound HRP, in which the oxidation of luminol to an excited state was catalyzed by HRP in the presence of hydrogen peroxide. Afterwards, the light emission, resulting from the decay of luminol from excited state to ground state, was measured by CCD sensor (Yang et al. 2009).

3.3 *Analysis of Common Food Adulterants*

3.3.1 Melamine Adulteration

Melamine, a trimer of cyanamide with a 1,3,5-triazine skeleton, existing as a white crystalline solid, is a well-studied chemical as one of the major food adulterants in many countries (Rai and Banerjee 2017; Singh and Kumar 2009). Depending on its high nitrogen content (66%), melamine is added illegally into various food products -most commonly in raw milk- in order to falsely increase their protein contents (Rai and Banerjee 2017; Kim et al. 2010b; Gossner et al. 2009). Melamine-added food products can give high protein readings in traditional methods applied for protein analysis since these methods do not distinguish between nitrogen from protein sources and nitrogen from non protein sources (Gossner et al. 2009). Melamine adulteration has been indicated to cause the formation of insoluble melamine-cyanuric acid crystals in kidneys resulting in renal failure. Reported outbreaks of renal failure related to melamine adulteration included infants in China that was traced to the contamination of infant formula with melamine, and pets in the Republic of Korea and in the United States as a result of melamine addition to pet food ingredients (Brown et al. 2007).

Nanosensors are among the modern instrumental methods that are currently utilized for melamine determination in food. Li et al. (2010) worked on the use of AuNPs as probe in establishing a colorimetric method for the detection of melamine in raw milk samples. The principle of this assay relies on the distance-dependent optical properties of AuNPs. In an aqueous solution, AuNPs are known to be stabilized against aggregation due to the existing citrate ions' electrostatic repulsion against van der Waals attraction between AuNPs (Li and Rothberg 2004). The interparticle distances give rise to a particular red colour in AuNPs solution by moderating the collective oscillations of the surface electrons induced by visible light at the suitable wavelength (Takeuchi et al. 1996). Whereas, in the presence of melamine, a red-to-blue color change takes place, as a result of binding of melamine onto the surface of AuNPs through its amine group which leads to the shortening of interparticle distances (Li et al. 2010; Zhong et al. 2004).

In another study, dopamine-stabilized AgNPs were utilized as a colorimetric reader for melamine detection in raw milk (Ma et al. 2011). Dopamine, as a naturally bifunctional molecule, can reduce Ag^{+} to AgNPs and can further functionalize the produced AgNPs to form monodispersed AgNPs, giving typical bright yellow color in their solution. Whereas, in the presence of melamine, the dopamine-functionalized AgNPs become aggregated leading to a bright yellow-to-brown color change in the suspension. Due to this color change of the solution, melamine could be determined using dopamine-modified AgNPs (Ma et al. 2011). Similarly, Song et al. (2015) established a colorimetric assay for melamine detection in milk samples using sulfanilic acid-modified silver nanoparticles (SAA-AgNPs). In the principle of this assay, the affinity of exocyclic amine group of melamine towards the functional groups of SAA led to the aggregation SAA-AgNPs resulting in a color change from bright yellow to blue-green that could be monitored by a UV-visible spectrophotometer (Song et al. 2015).

Aptamer-modified AuNPs were used for monitoring melamine in whole milk, where aptamer was utilized as a melamine recognition element and AuNPs were utilized as the color indicator (Yun et al. 2014). In the mechanism of this method, negative-charged citrate ions coated on the surface of AuNPs prevents the particles from aggregation (Li and Rothberg 2004); while the negative charge of citrate could be neutralized with high salt concentration, resulting in the aggregation of AuNPs with a simultaneous wine red-to-blue color change. Aptamer adsorption onto AuNPs enhances the stability of the particles against NaCl-induced aggregation by the electrostatic interactions between the positively charged bases of aptamers and negatively charged AuNPs (Wei et al. 2007; Wang et al. 2006). In the absence of melamine, AuNPs stay stable by the adsorbed aptamers; whereas in the presence of melamine, aptamers preferably bind with their target, resulting in the subsequent aggregation of AuNPs (as their salt tolerance decreases) and a color change from wine red to blue (Yun et al. 2014).

3.3.2 Meat Adulteration

In meat processing sector, substitution of high-priced meats by lower commercial value species is a serious adulteration problem that challenges consumers' confidence in the sense of economical, ethical, religious, and safety issues. The inclusion of pork in other processed meats has been reported as a common adulteration method in the global Halal food market where Islamic dietary guidelines are applicable (Kuswandi et al. 2015). Ali et al. (2011) used AuNPs as the colorimetric nanobiosensor for determining pork adulteration in beef and chicken meatballs. AuNPs can be protected against salt-induced aggregation by the adsorption of single stranded (ss) DNA. In the absence of pork DNA, the ssDNA probes were available to interact with AuNPs to protect them from salt-induced aggregation; whereas the presence of pork DNA consumed the probes totally by hybridization leading to the aggregation of the AuNPs that resulted in a color change from pinkish-red to gray-purple. This differential aggregation profile of AuNPs in the presence or absence of specific target hybridization was also utilized for monitoring horse adulteration in a number of meat products including fresh meat cuts, slices of roasted beef and pork, ready to eat meatballs, and sausages (Houhoula et al. 2017). In this work, thiolated oligonucleotide-functionalized AuNPs were used for specifically identifying a 69 bp fragment of the horse cytochrome b gene. Upon HCl acid addition, the aggregation of AuNPs in the AuNP-oligonucleotide reaction mixture was prevented in the presence of specific horse DNA target, retaining the original pink colour of the colloidal particles; whereas in the absence of this DNA target, the aggregation of AuNP-oligonucleotide probes was enhanced with the addition of HCl that resulted in a pink-to-purple colour change in the reaction mixture (Houhoula et al. 2017).

3.4 Use of Nanosensors in Food Packaging

Food packages equipped with nanosensors are being developed to track internal and external conditions to which the food products are exposed (Patolsky and Lieber 2005). While external conditions include atmospheric effects, internal conditions stand for the microbial growth and chemical detection inside the package (Ileš et al. 2011). Nanosensors can provide monitoring the quality of the food product throughout food processing chain and the logistic process to the final consumer (Neethirajan and Jayas 2011).

The extensively preferential nanosensors in food packaging are the time – temperature or humidity integrators, which ensure relevant information about these conditions. Moreover, nanosensors used in food packaging can monitor any gases being generated by foods owing to spoilage (Wesley et al. 2014). Additionally, they have also been utilized for the detection of microorganisms, toxic substances and other spoilage factors in foods (Nachay 2007).

Due to the increased demand for high quality, safe and fresh-like foods, progress for the innovative packaging is gaining popularity. A novel humidity nanosensor is one of the indicator system which has been designed by Zhang et al. (2010). The humidity nanosensor based on ZnO/NaCl electrospun composite fiber membrane was prepared *via* electrospinning and calcination. It was found that the whole relative humidity (11%–95%) measurement took 3 s and 6 s recovery time, respectively. The observed results indicate that this newly designed humidity nanodetector exhibits rapid response and recovery behavior, and good reproducibility. Another nanosensor, that consists of a film including carbon-coated copper nanoparticles, has been developed for moisture content analysis. Swelling of the polymer matrix in humid environments causes sensor strips to reflect or absorb different colors of light which can be useful for determination of package moisture levels (Luechinger et al. 2007).

An electronic tongue consisting of an array of nanosensors was used in a packaging material to monitor gases released as a result of food spoilage, by measuring color change (Joseph and Morrison 2006). In another study, an optical sensor based on polyetherimide and Ag nanoparticles has been manufactured for in-package monitoring of food freshness through the measurement of color change. The real-time meat (salmon, chicken, turkey and beef) freshness measurement was conducted by monitoring the colorimetric and UV-vis absorbance responses during 80 hours. While the released gases from salmon, chicken and turkey led to color change of the polyetherimide/Ag films after 20 hours, the gases from beef triggered this alteration after 60 hours of exposure. As a result, this nanosensor has a great potential to detect food spoilage in the packaging (Rvspayeva et al. 2018). With regard to fabrication of highly efficient hydrogen gas sensing ZnO-Si heterojunction gas nanosensors, it has been shown that ZnO nanoparticle size was decisive on the responsivity and sensitivity. The response and recovery times were found to be 8 s and 17 s, respectively, for the smaller ZnO nanoparticles; whereas 60 s and 100 s, respectively, for the larger nanoparticles (Iliadis and Ali 2010). Moreover, light sensitive nanoparticles containing intelligent ink for labels of any foods has been generated to detect oxygen. Additionally, this inexpensive ink could also be utilized to determine the originality of the modified atmosphere inside a package (Dunn 2004). Mills and Hazafy (2008), Lawrie et al. (2013) and Lee et al. (2005) have also developed a nanosized TiO_2 or SnO_2 particles and a redox-active dye based photoactivated indicator ink for in-package oxygen detection.

Various gas/water vapor permeability levels could be reached by different nanostructured plastics. They could satisfy the requirements for the preservation of fruits, vegetables, beverages, and other foods. Packages with increased durability to light, stronger mechanical and thermal properties and less gas absorptive performance can be produced by adding nanoparticles (Asadi and Mousavi 2006). For instance; a transparent plastic film including clay nanoparticles was fabricated by Bayer Company. The nanoclay particles provided a lighter and stronger plastic material with increased heat resistance. Additionally, the clay nanoparticles included into the plastic packaging were reported to slow down oxygen, carbon dioxide, and moisture release from food (Brody 2006).

The nanostructured films have also been studied for their potential to protect the packaged food products from microbial deterioration (Asadi and Mousavi 2006). In a related study, a proteomic biosensor incorporated into a thin film has been improved for the optical detection of enteropathogenic Escherichia coli (Horner et al. 2006). Zinc oxide and magnesium oxide nanoparticles, that are known for their potent antimicrobial properties, might also be utilized in food packaging materials to inhibit the microbial growth (Patton 2006).

On the other hand, Kim et al. (2013) produced a multi-walled carbon nanotube-doped titania-Nafion composite modified glassy carbon electrode to detect bisphenol A (BPA). The authors reported this nanosensor as a promising tool for BPA detection in packaged samples due to its enhanced sensitivity, electrocatalytic activity and voltammetric responses as well as improved reproducibility properties.

3.5 *Monitoring Freshness and Traceability*

Ammonia and simple amine species, including methylamine, dimethylamine and trimethylamine, which contribute to the total volatile bases-nitrogen (TVB-N) values in raw and processed fish samples, are accepted as quality indicators of freshness and as indicators for the potential to spoil (Mohebi and Marquez 2015; Wu and Bechtel 2008). Graphene-based materials are applied as sensors because of its great mechanical, electronic, optical, chemical, and thermal features (Geim and Novoselov 2010). A recent study investigated the detection of fish spoilage by using fluoro-functionalized graphene nanosensor (Rouhani 2019). With regard to increasing electronic selectivity of the nanosensor towards TVBN species, fluorinated graphene has gained an attention because of its enhanced sensitivity, short recovery time and thermal features. A glance at the results reveals the fluorographene as a favorable candidate with more obvious signals (ΔEg; the difference of the highest occupied molecular orbital and lowest unoccupied molecular orbital = 0.036,

0.0308, 0.0499 and 0.0191 eV) compared to the graphene sheet (ΔEg = 0.0015, 0.0006, 0.0041 and 0.0022 eV) for ammonia, methylamine, dimethylamine and trimethylamine, respectively. Xanthine, an intermediary product of purine catabolism, is another chemical used for fish freshness detection. Based on this knowledge, a xanthine sensor has been developed by chitosan modified ZnO nanoparticles on a multiwall carbon nanotube scaffold within a polyaniline matrix with 0.1 mM detection limit (Devi et al. 2012). Another xanthine sensor has been developed through immobilization of xanthine oxidase onto silver nanoparticles (Devi et al. 2013).

Ziyaina et al. (2019) fabricated a silicon dioxide (SiO_2) and Schiff's reagent based colorimetric nanosensor to monitor the shelf-life of milk. It was found that the levels of volatile organic compounds, the products of spoilage bacteria, were in agreement with the color changes throughout the shelf-life period.

Bovine serum albumin (BSA) is an indicator of milk quality, thus the identification of BSA is crucial for food, biochemical and immunological sciences. Since the traditional techniques, that are applied for BSA detection, has several drawbacks (i.e., time-consuming, expensive, limited sensitivity), polydiacetylenes (PDAs) and conjugated polymers have been developed as alternative ways. Although PDAs are appropriate vesicles to be used in sensor systems owing to their chromic properties (Jiang et al. 2015), they have low stability in extended periods of usage and also have restricted colorimetric transition in some conditions (Lee and Kim 2012). De Paula Rezende et al. (2017) studied on developing nanoaggregates by mixing two polymers, PDA and triblock copolymer L64, to overcome the limitations of PDA vesicles. Their results highlighted the PCDA/L64 nanostructures as simple and low-cost analytical tools for rapid BSA detection.

Nanosensors, in which a taste receptor is immobilized on the surface of plypyrrole modified carbon nanotubes, have been developed for assessing the quality of the final product. In this study, reorientation of the polypyrrole, as a result of the presence of bitter compounds in the food sample, was identified by an electrical signal (Song et al. 2012b).

The use of traceability systems for ensuring food safety and quality, for process standardization, and for rapid and effective recalls can be maintained by linking of global positioning, bar/chip coding and hazard analysis critical control point (HACCP) management systems with nanotechnology–based marker assays (Lachance 2008). Nam et al. (2003) have produced gold and nickel based nanodisks to be used as biological labels and tags for tracking food products by combining with chromophores to be illuminated with a laser beam. In another study, a nanobarcode system which fluoresces under ultraviolet light and recognized by a computer scanner has been made by Li et al. (2007).

3.6 *Aroma and Dye Analysis Using Nanosensors*

An electronic nose, arising from aroma detection techniques, can replace mammalian olfactory system. This device consists of groups of gas sensors with an ability to analyze the change in type, quality and quantity of odor/flavor (Dasgupta et al.

2015). In this regard, Del Carlo et al. (2014) developed an electronic nose including an array of seven quartz crystal microbalance sensors modified with gold nanoparticles to differentiate the quality of the olive oil samples according to their sensory profiles. In another study, a nanovesicle-based bioelectronic nose derived from single-walled carbon nanotube-based field effect transistor, which includes human olfactory receptors and calcium ion signal pathways, to analyze amylbutyrate. The results of this study provided a promising perspective for the use of this electronic nose that showed 100 times better sensitivity than the electronic nose that was based on only olfactory receptor protein and carbon nanotube transistors (Jin et al. 2012). Moreover, Park et al. (2012) designed an olfactory-nanovesicle-fused carbon-nanotube-transistor biosensor that could detect the hexanal in the spoiled milk with a superior selectivity and sensitivity. Furthermore, surface plasmon resonance-based immunosensor with the use of gold nanoparticles has been fabricated for the detection of trace amount of benzaldehyde, a fragrant compound in peach products. This immunosensor showed high sensitivity to benzaldehyde and could detect it in the concentration range of 0.1–80 ng/mL (ppb) (Gobi et al. 2008).

Currently, coloring agents/dyes have been applied to the food products as food additives, in order to improve some specific characteristics of the final product and to further promote consumption (Kobylewski and Jacobson 2010). In food industry, the use of synthetic food dyes is much higher than the use of natural dyes due to their superior color, better stability, lower cost and easy color formulation (Dotto et al. 2015). However, adverse health effects, such as allergy and asthmatic reaction (Combes and Haveland-Smith 1982), DNA damage (Hassan 2010) and hyperactivity (Directive 1994), have been linked to the consumption of these synthetic dyes. Thus, several nanosensors have been developed to monitor these synthetic dyes, such as sunset yellow and tartrazine. For example; electrochemical platform based on graphene oxide and multi-walled carbon nanotubes nanocomposite (Qiu et al. 2016), an ionic liquid of 1-allyl-3-methylimidazolium chloride functionalized reduced graphene oxide supported Au nanoparticle composite (Wang and Zhao 2015), β-cyclodextrin-coated poly (diallyldimethylammonium chloride) – functionalized graphene composite film (Ye et al. 2013) and molecularly imprinted polymer polypyrrole sensor (Jiang et al. 2014) etc.

4 Conclusion

Nanotechnology and nanomaterials provide important advantages in the food industry. However, the potential toxicities of nanomaterials are not fully understood and further studies are required, especially in case of their use in the food sector. On the other hand, nanosensors are important tools for ensuring safety and hygiene of foodstuffs by monitoring microbes, toxins and contaminants through all steps of the production and consumption chains. They also provide information for food adulterants, freshness and traceability of the food products as well as internal and external conditions to which the food products are exposed to by means of food packages equipped with nanosensors. For the present situation, nanosensors are relatively expensive with high manufacturing costs. However, after achieving high production volumes and low-cost products, nanosensors may find more extensive use in different sectors including food, medicine, health care, environment, etc., and indeed new nanosensors can be designed to meet special requirements.

References

Aigbogun IE, Mohammed SSD, Orukotan AA, Tanko JD (2017) The role of nanotechnology in food industries - a review. Journal of Advances in Microbiology 7(4):1–9. https://doi.org/10.9734/JAMB/2017/38175

Akhavan O, Ghaderi E (2009) Enhancement of antibacterial properties of Ag nanorods by electric field. Sci Technol Adv Mater 10:015003. https://doi.org/10.1088/1468-6996/10/1/015003

Alfadul SM, Elneshwy AA (2010) Use of nanotechnology in food processing, packaging and safety – review. Afr J Food Agric Nutr Dev 10(6):2719–2739. ISSN:1684 5374

Ali ME, Hashim U, Mustafa S, Che Man YB, Islam KN (2011) Gold nanoparticle sensor for the visual detection of pork adulteration in meatball formulation. J Nanomater 2012:1–7. https://doi.org/10.1155/2012/103607

Andreescu S, Marty JL (2006) Twenty years research in cholinesterase biosensors: from basic research to practical applications. Biomol Eng 23(1):1–15. https://doi.org/10.1016/j.bioeng.2006.01.001

Ansari AA, Kaushik A, Solanki PR, Malhotra BD (2010) Nano structured zinc oxide platform for mycotoxin detection. Bioelectrochemistry 77(2):75–81. https://doi.org/10.1016/j.bioelechem.2009.06.014

Asadi G, Mousavi M (2006) Application of nanotechnology in food packaging. In: Proceedings of 13th IUFoST World Congress 2006

Asao T, Kumeda Y, Kawai T, Shibata T, Oda H, Haruki K, Nakazawa H, Kozaki S (2003) An extensive outbreak of staphylococcal food poisoning due to low-fat milk in Japan: estimation of enterotoxin A in the incriminated milk and powdered skim milk. Epidemiol Infect 130(1):33–40. https://doi.org/10.1017/s0950268802007951

Assadpour E, Jafari SM (2019) A systematic review on nanoencapsulation of food bioactive ingredients and nutraceuticals by various nanocarriers. Crit Rev Food Sci Nutr 59(19):3129–3151. https://doi.org/10.1080/10408398.2018.1484687

Atafar Z, Mesdaghinia A, Nouri J, Homaee M, Yunesian M, Ahmadimoghaddam M, Mahvi AH (2010) Effect of fertilizer application on soil heavy metal concentration. Environ Monit Assess 160(1–4):83–89. https://doi.org/10.1007/s10661-008-0659-x

Baba A, Advincula RC, Knoll W (2001) Simultaneous observation of the electropolymerisation process of conducting polymers by surface plasmon resonance spectroscopy, surface plasmon enhanced light scattering and cyclic voltammetry. In: Möbius D, Miller R (eds) Studies in interface science: novel methods to study interfacial layers. Elsevier Science, New York, USA. https://doi.org/10.1016/S1383-7303(01)80016-8

Bagheri H, Afkhami A, Khoshsafar H, Rezaei M, Shirzadmehr A (2013) Simultaneous electrochemical determination of heavy metals using a triphenylphosphine/MWCNTs composite carbon ionic liquid electrode. Sensors Actuators B Chem 186:451–460. https://doi.org/10.1016/j.snb.2013.06.051

Barry S, O'Riordan A (2016) Electrochemical nanosensors: advances and applications. Reports in Electrochemistry 6:1–14. https://doi.org/10.2147/RIE.S80550

Biju V, Itoh T, Anas A, Sujith A, Ishikawa M (2008) Semiconductor quantum dots and metal nanoparticles: syntheses, optical properties, and biological applications. Analytical and Bioanalytical Chemistry, 391, 2469–2495

Brody AL (2003) Nano food packaging technology. Food Technol 57(12):52–54
Brody AL (2006) Nano and food packaging technologies converge. Food Technol 60(3):92–93
Brown CA, Jeong KS, Poppenga RH, Puschner B, Miller DM, Ellis AE (2007) Outbreaks of renal failure associated with melamine and cyanuric acid in dogs and cats in 2004 and 2007. J Vet Diagn Investig 19(5):525–531. https://doi.org/10.1177/104063870701900510
Brunner T, Wick P, Pius P M, Spohn P, Grass R, Limbach L, Ruinink AB, Stark W (2006) In vitro cytotoxicity of oxide nanoparticles: comparison to Asbestos, silica, and the effect of particle solubility. Environ Sci Technol 40:4374–4381. https://doi.org/10.1021/es052069i
Buck SM, Koo Y-EL, Park E, Xu H, Philbert MA, Brasuel MA, Kopelman R (2004a) Optochemical nanosensor PEBBLEs: photonic explorers for bioanalysis with biologically localized embedding. Curr Opin Chem Biol 8:540–546. https://doi.org/10.1016/j.cbpa.2004.08.011
Buck SM, Xu H, Brasuel M, Philbert MA, Kopelman R (2004b) Nanoscale probes encapsulated by biologically localized embedding (PEBBLEs) for ion sensing and imaging in live cells. Talanta 63:41–59. https://doi.org/10.1016/j.talanta.2003.12.048
Bülbül G, Hayat A, Andreescu S (2015) Portable nanoparticle-based sensors for food safety assessment. Sensors 15(12):30736–30758. https://doi.org/10.3390/s151229826
Caetano J, Machado SAS (2008) Determination of carbaryl in tomato "in natura" using an amperometric biosensor based on the inhibition of acetylcholinesterase activity. Sens Actuators B Chem 129(1):40–46. https://doi.org/10.1016/j.snb.2007.07.098
Centers for Disease Control and Prevention (CDC). 2010. Available online: http://www.cdc.gov/foodborneburden/2011-foodborne-estimates.html (accessed on 5 November 2015)
Cesarino I, Moraes FC, Lanza MR, Machado SA (2012) Electrochemical detection of carbamate pesticides in fruit and vegetables with biosensor based on acetylcholinesterase immobilized on a composite of polyaniline-carbon nanotubes. Food Chem 135(3):873–879. https://doi.org/10.1016/j.foodchem.2012.04.147
Chao G, Zheng G, Jin-Huai L, Xing-Jiu H (2012) The new age of carbon nanotubes: An updated review of functionalized carbon nanotubes in electrochemical sensors. Nanoscale, 4(6):1948–1963
Chaubey A, Malhotra BD (2002) Mediated biosensors. Biosens Bioelectron 17:441–456. https://doi.org/10.1016/s0956-5663(01)00313-x
Cheli F, Pinotti L, Campagnoli A, Fusi E, Rebucci R, Baldi A (2008) Mycotoxin analysis, mycotoxin-producing fungi assays and mycotoxin toxicity bioassays in food mycotoxin monitoring and surveillance. Ital J Food Sci 20(4):447–462
Chen H, Hu W, Li CM (2015) Colorimetric detection of mercury(II) based on 2,2-bipyridyl induced quasi-linear aggregation of gold nanoparticles. Sensors Actuators B Chem 215:421–427. https://doi.org/10.1016/j.snb.2015.03.083
Chen H, Rim RS, Wang IC, Li C, Zhu B, Sun M, Goorsky MS, He X, Yang Y (2017) Quasi-two-dimensional metal oxide semiconductors based ultrasensitive potentiometric biosensors. ACS Nano 11:4710–4718. https://doi.org/10.1021/acsnano.7b00628
Chen M, von Mikecz A (2005) Formation of nucleoplasmic protein aggregates impairs nuclear function in response to SiO_2 nanoparticles. Exp Cell Res 305:51–62. https://doi.org/10.1016/j.yexcr.2004.12.021
Chen SH, Wu VCH, Chuang YC, Lin CS (2008) Using oligonucleotide-functionalized au nano particles to rapidly detect food borne pathogens on a piezoelectric biosensor. J Microbiol Methods 73(1):7–17. https://doi.org/10.1016/j.mimet.2008.01.004
Cheraghi S, Taher MA, Karimi-Maleh H (2017) Highly sensitive square wave voltammetric sensor employing CdO/SWCNTsand room temperature ionic liquid for analysis of vanillin and folic acid infood samples. J Food Compos Anal 62:254–259. https://doi.org/10.1016/j.jfca.2017.06.006
Clark HA, Barker SLR, Brasuel M, Miller MT, Monson E, Parus S, Shi Z-Y, Song A, Thorsrud B, Kopelman R, Ade A, Meixner W, Athey B, Hoyer M, Hill D, Lightle R, Philbert MA (1998) Subcellular optochemical nanobiosensors: probes encapsulated by biologically

localised embedding (PEBBLEs). Sensors Actuators B 51:12–16. https://doi.org/10.1016/S0925-4005(98)00212-3

Combes R, Haveland-Smith R (1982) A review of the genotoxicity of food, drug and cosmetic colours and other azo, triphenylmethane and xanthene dyes. Mutat Res/Rev Genet Toxicol 98:101–243. https://doi.org/10.1016/0165-1110(82)90015-X

Dasgupta N, Ranjan S, Mundekkad D, Ramalingam C, Shanker R, Kumar A (2015) Nanotechnology in agro-food: from field to plate. Food Res Int 69:381–400. https://doi.org/10.1016/j.foodres.2015.01.005

de Paula Rezende J, Ferreira GMD, Ferreira GMD, da Silva LHM, da MdCH S, Pinto MS, dos Santos Pires AC (2017) Polydiacetylene/triblock copolymer nanosensor for the detection of native and free bovine serum albumin. Mater Sci Eng C 70:535–543. https://doi.org/10.1016/j.msec.2016.09.009

Del Carlo M, Fusella G, Pepe A, Sergi M, Di Martino M, Mascini M, Martino G, Cichelli A, Di Natale C, Compagnone D (2014) Novel oligopeptides based e-nose for food quality control: application to extra-virgin olive samples. Qual Assur Saf Crops Food 6:309–317. https://doi.org/10.3920/QAS2013.0377

DeMarco DR, Saaski EW, McCrae DA, Lim DV (1999) Rapid detection of *Eschrichia coli* O157:H7 in ground beef using a fiber-optic biosensor. J Food Prot 62(7):711–716. https://doi.org/10.4315/0362-028X-62.7.711

Devi R, Batra B, Lata S, Yadav S, Pundir C (2013) A method for determination of xanthine in meat by amperometric biosensor based on silver nanoparticles/cysteine modified Au electrode. Process Biochem 48:242–249. https://doi.org/10.1016/j.procbio.2012.12.009

Devi R, Yadav S, Pundir C (2012) Amperometric determination of xanthine in fish meat by zinc oxide nanoparticle/chitosan/multiwalled carbon nanotube/polyaniline composite film bound xanthine oxidase. Anal 137:754–759. https://doi.org/10.1039/C1AN15838D

Directive HAT (1994) European Parliament and council Directive 94/36/EC of 30 June 1994 on colours for use in foodstuffs. O J L 237:0013–0029

Dominguez RB, Alonso GA, Muñoz R, Hayat A, Marty J-L (2015) Design of a novel magnetic particles based electrochemical biosensor for organophosphate insecticide detection in flow injection analysis. Sens Actuat B Chem 208:491–496. https://doi.org/10.1016/j.snb.2014.11.069

Dotto G, Pinto L, Hachicha M, Knani S (2015) New physicochemical interpretations for the adsorption of food dyes on chitosan films using statistical physics treatment. Food Chem 171:1–7. https://doi.org/10.1016/j.foodchem.2014.08.098

Duncan TV (2011) Applications of nanotechnology in food packaging and food safety: barrier materials, antimicrobials and sensors. J Colloid Interface Sci 363:1–24. https://doi.org/10.1016/j.jcis.2011.07.017

Dunn J (2004) A mini revolution. Food Manuf 79:49–50. London, UK. http://www.foodmanufacture.co.uk/news/fullstory.php/aid/472/A%20mini%20revolution.htm Accessed 23 April 2009

Düzgün A, Zelada-Guillén GA, Crespo GA, Macho S, Riu J, Rius FX (2011) Nanostructured materials in potentiometry. Anal Bioanal Chem 399:171–181. https://doi.org/10.1007/s00216-010-3974-3

Elyasi M, Khalilzadeh MA, Karimi-Maleh H (2013) High sensitive voltammetric sensor based on Pt/CNTs nanocompositemodified ionic liquid carbon paste electrode for determination of Sudan I in food samples. Food Chem 141:4311–4317. https://doi.org/10.1016/j.foodchem.2013.07.020

Farhadi K, Forough M, Molaei R, Hajizadeh S, Rafipour A (2012) Highly selective Hg^{2+} colorimetric sensor using green synthesized and unmodified silver nanoparticles. Sensors Actuators B Chem 161:880–885. https://doi.org/10.1016/j.snb.2011.11.052

FDA (2009) U.S. Food and Drug Administration, Fed. Regist. 74 (2009) 11476

Fedorova MD, Andreeva IP, Vilegzhanina ES, Komarov AA, Rubtsova YM, Samsonova JV, Egorov AM (2010) Enzyme-linked immunosorbent assay of chlorampenicol in foodstuff. Appl Biochem Microbiol 46(8):795–801. https://doi.org/10.1134/S0003683810080107

Fernández-Cruz ML, Mansilla ML, Tadeo JL (2010) Mycotoxins in fruits and their processed products: analysis, occurrence and health implications. J Adv Res 1(2):113–122. https://doi.org/10.1016/j.jare.2010.03.002

Fleischmann M, Hendra PJ, McQuillan AJ (1974) Raman spectra of pyridine adsorbed at a silver electrode. Chem Phys Lett 26(2):163–166. https://doi.org/10.1016/0009-2614(74)85388-1

Frasconi M, Tel-Vered R, Riskin M, Willner I (2010) Surface plasmon resonance analysis of antibiotics using imprinted boronic acid-functionalized au nanoparticle composites. Anal Chem 82(6):2512–2519. https://doi.org/10.1021/ac902944k

Fu J, Park B, Siragusa G, Jones L, Tripp R, Zhao Y, Cho Y-J (2008) An Au/Si hetero nanorod-based biosensor for Salmonella detection. Nanotechnology 19:1–7. https://doi.org/10.1088/0957-4484/19/15/155502

Fuertes G, Sánchez-Muñoz OL, Pedrueza E, Abderrafi K, Salgado J, Jiménez E (2011) Switchable bactericidal effects from novel silica-coated silver nanoparticles mediated by light irradiation. Langmuir 26:2826–2833. https://doi.org/10.1021/la1045282

Galvan DD, Yu Q (2018) Surface-enhanced Raman scattering for rapid detection and characterization of antibiotic-resistant Bacteria. Adv Healthc Mater 7(13):1701335

Gandhi M, Chikindas ML (2007) Listeria: a foodborne pathogen that knows how to survive. Int J Food Microbiol 113(1):1–15. https://doi.org/10.1016/j.ijfoodmicro.2006.07.008

García M, Aleixandre M, Horrillo MC (2005) Electronic nose for the identification of spoiled Iberian hams, in: Conference on Electron Devices Conference on Electron Devices, 2005 Spanish

Gardner JW, Bartlett PN (2000) Electronic noses. Principles and applications. Meas Sci Technol 11(7):1087

Geim, AK, Novoselov, KS (2010) The rise of graphene. In Nanoscience and Technology: A Collection of Reviews from Nature Journals 11–19. doi: https://doi.org/10.1142/9789814287005_0002

Ghaedi M, Montazerozohori M, Biyareh MN, Mortazavi K, Soylak M (2013) Chemically bonded multiwalled carbon nanotubes as efficient material for solid phase extraction of some metal ions in food samples. Int J Environ An Ch 93(5):528–542. https://doi.org/10.1080/03067319.2011.629351

Gobi KV, Matsumoto K, Toko K, Miura N (2008) Highly regenerable and storageable all-chemical based PEG-immunosensor chip for SPR detection of ppt levels of fragrant compounds from beverage samples. Sens & Instrumen Food Qual 2:225. https://doi.org/10.1007/s11694-008-9033-5

Goodrace R, Shann B, Gilbert RJ, Timmins EM, McGovern AC, Alsberg BK, Kell DB Logan NA (2000) Detection of the dipicolinic acid biomarker in *Bacillus* spores using curie-point pyrolysis mass spectrometry and Fourier transform infrared spectroscopy. Anal Chem. 72(1): 119–127. doi: https://doi.org/10.1021/ac990661i

Gossner CME, Schlundt J, Embarek PB, Hird S, Lo-Fo-Wong D, Javier J, Beltran O, Teoh KN, Tritscher A (2009) The melamine incident: implications for international food and feed safety. Environ Health Perspect 117(12):1803–1808. https://doi.org/10.1289/ehp.0900949

Gu B, Ruan C, Wang W (2009) Perchlorate detection at nanomolar concentrations by surface-enhanced Raman scattering. Appl Spectrosc 63(1):98–102. https://doi.org/10.1366/000370209787169894

Guo J, Chai Y, Yuan R, Song Z, Zou Z (2011b) Lead (II) carbon paste electrode based on derivatized multi-walled carbon nanotubes: application to lead content determination in environmental samples. Sensors Actuators B Chem 155(2):639–645. https://doi.org/10.1016/j.snb.2011.01.023

Guo S, Wang E (2011) Noble metal nanomaterials: controllable synthesis and application in fuel cells and analytical sensors. Nano Today 6(3):240–264. https://doi.org/10.1016/j.nantod.2011.04.007

Guo Z, Ren J, Wang J, Wang E (2011a) Single-walled carbon nanotubes based quenching of free FAM-aptamer for selective determination of ochratoxin A. Talanta 85(5):2517–2521. https://doi.org/10.1016/j.talanta.2011.08.015

Hassan G (2010) Effects of some synthetic coloring additives on DNA damage and chromosomal aberrations of rats. Arab J Biotech 13:13–24

He L, Liu Y, Mustapha A, Lin M (2011) Antifungal activity of zinc oxide nanoparticles against *Botrytis cinerea* and *Penicillium expansum*. Microbiol Res 166:207–215. https://doi.org/10.1016/j.micres.2010.03.003

He X, Hwang H-M (2016) Nanotechnology in food science: functionality, applicability, and safety assessment. J Food Drug Anal 24:671–681. https://doi.org/10.1016/j.jfda.2016.06.001

He Y, Xu B, Li W, Yu H (2015) Silver nanoparticle-based chemiluminescent sensor array for pesticide discrimination. J Agric Food Chem 63(11):2930–2934. https://doi.org/10.1021/acs.jafc.5b00671

Health and Safety Executive (2009) What Are Pesticides? Available via: http://www.pesticides.gov.uk/about_pesticides.asp?id0219

Heo DN, Yang DH, Moon H, Lee JB, Bae MS, Lee SC, Lee WJ, Sun I, Kwon K (2012) Gold nanoparticles surface-functionalized with paclitaxel drug and biotin receptor as theranostic agents for cancer therapy. Biomaterials 33(3):856–866. https://doi.org/10.1016/j.biomaterials.2011.09.064

Homola J (2008) Surface Plasmon resonance sensors for detection of chemical and biological species. Chem Rev 108(2):462–493. https://doi.org/10.1021/cr068107d

Homola J, Yee SS, Gaugliz G (1999) Surface plasmon resonance sensors: review. Sensors Actuators B Chem 54(1):3–15. https://doi.org/10.1016/S0925-4005(98)00321-9

Horner SR, Mace CR, Rothberg LJ, Miller BL (2006) A proteomic biosensor for enteropathogenic E. coli. Biosens Bioelectron 21:1659–1663. https://doi.org/10.1016/j.bios.2005.07.019

Hosseini M, Khabbaz H, Dadmehr M, Ganjali MR, Mohamadnejad J (2015) Aptamer-based colorimetric and chemiluminescence detection of aflatoxin b1 in foods samples. Acta Chim Slov 62(3):721–728. https://doi.org/10.17344/acsi.2015.1358

Huang YF, Wang YF, Yan XP (2010) Amine-functionalized magnetic nano particles for rapid capture and removal of bacterial pathogens. Environ Sci Technol 44(20):7908–7913. https://doi.org/10.1021/es102285n

Huet AC, Charlier C, Weigel S, Godefroy SB, Delahaut PH (2009) Validation of an optical surface plasmon resonance biosensor assay for screening (fluoro)quinolones in egg, fish, and poultry meat. Food Addit Contam A Chem 26(10):1341–1347. https://doi.org/10.1080/02652030903013328

Hussein HS, Brasel JM (2001) Toxicity, metabolism, and impact of mycotoxins on humans and animals. Toxicology 167(2):101–134. https://doi.org/10.1016/s0300-483x(01)00471-1

Ileš D, Martinović G, Kozak D (2011) Review of potential use, benefits and risks of nanosensors and nanotechnologies in food. Strojarstvo 53:127–136

Iliadis AA, Ali HA (2010) Properties of fast response room temperature ZnO-Si heterojunction gas nanosensors. IEEE Trans Nanotechnol 10:652–656. https://doi.org/10.1109/TNANO.2010.2089637

Jadzinsky PD, Calero G, Ackerson CJ, Bushnell DA, Kornberg RD (2007) Structure of a thiol monolayer-protected gold nanoparticle at 1.1 a resolution. Science 318(5849):430–433. https://doi.org/10.1126/science.1148624

Jain S, Singh SR, Horn DW, Davis VA, Ram MK, Pillai S (2012) Development of an antibody functionalized carbon nano tube biosensor for food borne bacterial pathogens. J Biosens Bioelectron (S11):1–7. https://doi.org/10.4172/2155-6210.S11-002

Ji X, Zheng J, Xu J, Rastogi VK, Cheng TC, DeFrank JJ, Leblanc RM (2005) (CdSe)ZnS quantum dots and Organophosphorus hydrolase bioconjugate as biosensors for detection of Paraoxon. J Phys Chem B 109(9):3793–3799. https://doi.org/10.1021/jp044928f

Jiang L, Luo J, Dong W, Wang C, Jin W, Xia Y, Wang H, Ding H, Jiang L, He H (2015) Development and evaluation of a polydiacetylene based biosensor for the detection of H5 influenza virus. J Virol Methods 219:38–45. https://doi.org/10.1016/j.jviromet.2015.03.013

Jiang S, Xu J, Xu P, Liu L, Chen Y, Qiao C, Yang S, Sha Z, Zhang J (2014) A novel molecularly imprinted sensor for direct tartrazine detection. Anal Lett 47:323–330. https://doi.org/10.1080/00032719.2013.834442

Jin HJ, Lee SH, Kim TH, Park J, Song HS, Park TH, Hong S (2012) Nanovesicle-based bioelectronic nose platform mimicking human olfactory signal transduction. Biosens Bioelectron 35:335–341. https://doi.org/10.1016/j.bios.2012.03.012

Jokerst JC, Adkins JA, Bisha B, Mentele MM, Goodridge LD, Henry CS (2012) Development of a paper-based analytical device for colorimetric detection of select. Foodborne Pathog Anal Chem 84(6):2900–2907. https://doi.org/10.1021/ac203466y

Joo J, Yim C, Kwon D, Lee J, Shin HH, Cha HJ, Jeon S (2012) A facile and sensitive detection of pathogenic bacteria using magnetic nanoparticles and optical nano crystal probes. Analyst 137(16):3609–3612. https://doi.org/10.1039/c2an35369e

Joseph T, Morrison M (2006) Nanotechnology in agriculture and food, Institute of nanotechnology. Website: http://www.nanoforum.org/dateien/temp/nanotechnology%20in%20agriculture%20and%20food.pdf?20032007152346

Jung BY, Jung SC, Kweon CH (2005) Development of a rapid immunochromatographic strip for detection of Escherichia coli O157. J Food Prot 68(10):2140–2143. https://doi.org/10.4315/0362-028x-68.10.2140

Kamer G, Demirata B, Bayraktar R, Ozyurt D, Apak R (2019) Nanoceria-based reactive species scavenging activity of antioxidants using N,N-dimethyl-*p*-phenylenediamine (DMPD) probe. Anal Methods 11:1908–1915. https://doi.org/10.1039/C8AY02561D

Kara M, Uzun L, Kolaylı S, Denizli A (2013) Combining molecular imprinted nanoparticles with surface plasmon resonance nanosensor for chloramphenicol detection in honey. J Appl Polym Sci 129(4):2273–2279. https://doi.org/10.1002/app.38936

Karimi-Maleh H, Moazampour M, Yoosefian M, Sanati AL, Tahernejad-Javazmi F, Mahani M (2014) An electrochemical Nanosensor for simultaneous Voltammetric determination of ascorbic acid and Sudan I in food samples. Food Anal Methods 7:2169–2176. https://doi.org/10.1007/s12161-014-9867-x

Keiji S, Zhong-You S, Raoul K, Hiroshi M (1996) Three-dimentional pH microprobing with an optically-manipulated fluorescent particle. Chem Lett 25(2):141–142. https://doi.org/10.1246/cl.1996.141

Kim BK, Kim JY, Kim DH, Choi HN, Lee WY (2013) Electrochemical determination of bisphenol a at carbon nanotube-doped titania-nafion composite modified electrode. Bull Kor Chem Soc 34:1065–1069. https://doi.org/10.5012/bkcs.2013.34.4.1065

Kim JS, Taitt CR, Ligler FS, Anderson GP (2010a) Multiplexed magnetic microsphere immunoassays for detectionof pathogens in foods. Sens & Instrumen Food Qual 4(2):73–81. https://doi.org/10.1007/s11694-010-9097-xs

Kim CW, Yun JW, Bae IH, Lee JS, Kang HJ, Joo KM, Jeong HJ, Chung JH, Park YH, Lim KM (2010b) Determination of spatial distribution of melamineecyanuric acid crystals in rat kidney tissue by histology and imaging matrix assisted laser desorption/ionization quadrupole time-of-flight mass spectrometry. Chem Res Toxicol 23(1):220–227. https://doi.org/10.1021/tx900354z

Knoll W (1998) Interfaces and thin films as seen by bound electromagnetic waves. Annu Rev Phys Chem 49(1):569–638. https://doi.org/10.1146/annurev.physchem.49.1.569

Kobylewski S, Jacobson MF (2010) Food dyes: a rainbow of risks center for science in the public interest, Washington, DC. Available at: http://cspinet.org/new/pdf/food-dyes-rainbow-of-risks.pdf

Kumar SV, Fareedullah M, Sudhakar Y, Venkateswarlu B, Kumar EA (2010) Current review on organophosphorus poisoning. Arch Appl Sci Res 2(4):199–215

Kumar VV, Anthony SP (2014) Silver nanoparticles based selective colorimetric sensor for Cd^{2+}, Hg^{2+} and Pb^{2+} ions: tuning sensitivity and selectivity using co-stabilizing agents. Sensors Actuators B Chem 191:31–36. https://doi.org/10.1016/j.snb.2013.09.089

Kuswandi B, Cendekiawan KA, Kristiningrum N, Ahmad M (2015) Pork adulteration in commercial meatballs determined by chemometric analysis of NIR spectra. J Food Meas Charact 9(3):313–323. https://doi.org/10.1007/s11694-015-9238-3

Kuswandi B, Fikriyah CI, Gani AA (2008) An optical fiber biosensor for chlorpyrifos using a single solegel film containing acetylcholinesterase and bromothymol blue. Talanta 74(4):613–622. https://doi.org/10.1016/j.talanta.2007.06.042

Kuswandi B, Futra D, Heng LY (2017) Nanosensors for the detection of food contaminants. In: Oprea AE, Grumezescu AM (eds) Nanotechnology applications in food: flavor, stability, nutrition and safety. Elsevier Academic Press, Massachusetts, USA. https://doi.org/10.1016/B978-0-12-811942-6.00015-7

Kuswandi B, Mascini M (2005) Enzyme inhibition based biosensors for environmental monitoring. Curr Enzym Inhib 1(3):11–21. https://doi.org/10.2174/1573408054580484

Kuswandi B, Suwandari NW (2007) Simple and sensitive flow injection optical fibre biosensor based on immobilised enzyme for monitoring of pesticides. Sens Trans 76(2):978–986

Lachance PA (2008) Nutraceutical/drug/anti-terrorism safety assurance through traceability. In: Nutraceutical and functional food regulations in the United States and around the world. Elsevier, pp 109–112. https://doi.org/10.1016/B978-012373901-8.00008-1

Law JW, Ab Mutalib NS, Chan KG, Lee LH (2015) Rapid methods for the detection of foodborne bacterial pathogens: principles, applications, advantages and limitations. Front Microbiol 5:770. https://doi.org/10.3389/fmicb.2014.00770

Lawrie K, Mills A, Hazafy D (2013) Simple inkjet-printed, UV-activated oxygen indicator. Sens Actuators B Chem 176:1154–1159. https://doi.org/10.1016/j.snb.2012.10.125

Lebrun M, Plotto A, Goodner K, Ducamp MN, Baldwin E (2008) Discrimination of mango fruit maturity by volatiles using the electronic nose and gas chromatography. Postharvest Biology Technology 48:122–131. https://doi.org/10.1016/j.postharvbio.2007.09.010

Lee J, Kim J (2012) Multiphasic sensory alginate particle having polydiacetylene liposome for selective and more sensitive multitargeting detection. Chem Mater 24:2817–2822. https://doi.org/10.1021/cm3015012

Lee SK, Sheridan M, Mills A (2005) Novel UV-activated colorimetric oxygen indicator. Chem Mater 17:2744–2751. https://doi.org/10.1021/cm0403863

Li H, Rothberg L (2004) Label-free colorimetric detection of specific sequences in genomic DNA amplified by the polymerase chain reaction. J Am Chem Soc 126(35):10958–10961. https://doi.org/10.1021/ja048749n

Li L, Li B, Cheng D, Mao L (2010) Visual detection of melamine in raw milk using gold nanoparticles as colorimetric probe. Food Chem 122(3):895–900. https://doi.org/10.1016/j.foodchem.2010.03.032

Li X, Jin Z, Wang J (2007) Complexation of allyl isothiocyanate by α-and β-cyclodextrin and its controlled release characteristics. Food Chem 103:461–466. https://doi.org/10.1016/j.foodchem.2006.08.017

Li Y-S, Church JS (2014) Raman spectroscopy in the analysis of food and pharmaceutical nanomaterials. J Food Drug Anal 22(1):29–48. https://doi.org/10.1016/j.jfda.2014.01.003

Li Z, Sheng C (2014) Nanosensors for food safety. J Nanosci Nanotechno 14(1):905–912. https://doi.org/10.1166/jnn.2014.8743

Liana DD, Raguse B, Wieczorek L, Baxter GR, Chuah K, Gooding JJ, Chow E (2013) Sintered gold nanoparticles as an electrode material for paper-based electrochemical sensors. RSC Adv 3:8683–8691. https://doi.org/10.1039/c3ra00102d

Liang M, Fan K, Pan Y, Jiang H, Wang F, Yang D, Lu D, Feng J, Zhao J, Yang L, Yan X (2013) Fe3O4 magnetic nanoparticle peroxidase mimetic-based colorimetric assay for the rapid detection of Organophosphorus pesticide and nerve agent. Anal Chem 85(1):308–312. https://doi.org/10.1021/ac302781r

Lin X, Ni Y, Kokot S (2013) Glassy carbon electrodes modified with gold nanoparticles for the simultaneous determination of three food antioxidants. Anal Chim Acta 765:54–62. https://doi.org/10.1016/j.aca.2012.12.036

Lin Y, Hamme L (2014) Targeted highly sensitive detection/eradication of multi-drug resistant Salmonella DT104 through gold nanoparticle-SWCNT bioconjugated nanohybrids. Analyst 139:3702–3705. https://doi.org/10.1039/C4AN00744A

Liu MM, Xu L, Cheng WQ, Zeng Y, Yan ZY (2008b) Surface modified CdS quantum dots as luminescent probes for sulfadiazine determination. Spectrochim Acta A 70(5):1198–1202. https://doi.org/10.1016/j.saa.2007.10.040

Liu S, Yuan L, Yue X, Zheng Z, Tang Z (2008a) Recent advances in nanosensors for organophosphate pesticide detection. Adv Powder Technol 19(5):419–441. https://doi.org/10.1016/S0921-8831(08)60910-3

Liu TT, Lin Y-H, Hung C-S, Liu T-J, Chen Y, Huang Y-C, Tsai T-H, Wang H-H, Wang D-W, Wang J-K, Wang Y-L, Lin C-H (2009) A high speed detection platform based on surface-enhanced Raman scattering for monitoring antibiotic-induced chemical changes in bacteria cell wall. PLoS One 4(5):e5470. https://doi.org/10.1371/journal.pone.0005470

Long TC, Saleh N, Tilton RD, Lowry GV, Veronesi B (2006) Titanium dioxide (P25) produces reactive oxygen species in immortalized brain microglia (BV2): implications for nanoparticle neurotoxicity. Environ Sci Technol 40(14):4346–4352. https://doi.org/10.1021/es060589n

Luechinger NA, Loher S, Athanassiou EK, Grass RN, Stark WJ (2007) Highly sensitive optical detection of humidity on polymer/metal nanoparticle hybrid films. Langmuir 23:3473–3477. https://doi.org/10.1021/la062424y

Ma Y, Niu H, Zhang X, Cai Y (2011) One-step synthesis of silver/dopamine nanoparticles and visual detection of melamine in raw milk. Analyst 136(20):4192–4196. https://doi.org/10.1039/c1an15327g

Malik P, Katyal V, Malik V, Asatkar A, Inwati G, Mukherjee TK (2013) Nanobiosensors: Concepts and Variations, ISRN Nanomaterials, Article ID 327435. doi: https://doi.org/10.1155/2013/327435

Martin-Orozco N, Touret N, Zaharik ML, Park E, Kopelman R, Miller S, Finlay BB, Gros P, Grinstein S (2006) Visualization of Vacuolar acidification-induced transcription of genes of pathogens inside macrophages. Mol Biol Cell 17(1):498–510. https://doi.org/10.1091/mbc.e04-12-1096

Matlock-Colangelo L, Baeumner AJ (2012) Recent progress in the design of nanofiber-based biosensing devices. Lab Chip 12(15):2612–2620. https://doi.org/10.1039/c2lc21240d

Maurer EI, Comfort KK, Hussain SM, Schlager JJ, Mukhopadhyay SM (2012) Novel platform development using an assembly of carbon nanotube, nano gold and immobilized RNAcapture element towards rapid, selective sensing of bacteria. Sensor 12(6):8135–8144. https://doi.org/10.3390/s120608135

Mazumdar SD, Hartmann M, Kämpfer P, Keusgen M (2007) Rapid method for detection of Salmonella in milk by surface plasmon resonance (SPR). Biosens Bioelectron 22(9–10):2040–2046. https://doi.org/10.1016/j.bios.2006.09.004

Men H, Chen D, Zhang X, Liu J, Ning K (2014) Data Fusion of Electronic Nose and Electronic Tongue for Detection of Mixed Edible-Oil, Journal of Sensors, Article ID 840685, doi: https://doi.org/10.1155/2014/840685

Mercante LA, Scagion VP, Migliorini FL, Mattoso LHC, Correa DS (2017) Electrospinning-based (bio)sensors for food and agriculturalapplications: a review. Trens in Anal Chem 91:91–103. https://doi.org/10.1016/j.trac.2017.04.004

Mills A, Hazafy D (2008) A solvent-based intelligence ink for oxygen. Anal 133:213–218. https://doi.org/10.1039/B713450A

Minke WE, Roach C, Hol WGJ, Verlinde CLMJ (1999) Structure-based exploration of the ganglioside GM1 binding sites of E. coli heat-labile enterotoxin and cholera toxin for the discovery of receptor antagonists. Bio Chem 38(18):5684–5692. https://doi.org/10.1021/bi982649a

Miranda OR, Li X, Garcia-Gonzalez L, Zhu ZJ, Yan B, Bunz UH, Rotello VM (2011) Colorimetric bacteria sensing using a supramolecular enzyme-nanoparticle biosensor. J Am Chem Soc 133(25):9650–9653. https://doi.org/10.1021/ja2021729

Mohebi E, Marquez L (2015) Intelligent packaging in meat industry: an overview of existing solutions. J Food Sci Technol 52:3947–3964. https://doi.org/10.1007/s13197-014-1588-z

Müller C, David L, Chiş V, Pinzaru SC (2014) Detection of thiabendazole applied on citrus fruits and bananas using surface enhanced Raman scattering. Food Chem 145:814–820. https://doi.org/10.1016/j.foodchem.2013.08.136

Murphy C, Carroll C, Jordan KN (2006) Environmental survival mechanisms of the food-borne pathogen campylobacter jejuni. J Appl Microbiol 100(4):623–632. https://doi.org/10.1111/j.1365-2672.2006.02903.x

Murphy CJ (2002) Optical sensing with quantum dots. Anal Chem 74(19):520 A–526 A. https://doi.org/10.1021/ac022124v

Nachay K (2007) Analyzing nanotechnology. Food Technol 61(1):34–36

Nagarajan R (2008) Nanoparticles: Building Blocks for Nanotechnology. In: Nanoparticles: Synthesis, Stabilization, Passivation, and Functionalization; Nagarajan, R., et al.; ACS Symposium Series; American Chemical Society; Washington, DC.

Nam JM, Thaxton CS, Mirkin CA (2003) Nanoparticle-based bio-bar codes for the ultrasensitive detection of proteins. Sci 301:1884–1886. https://doi.org/10.1126/science.1088755

Nanduri V, Kim G, Morgan MT, Ess D, Hahm B-K, Kothapalli A, Valadez A, Geng T, Bhunia AK (2006) Antibody immobilization on waveguides using a flow–through system shows improved Listeria monocytogenes detection in an automated Fiber optic biosensor: RAPTORTM. Sensors 6(8):808–822. https://doi.org/10.3390/s6080808

Navarro E, Piccapietra F, Wagner B, Marconi F, Kaegi R, Odzak N, Sigg L, Behra R (2008) Toxicity of silver nanoparticles to Chlamydomonas reinhardtii. Environ Sci Technol 42(23):8959–8964. https://doi.org/10.1021/es801785m

Neethirajan S, Jayas DS (2011) Nanotechnology for the food and bioprocessing industries. Food Bioprocess Technol 4(1):39–47. https://doi.org/10.1007/s11947-010-0328-2

Nesakumar N, Sethuraman S, Krishnan UM, Rayappan JB (2016) Electrochemical acetylcholinesterase biosensor based on ZnO nanocuboids modified platinum electrode for the detection of carbosulfan in rice. Biosens Bioelectron 77:1070–1077. https://doi.org/10.1016/j.bios.2015.11.010

Nu Mega Personal communication with company representative November 2007

Ornatska M, Sharpe E, Andreescu D, Andreescu S (2011) Paper bioassay based on ceria nanoparticles as colorimetric probes. Anal Chem 83(11):4273–4280. https://doi.org/10.1021/ac200697y

Ozdemir Olgun FA, Üzer A, Demirata Ozturk B, Apak R (2018) A novel cerium oxide nanoparticles–based colorimetric sensor using tetramethyl benzidine reagent for antioxidant activity assay. Talanta 182:55–61. https://doi.org/10.1016/j.talanta.2018.01.047

Houhoula DP, Kouzilou M, Tzogias C, Kyrana V, Sflomos C, Tsaknis J, Pavlos Lougovois V (2017) Effectual gold nanoprobe sensor for screening horse adulteration in meat products. J Food Res 6(4):34–39. https://doi.org/10.5539/jfr.v6n4p34

Park J, Lim JH, Jin HJ, Namgung S, Lee SH, Park TH, Hong S (2012) A bioelectronic sensor based on canine olfactory nanovesicle–carbon nanotube hybrid structures for the fast assessment of food quality. Anal 137:3249–3254. https://doi.org/10.1039/C2AN16274A

Park SS, Worobo RW, Durst RA (2001) Escherichia coli O157: H7 as an emerging food-borne pathogen: a literature review. Crit Rev Biotechnol 21(1):27–48. https://doi.org/10.1080/20013891081674

Patolsky F, Lieber CM (2005) Nanowire nanosensors. Mater Today 8:20–28. https://doi.org/10.1016/S1369-7021(05)00791-1

Patton D (2006) Australian nanotech firm promises better food packaging film, Website. http://www.foodproductiondaily.com/news

Peng C, Zheng L, Chen Q, Shen M, Guo R, Wang H, Cao X, Zhang G, Shi X (2012) PEGylated dendrimer-entrapped gold nanoparticles for in vivo blood pool and tumor imaging by computed tomography. Biomaterials 33(4):1107–1119. https://doi.org/10.1016/j.biomaterials.2011.10.052

Perez-Lopez B, Merkoçi A (2011) Nanomaterials based biosensors for food analysis applications. Trends Food Sci Technol 22(11):625–639. https://doi.org/10.1016/j.tifs.2011.04.001

Peris M, Escuder-Gilabert L (2009) A 21st century technique for food control: electronic noses. Anal Chim Acta 638(1):1–15. https://doi.org/10.1016/j.aca.2009.02.009
Phillips RL, Miranda OR, You CC, Rotello VM, Bunz UHF (2008) Rapid and efficient identification of Bacteria using gold-nanoparticle–poly (para phenyleneethynylene) constructs. Angew Chem Int Ed Engl 47(14):2590–2594. https://doi.org/10.1002/anie.200703369
Pinheiro C, Rodrigues CM, Schäfer T, Crespo JG (2002) Monitoring the aroma production during wine-must fermentation with an electronic nose. Biotechnol Bioeng 77(6):632–640. https://doi.org/10.1002/bit.10141
Poltronieri P, Mezzolla V, Primiceri E, Maruccio G (2014) Biosensors for the detection of food pathogens. Foods 3(3):511–526. https://doi.org/10.3390/foods3030511
Pope CN (1999) Organophosphorus pesticides: do they all have the same mechanism of toxicity? J Toxicol Environ Health B Crit Rev 2(2):161–181. https://doi.org/10.1080/109374099281205
Qiu X, Lu L, Leng J, Yu Y, Wang W, Jiang M, Bai L (2016) An enhanced electrochemical platform based on graphene oxide and multi-walled carbon nanotubes nanocomposite for sensitive determination of sunset yellow and Tartrazine. Food Chem 190:889–895. https://doi.org/10.1016/j.foodchem.2015.06.045
Qu W, Liu Y, Liu D, Wang Z, Jiang X (2011) Copper-mediated amplification allows readout of immunoassays by the naked eye. Angew Chem Int Ed Engl 50(15):3442–3445. https://doi.org/10.1002/anie.201006025
Rai N, Banerjee D (2017) Melamine adulteration of food: detection by point-of-care testing tool. Curr Sci 112(3):454–456
Ramgir NS (2013) Electronic Nose Based on Nanomaterials: Issues, Challenges, and Prospects, International Scholarly Reseach Notices (ISRN) Nanomaterials, Article ID 941581. doi: https://doi.org/10.1155/2013/941581
Ravelo-Perez LM, Hernandez-Borges J, Rodriguez-Delgado MA (2008) Multi-walled carbon nanotubes as efficient solid-phase extraction materials of organophosphorus pesticides from apple, grape, orange and pineapple fruit juices. J Chromatogr A 1211(1–2):33–42. https://doi.org/10.1016/j.chroma.2008.09.084
Ravindran A, Elavarasi M, Prathna TC, Raichur AM, Chandrasekaran N, Mukherjee A (2012) Selective colorimetric detection of nanomolar Cr (VI) in aqueous solutions using unmodified silver nanoparticles. Sensors Actuators B Chem 166-167:365–371. https://doi.org/10.1016/j.snb.2012.02.073
Ray SS, Yamada K, Okamoto M, Ueda K (2002) Polylactide-layered silicate Nanocomposite: a novel biodegradable material. Nano Lett 2(10):1093–1096. https://doi.org/10.1021/nl0202152
Rezaei A, Nasirpour A, Fathi M (2015) Application of cellulosic Nanofibers in food science using electrospinning and its potential risk. Compr Rev Food Sci Food Saf 14:269–284. https://doi.org/10.1111/1541-4337.12128
Rivett J, Speer DV (2009) Oxygen scavenging film with good interplay adhesion., US Patent 75141512
Roco MC (2003) Nanotechnology: convergence with modern biology and medicine. Curr Opin Biotechnol 14(3):337–346. https://doi.org/10.1016/S0958-1669(03)00068-5
Roco MC, Bainbridge WS (eds) (2001) National Science Foundation, NSET workshop report, societal implications of nanoscience and nanotechnology. Kluwer Academic Publishers, Dordrecht
Rouhani M (2019) Fluoro-functionalized graphene as a promising nanosensor in detection of fish spoilage: a theoretical study. Chem Phys Lett 719:91–102. https://doi.org/10.1016/j.cplett.2019.02.001
Rvspayeva A, Jones TD, Hughes PA, Esfahani MN, Shuttleworth MP, Harris RA, Kay RW, Desmulliez MP, Marques-Hueso J (2018) PEI/Ag as an optical gas nano-sensor for intelligent food packaging, 2018 IEEE 18th International Conference on Nanotechnology (IEEE-NANO), IEEE. pp. 1–4
Sack DA, Sack RB, Nair GB, Siddique AK (2004) Cholera. Lancet 363(9404):223–233. https://doi.org/10.1016/s0140-6736(03)15328-7

Saha B, Evers TH, Prins MWJ (2014) How antibody surface coverage on nanoparticles determines the activity and kinetics of antigen capturing for biosensing. Anal Chem 86(16):8158–8166. https://doi.org/10.1021/ac501536z

Sarkar P, Panigrahi SS, Roy E, Banerjee P (2017) In book: Portable biosensors and point-of-care systems, Chapter: Nanosensors in food safety, Editor: Spyridon E. Kintzios, e-book, January 2017, doi: https://doi.org/10.1049/PBHE003E_ch10

Saute B, Narayanan R (2011) Solution-based direct readout surface enhanced Raman spectroscopic (SERS) detection of ultra-low levels of thiram with dogbone shaped gold nanoparticles. Analyst 136(3):527–532. https://doi.org/10.1039/C0AN00594K

Schofield CL, Field RA, Russell DA (2007) Glyconanoparticles for the colorimetric detection of cholera toxin. Anal Chem 79(4):1356–1361. https://doi.org/10.1021/ac061462j

Sharma A, Matharu Z, Sumana G, Solanki PR, Kim CG, Malhotra BD (2010) Antibody immobilized cysteamine functionalized-gold nanoparticles for aflatoxin detection. Thin Sol Films 519(3):1213–1218. https://doi.org/10.1016/j.tsf.2010.08.071

Sharp D, Burkitt R (2015) Carbon materials for analytical electrochemistry: printed carbon materials and composites. Mater Technol 30(B3):B155–B162. https://doi.org/10.1179/1753555714Y.0000000213

Sharpe E, Frasco T, Andreescu D, Andreescu S (2013) Portable ceria nanoparticle-based assay for rapid detection of food antioxidants (NanoCerac). Analyst 138(1):249–262. https://doi.org/10.1039/c2an36205h

Shikha S, Salafi T, Cheng J, Zhang Y (2017) Versatile design and synthesis of nano-barcodes. Chem Soc Rev 46:7054–7093. https://doi.org/10.1039/C7CS00271H

Singh A, Choudhary M, Singh MP, Verma HN, Singh SP, Arora K (2015) DNA functionalized direct electro-deposited gold nanoaggregates for efficient detection of salmonella typhi. Bioelectrochemistry 105:7–15. https://doi.org/10.1016/j.bioelechem.2015.03.005

Singh BK, Walker A (2006) Microbial degradation of organophosphorus compounds. FEMS Microbiol Rev 30(3):428–471. https://doi.org/10.1111/j.1574-6976.2006.00018.x

Singh M, Kumar V (2009) Preparation and characterization of melamine–formaldehyde polyvinylpyrrolidone polymer resin for better industrial uses over melamine resins. J Appl Polym Sci 114:1870–1878. https://doi.org/10.1002/app.30805

Sinha R, Kulldorff M, Gunter MJ, Strickland P, Rothman N (2005) Dietary Benzo[a]Pyrene intake and risk of colorectal adenoma. Cancer Epidemiol Biomark Prev 14(8):2030–2034. https://doi.org/10.1158/1055-9965.EPI-04-0854

Soh JH, Lin Y, Rana S, Ying JY, Stevens MM (2015) Colorimetric detection of small molecules in complex matrixes via target-mediated growth of aptamer-functionalized gold nanoparticles. Anal Chem 87(15):7644–7652. https://doi.org/10.1021/acs.analchem.5b00875

Solsana F, Méndez JP (2003) Water Disinfection, Pan American Center for Sanitary Engineering and Environmental Sciences, Pan American Health Organization, Lima, Peru, 2003, http://whqlibdoc.who.int/paho/2003/a85637.pdf

Somerset V, Baker P, Iwuoha E (2009) Mercaptobenzothiazole-on-gold organic phase biosensor systems: 1. Enhanced organosphosphate pesticide determination. J Environ Sci Health B 44(2):164–178. https://doi.org/10.1080/03601230802599092

Somerset VS, Klink MJ, Baker PGL, Iwuoha EI (2007) Acetylcholinesterase polyaniline biosensor investigation of organophosphate pesticides in selected organic solvents. J Environ Sci Health B 42(3):297–304. https://doi.org/10.1080/03601230701229288

Sonawane SK, Arya SS, LeBlanc JG, Jha N (2014) Use of nanomaterials in the detection of food contaminants. European J Nutr Food Saf 4(4):301–317. https://doi.org/10.9734/ejnfs

Song KM, Jeong E, Jeon W, Cho M, Ban C (2012a) Aptasensor for ampicillin using gold nanoparticle based dual fluorescenceecolorimetric methods. Anal Bioanal Chem 402(6):2153–2161. https://doi.org/10.1007/s00216-011-5662-3

Song HS, Kwon OS, Lee SH, Park SJ, Kim UK, Jang J, Park TH (2012b) Human taste receptor-functionalized field effect transistor as a human-like nanobioelectronic tongue. Nano Lett 13:172–178. https://doi.org/10.1021/nl3038147

Song J, Wu F, Wan Y, Ma L (2015) Colorimetric detection of melamine in pretreated milk using silver nanoparticles functionalized with sulfanilic acid. Food Control 50:356–361. https://doi.org/10.1016/j.foodcont.2014.08.049

Sozer N, Kokini JL (2012) In: Pico Y (ed) The applications of nanotechnology in chemical analysis of food: techniques and applications. Academic, San Diego, California, USA

Srivastava AK, Dev A, Karmakar S (2018) Nanosensors and nanobiosensors in food and agriculture. Environ Chem Lett 16(1):161–182. https://doi.org/10.1007/s10311-017-0674-7

Subbiah T, Bhat GS, Tock RW, Parameswaran S, Ramkumar SS (2005) Electrospinning of Nanofibers. J Appl Polym Sci 96(2):557–569. https://doi.org/10.1002/app.21481

Sun X, Du S, Wang X (2012) Amperometric immunosensor for carbofuran detection based on gold nano particles and PB-MWCNTs-CTS composite film. Eur Food Res Technol 235(3):469–477. https://doi.org/10.1007/s00217-012-1774-z

Takeuchi Y, Ida T, Kimura K (1996) Temperature effect on gold nano dispersion in organic liquids. Surf Rev Lett 3(1):1205–1208. https://doi.org/10.1142/S0218625X96002175

Talley CE, Jusinski L, Hollars CW, Lane SM, Huser T (2004) Intracellular pH sensors based on surface-enhanced Raman scattering. Anal Chem 76(23):7064–7068. https://doi.org/10.1021/ac049093j

Tang X, Liu Y, Hou H, You T (2011) A nonenzymatic sensor for xanthine based on electrospun carbon nanofibers modified electrode. Talanta 83:1410–1414. https://doi.org/10.1016/j.talanta.2010.11.019

Torri L, Piochi M (2016) Sensory methods and electronic nose as innovative tools for the evaluation of the aroma transfer properties of food plastic bags. Food Res Int 85:235–243. https://doi.org/10.1016/j.foodres.2016.05.004

Ubbink J, Krüger J (2006) Physical approaches for the delivery of active ingredients in foods. Trends Food Sci Technol 17(5):244–254. https://doi.org/10.1016/j.tifs.2006.01.007

US DOA. Nanoscale science and engineering for agriculture and food systems: a report submitted to cooperative state research, education and extension service. Washington, DC: Department of Agriculture; 2003. the United States Department of Agriculture: National Planning Workshop

Valadez AM, Lana CA, Tu S-I, Morgan MT, Bhunia AK (2009) Evanescent wave Fiber optic biosensor for *Salmonella* detection in food. Sensors 9(7):5810–5824. https://doi.org/10.3390/s90705810

Vilela D, González MC, Escarpa A (2012) Sensing colorimetric approaches based on gold and silver nanoparticles aggregation: chemical creativity behind the assay. Anal Chim Acta 751:24–43. https://doi.org/10.1016/j.aca.2012.08.043

Vinayaka AC, Basheer S, Thakur MS (2009) Bioconjugation of CdTe quantum dot for the detection of 2,4-dichlorophenoxyacetic acid by competitive fluoroimmunoassay based biosensor. Biosens Bioelectron 24(6):1615–1620. https://doi.org/10.1016/j.bios.2008.08.042

Virolainen NE, Pikkemaat MG, Elferink JWA, Karp MT (2008) Rapid detection of tetracyclines and their 4-epimer derivatives from poultry meat with bioluminescent biosensor bacteria. J Agric Food Chem 56(23):11065–11070. https://doi.org/10.1021/jf801797z

Vogel V (2019) Nanosensors and particles: a technology frontier with pitfalls. J Nanobiotechnol 17:111. https://doi.org/10.1186/s12951-019-0542-7

Vollmer F, Arnold S (2008) Whispering-gallery-mode biosensing: label-free detection down to single molecules. Nat Methods 5(7):591–596. https://doi.org/10.1038/NMETH.1221

Wan H, Sun Q, Li H, Sun F, Hu N, Wang P (2015) Screen-printed gold electrode with gold nanoparticles modification for simultaneous electrochemical determination of lead and copper. Sensors Actuators B Chem 209:336–342. https://doi.org/10.1016/j.snb.2014.11.127

Wang L, Liu X, Hu X, Song S, Fan C (2006) Unmodified gold nanoparticles as a colorimetric probe for potassium DNA aptamers. Chem Commun 28(36):3780–3782. https://doi.org/10.1039/b607448k

Wang M, Zhao J (2015) Facile synthesis of au supported on ionic liquid functionalized reduced graphene oxide for simultaneous determination of sunset yellow and tartrazine in drinks. Sens Actuators B Chem 216:578–585. https://doi.org/10.1016/j.snb.2015.04.053

Wang Y, Duncan TV (2017) Nanoscale sensors for assuring the safety of food products. Curr Opin Biotechnol 44:74–86. https://doi.org/10.1016/j.copbio.2016.10.005

Wang Y, Hsieh Y-L (2004) Enzyme immobilization to ultra-fine cellulose fibers via Amphiphilic polyethylene glycol spacers. J Polym Sci Part a-Polym Chem 42:4289–4299. https://doi.org/10.1002/pola.20271

Warriner K, Reddy SM, Namvar A, Neethirajan S (2014) Developments in nanoparticles for use in biosensors to assess food safety and quality. Trends Food Sci Technol 40(2):183–199. https://doi.org/10.1016/j.tifs.2014.07.008

Wei H, Li B, Li J, Dong S, Wang E, Dong S (2007) Simple and sensitive aptamer-based colorimetric sensing of protein using unmodified gold nanoparticle probes. Chem Commun 28(36):3735–3737. https://doi.org/10.1039/b707642h

Wesley SJ, Raja P, Raj AA, Tiroutchelvamae D (2014) Review on - nanotechnology applications in food packaging and safety. International Journal of Engineering Research 3(11):645–651. https://doi.org/10.17950/ijer/v3s11/1105

WHO Food safety—a worldwide public health issue. World Health Organization, Geneva. Available: http://www.who.int/fsf/fctshtfs.htm [accessed 15 August 2002]

Willander M, Khun Z, Ibupoto ZH (2014) ZnO based potentiometric and Amperometric nanosensors. J Nanosci Nanotechnol 14:1–12. https://doi.org/10.1166/jnn.2014.9349

Wu S, Duan N, Zhu C, Ma X, Wang M, Wang Z (2011) Magnetic nanobead-based immunoassay for the simultaneous detection of aflatoxin B1 and ochratoxin a using up conversion nanoparticles as multicolor labels. Biosens Bioelectron 30(1):35–42. https://doi.org/10.1016/j.bios.2011.08.023

Wu TH, Bechtel PJ (2008) Ammonia, dimethylamine, trimethylamine, and trimethylamine oxide from raw and processed fish by-products. J Aquat Food Prod Technol 17(1):27–38. https://doi.org/10.1080/10498850801891140

Xiao R, Wang D, Lin Z, Qiu B, Liu M, Guo L, Chen G (2015) Disassembly of gold nanoparticle dimers for colorimetric detection of ochratoxin a. Anal Methods 7(3):842–825. https://doi.org/10.1039/C4AY02970D

Xie Y, Xu L, Wang Y, Shao J, Wang L, Wang H, Qian H, Yao W (2013) Label-free detection of the foodborne pathogens of Enterobacteriaceae by surface enhanced Raman spectroscopy. Anal Methods 5:946–952. https://doi.org/10.1039/C2AY26107C

Xiong D, Chen M, Li H (2008) Synthesis of Para-sulfonatocalix[4]arene-modified silver nanoparticles as colorimetric histidine probes. Chem Commun (7):880–882. https://doi.org/10.1039/B716270G

Yaktine A, Pray L (2009) Nanotechnology in food products: workshop summary. National Academies Press, Washington, DC, USA. https://doi.org/10.17226/12633

Yan J, Guan H, Yu J, Chi D (2013) Acetylcholinesterase biosensor based on assembly of multiwall carbon nanotubes onto liposome bioreactors for detection of organophosphates pesticides. Pestic Biochem Physiol 105(3):197–202. https://doi.org/10.1016/j.pestbp.2013.02.003

Yang M, Kostov Y, Bruck HA, Rasooly A (2009) Gold nanoparticle-based enhanced chemiluminescence immunosensor for detection of staphylococcal enterotoxin B(SEB) in food. Int J Food Microbiol 133(3):265–271. https://doi.org/10.1016/j.ijfoodmicro.2009.05.029

Yao JN, Hashimoto K, Fujishima A (1992) Photochromism induced in an electrolytically pretreated MoO_3 thin film by visible light. Nature 355:624–626. https://doi.org/10.1038/355624a0

Yavuz S, Erkal A, Kariper IA, Solak AO, Jeon S, Mulazimoglu IE, Ustundag Z (2016) Carbonaceous materials-12: a novel highly sensitive graphene oxide-based carbon electrode: preparation, characterization, and heavy metal analysis in food samples. Food Anal Methods 9(2):322–331. https://doi.org/10.1007/s12161-015-0198-3

Ye X, Du Y, Lu D, Wang C (2013) Fabrication of β-cyclodextrin-coated poly (diallyldimethylammonium chloride)-functionalized graphene composite film modified glassy carbon-rotating disk electrode and its application for simultaneous electrochemical determination colorants of sunset yellow and tartrazine. Anal Chim Acta 779:22–34. https://doi.org/10.1016/j.aca.2013.03.061

Yılmaz F, Saylan Y, Akgönüllü S, Çimen D, Derazshamshir A, Bereli N, Denizli A (2017) Surface plasmon resonance based nanosensors for detection of triazinic pesticides in agricultural foods. In: Book: new pesticides and soil sensors Ed: Alexandru Grumezescu, pp 679–718. https://doi.org/10.1016/B978-0-12-804299-1.00019-9

Yun W, Li H, Chen SQ, Tu DW, Xie WY, Huang Y (2014) Aptamer-based rapid visual biosensing of melamine in whole milk. Eur Food Res Technol 238(6):989–995. https://doi.org/10.1007/s00217-014-2166-3

Zhang H, Li Z, Wang W, Wang C, Liu L (2010) Na+–doped zinc oxide nanofiber membrane for high speed humidity sensor. J Am Ceram Soc 93:142–146. https://doi.org/10.1111/j.1551-2916.2009.03341.x

Zhang L, Xu J, Mi L, Gong H, Jiang S, Yu Q (2012) Multifunctional magnetic–plasmonic nanoparticles for fast concentration and sensitive detection of bacteria using SERS. Biosens Bioelectron 31:130–136. https://doi.org/10.1016/j.bios.2011.10.006

Zhang X, Young MA, Lyandres O, Van Duyne RP (2005) Rapid detection of an anthrax biomarker by surface-enhanced Raman spectroscopy. Journal of American Chemical Society 127:4484–4489. https://doi.org/10.1021/ja043623b

Zhao X, Hilliard LR, Mechery SJ, Wang Y, Bagwe RP, Jin S, Tan W (2004) A rapid bioassay for single bacterial cell quantitation using bioconjugated nanoparticles. Proc Natl Acad Sci U S A 101(42):15027–15032. https://doi.org/10.1073/pnas.0404806101

Zhong Z, Patskovskyy S, Bouvrette P, Luong JHT, Gedanken A (2004) The surface chemistry of au colloids and their interactions with functional amino acids. J Phys Chem B 108(13):4046–4052. https://doi.org/10.1021/jp037056a

Zhou Q, Liu N, Qie Z, Wang Y, Ning B, Gao Z (2011a) Development of gold nano particle based rapid detection kit for melamine in milk products. J Agric Food Chem 59(22):12006–12011. https://doi.org/10.1021/jf202919a

Zhou Y, Dong H, Liu L, Li M, Xiao K, Xu M (2014) Selective and sensitive Colorimetric sensor of mercury (II) based on gold nanoparticles and 4-mercaptophenylboronic acid. Sensors Actuators B Chem 196:106–111. https://doi.org/10.1016/j.snb.2014.01.060

Zhou Y, Zhao H, He Y, Ding N, Cao Q (2011b) Colorimetric detection of Cu2þ using 4-mercaptobenzoic acid modified silver nanoparticles. Colloids Surf A Physicochem Eng Asp 391(1–3):179–183. https://doi.org/10.1016/j.colsurfa.2011.07.026

Ziyaina M, Rasco B, Coffey T, Ünlü G, Sablani SS (2019) Colorimetric detection of volatile organic compounds for shelf-life monitoring of milk. Food Control 100:220–226. https://doi.org/10.1016/j.foodcont.2019.01.018

Index

U. Hebbar et al. (eds.), *Nano-food Engineering*, Food Engineering Series,
https://doi.org/10.1007/978-3-030-44552-2

Q

The manufacturer's authorised representative in the EU is Springer Nature Customer Service Centre GmbH, Europaplatz 3, 69115 Heidelberg, Germany. If you have any concerns regarding our products, please contact ProductSafety@springernature.com

Printed and bound by CPI Group (UK) Ltd, Croydon, CR0 4YY
30/09/2026
02225503-0013